Annals of Mathematics Studies
Number 179

Fréchet Differentiability of Lipschitz Functions and Porous Sets in Banach Spaces

Joram Lindenstrauss
David Preiss
Jaroslav Tišer

PRINCETON UNIVERSITY PRESS

PRINCETON AND OXFORD

2012

Published by Princeton University Press
41 William Street, Princeton, New Jersey 08540

In the United Kingdom: Princeton University Press
6 Oxford Street, Woodstock, Oxfordshire OX20 1TW

Library of Congress Cataloging-in-Publication Data

Lindenstrauss, Joram, 1936–
  Fréchet differentiability of Lipschitz functions and porous sets in Banach spaces /
Joram Lindenstrauss, David Preiss, Jaroslav Tišer.
    p. cm. – (Annals of mathematics studies; no. 179)
    Includes bibliographical references and index.
    ISBN 978-0-691-15355-1 (hardcover) – ISBN 978-0-691-15356-8 (pbk.) 1. Banach
spaces. 2. Calculus of variations. 3. Functional analysis. I. Preiss, David. II. Tišer,
Jaroslav, 1957– III. Title.
    QA322.2.L564 2012
    515'.88–dc23
                                2011037616

British Library Cataloging-in-Publication Data is available

This book has been composed in LaTeX.
The publisher would like to acknowledge the authors of this volume for providing the
camera-ready copy from which this book was printed.

press.princeton.edu

10 9 8 7 6 5 4 3 2 1

The authors would like to thank Naomi Lindenstrauss for her help and the Hebrew University for its support.

This work was also partially supported by the grants EPSRC (UK) EP/D053099/1 (D. Preiss) and GAČR 207/07/0394 and MSM 6840770010 (J. Tišer).

Contents

1 Introduction   **1**
  1.1   Key notions and notation   9

2 Gâteaux differentiability of Lipschitz functions   **12**
  2.1   Radon-Nikodým property   12
  2.2   Haar and Aronszajn-Gauss null sets   13
  2.3   Existence results for Gâteaux derivatives   15
  2.4   Mean value estimates   16

3 Smoothness, convexity, porosity, and separable determination   **23**
  3.1   A criterion of differentiability of convex functions   23
  3.2   Fréchet smooth and nonsmooth renormings   24
  3.3   Fréchet differentiability of convex functions   28
  3.4   Porosity and nondifferentiability   31
  3.5   Sets of Fréchet differentiability points   33
  3.6   Separable determination   37

4 ε-Fréchet differentiability   **46**
  4.1 ε-differentiability and uniform smoothness   46
  4.2   Asymptotic uniform smoothness   51
  4.3 ε-Fréchet differentiability of functions on asymptotically smooth spaces   59

5 Γ-null and Γ_n-null sets   **72**
  5.1   Introduction   72
  5.2 Γ-null sets and Gâteaux differentiability   74
  5.3   Spaces of surfaces, and Γ- and Γ_n-null sets   76
  5.4 Γ- and Γ_n-null sets of low Borel classes   81
  5.5   Equivalent definitions of Γ_n-null sets   87
  5.6   Separable determination   93

6 Fréchet differentiability except for Γ-null sets   **96**
  6.1   Introduction   96
  6.2   Regular points   97
  6.3   A criterion of Fréchet differentiability 100
  6.4   Fréchet differentiability except for Γ-null sets 114

7 Variational principles   **120**
  7.1 Introduction 120
  7.2 Variational principles via games 122
  7.3 Bimetric variational principles 127

8 Smoothness and asymptotic smoothness   **133**
  8.1 Modulus of smoothness 133
  8.2 Smooth bumps with controlled modulus 141

9 Preliminaries to main results   **156**
  9.1 Notation, linear operators, tensor products 156
  9.2 Derivatives and regularity 157
  9.3 Deformation of surfaces controlled by ω_n 161
  9.4 Divergence theorem 164
  9.5 Some integral estimates 165

10 Porosity, Γ_n- and Γ-null sets   **169**
  10.1 Porous and σ-porous sets 169
  10.2 A criterion of Γ_n-nullness of porous sets 173
  10.3 Directional porosity and Γ_n-nullness 186
  10.4 σ-porosity and Γ_n-nullness 189
  10.5 Γ_1-nullness of porous sets and Asplundness 192
  10.6 Spaces in which σ-porous sets are Γ-null 198

11 Porosity and ε-Fréchet differentiability   **202**
  11.1 Introduction 202
  11.2 Finite dimensional approximation 203
  11.3 Slices and ε-differentiability 208

12 Fréchet differentiability of real-valued functions   **222**
  12.1 Introduction and main results 222
  12.2 An illustrative special case 225
  12.3 A mean value estimate 230
  12.4 Proof of Theorems 12.1.1 and 12.1.3 234
  12.5 Generalizations and extensions 261

13 Fréchet differentiability of vector-valued functions   **262**
  13.1 Main results 262
  13.2 Regularity parameter 263
  13.3 Reduction to a special case 269
  13.4 Regular Fréchet differentiability 289
  13.5 Fréchet differentiability 304
  13.6 Simpler special cases 317

14 Unavoidable porous sets and nondifferentiable maps   **319**
    14.1 Introduction and main results 319
    14.2 An unavoidable porous set in ℓ_1 325
    14.3 Preliminaries to proofs of main results 332
    14.4 The main construction, Part I 339
    14.5 The main construction, Part II 344
    14.6 Proof of Theorem 14.1.3 347
    14.7 Proof of Theorem 14.1.1 351

15 Asymptotic Fréchet differentiability   **355**
    15.1 Introduction 355
    15.2 Auxiliary and finite dimensional lemmas 359
    15.3 The algorithm 363
    15.4 Regularity of f at x_∞ 372
    15.5 Linear approximation of f at x_∞ 380
    15.6 Proof of Theorem 15.1.3 389

16 Differentiability of Lipschitz maps on Hilbert spaces   **392**
    16.1 Introduction 392
    16.2 Preliminaries 394
    16.3 The algorithm 396
    16.4 Proof of Theorem 16.1.1 403
    16.5 Proof of Lemma 16.2.1 403

Bibliography   **415**

Index   **419**

Index of Notation   **423**

Chapter One

Introduction

The notion of a derivative is one of the main tools used in analyzing various types of functions. For vector-valued functions there are two main versions of derivatives: Gâteaux (or weak) derivatives and Fréchet (or strong) derivatives. For a function f from a Banach space X into a Banach space Y the Gâteaux derivative at a point $x_0 \in X$ is by definition a bounded linear operator $T \colon X \longrightarrow Y$ such that for every $u \in X$,

$$\lim_{t \to 0} \frac{f(x_0 + tu) - f(x_0)}{t} = Tu. \tag{1.1}$$

The operator T is called the Fréchet derivative of f at x_0 if it is a Gâteaux derivative of f at x_0 and the limit in (1.1) holds uniformly in u in the unit ball (or unit sphere) in X. An alternative way to state the definition is to require that

$$f(x_0 + u) = f(x_0) + Tu + o(\|u\|) \text{ as } \|u\| \to 0.$$

Thus T defines the natural linear approximation of f in a neighborhood of the point x_0. Sometimes T is called the first variation of f at the point x_0.

Clearly, for both notions of derivatives we have only to require that f be defined in a neighborhood of x_0.

The existence of a derivative of a function f at a point x_0 is not obvious. The question of existence of a derivative for functions from $\mathbb{R}$ to $\mathbb{R}$ was the subject of research and much discussion among mathematicians for a long time, mainly in the nineteenth century. If $f \colon \mathbb{R} \longrightarrow \mathbb{R}$ has a derivative at x_0 then it must be continuous at x_0. While it is obvious how to construct a continuous function $f \colon \mathbb{R} \longrightarrow \mathbb{R}$ which fails to have a derivative at a given point, the problem of finding such a function which is nowhere differentiable is not easy. The first to do this was the Czech mathematician Bernard Bolzano in an unpublished manuscript about 1820. He did not supply a full proof that his function had indeed the desired properties. Later, around 1850, Bernhard Riemann mentioned in passing such an example. It was found out later that his example was not correct. The first one who published such an example with a valid proof was Karl Weierstrass in 1875. The first general result on existence of derivatives for functions $f \colon \mathbb{R} \longrightarrow \mathbb{R}$ was found by Henri Lebesgue in his thesis (around 1900). He proved that a monotone function $f \colon \mathbb{R} \longrightarrow \mathbb{R}$ is differentiable almost everywhere. As a consequence it follows that every Lipschitz function $f \colon \mathbb{R} \longrightarrow \mathbb{R}$, that is, a function which satisfies

$$|f(s) - f(t)| \le C|s - t|$$

for some constant C and every $s, t \in \mathbb{R}$, has a derivative a.e. This result is sharp in the sense that for every $A \subset \mathbb{R}$ of measure zero there is a Lipschitz (even monotone) function $f \colon \mathbb{R} \longrightarrow \mathbb{R}$ which fails to have a derivative at any point of A.

Lebesgue's result was extended to Lipschitz functions $f \colon \mathbb{R}^n \longrightarrow \mathbb{R}$ by Hans Rademacher, who showed that in this case f is also differentiable a.e. However, this result is not as sharp as Lebesgue's: there are planar sets of measure zero that contain points of differentiability of all Lipschitz functions $f \colon \mathbb{R}^2 \longrightarrow \mathbb{R}$. This can be seen by detailed inspection of our arguments in Chapter 12 (more details are in [39]). Questions related to sharpness of Rademacher's theorem have recently received considerable attention. See, for example, [2] or [15]. We do not cover this development here since its main interest and deepest results are finite dimensional, whereas our aim is to contribute to the understanding of the infinite dimensional situation.

The concept of a Lipschitz function makes sense for functions between metric spaces. Consequently, this gives rise to the study of derivatives of Lipschitz functions between Banach spaces X and Y. It is easy to see that in view of the compactness of balls in finite dimensional Banach spaces both concepts of a derivative, defined above, coincide if $\dim X < \infty$ and f is Lipschitz. However, if $\dim X = \infty$ easy examples show that there is a big difference between Gâteaux and Fréchet differentiability even for simple Lipschitz functions.

In the formulation of Lebesgue's theorem there appears the notion of a.e. (almost everywhere). If we consider infinite dimensional spaces and want to extend Lebesgue's theorem to functions on them, we have first to extend the notion of a.e. to such spaces. In other words, we have to define in a reasonable way a family of negligible sets on such spaces. (These sets are also often called exceptional or null.) The negligible sets should form a proper σ-ideal of subsets of the given space X, that is, be closed under subsets and countable unions, and should not contain all subsets of X. Since sets that are involved in differentiability problems are Borel, we can equivalently consider σ-ideals of Borel subsets of X, that is, families of Borel sets, closed under taking Borel subsets and countable unions. It turns out that this can be done in several nonequivalent ways (in our study below we were led to an infinite family of such σ-ideals). Thus the study of derivatives of functions defined on Banach spaces leads in a natural way to questions of descriptive set theory.

In the study of differentiation of Lipschitz functions on Banach spaces, one obstacle has been apparent from the outset. It was recognized already in 1930. The isometry $t \rightarrow \mathbf{1}_{[0,t]}$ (the indicator function of the interval $[0, t]$) from the unit interval to $L_1[0, 1]$ does not have a Gâteaux derivative at a single point. The class of Banach spaces where this pathology does not appear was singled out already in the 1930s and characterized in various ways. The "good" Banach spaces (i.e., spaces X so that Lipschitz maps from $\mathbb{R}$ to X have a derivative a.e.) are now called spaces with the Radon-Nikodým property (or RNP spaces). The reason for this terminology is that one of their characterizations is that a version of the Radon-Nikodým theorem holds for measure with values in them. A detailed study of this class of spaces is presented in the books [4], [9], and [14]. All separable conjugate (in particular reflexive) spaces are RNP spaces. As we have just noted, this class does not include Banach spaces having $L_1[0, 1]$ as a subspace. A similar easy argument shows that an RNP space cannot contain c_0 as a subspace. More sophisticated arguments are needed to show that there are separable Banach spaces with the RNP which are not subspaces of separable conjugate spaces or that there are spaces which fail to have the RNP but do not contain subspaces isomorphic to $L_1[0,1]$ or c_0. Such examples are presented in detail in [4].

The theorem of Lebesgue can be extended to Gâteaux differentiability of Lipschitz functions from an open subset of a separable Banach space into Banach spaces with the RNP. This was done by various authors independently in the 1970s by using different σ-ideals of negligible sets. The proofs are not difficult, and again all the details may be found in [4]. The situation concerning the existence of Gâteaux derivatives is at present quite satisfactory. On the other hand, the question of existence of Fréchet derivatives seems to be deep and our current knowledge concerning it is rather incomplete. This book is devoted to the study of this topic. Most of it consists of new material. We also recall the known results concerning this question and mention several of the problems which are still open. The proofs of most known results in this direction are at present quite difficult. It is not clear to us whether they can be considerably simplified. We present the proofs of the main results with all details and often accompany them with some words of motivation. Some examples we present seem to indicate that the fault in the difficulty lies mainly in the nature of things.

In dealing with Fréchet differentiability it turns out quite soon that we have to restrict the Banach spaces that can serve as domain spaces. The function $x \to \|x\|$ from X to the reals is obviously continuous and convex and thus Lipschitz. If $X = \ell_1$ this function is easily seen not to be Fréchet differentiable at a single point. A similar situation may occur whenever X is separable but X^* is not. A separable Banach space X is called an Asplund space if X^* is again separable. The reason for this terminology is that Asplund was the first to prove that in such spaces real-valued convex and continuous functions have many points of Fréchet differentiability. "Many points" means here a set whose complement is a set of the first category (i.e., small in the sense of category and not in general in the sense of measure). This shows again why the study of Fréchet differentiability is strongly connected to descriptive set theory. Thus our real subject of study in this book is the existence of Fréchet derivatives of Lipschitz functions from X to Y, where X is an Asplund space and Y has the RNP.

Perhaps the best known open question about differentiability of Lipschitz mappings is whether every countable collection of real-valued Lipschitz functions on an Asplund space has a common point of Fréchet differentiability.

Optimistic conjectures would assert Fréchet differentiability of Lipschitz functions almost everywhere with respect to a suitable proper σ-ideal of exceptional sets (or null sets). Based on what we currently know (including the results proved here), an optimistic differentiability conjecture may be stated in the following way.

Conjecture. In every Asplund space X there is a nontrivial notion of exceptional sets such that, for every locally Lipschitz map f of an open subset $G \subset X$ into a Banach space Y having the RNP:

(C1) f is Gâteaux differentiable almost everywhere in G.

(C2) If $S \subset G$ is a set with null complement such that f is Gâteaux differentiable at every point of S, then $\mathrm{Lip}(f) = \sup_{x \in S} \|f'(x)\|$.

(C3) If the set of Gâteaux derivatives of f attained on some $E \subset G$ is norm separable, then f is Fréchet differentiable at almost every point of E.

There is very little evidence for validity of this Conjecture in the generality given above. On the positive side it holds if $Y = \mathbb{R}$ and, as we shall see in Chapter 6, it also holds for some infinite dimensional spaces X. On the negative side, it is unknown even whether every three real-valued Lipschitz functions on a Hilbert space have a common point of Fréchet differentiability. (The fact that every two real-valued Lipschitz functions on a Hilbert space have a common point of Fréchet differentiability is one of the new results we prove here.) Moreover, as far as we know, the Conjecture fails with any known nontrivial σ-ideal of subsets of infinite dimensional Hilbert spaces.

In any detailed study of Fréchet differentiability one immediately encounters the notion of porous sets (and of σ-porous sets that are their countable unions). We will give their usual definition later, since at this moment the only relevant fact is that a set $E \subset X$ is porous if an only if the function $x \to \mathrm{dist}(x, E)$ is Fréchet nondifferentiable at any point of E. It follows that porous sets have to belong to the σ-ideal hoped for in the above Conjecture. Of course, it also has to contain the sets of Gâteaux nondifferentiability of Lipschitz maps to RNP spaces. Denoting just for the purpose of this discussion by $\mathcal{I}$ the σ-ideal of subset of X generated by the porous sets and sets of Gâteaux nondifferentiability of Lipschitz maps from X into RNP spaces, one may hope that the Conjecture holds with exceptional sets being defined as elements of $\mathcal{I}$. Both subproblems of this variant of the Conjecture are open: it is unknown whether it holds with these exceptional sets, and it is unknown whether $\mathcal{I}$ is nontrivial.

A weaker variant of the Conjecture than the one at which we arrived above is in fact true. For the Γ-null sets (which will be defined in Chapter 5) we show that (C1) and (C2) hold, and that (C3) holds for the given space X if and only if every porous subset in X is Γ-null. This result coming from [28] was the first showing that the problem of smallness of porous sets is related to the problem of existence of Fréchet differentiability points of Lipschitz functions. One of the contributions of this text is to bring better understanding of this relation.

The statement (C2) is a weak form of the mean value estimate. Although it is stated for vector-valued functions, it can be equivalently asked only for real-valued functions (as the general case follows by considering $x^* \circ f$ for a suitable $x^* \in X^*$). One can argue that without the validity of this statement a differentiability result would not be very useful. For vector-valued functions there is, however, a stronger mean value estimate, the one that one would obtain by estimating in the Gauss-Green divergence theorem the integral of the divergence by its supremum. We will explain this concept, which we call a multidimensional mean value estimate, in detail in the last section of Chapter 2. The main results of this book give a fairly complete answer to the question under what conditions all Lipschitz mapping of X to finite dimensional spaces not only possess points of Fréchet differentiability, but possess so many of them that even the multidimensional mean value estimate holds. It turns out that this property is much stronger that mere existence of points of Fréchet differentiability: for example, for mappings on Hilbert spaces it holds if the target is two-dimensional, but fails if it is three-dimensional.

We now describe some of the contents of this book in more detail. Every chapter starts with a brief information about its content and basic relation to results proved elsewhere. In most cases this is followed by an introductory section, which may also state the main results and prove their most important corollaries. However, some chapters contain rather diverse sets of results, in which case their statements are often deferred to the section in which they are proved. The key notions and notation are introduced at the end of this Introduction; more specialized notions and notation are given only when they are needed. The index and index of notation at the end should help the reader to find the definitions quickly.

The main point of the starting chapters is to revise some basic notions and results, although they also contain new results or concepts. Proofs that are well covered in the main reference [4] are not repeated here.

Chapter 2 recalls the notion of the Radon-Nikodým property and main results on Gâteaux differentiability of Lipschitz functions and related notions of null sets. Throughout the text, we will be interested not only in mere existence of points of Fréchet differentiability, but also, and often more important, in validity of the mean value estimates. We therefore explain this concept here in some detail. In particular, we spend some time on explaining the meaning of multidimensional mean value estimates, as this seems to be the concept behind nearly all positive results as well as the main counterexamples.

In Chapter 3 we meet some of the most basic concepts that we will use in the rest of the book. The existence of a Fréchet smooth equivalent norm on a Banach space with separable dual is of such fundamental importance that we prove it here even though it is treated in [4] as well. In separable Banach spaces with nonseparable dual we construct an equivalent norm which not only is rough in the usual sense, but has roughness directions inside every finite codimensional subspace (and even inside every subspace for which the quotient has separable dual). We also establish here the basic relations between porosity and differentiability, many of which will be further developed in the following text. Some of our finer results are deeply ingrained in descriptive theoretical complexity of sets related to differentiability. We therefore show here also that the set of points of Fréchet differentiability of maps between Banach spaces is of type $F_{\sigma\delta}$. In this book we normally assume that the Banach spaces we work with are separable (although we sometimes remark what happens in the nonseparable situation). The last section of this chapter justifies these assumptions: it describes how one can show that many differentiability type results hold in nonseparable spaces provided they hold in separable ones.

In Chapter 4 we treat results on ε-Fréchet differentiability of Lipschitz functions in asymptotically smooth spaces. In our context, these results are highly exceptional in the sense that they are the only differentiability results in which we do not prove that the multidimensional version of the mean value estimate holds; in fact we will show later that it may be false. The chapter, however, is not isolated from the others: some of the concepts and techniques introduced here will be used in what follows.

Chapter 5 introduces the notions of Γ- and Γ_n-null sets. The former is key for the strongest known general Fréchet differentiability results in Banach spaces; the latter presents a new, more refined concept. The reason for these notions comes from an (imprecise) observation that differentiability problems are governed by measure in finite dimension, but by Baire category when it comes to behavior at infinity. Γ_n-null sets are thus defined as those (Borel) sets that are null on residually many n-dimensional surfaces; to define Γ-null sets we use the same idea with $n = \infty$. The results connecting Γ- and Γ_n-null sets will later lead to finding a new class of spaces for which the strong Fréchet differentiability result holds.

The results of Chapter 6 are based on the already mentioned simple but important observation that for the above Conjecture to hold, porous sets have to be negligible. We show here that for Γ-null sets the converse is also true: the Conjecture holds (for the given space X) with Γ-null sets provided all sets porous in X are Γ-null. In the proof we meet for the first time the concept of regular differentiability. It was observed a long time ago that if f is Fréchet differentiable at x, then for any $c > 0$ and any direction u, $f(y + tu) - f(y)$ is well approximated by $tf'(x)(u)$ provided t is small and $\|y - x\| \le ct$. (This observation was crucial, for example, in Zahorski's deep study [47] of derivatives of real-valued functions of one real variable.) The validity of this observation or of its variants is called regular (Fréchet) differentiability. In general, it is weaker than Fréchet differentiability. From this moment on, proofs of Fréchet differentiability become two-stage: we show first that the function is, at a particular point, regularly differentiable, and only then proceed to the proof of differentiability. In this chapter the first stage is based on the relatively simple result that the set of points of irregular differentiability is σ-porous. In what follows, we will have to use considerably more sophisticated methods to handle this stage. Much of the material in this chapter follows ideas from [28], but the study of spaces in which porous sets are Γ-null is based on new results connecting Γ- and Γ_n-null sets that we will show later (in Chapter 10).

Chapters 7–14 contain our main new results. The first three contain new, sometimes technical, methods on which these results are based.

To explain the direction we have taken, we recall one of the facts that is in the background of the proofs of existing general Fréchet differentiability results in Banach spaces: for real-valued Lipschitz functions on spaces with smooth norms, attainment of the (local) Lipschitz constant by some directional derivative $f'(x; u)$, where $\|u\| = 1$, implies Fréchet differentiability of f at x. In this form, this idea was used in [38] to prove that everywhere Gâteaux differentiable Lipschitz functions on Asplund spaces have points of Fréchet differentiability. Fitzpatrick [18] replaced attainment of the Lipschitz constant by attainment of what he called a modified version of the local Lipschitz constant, which is related to the notion of regularity mentioned above, and applied it to the problem of Fréchet differentiability of distance functions. The first Fréchet differentiability result for Lipschitz functions in [39] replaced the attainment of the Lipschitz constant by the requirement that the function $(x, u) \mapsto f'(x; u)$, perturbed in a suitable way, attains its maximum on a suitable set of pairs (x, u). These perturbations are similar to those used in the proof of the Bishop-Phelps theorem. One of the contributions of the present work is the recognition that this similarity can be carried much farther, to the use of a variational principle of Ekeland's type (see [16]). We therefore devote Chapter 7 to the study in some detail of such principles, in particular, of smooth variational principles of the type found in [7]. We describe these principles as infinite two-player games, which allows some unusual applications later. The material of this chapter should be relevant to other areas of nonlinear analysis, and readers interested in such applications may find it useful that the chapter can be read independently of the others.

As we have already mentioned in description of Chapter 4, asymptotic smoothness of the norm can be successfully used to prove a weaker form of differentiability. We intend to extend this idea to showing that higher order asymptotic smoothness of the norm implies existence of common points of differentiability of more functions. In Chapter 8 we study general forms of such smoothness assumptions, which we make on bump functions rather than on norms. (This is more a matter of technical convenience than a serious generalization, and so in the following description of results we will speak only about asymptotic smoothness of norms.) Unlike ordinary smoothness, asymptotic smoothness allows arbitrary moduli, and one of the main contributions of our work is in isolating the "right" modulus of asymptotic uniform smoothness to guarantee existence of many common points of Fréchet differentiability of n real-valued Lipschitz functions, namely, $o(t^n \log^{n-1}(1/t))$.

Chapter 9, as its title "Preliminaries to main results" indicates, gives a number of results and notions that will be needed in the sequel. In particular, it treats the important notion of regular differentiability in some detail and proves inequalities controlling the increment of functions by the integral of their derivatives. Another very important technical tool introduced here describes a particular deformation of n-dimensional surfaces. The idea is to deform a flat surface passing through a point x (along which we imagine that a certain function f is almost affine) to a surface passing through a point witnessing that f is not Fréchet differentiable at x. This is done in such a way that certain "energy" associated to surfaces increases less than the "energy" of the function f along the surface, which is precisely what enables the use of the variational principle to conclude that f is in fact Fréchet differentiable at x.

The ideal goal of Chapter 10 would be to show that in spaces with modulus of asymptotic uniform smoothness $o(t^n \log^{n-1}(1/t))$ porous sets are Γ_n-null. We prove that this is the case for $n = 1, 2$, but for $n \geq 3$ we were unable to decide whether this holds in finite dimensional spaces. Nevertheless, we decompose every porous set in such a space into a union of its "finite dimensional" part (which is, for example, Haar null) and of a set porous "at infinity." We show that the latter set is Γ_n-null. In fact, we use a condition weaker than asymptotic uniform smoothness $o(t^n \log^{n-1}(1/t))$. The main advantage of this is that in the case $n = 1$ our results imply that a separable space has separable dual if and only if all its porous subsets are Γ_1-null. Finally, we establish a (not completely straightforward) relation between Γ_n-nullness for infinitely many values of n and Γ-nullness. This produces a new class of spaces for which the Conjecture holds, containing all spaces for which it was known before.

In Chapter 11 we investigate whether the result of Chapter 6 that Γ-nullness of porous sets implies the Conjecture has a meaningful analogy for Γ_n-null sets. One of the difficulties is that for $n \geq 3$ there are no (infinite dimensional) spaces in which we are able to verify the assumption that the porous sets are Γ_n-null. From the previous chapter we know only that in spaces admitting an appropriately asymptotically smooth norm every porous set can be covered by a union of a Haar null set and a Γ_n-null G_δ set. We therefore consider a space X having this property and show that the set of Gâteaux derivatives of a Lipschitz function $f\colon X \longrightarrow Y$, where $\dim Y = n$, is, for every $\varepsilon > 0$, contained in the closed convex hull of the "ε-Fréchet derivatives," that is, of those $f'(x)$ such that

$$\|f(x + u) - f(x) - f'(x)(u)\| \leq \varepsilon\|u\|$$

for all u with $\|u\|$ small enough.

In Chapter 12 we use the ideas developed in the previous chapters to improve known results in the classical case of real-valued functions on spaces with separable dual. Here the main novelty is that the variational principle is used to organize the arguments in a way that leads to two new results. First, for Lipschitz functions we prove not only the known result that they have many points of Fréchet differentiability, but also that they have many points of Fréchet differentiability outside any given σ-porous set. Second, we show that even cone-monotone functions have points of Fréchet differentiability, and again such points may be found outside any given σ-porous set.

Chapter 13 contains perhaps the most surprising results of this text. We show that if a Banach space with a Fréchet smooth norm is asymptotically smooth with modulus $o(t^n \log^{n-1}(1/t))$ then every Lipschitz map of X to a space of dimension not exceeding n has many points of Fréchet differentiability. In particular, we get that two real-valued Lipschitz functions on a Hilbert space have a common point of Fréchet differentiability and, more generally, that every collection of n real-valued Lipschitz functions on ℓ^p has a common point of Fréchet differentiability provided $1 < p < \infty$ and $n \leq p$. The argument combines a significant extension of methods developed in Chapter 12 with understanding of the role of a particular porous set: the set of points of irregular differentiability. However, the results about porosity do not enter the proof directly.

In Chapter 14 we explain the need for the rather strong smoothness assumptions made in Chapters 10 and 13 to show Γ_n-nullness of sets porous "at infinity" and/or existence of many points of Fréchet differentiability of Lipschitz maps into n-dimensional spaces. Although it is probable that our assumptions are not optimal, we show that they are rather close to being so. As a corollary we obtain that Lipschitz maps of Hilbert spaces to $\mathbb{R}^3$ and, more generally, maps of ℓ^p to $\mathbb{R}^n$ for $n > p$ do not have many points of Fréchet differentiability. Here, of course, the word "many" (in the "mean value" sense explained above) is crucial; without it we would have disproved the Conjecture.

The two last chapters do not follow the main direction of the book in the sense that they return to the iterative procedure for finding points of Fréchet differentiability. Chapter 15 actually presents the current development of our first, unpublished proof of existence of points Fréchet differentiability of Lipschitz mappings to two-dimensional spaces. Unlike in Chapter 13, for functions into higher dimensional spaces the method does not lead to a point of Gâteaux differentiability but constructs points of so-called asymptotic Fréchet differentiability. Although some of these results follow from Chapter 13, in full generality they are new. The proof does not use the variational approach;

indeed, it uses perturbations that are not additive. It nevertheless still provides (asymptotic) Fréchet derivatives in every slice of Gâteaux derivatives, and so by results of Chapter 14 it also cannot be used to show existence of points of Fréchet differentiability of Lipschitz mappings of Hilbert spaces to three-dimensional spaces. Readers interested in attempts to find such a point by modifying the iterative (or variational) argument should probably start by checking in what way their approach differs from ours, and why the difference may allow for slices of Gâteaux derivatives without Fréchet derivatives.

The final Chapter 16 gives a separate, essentially self-contained, nonvariational proof of existence of points of Fréchet differentiability of $\mathbb{R}^2$-valued Lipschitz maps on Hilbert spaces. It is mainly directed toward readers whose main interest is in Hilbert spaces. The structure of the Hilbert space is heavily used, but even then some readers interested in the much stronger results of Chapter 13 may find useful that it covers some of the basic ideas in a substantially simpler way.

1.1   KEY NOTIONS AND NOTATION

We finish the Introduction by quickly recapitulating the main notions defined above and introducing some notation that will be used throughout or for which there was no other natural place. More specific notions and notation will be given only when needed (and can be found in the Index or Index of Notation at the end).

First, we should remind the reader that not much would be lost by assuming that the Banach spaces in this book are separable. We tend to give separability assumptions only in key statement and notions and only when there is a significant difference between the separable and nonseparable case. For example, in the definition of Haar null sets in Chapter 2 we assume that the space is separable, although there is a natural definition also in the nonseparable situation. The reason is that for separable spaces we may use in this definition Borel measures (which we call simply measures) while for nonseparable spaces we have to use Radon measures.

We will use the following notation for derivatives of functions between Banach spaces X and Y. The *directional derivative of a function f at $x \in X$ in the direction $u \in X$* is defined by

$$f'(x; u) = \lim_{t \to 0} \frac{f(x + tu) - f(x)}{t},$$

provided the limit exists. To avoid any misunderstanding, we agree that by saying that "a limit exists" we understand that it is finite.

The Gâteaux derivative of f at x (defined at the beginning of the Introduction) will be denoted by $f'(x)$. So, provided f is Gâteaux differentiable at x, we have two ways of writing the directional derivatives at x: $f'(x)(u)$ and $f'(x; u)$. The second expression may be meaningful even if the first is not. We will not use any particular notation for the Fréchet derivative (also defined at the beginning of the Introduction), since its value is the same as that of the Gâteaux derivative and so we may use $f'(x)$ if needed. (But this expression by itself does not say that f is Fréchet differentiable at x.)

Although Fréchet differentiability is the main problem we study, we will also have to use various other notions of differentiability. Gâteaux or Fréchet differentiability of a function $f\colon X \longrightarrow Y$ in the direction of a subspace Z of X is defined in the obvious way as Gâteaux or Fréchet differentiability of the function $z \in Z \to f(x + z)$ at the point $z = 0$. We will use without any reference that the two notions of differentiability in the direction of Z are equivalent when f is Lipschitz and Z is finite dimensional; in this case we will often speak just about differentiability or derivative. Other, more special notions such as regular differentiability and differentiability in directions of linear maps will be introduced only when needed, most of them in Definition 9.2.1.

Since we will work with a number of (pseudo)norms, we fix basic notation for them. (The only difference between a norm and a pseudonorm is that the latter may be zero for some nonzero vectors. Similarly for pseudometrics, although for them we will occasionally allow also infinite values.) The symbol $\|\cdot\|$ is used for most norms: for example, it may denote the norm on the Banach space X we study or the usual operator norm on the space $L(X, Y)$ of bounded linear operators from X to Y. However, since the Euclidean norm in $\mathbb{R}^n$ will often be used at the same time as other norms (on other spaces), we will denote it by $|\cdot|$. When needed, we add an index to give more information about a particular norm. By $\|\cdot\|_\infty$ we denote the supremum norm on the space of (Banach space–valued) functions. More special norms that we will use include $\|\gamma\|_{C^1} = \|\gamma\|_{C^1(S)} = \max\{\|\gamma\|_\infty, \|\gamma'\|_\infty\}$ on the space of (Banach space–valued) C^1 functions on a subset S of $\mathbb{R}^n$ and the norms $\|\gamma\|_k$ and $\|\gamma\|_{\leq k}$ on the space of surfaces introduced in Chapter 5.

By $B(x, r)$ we will denote the open ball with center x and radius r; in situations when the space or the norm is not clear, we invent a way of indicating it.

After derivatives, the notion of σ-porosity is perhaps the most often occurring notion in this book. The underlying notion of porosity comes in several variants, of which we will use the following three.

A set E in a Banach space X (or in a general metric space X) is called *porous* if there is $0 < c < 1$ such that for every $x \in E$ and every $\varepsilon > 0$ there is a $y \in X$ with $0 < \operatorname{dist}(x, y) < \varepsilon$ and
$$B\big(y, c\operatorname{dist}(x, y)\big) \cap E = \emptyset.$$

In this situation we also say that E is *porous with constant c*.

If Y is a subspace of X, then E is called *porous in the direction of Y* if there is $0 < c < 1$ such that for every $x \in E$ and $\varepsilon > 0$ there is $y \in Y$ so that $0 < \|y\| < \varepsilon$ and

$$B(x + y, c\|y\|) \cap E = \emptyset.$$

Instead of "porous in the direction of the linear span of a vector u" we say "porous in the direction u"; we also say that $E \subset X$ is *directionally porous* if it is porous in some direction.

The sets that will be really important for us are not the porous sets we have just defined, but their countable unions. A subset of X is termed *σ-porous, σ-porous in the direction of Y*, or *σ-directionally porous* if it is a union of countably many porous, porous in the direction of Y, or directionally porous sets, respectively.

It remains to introduce some notation which is used more often but for which there is no reasonable place elsewhere. The symbol $\fint$ means the average integral: provided $0 < \mu A < \infty$,

$$\fint_A f \, d\mu = \frac{1}{\mu A} \int_A f \, d\mu.$$

In this book we will never encounter serious measurability problems, but to make some assumptions more compact we will identify measures with outer measures. So, for example, $\mathscr{L}^n$ denotes the outer Lebesgue measure in $\mathbb{R}^n$. The Lebesgue measure of the Euclidean unit ball in $\mathbb{R}^n$ is $\boldsymbol{\alpha}_n$.

As one of our main goals is finding points of Fréchet differentiability in arbitrary slices of the set of Gâteaux derivatives, we recall the relevant notions here. In general, a *slice* of a subset M of a Banach space X is any nonempty set of the form

$$S = \{x \in M \mid x^*(x) > c\},$$

where $x^* \in X^*$ and $c \in \mathbb{R}$. However, for us the more pertinent concept is that of w^*-*slice*, which is defined for a subset M of the dual space X^* as any nonempty set of the form

$$S = \{x^* \in M \mid x^*(x) > c\},$$

where $x \in X$ and $c \in \mathbb{R}$. More generally, we will use the concept of w^*-slices also in the space of operators $L(X, Y)$. A w^*-slice S of a subset $M \subset L(X, Y)$ is any nonempty set of the form

$$S = \left\{ L \in M \,\middle|\, \sum_{i=1}^{m} y_i^* L x_i > c \right\},$$

where $m \in \mathbb{N}$, $x_1, \ldots, x_m \in X$, $y_1^*, \ldots, y_m^* \in Y^*$ and $c \in \mathbb{R}$. Since sometimes the number m of points and functionals determining the given slice plays a role, we will call the least such number the *rank of the slice S*.

Finally, we make agreements about the use of some standard terms. We will use the terms increasing and strictly increasing in the sense for which some authors use nondecreasing and increasing, respectively. Similarly, we use the terms decreasing and strictly decreasing, positive and strictly positive. However, we will not be pedantic in the use of these terms; in particular, we may say non-negative instead of positive and positive instead of strictly positive, the former since it may be clearer and the latter when the meaning is obvious from the context. In most situations it does not matter whether the set of natural numbers $\mathbb{N}$ starts from zero or one, but when it does, we assume it starts from one.

Chapter Two

Gâteaux differentiability of Lipschitz functions

We start by quickly recalling some basic notions and results that are well covered in [4]: the Radon-Nikodým property, main results on Gâteaux differentiability of Lipschitz functions, and related notions of null sets. We also discuss what is meant by validity of the mean value estimates, since this concept is deeply related to most of what is done in this book.

2.1 RADON-NIKODÝM PROPERTY

Lipschitz maps even from the real line into a Banach space need not have a single point of differentiability. A simple and well-known example is $f\colon [0,1] \longrightarrow L_1([0,1])$ defined by

$$f(t) = \mathbf{1}_{[0,t]} \,.$$

This map is even an isometry into $L_1([0,1])$. For $0 \leq s < t \leq 1$

$$\frac{f(t) - f(s)}{t - s} = \frac{\mathbf{1}_{[s,t]}}{t - s},$$

and this expression does not tend to any limit as $t \to s$. It is known (see [4, Theorem 5.21]) that the following two properties of a Banach space X are equivalent:

(i) Every Lipschitz map $f\colon \mathbb{R} \longrightarrow X$ is differentiable a.e.

(ii) Every Lipschitz map $f\colon \mathbb{R} \longrightarrow X$ has at least one point of differentiability.

We say that a Banach space X which has these equivalent properties has the *Radon-Nikodým property* (RNP). Many other equivalent ways of defining the RNP can be found in literature. Several geometric properties of RNP spaces, including the validity of a version of the Radon-Nikodým theorem for vector measures, are discussed in detail in [4]. We refer to this book for further details and references to the original literature.

We have just noted above that $L_1([0,1])$ fails to have the RNP. Another simple argument shows that c_0 fails to have RNP as well. Indeed, consider, for example, the 1-Lipschitz map $f\colon [0,1] \longrightarrow c_0$ given as

$$f(t) = (a_1, a_2, \ldots), \quad \text{where } a_n = \int_0^t \sin(2n\pi s)\, ds.$$

The Riemann-Lebesgue lemma (or a direct calculation of the integral) guarantees that f is a c_0-valued function, but one can directly verify that $f'(t)$ does not exist at any point $t \in [0, 1]$. On the other hand, it is well known that a separable conjugate space (and hence every reflexive Banach space) has the RNP. Clearly, any space isomorphic to a subspace of a space with RNP has the RNP.

These facts, however, do not give a complete picture of the situation. Somewhat more sophisticated proofs show that there are separable spaces with the RNP which do not embed isomorphically into a separable conjugate space, and that there are separable spaces which fail to have RNP but do not contain subspaces isomorphic to $L_1([0, 1])$ or c_0. (See [4, Examples 5.25 and 5.30].)

2.2 HAAR AND ARONSZAJN-GAUSS NULL SETS

On an infinite dimensional Banach space there is no translation invariant σ-finite Borel measure (except the trivial and useless case of identically zero measure). In our setting, as well as in many others, it is essential to have an appropriate notion of "almost everywhere" in a Banach space. There is no canonical notion for this; there are several reasonable notions that are mutually nonequivalent. The simplest and most often used ones are Haar null sets and Gauss null sets.

Definition 2.2.1. Let X be a separable Banach space. A Borel set $A \subset X$ is said to be *Haar null* if there is a Borel probability measure μ on X such that

$$\mu(A + x) = 0$$

for every $x \in X$. A possibly non-Borel set is called Haar null if it is contained in a Borel Haar null set.

This notion of Haar null sets was introduced by Christensen in [11]. In finite dimensional spaces Haar null sets are exactly the sets of Lebesgue (i.e., Haar) measure zero. A simple example of a Haar null set A in an infinite dimensional Banach space X is a set for which we can find a line p in X such that the set $(p+x) \cap A$ has one-dimensional measure zero for every $x \in X$. Then we can take as μ any probability measure which is supported on p and equivalent to the Lebesgue measure on this line. Also, if A is a Borel set which is not Haar null one can show that $A - A$ is a neighborhood of the origin. Hence in any infinite dimensional Banach space every norm compact set is necessarily Haar null. The class of Haar null sets in a Banach space forms a σ-ideal.

Another often used class of null sets is the class of Gauss null sets. A Borel probability measure μ on a separable Banach space X is called *Gaussian* if for every $x^* \in X$ the measure $\nu = x^*\mu$ on $\mathbb{R}$ has a Gaussian distribution. The Gaussian measure μ is called nondegenerate if for every $x^* \neq 0$ the measure $\nu = x^*\mu$ has a positive variance or equivalently, the measure μ is not supported on a proper closed hyperplane in X.

Definition 2.2.2. A Borel set $A \subset X$ is said to be *Gauss null* if $\mu A = 0$ for every nondegenerate Gaussian measure μ on X.

For reasons that will be briefly touched upon after the statement of Theorem 2.3.1, Gauss null sets are also called Aronszajn null. Again, in finite dimensional spaces these null sets coincide with Lebesgue null sets. Since the translate of any Gaussian measure is again a Gaussian measure, every Gauss null set is a Haar null set. The converse fails in every infinite dimensional Banach space. For example, in ℓ_2 the Hilbert cube is Haar null since it is norm compact but fails to be Gauss null. For a detailed discussion of these notions and results we refer to [4]. However, the following simple property of Haar null sets is so useful that we also prove it here. Notice that, since the family of Gauss null sets is smaller, the statement also holds for Gauss null sets.

Proposition 2.2.3. *Let N be a Haar null set in a separable Banach space X and γ a Borel measurable mapping of a Borel subset Ω of $\mathbb{R}^k$ to X. Then the set of $x \in X$ for which $\mathscr{L}^k\{u \in \Omega \mid x + \gamma(u) \in N\} = 0$ is dense in X.*

Proof. Define a σ-finite Borel measure ν on X by $\nu(E) = \mathscr{L}^k(\gamma^{-1}(E))$. Assuming, as we may, that N is Borel, we find a Borel probability measure μ on X such that $\mu(N - x) = 0$ for every $x \in X$. Convolving μ with a measure whose support is the whole of X, we may without loss of generality assume that the support of μ is X. Then

$$\int_X \nu(N - x)\, d\mu = \mu \star \nu(N) = \int_X \mu(N - x)\, d\nu = 0,$$

showing that $\mathscr{L}^k\{u \in \Omega \mid x + \gamma(u) \in N\} = \nu(N - x) = 0$ for μ-almost all $x \in X$, and hence for a dense set of $x \in X$. □

Using this proposition for an affine map of $\mathbb{R}^n$ onto a k-dimensional subspace of X, we get

Corollary 2.2.4. *Let N be a Haar null set in a separable Banach space X and V be a k-dimensional subspace of X. Then there is a dense subset D of X such that for every $x \in D$ the k-dimensional measure of $(V - x) \cap N$ is zero.*

Example 2.2.5. Let E be a Borel set in X which is (Lebesgue) null on every line in the direction of a fixed vector $0 \neq u \in X$. Then E is Haar null. In particular, σ-directionally porous sets are Haar null.

Proof. Fix a line L in the direction of u and any Borel probability measure on X that is concentrated on L and absolutely continuous with respect to the Lebesgue measure on L. Then any translate of E has measure zero, showing that E is Haar null. A set porous in direction of u is (for example, by the Lebesgue density theorem) null on every line in direction of u, and the additional statement follows. □

Remark 2.2.6. If E is a directionally porous set, the function $x \to \operatorname{dist}(x, E)$ is not Gâteaux differentiable at any point of E. Hence σ-directionally porous sets belong to any σ-ideal of subsets of X with respect to which every real-valued Lipschitz function is Gâteaux differentiable almost everywhere. As we will recall shortly, this condition holds for Gauss null sets, and so σ-directionally porous sets are even Gauss null.

2.3 EXISTENCE RESULTS FOR GÂTEAUX DERIVATIVES

The theorem of Lebesgue that every Lipschitz function $f \colon \mathbb{R} \longrightarrow \mathbb{R}$ is differentiable a.e. is the basic result of the theory of differentiation. This theorem was generalized by Rademacher, who proved that every Lipschitz function $f \colon \mathbb{R}^n \longrightarrow \mathbb{R}$ ($n \geq 2$) is differentiable a.e. Our definition of the RNP easily leads to the extension of Rademacher's theorem to Lipschitz maps $f \colon \mathbb{R}^n \longrightarrow Y$, where Y is a Banach space having the RNP.

The main existence result for Gâteaux derivatives reads as follows.

Theorem 2.3.1. *Every Lipschitz map f from a separable Banach space X into a space Y with the RNP is Gâteaux differentiable almost everywhere.*

This theorem was proved in different levels of generality independently by various authors. Mankiewicz, Christensen, and Aronszajn obtained the above result, each with a different interpretation of the notion "almost everywhere." Later Phelps noted that the exceptional sets defined by Aronszajn are Gauss null. More recently, Csörnyei proved that the exceptional sets used by Mankiewicz and those defined by Aronszajn both actually coincide with the Gauss null sets. The result of Csörnyei is, however, not the end of the story. We clearly get a stronger version of the theorem once the σ-ideal used as exceptional sets to define a.e. is smaller. In [41] Preiss and Zajíček defined several examples of σ-ideals which can be used as exceptional sets in Theorem 2.3.1. The smallest of the classes defined in [41] is strictly smaller than the class of Gauss null sets. This class, called there $\widetilde{\mathcal{A}}$, is defined as follows. Let $x \in X$ and $\varepsilon > 0$. We let $\widetilde{\mathcal{A}}(x, \varepsilon)$ be the class of all Borel sets B such that

$$\mathscr{L}^1\{t \in \mathbb{R} \mid \gamma(t) \in B\} = 0$$

whenever $\gamma \colon \mathbb{R} \longrightarrow X$ is such that the Lipschitz constant of $\gamma(t) - tx$ is at most ε. We denote by $\widetilde{\mathcal{A}}(x)$ the family of all Borel sets B which can be written as $B = \bigcup_{k=1}^{\infty} B_k$, where $B_k \in \widetilde{\mathcal{A}}(x, \varepsilon_k)$ for suitable $\varepsilon_k > 0$. Finally, we define $\widetilde{\mathcal{A}}$ as the class of all Borel sets B which can be, for every sequence $(x_n)_{n=1}^{\infty}$ whose closed linear span is X, written in the form

$$B = \bigcup_{n=1}^{\infty} B_n \quad \text{with} \quad B_n \in \widetilde{\mathcal{A}}(x_n).$$

This definition is inspired by Aronszajn's definition of null sets; see [3] or [4, Definition 6.23(ii)]. It is an open question whether $\widetilde{\mathcal{A}}$ is the smallest possible σ-ideal of Borel sets which can be used in Theorem 2.3.1. We shall not use this σ-ideal in this volume.

Even if we consider Lipschitz maps from $\mathbb{R}^n$ to $\mathbb{R}^m$, the question of structure of the sets of nondifferentiability points is delicate and not completely resolved. It is a classical fact that Lebesgue's theorem is sharp. (See, for example, [4, page 165].) A more precise result was proved by Zahorski [46]. A set $A \subset \mathbb{R}$ is a $G_{\delta\sigma}$ set of Lebesgue measure zero if and only if there is a Lipschitz function $f \colon \mathbb{R} \longrightarrow \mathbb{R}$ which is differentiable exactly at points of $\mathbb{R} \setminus A$. (For a more up-to-date proof see [19].) However, while considering questions of Fréchet differentiability of Lipschitz functions from a Banach space to $\mathbb{R}$, Preiss proved in [39] that Rademacher's theorem is not always sharp. There is a Borel set $A \subset \mathbb{R}^2$ of Lebesgue measure zero such that every Lipschitz function $f \colon \mathbb{R}^2 \longrightarrow \mathbb{R}$ is differentiable somewhere in A. This topic is studied in detail in [2], and the situation is still not completely clear.

Theorem 2.3.1 holds also if f is defined just on an open set in X. The same remark applies to all proofs of existence of derivatives in this volume. For the discussion and proof of Theorem 2.3.1 and references to the original literature we refer to [4]. Here we just give a simple corollary, which we need for future reference.

Lemma 2.3.2. *Let X be a separable Banach space, $x_0 \in X$, and Y, Z be subspaces of X such that $X = Y + Z$. If E is a Banach space with the RNP, f is a Lipschitz map from an open subset G of X to E and $T \in L(\mathbb{R}^p, X)$, then there is a dense set of points $(y, z) \in Y \times Z$ such that f is Gâteaux differentiable at the point $x_0 + z + ty + Tu$ for almost all $(t, u) \in \mathbb{R} \times \mathbb{R}^p$ for which this point belongs to G.*

Proof. Define $\Phi \colon \mathbb{R} \times \mathbb{R}^p \times Y \times Z \longrightarrow X$ by

$$\Phi(t, u, y, z) = x_0 + z + ty + Tu.$$

Then $g := f \circ \Phi$ is locally Lipschitz on $\Phi^{-1}(G)$, so it is Gâteaux differentiable except for a Gauss null set by Theorem 2.3.1. Hence given an arbitrary $(\tilde{y}, \tilde{z}) \in Y \times Z$ we can find $(y, z) \in Y \times Z$ as close to $(\tilde{y}, \tilde{z})$ as we wish so that g is Gâteaux differentiable at (t, u, y, z), for almost every $(t, u) \in \mathbb{R} \times \mathbb{R}^p$ for which $\Phi(t, u, y, z) \in G$. To see that such pairs y, z have the required properties, we just need to realize that Gâteaux differentiability of g at a point (t, u, y, z), where $t \neq 0$, implies Gâteaux differentiability of f at $x_0 + z + ty + Tu$. Indeed, the derivative of f at $x_0 + z + ty + Tu$ in the direction $x = v + w$, where $v \in Y$ and $w \in Z$, is

$$f'\big(x_0 + z + ty + Tu; x\big) = g'\big((t, u, y, z); (0, \ldots, 0, v/t, w)\big),$$

and this expression is linear in $x \in X$. $\qquad\square$

2.4   MEAN VALUE ESTIMATES

Throughout this work, results on existence of derivatives are intimately connected with (usually) validity or (mainly in Chapter 14) nonvalidity of an appropriate form of the mean value estimates. For Gâteaux derivatives, any of the notions of null sets with which Theorem 2.3.1 has been proved leads to such estimates. We first treat the one-dimensional case and then use this opportunity to explain what is behind the higher dimensional case. In the additional statement we use for definiteness the Haar null sets (as then the statement with the Gauss null sets follows), but the statement holds also for other null sets, with a similarly easy proof.

Proposition 2.4.1. *Let f be a real-valued locally Lipschitz function on an open subset G of a separable Banach space X, and let $a, b \in G$ be such that the straight segment $[a, b]$ is contained in G. Then for every $\varepsilon > 0$ there is a point $x \in G$ at which f is Gâteaux differentiable and*

$$f'(x)(b - a) > f(b) - f(a) - \varepsilon. \tag{2.1}$$

Consequently, for any Lipschitz map g from an open convex set G in a separable Banach space to an RNP space and $\varepsilon > 0$, there is a point $x \in G$ at which g is Gâteaux differentiable and

$$\|g'(x)\| > \mathrm{Lip}(g) - \varepsilon. \tag{2.2}$$

If, in addition, a Haar null set $N \subset X$ is given, we may find the point x satisfying (2.1) (resp. (2.2)) belonging to $G \setminus N$.

Proof. We prove directly the first part of the last statement, assuming, as we may, that $a \neq b$, $\mathrm{Lip}(f) = 1$ and N contains the points of Gâteaux nondifferentiability of f. Let L be a line in the direction $b - a$. By Corollary 2.2.4 find $y \in X$ with $\|y\| < \frac{1}{2}\varepsilon$ such that $L - y$ meets N in a set of one-dimensional measure zero. We may take $\|y\|$ small enough that the segment $[a - y, b - y]$ lies entirely in G. Since for almost every $t \in \mathbb{R}$, $a - y + t(b - a) \notin N$, and since

$$\int_0^1 \frac{d}{dt} f\big(a - y + t(b - a)\big)\, dt = f(b - y) - f(a - y) > f(b) - f(a) - \varepsilon, \tag{2.3}$$

we infer that there is $t_0 \notin N$ with

$$f(b) - f(a) - \varepsilon < \frac{d}{dt} f\big(a - y + t_0(b - a)\big) = f'\big(a - y + t_0(b - a)\big)(b - a).$$

For the second statement, we again assume $\mathrm{Lip}(g) = 1$ and find $a, b \in G$, $a \neq b$ such that $\|g(b) - g(a)\| > \|b - a\| - \varepsilon\|b - a\|$. Then choose $x^* \in X^*$ with norm one such that $x^*(g(b) - g(a)) = \|g(b) - g(a)\|$. We use the first part of the proof with N containing set of points of Gâteaux nondifferentiability of g and with $f = x^* \circ g$ to find $x \in G \setminus N$ at which

$$\|g'(x)\| \geq \frac{(x^* \circ g)'(x)(b - a)}{\|b - a\|} > \frac{x^*(g(b) - g(a)) - \varepsilon\|b - a\|}{\|b - a\|} \geq 1 - \varepsilon. \qquad \square$$

We use this opportunity to explain how we understand the validity of the mean value estimate for other derivatives. In the general setting, we are given a map $f \colon G \longrightarrow Y$ and a set $D \subset G$ contained in the set of points of Gâteaux differentiability of f (for example, D may be the set of points of Fréchet differentiability of f). We wish to get inequalities similar to those in Proposition 2.4.1. To be more precise, we start with the case when f is real-valued. Then the most natural meaning of the mean value estimate is the following.

- For every $a, b \in G$ such that $[a, b] \subset G$,

$$\inf_{x \in D} f'(x; b - a) \leq f(b) - f(a) \leq \sup_{x \in D} f'(x; b - a). \tag{2.4}$$

If X is separable (which we assume here), we already know that this statement holds with Gâteaux derivatives, and so we may reformulate this inequality as a comparison between derivatives achieved in D and all Gâteaux derivatives. Again the example of Fréchet derivatives may explain it best: we may equivalently say that a real-valued locally Lipschitz function f on an open subset G of a separable Banach space X *satisfies the mean value estimate for Fréchet derivatives* if the following condition holds.

- For every $x \in G$ at which f is Gâteaux differentiable and for every $u \in X$ and $\varepsilon > 0$, there are $y, z \in G$ such that f is Fréchet differentiable at both y and z and $f'(y)(u) + \varepsilon > f'(x)(u) > f'(z)(u) - \varepsilon$.

Another easy reformulation of the same condition, in spirit closer to the theory of sub-differentials, is the following.

- For any open $H \subset G$ the w^*-closed convex hull of the sets of Fréchet and Gâteaux derivatives attained at points of H coincide.

Yet another easy reformulation, this time in more geometric terms:

- For any open $H \subset G$ every w^*-slice S of the set of Gâteaux derivatives of f attained in H contains a Fréchet derivative.

Recall that a w^*-slice S of a subset $M \subset X^*$ is any nonempty set of the form

$$S = \{x^* \in M \mid x^*(x) > c\},$$

where $x \in X$ and $c \in \mathbb{R}$.

It is rarely interesting to know only that a particular function satisfies these esti-mates. Usually, one wishes to know that (2.4) holds for every function in a given class. Obviously, if this class is closed under changing sign, then either of the inequalities from (2.4) is sufficient. However, in Chapter 12 we prove a Fréchet differentiability result also for the class of so-called cone-monotone functions, for which only one of the inequalities from (2.4) holds.

For vector-valued f there are several possibilities. We may require the rather weak condition that the second statement of Proposition 2.4.1 holds. (Notice that this implies the weakest form of the mean value estimates, namely, that the function is constant pro-vided its derivative vanishes at every considered point.) Alternatively, we may require the perhaps strongest condition, the validity of the first statement of Proposition 2.4.1 for every $y^* \circ f$, $y^* \in Y^*$. (Various conditions between these two have also been used.) Here we show that under relatively mild assumptions these two conditions are equivalent. There are clearly many variants of this statement; we have chosen the one we will actually use. Notice, however, that easy examples show these conditions are not equivalent in general. The proof has additional interest since it explains the rea-son behind various "change of norm" arguments that have been employed in finding a point of Fréchet differentiability. While the algorithmic procedure kept choosing points x_k and directions h_k with the directional derivative $f'(x_k, h_k)$ as large as pos-sible, to assure convergence one needed that the new direction be close to the old one and the directional derivative at the new point in the old direction be close to the di-rectional derivative at the old point in the old direction; i.e. that $\|h_{k+1} - h_k\|$ and $|f'(x_{k+1}, h_k) - f'(x_k, h_k)|$ are small. This is essentially what we do in the argument below.

Observation 2.4.2. *Let f be a locally Lipschitz map of an open subset G of a Banach space X to a Banach space Y and let $D \subset G$ be a set of points at each of which f*

is Gâteaux differentiable. Suppose further that when X and Y are equipped with any norms of the form

$$\|x\| = \max\{\|x\|, C\|x - e^*(x)e\|\} \text{ and } \|y\| = \|y\| + |y^*(y)|,$$

respectively, where $C > 0$, $e \in X$, $e^* \in X^*$, $\|e^*\| = \|e\| = e^*(e) = 1$ and $y^* \in Y^*$, then for every $T \in L(X,Y)$ there is $x \in D$ satisfying

$$\|f'(x) + T\| > \mathrm{Lip}_{\|\cdot\|}(f + T) - 1.$$

Then for every $\varepsilon > 0$, $y^* \in Y^*$, and $a, b \in G$ such that $[a, b] \subset G$ there is a point $x \in D$ such that

$$(y^* \circ f)'(x)(b - a) > (y^* \circ f)(b) - (y^* \circ f)(a)) - \varepsilon.$$

Proof. Let $\varepsilon > 0$, $y^* \in Y^*$, and $a, b \in G$. We may assume that $\|y^*\| = 1$ and $b \neq a$. Choose a large constant $C > 1$; particular requirements will become clear as the argument develops. Let $e \in X$ be the unit vector in the direction of $b - a$, choose $e^* \in X^*$ such that $\|e^*\| = e^*(e) = 1$, and also choose $y_0 \in Y$ such that $\|y_0\| = 1$ and $y^*(y_0) > 1 - 1/C^2$. We introduce new norms

$$\|x\| = \max\{\|x\|, C^2\|x - e^*(x)e\|\} \text{ and } \|y\| = \|y\| + C|y^*(y)|$$

and the operator $T \in L(X,Y)$ by

$$Tx = Ce^*(x)y_0 - e^*(x)w, \text{ where } w = \frac{f(b) - f(a)}{\|b - a\|}.$$

By assumption, there is $x \in D$ such that $\|f'(x) + T\| > \mathrm{Lip}_{\|\cdot\|}(f + T) - 1$.

We show that x is the point we require. This just needs several estimates. First notice that

$$\begin{aligned}
\mathrm{Lip}_{\|\cdot\|}(f + T) &\geq \frac{\|f(b) - f(a) + T(b - a)\|}{\|b - a\|} \\
&= C\|y_0\| \\
&= C(1 + C|y^*(y_0)|) > C^2 + C - 1.
\end{aligned}$$

Let $u \in X$ be such that $\|u\| = 1$ and $\|(f'(x) + T)(u)\| > C^2 + C - 1$. We denote $v = u - e^*(u)e$ for short. Clearly, $e^*(v) = 0$ and $C^2\|v\| \leq \|u\| = 1$. Hence

$$\begin{aligned}
\|(f'(x) + T)(v)\| &= \|f'(x)(v)\| \\
&\leq (1 + C)\|f'(x)(v)\| \\
&\leq (1 + C)\,\mathrm{Lip}_{\|\cdot\|}(f)\,\|v\| \leq \frac{1 + C}{C^2}\,\mathrm{Lip}_{\|\cdot\|}(f) \leq 1,
\end{aligned}$$

provided C is big enough. Since also $|e^*(u)| \leq \|u\| \leq \|u\| = 1$, we obtain

$$\begin{aligned}
\|(f'(x) + T)(e)\| &\geq \|(f'(x) + T)(e^*(u)e)\| \\
&\geq \|(f'(x) + T)(u)\| - \|(f'(x) + T)(v)\| \\
&> C^2 + C - 2.
\end{aligned}$$

For the upper estimate of $\|(f'(x) + T)(e)\|$ we use the fact that

$$
\begin{aligned}
y^*(f'(x)(e) + Te) &= y^*(f'(x)(e) + Cy_0 - w) \\
&\geq Cy^*(y_0) - 2\operatorname{Lip}_{\|\cdot\|}(f) \\
&\geq C - 1/C - 2\operatorname{Lip}_{\|\cdot\|}(f) \geq 0,
\end{aligned}
$$

assuming again that C is big. It now follows

$$
\begin{aligned}
\|(f'(x) + T)(e)\| &= \|(f'(x) + T)(e)\| + Cy^*(f'(x)(e) + Te) \\
&\leq C + 2\operatorname{Lip}_{\|\cdot\|}(f) + C^2 + Cy^*(f'(x)(e) - w).
\end{aligned}
$$

Hence $C + 2\operatorname{Lip}_{\|\cdot\|}(f) + C^2 + Cy^*(f'(x)(e) - w) > C^2 + C - 2$, implying that

$$
y^*(f'(x)(e) - w)) > -\frac{2}{C}(1 + \operatorname{Lip}_{\|\cdot\|}(f)),
$$

which, multiplied by $\|b - a\|$ and for C large enough, gives the statement. $\qquad\square$

So far we have considered what could be termed one-dimensional mean value estimates: although the range could be even infinite dimensional, the estimate involved only derivative in a single direction. Arguments similar to those of Proposition 2.4.1 also lead to higher dimensional mean value estimates for Gâteaux derivatives: just use the divergence theorem (or the Stokes theorem) instead of (2.3). The natural formulation may then sound somewhat awkward, but we may follow the same reasoning as above. If, instead of a line segment, we imagine an n-dimensional (say C^1) surface in X, then for any Lipschitz $f\colon X \longrightarrow Y$ and $u_1^*, \ldots, u_n^* \in Y^*$ the appropriate integral over the boundary of the surface will be majorized by the supremum of expressions

$$
\sum_{k=1}^{n} (u_k^* \circ f)'(x; u_k),
$$

where x lies on the surface, $u_1, \ldots, u_n$ are suitable tangent vectors to the surface, and $u_k^* \circ f$ are differentiable in the direction of the linear span of the vectors $u_1, \ldots, u_n$. We may therefore define the *validity of the multidimensional mean value estimate for Gâteaux (resp. Fréchet) derivatives* of a locally Lipschitz function from an open subset G of a Banach space X to a Banach space Y by the following requirement.

- For every $x \in G$, $n \in \mathbb{N}$, every choice of vectors $u_1, \ldots, u_n \in X$ and functionals $u_1^*, \ldots, u_n^* \in Y^*$ such that the functions $u_i^* \circ f$ are differentiable in the direction of the linear span of $u_1, \ldots, u_n$, and for every $\varepsilon > 0$, there is $y \in G$, arbitrarily close to x, such that f is Gâteaux (resp. Fréchet) differentiable at y and

$$
\sum_{k=1}^{n} u_k^*(f'(y; u_k)) > \sum_{k=1}^{n} (u_k^* \circ f)'(x; u_k) - \varepsilon.
$$

To avoid possible misunderstanding, we point out that the assumption of differentiability of $u_i^* \circ f$ in the direction of $\operatorname{span}\{u_1, \ldots, u_n\}$ cannot be replaced by the existence of all directional derivatives $(u_k^* \circ f)'(x; u_k)$, even though only these derivatives appear in the mean value formula. For example, let $f : \mathbb{R}^2 \longrightarrow \mathbb{R}^2$ be defined by

$$f(s,t) = \big(\max\{|s| - |t|, 0\}\operatorname{sign}(s), \max\{|t| - |s|, 0\}\operatorname{sign}(t)\big).$$

On a neighborhood of every point at which f is differentiable one of the components of the maximum is zero, which implies $\frac{\partial f_1}{\partial s} + \frac{\partial f_2}{\partial t} = 1$; but at the origin this expression has value 2.

We show the simple but important fact that in the situation when we know that Lipschitz functions have enough points of Gâteaux differentiability, the multidimensional mean value estimate for Gâteaux derivatives holds.

Proposition 2.4.3. *Let f be a locally Lipschitz function from an open subset G of a separable Banach space X to a Banach space Y with the RNP. Then for every $x \in G$, $n \in \mathbb{N}$, every choice of vectors $u_1, \ldots, u_n \in X$ and functionals $u_1^*, \ldots, u_n^* \in Y^*$ such that the functions $u_i^* \circ f$ are differentiable in the direction of the linear span of $u_1, \ldots, u_n$, and every $\varepsilon > 0$, there is $z \in G$, arbitrarily close to x, such that f is Gâteaux differentiable at z and*

$$\sum_{k=1}^n u_k^*(f'(z; u_k)) > \sum_{k=1}^n (u_k^* \circ f)'(x; u_k) - \varepsilon. \tag{2.5}$$

Moreover, if $N \subset X$ is Haar null, the point z may be found in $G \setminus N$.

Proof. It suffices to prove the additional statement assuming also that the Haar null set N contains all points at which f is not Gâteaux differentiable. Define $T \colon \mathbb{R}^n \longrightarrow X$ and $L \colon Y \longrightarrow \mathbb{R}^n$ by

$$Tw = x + \sum_{k=1}^n w_k u_k \quad \text{and} \quad Ly = (u_1^*(y), \ldots, u_k^*(y)).$$

Let $g \colon \mathbb{R}^n \longrightarrow \mathbb{R}^n$ be the composition $g = L \circ f \circ T$. Observe that since f is differentiable in the direction of the image of T, g is differentiable at 0.

Let $\eta = \varepsilon/(2n)$ and suppose that $\delta_0 > 0$ is such that $B(x, \delta_0) \subset G$ and f is Lipschitz on $B(x, \delta_0)$ with constant, say, C. Find $0 < \delta < \frac{1}{2}\delta_0$ such that $C\|L\|\delta < \eta$, and choose $0 < r < 1$ such that, letting $\Omega := \overline{B_{\mathbb{R}^n}(0, r)}$, we have $T(\Omega) \subset B(x, \frac{1}{2}\delta_0)$ and

$$|g(w) - g(0) - g'(0)(w)| \le \eta|w|$$

for every $w \in \Omega$. By Proposition 2.2.3 there is $\tilde{x} \in X$ with $\|\tilde{x}\| < \delta r$ such that the set $M := \{w \in \mathbb{R}^n \mid \tilde{x} + Tw \in N\}$ has Lebesgue measure zero. If we denote for a moment by h the function $h \colon \mathbb{R}^n \longrightarrow \mathbb{R}^n$ given by

$$h(w) = L \circ f(\tilde{x} + Tw),$$

then

$$|h(w) - g(0) - g'(0)(w)| \le |h(w) - g(w)| + |g(w) - g(0) - g'(0)(w)|$$
$$\le C\|L\|\|\tilde{x}\| + \eta r < 2\eta r$$

for $w \in \Omega$. Denoting by ν_Ω the outer unit normal vector to Ω, and recalling that $\mathscr{H}^{n-1}\partial\Omega = n\alpha_n r^n$, we obtain by the divergence theorem,

$$\int_\Omega \operatorname{div}\big(h(w) - g'(0)(w)\big)\, d\mathscr{L}^n(w)$$
$$= \int_\Omega \operatorname{div}\big(h(w) - g(0) - g'(0)(w)\big)\, d\mathscr{L}^n(w)$$
$$= \int_{\partial\Omega} \big(h(w) - g(0) - g'(0)(w)\big) \cdot \nu_\Omega\, d\mathscr{H}^{n-1}(w)$$
$$> -2\eta n\alpha_n r^n.$$

This implies that there is $w \in \Omega \setminus M$ such that

$$\operatorname{div} h(w) > \operatorname{div}(g'(0)(w)) - 2\eta n = \sum_{k=1}^n (u_k^* \circ f)'(x; u_k) - \varepsilon. \qquad (2.6)$$

Let $z = \tilde{x} + Tw$. Then $z \notin N$, $\|z - x\| \le \|\tilde{x}\| + \|Tw - x\| < \delta_0$, f is Gâteaux differentiable at z, and (2.6) is precisely (2.5). $\qquad\square$

It follows that for locally Lipschitz maps of separable Banach spaces to spaces with the RNP the condition for the validity of the multidimensional mean value estimate may be simplified to:

- For every $x \in G$ at which f is Gâteaux differentiable, $n \in \mathbb{N}$, every choice of $u_1, \dots, u_n \in X$ and $u_1^*, \dots, u_n^* \in Y^*$, and every $\varepsilon > 0$ there is $y \in G$, arbitrarily close to x, such that f is Fréchet differentiable at y and

$$\sum_{k=1}^n u_k^*(f'(y; u_k)) > \sum_{k=1}^n u_k^*(f'(x; u_k)) - \varepsilon.$$

Equivalent conditions similar to the ones we stated for the real-valued function may again be given. Perhaps the most natural one is in the language of slices. It would say exactly same as the one for real-valued functions:

- For any open $H \subset G$ every w^*-slice S of the set of Gâteaux derivatives of f attained in H contains a Fréchet derivative.

But this time, a w^*-slice S of a subset $M \subset L(X, Y)$ is any nonempty set of the form

$$S = \Big\{ L \in M \,\Big|\, \sum_{i=1}^m y_i^* L x_i > c \Big\},$$

where $m \in \mathbb{N}$, $x_1, \dots, x_m \in X$, $y_1^*, \dots, y_m^* \in Y^*$, and $c \in \mathbb{R}$.

Chapter Three

Smoothness, convexity, porosity, and separable determination

In this chapter we prove some results that will be crucial in what follows; in particular, we show that spaces with separable dual admit a Fréchet smooth norm. For the first time, we meet the σ-porous sets and see their relevance for differentiability: the set of points of Fréchet nondifferentiability of continuous convex functions forms a σ-porous set. We also give some basic facts about σ-porous sets and show that they are contained in sets of Fréchet nondifferentiability of real-valued Lipschitz functions. These results, and their proofs, are important in order to understand some of the development that follows. So, although much of this material is discussed in greater detail in [4], and we refer again to this book for the original references, we present here the most relevant results and arguments. Finally, we discuss a new way of treating separable reduction arguments that can be used to extend some of our results to nonseparable setting.

3.1 A CRITERION OF DIFFERENTIABILITY OF CONVEX FUNCTIONS

Continuous convex functions on Banach spaces are easily seen to be locally Lipschitz (see [4, Proposition 4.6]). Thus the differentiability of such functions is naturally included, as a much simpler case, in the study of differentiability of Lipschitz functions. The key to the simplicity is that convexity implies "differentiability from below," which is usually expressed more precisely as nonemptiness of its subdifferential $\partial f(x_0)$ of f at x_0.

Definition 3.1.1. Let f be a convex function defined on a convex set C and $x_0 \in C$. The *subdifferential* $\partial f(x_0)$ *of* f *at* x_0 is the set of functionals $x^* \in X^*$ satisfying

$$x^*(x - x_0) \leq f(x) - f(x_0) \quad \text{for all } x \in C.$$

Geometrically, this is the set of all linear functionals which support the convex epigraph of f at $(x_0, f(x_0))$. Hence, by the Hahn-Banach theorem, in the interior points of C the subdifferential of a continuous convex function is nonempty.

Proposition 3.1.2. *Let* f *be a continuous convex function on an open convex subset* $C \subset X$ *of a Banach space* X, *and let* $x \in C$. *Then for every* $x^* \in \partial f(x)$ *and every* u *such that* $x \pm u \in C$,

$$0 \leq f(x + u) - f(x) - x^*(u) \leq f(x + u) + f(x - u) - 2f(x).$$

Proof. The first inequality comes from the definition of the subdifferential. The same definition implies that $f(x-u) - f(x) + x^*(u) \geq 0$, and the second inequality follows by adding it to $f(x+u) - f(x) - x^*(u)$. □

An immediate corollary of this statement is a useful criterion for Fréchet or Gâteaux differentiability of convex functions on Banach spaces. Its main point is that differentiability may be easily proved without actually finding the derivative. Although one can prove criteria of differentiability avoiding the value of the derivative also in more general situations, they become too complicated and their usefulness seems to be limited to special tasks (see, for example, Section 3.6).

Proposition 3.1.3. *Let f be a continuous convex function on an open convex subset C of a Banach space X and let $x \in C$. Then*

(i) *f is Gâteaux differentiable at x if and only if for any $u \in X$*

$$f(x + tu) + f(x - tu) - 2f(x) = o(t) \ \text{as } t \searrow 0;$$

(ii) *f is Fréchet differentiable at x if and only if*

$$f(x + u) + f(x - u) - 2f(x) = o(\|u\|) \ \text{as } u \to 0.$$

Proof. From Proposition 3.1.2 we immediately see that any element of the subdifferential of f at x is its Gâteaux (resp. Fréchet) derivative. □

We may notice that the above criterion did not use the full strength of convexity. For example, for the validity of (ii) we just needed nonemptiness of the "Fréchet subdifferential" of f at x, that is, existence of $x^* \in X^*$ such that

$$\liminf_{u \to 0} \frac{f(x + u) - f(x) - x^*(u)}{\|u\|} \geq 0.$$

We also remark that Proposition 3.1.2 shows that if a continuous convex function f is Gâteaux differentiable at x then its subdifferential at x has a single element, namely, its Gâteaux derivative at x. Conversely, if $\partial f(x)$ consists of a single point, f is Gâteaux differentiable at x. This may be seen by an application of the Hahn-Banach theorem. Details, and much additional information, may be found in [4, Chapter 4].

3.2 FRÉCHET SMOOTH AND NONSMOOTH RENORMINGS

An important example of a continuous convex function on a Banach space is its norm. There are several known and for us very useful results on existence of interesting norms in rather general Banach space. The first result of this nature which we present here is due to Klee and Kadec independently.

Theorem 3.2.1. *Let X be a Banach space such that X^* is separable. Then there exists an equivalent norm on X which is Fréchet differentiable at every point $x \neq 0$.*

Such a norm is called a *Fréchet smooth* norm, or just *smooth* norm. Observe that no norm is (even Gâteaux) differentiable at the origin.

The proof of this theorem is based on the following simple yet important lemma. To state it, recall that the space X is said to be *locally uniformly convex* if $\|x_n - x\| \to 0$ whenever x, x_n are unit vectors and $\|x + x_n\| \to 2$. Naturally, such norms are called locally uniformly convex.

Lemma 3.2.2. *Assume that X^* is locally uniformly convex. Then the norm in X is smooth.*

Proof. Let $x \in X$ with $\|x\| = 1$. Find $x^* \in X^*$ such that $\|x^*\| = x^*(x) = 1$. For every $u \in X$ we choose a norm one functional y_u^* in X^* such that $y_u^*(x+u) = \|x+u\|$. Then

$$\|y_u^* + x^*\| \geq y_u^*(x) + x^*(x) = \|x + u\| - y_u^*(u) + 1.$$

Hence $\|y_u^* + x^*\| \to 2$ as $\|u\| \to 0$. By the local uniform convexity of X^* we deduce that $\|y_u^* - x^*\| \to 0$ as $\|u\| \to 0$. It follows that

$$\begin{aligned}
0 \leq \|x + u\| + \|x - u\| - 2 &= y_u^*(x + u) + y_{-u}^*(x - u) - 2 \\
&= y_u^*(x) + y_{-u}^*(x) - 2 + (y_u^* - y_{-u}^*)(u) \\
&\leq \|y_u^* - y_{-u}^*\| \, \|u\| = o(\|u\|). \qquad \square
\end{aligned}$$

Proof of Theorem 3.2.1. Since both X and X^* are separable there are a dense sequence $(x_n)_{n=1}^\infty$ in the unit sphere of X and an increasing sequence of finite dimensional subspaces $(F_n)_{n=1}^\infty$ of X^* such that

$$X^* = \overline{\bigcup_{n \geq 1} F_n}.$$

Consider now the following maps from X^* into ℓ_2:

$$Tx^* = \left(x^*(x_1), \tfrac{1}{2} x^*(x_2), \ldots, \tfrac{1}{n} x^*(x_n), \ldots \right),$$

$$Sx^* = \left(\mathrm{dist}(x^*, F_1), \tfrac{1}{2} \mathrm{dist}(x^*, F_2), \ldots, \tfrac{1}{n} \mathrm{dist}(x^*, F_n), \ldots \right).$$

The function on X^* given as $x^* \to \|Tx^*\|$ is w^*-continuous on bounded sets and the function $x^* \to \|Sx^*\|$ is w^*-lower semicontinuous on bounded sets. Also

$$\|T(x^* + y^*)\| \leq \|Tx^*\| + \|Ty^*\| \quad \text{and} \quad \|S(x^* + y^*)\| \leq \|Sx^*\| + \|Sy^*\|.$$

Hence we can define the equivalent norm $\|\cdot\|$ on X^* by

$$\|x^*\|^2 := \|x^*\|^2 + \|Tx^*\|^2 + \|Sx^*\|^2. \tag{3.1}$$

Since all three summands are w^*-lower semicontinuous the unit ball in this norm is w^*-closed. Thus $\|\cdot\|$ is norm dual to an equivalent norm on X. By Lemma 3.2.2 it suffices to prove that $(X^*, \|\cdot\|)$ is locally uniformly convex.

Assume that $\|x^*\| = \|x_k^*\| = 1$ for all $k \in \mathbb{N}$ and

$$\lim_{k \to \infty} \|x^* + x_k^*\| = 2.$$

Since each of the expressions $\|x^*\|$, $\|Tx^*\|$, and $\|Sx^*\|$ in (3.1) satisfies the triangle inequality, we obtain that

$$\left(\|x_k^*\| - \|x^*\|\right)^2 + \left(\|Tx_k^*\| - \|Tx^*\|\right)^2 + \left(\|Sx_k^*\| - \|Sx^*\|\right)^2$$
$$\leq \left(2\|x_k^*\|^2 + 2\|x^*\|^2 - \|x_k^* + x^*\|^2\right) \tag{3.2}$$
$$+ \left(2\|Tx_k^*\|^2 + 2\|Tx^*\|^2 - \|T(x_k^* + x^*)\|^2\right) \tag{3.3}$$
$$+ \left(2\|Sx_k^*\|^2 + 2\|Sx^*\|^2 - \|S(x_k^* + x^*)\|^2\right) \tag{3.4}$$
$$= 2\|x_k^*\|^2 + 2\|x^*\|^2 - \|x_k^* + x^*\|^2 \to 0.$$

Hence $\|x_k^*\| \to \|x^*\|$ (which we will not use) and, more important,

$$\|Tx_k^*\| \to \|Tx^*\| \quad \text{and} \quad \|Sx_k^*\| \to \|Sx^*\|.$$

Since each of the expressions (3.2), (3.3), and (3.4) is positive,

$$\|T(x^* + x_k^*)\| \to 2\|Tx^*\| \quad \text{and} \quad \|S(x^* + x_k^*)\| \to 2\|Sx^*\|.$$

By the (local) uniform convexity of ℓ_2 it follows that

$$\|T(x^* - x_k^*)\| \to 0 \quad \text{and} \quad \|S(x^* - x_k^*)\| \to 0,$$

which implies that for each n,

$$\lim_{k \to \infty} x_k^*(x_n) = x^*(x_n) \quad \text{and} \quad \lim_{k \to \infty} \operatorname{dist}(x_k^*, F_n) = \operatorname{dist}(x^*, F_n).$$

Since the sequence (x_n) is dense in the unit sphere in X and (x_k^*) is bounded, we infer that $x_k^* \to x^*$ in the w^*-topology. For a contradiction, suppose that for some $\varepsilon > 0$ there is a subsequence of (x_k^*) (which we may assume to be the original sequence) such that

$$\liminf_{k \to \infty} \|x_k^* - x^*\| > \varepsilon.$$

Pick n such that $\operatorname{dist}(x^*, F_n) < \frac{1}{3}\varepsilon$. Then for every k large enough there is a $z_k^* \in F_n$ with $\|z_k^* - x_k^*\| < \frac{1}{3}\varepsilon$. Since F_n is finite dimensional, a subsequence of (z_k^*) converges to some $z^* \in F_n$. Thus

$$\|x^* - z^*\| \leq \liminf_{k \to \infty} \|x_k^* - z^*\| \leq \liminf_{k \to \infty} \|x_k^* - z_k^*\| \leq \frac{\varepsilon}{3}$$

and we get the desired contradiction,

$$\varepsilon < \liminf_{k \to \infty} \|x_k^* - x^*\| \leq \|x^* - z^*\| + \lim_{k \to \infty} \|z^* - z_k^*\| + \limsup_{k \to \infty} \|z_k^* - x_k^*\| \leq \varepsilon. \qquad \square$$

In the space ℓ_1 it is easily seen that the norm is nowhere Fréchet differentiable. The next renorming result, whose first form is due to Leach and Whitfield, shows that this is a special case of a general phenomenon. We state and prove it in a rather strong form that will allow immediate applications to proving results showing that existence of suitable smooth functions implies separability of the dual.

Theorem 3.2.3. *Assume that $(X, \| \cdot \|)$ is separable but X^* is nonseparable. Then for every $0 < c < 1$ there is an equivalent norm $\| \cdot \|$ on X such that $\| \cdot \| \leq \| \cdot \|$, and for every $x \in X$, $r > 0$, and every subspace Y of X such that X/Y has separable dual,*

$$\sup_{y \in Y, \|y\| \leq r} \left(\|x + y\| - \|x\| \right) > cr.$$

Proof. We choose $1 > a > b > c$ and $\varepsilon, \tau > 0$ such that $(1 - \tau)b - \tau > c$. Let $\mathcal{E}$ be the family of norm separable subspaces of X^*. For every $E \in \mathcal{E}$ let

$$M_E = \{x^* \in X^* \mid \|x^*\| \leq 1, \ \text{dist}(x^*, E) > a\} \ \text{and}$$
$$p_E(x) = \sup_{x^* \in M_E} x^*(x).$$

Then p_E is a pseudonorm on X majorized by $\| \cdot \|$. Moreover, we notice that the pseudonorm p_E satisfies $p_E \leq \min\{p_{E_1}, p_{E_2}\}$ whenever $E \supset E_1 \cup E_2$. Thus the infimum

$$p(x) := \inf_{E \in \mathcal{E}} p_E(x)$$

is also a pseudonorm on X majorized by $\| \cdot \|$. Consequently,

$$\|x\| := \tau\|x\| + (1 - \tau)p(x)$$

is an equivalent norm X majorized by $\| \cdot \|$.

Suppose now that $x \in X$, $r > 0$ and Y is a subspace of X such that X/Y has separable dual. We let $\varepsilon = \frac{1}{2}(a - b)r$ and show that there is $y \in Y$ with $\|y\| = 1$ such that for every $E \in \mathcal{E}$ one can find $y_E^* \in M_E$ with

$$|y_E^*(x) - p(x)| < 2\varepsilon \quad \text{and} \quad y_E^*(y) > a. \tag{3.5}$$

Indeed, if this were not the case, then for every y belonging to a countable dense subset S of the unit sphere of Y we would find $E_y \in \mathcal{E}$ such that $y^*(y) \leq a$ for every $y^* \in M_{E_y}$ with $|y^*(x) - p(x)| < 2\varepsilon$. Choose $E \in \mathcal{E}$ containing $Y^\perp \cup \bigcup_{y \in S} E_y$ such that $|p_E(x) - p(x)| < \varepsilon$. Then by the definition of p_E there is $y^* \in M_E$ such that $y^*(x) > p_E(x) - \varepsilon$, and so

$$|y^*(x) - p(x)| \leq |y^*(x) - p_E(x)| + |p_E(x) - p(x)| < 2\varepsilon.$$

Since $M_E \subset \bigcap_{y \in S} M_{E_y}$ we see that y^* belongs to all $M_{E_y}, y \in S$, which implies that $\|y^*|_Y\| \leq a$. Hence

$$a \geq \|y^*|_Y\| = \text{dist}(y^*, Y^\perp) > a.$$

Having found $y \in Y$ and $y_E^* \in M_E$, $E \in \mathcal{E}$, such that $\|y\| = 1$ and (3.5) holds, we estimate

$$p_E(x + ry) \geq y_E^*(x + ry) = y_E^*(x) + ry_E^*(y) > p(x) - 2\varepsilon + ar.$$

Hence $p(x + ry) - p(x) \geq -2\varepsilon + ar = br$ and we are ready to finish the proof by concluding that

$$\|x + ry\| - \|x\| = \tau(\|x + ry\| - \|x\|) + (1 - \tau)(p(x + ry) - p(x))$$
$$\geq -\tau r + (1 - \tau)br > cr.$$
$\square$

Corollary 3.2.4. *Assume that X is separable but X^* is nonseparable. Then for every $0 < \varepsilon < 1$ there is an equivalent norm $\|\cdot\|$ on X such that for every x,*

$$\limsup_{\|y\| \to 0} \frac{\|x + y\| + \|x - y\| - 2\|x\|}{\|y\|} > \varepsilon.$$

Proof. Let $\|\cdot\|$ be the norm obtained in Theorem 3.2.3 with some $\varepsilon < c < 1$. Given $x \in X$, we find $x^* \in X^*$ with $\|x^*\| = 1$ and $x^*(x) = \|x\|$. Using the statement of Theorem 3.2.3 with Y the kernel of x^*, we find for any given $r > 0$ a point $y \in Y$ such that $\|y\| \leq r$ and

$$\|x + y\| - \|x\| > cr.$$

Since also $\|x - y\| - \|x\| \geq x^*(x - y) - x^*(x) = x^*(-y) = 0$, we have

$$\|x + y\| + \|x - y\| - 2\|x\| > cr \geq c\|y\|.$$
$\square$

Remark 3.2.5. Recall that, for $\varepsilon > 0$, a function $f: X \longrightarrow Y$ is called ε-Fréchet differentiable at x_0 if for some $T \in L(X, Y)$ and $\delta > 0$,

$$\|f(x_0 + x) - f(x_0) - Tx\| \leq \varepsilon \|x\|$$

whenever $\|x\|$ is small enough.

We will study this notion in detail in the next chapter (where it is defined more formally in Definition 4.1.1), but we notice already that, given any $0 < \varepsilon < \frac{1}{2}$, the norm $\|\cdot\|$ from Corollary 3.2.4 (used with 2ε instead of ε) is not ε-Fréchet differentiable at any x_0. Indeed, if $x^* \in X^*$ and $\left|\|x_0 + x\| - \|x_0\| - x^*(x)\right| \leq \varepsilon\|x\|$ for $\|x\| < \delta$, then

$$\|x_0 + x\| + \|x_0 - x\| - 2\|x_0\|$$
$$\leq \left|\|x_0 + x\| - \|x_0\| - x^*(x)\right| + \left|\|x_0 - x\| - \|x_0\| - x^*(x)\right| \leq 2\varepsilon\|x\|$$

for all such x, which is incompatible with the conclusion of Corollary 3.2.4.

3.3 FRÉCHET DIFFERENTIABILITY OF CONVEX FUNCTIONS

It is clear from Corollary 3.2.4 that positive results on existence of Fréchet derivatives of continuous convex functions $f: X \longrightarrow \mathbb{R}$, with X separable, may hold only if X^*

is separable. The same is true of course for general Lipschitz maps from X to any Banach space Y.

Positive results on existence of points of Fréchet differentiability of continuous convex functions on spaces with separable dual are indeed valid, and are even valid in the sense of Baire category. Lindenstrauss proved that if X is separable and reflexive then every continuous convex $f\colon X \longrightarrow \mathbb{R}$ is Fréchet differentiable on a dense G_δ set (hence outside a set of the first category). This result was extended by Asplund to every space with X^* separable. In view of this result, spaces X with X^* separable are called *Asplund spaces*. Notice that in the literature the term "Asplund space" often refers to Banach spaces X such that for every separable subspace $Z \subset X$ the space Z^* is separable. In this work, however, Asplund space will denote just spaces X with X^* separable. Only in comments or remarks (especially when we treat separable determination in Section 3.6) we may occasionally mention the general case.

A stronger result than Asplund's with a much simpler proof was obtained by Preiss and Zajíček. It seems to be the first result connecting the question of Fréchet differentiability with the notion of porous set. (See Section 1.1 of the Introduction for the definition.) It turned out to play a central role in the whole subject and in particular in this volume.

Theorem 3.3.1. *A continuous convex function f on an Asplund space X is Fréchet differentiable outside a σ-porous set.*

Proof. For each integer $m \in \mathbb{N}$ cover the space X^* by a sequence $(B_{k,m})_{k=1}^\infty$ of balls of radius $1/(6m)$. For $k, m, n \in \mathbb{N}$ consider the sets

$$A_{k,m,n} = \Big\{ x \in X \ \Big| \ \text{there is } u^* \in \partial f(x) \cap B_{k,m} \text{ with } \|u^*\| < n \text{ such that}$$

$$n > \limsup_{y \to 0} \frac{f(x+y) - f(x) - u^*(y)}{\|y\|} > \frac{1}{m} \Big\}.$$

The set where f is not Fréchet differentiable is $\bigcup_{k,m,n} A_{k,m,n}$. We show that each $A_{k,m,n}$ is porous with constant $1/(6mn)$.

Fix $x \in A_{k,m,n}$ and $\varepsilon > 0$. Diminishing ε if needed, we achieve that f is Lipschitz with constant $2n$ on $B(x, 2\varepsilon)$. Indeed, let $r, \tau > 0$ be such that

$$f(x+y) - f(x) - u^*(y) \le (n - \tau)\|y\|$$

whenever $\|y\| \le r$. If f were not $2n$-Lipschitz on a neighborhood of x, we could find points $y_i, z_i \in X$ such that $y_i \to x$, $z_i \to x$, and

$$f(z_i) - f(y_i) > 2n\|z_i - y_i\|.$$

Put $w_i = z_i + \frac{r}{2} \frac{z_i - y_i}{\|z_i - y_i\|}$. Then $w_i \in B(x, r)$ for large i. By the convexity of f we see that

$$f(w_i) - f(z_i) \ge \frac{f(z_i) - f(y_i)}{\|z_i - y_i\|} \|w_i - z_i\| > 2n \frac{r}{2} = nr.$$

On the other hand, for large i,

$$f(w_i) - f(z_i) \leq f(w_i) - f(x) - u^*(w_i - x) + (f(z_i) - f(x)) + u^*(w_i - x)$$
$$< (2n - \tau)\|w_i - x\| + f(z_i) - f(x).$$

Combining the last two inequalities we get

$$nr < (2n - \tau)\|w_i - x\| + f(w_i) - f(x),$$

which gives a contradiction for $i \to \infty$.

By the definition of $A_{k,m,n}$ we now find $u^* \in \partial f(x) \cap B_{k,m}$ and $y \in X$ with $\|y\| < \varepsilon$ such that $\|u^*\| < n$ and

$$f(x + y) - f(x) - u^*(y) > \frac{1}{m}\|y\|.$$

We claim that

$$A_{k,m,n} \cap B\big(x + y, \|y\|/(6mn)\big) = \emptyset.$$

Assume otherwise, that there is a point z in this intersection. Let v^* be the element in $\partial f(z)$ witnessing that $z \in A_{k,m,n}$. Then $v^*(x - z) \leq f(x) - f(z)$ and also

$$\|u^* - v^*\| \leq \frac{1}{3m}$$

because both belong to the same ball $B_{k,m}$. Noticing that

$$\|z - x\| \leq \|y\|\left(1 + \frac{1}{6mn}\right) \leq \frac{7}{6}\|y\|,$$

we obtain

$$f(x + y) - f(z) = f(x + y) - f(x) + f(x) - f(z)$$
$$> u^*(y) + \frac{\|y\|}{m} + v^*(x - z)$$
$$= \frac{\|y\|}{m} + u^*(x + y - z) + (v^* - u^*)(x - z)$$
$$\geq \frac{\|y\|}{m} - \|u^*\|\,\|x + y - z\| - \|v^* - u^*\|\,\|x - z\|$$
$$\geq \frac{\|y\|}{m} - \frac{n\|y\|}{6mn} - \frac{1}{3m}\frac{7}{6}\|y\| > \frac{\|y\|}{3m}.$$

On the other hand, we have already found that $\|z - x\| \leq \frac{7}{6}\|y\| < 2\varepsilon$, and so using that f is $2n$-Lipschitz we get

$$|f(x + y) - f(z)| \leq 2n\|x + y - z\| \leq \frac{\|y\|}{3m}.$$

This contradiction finishes the proof. $\qquad\qquad\square$

3.4   POROSITY AND NONDIFFERENTIABILITY

We first point out another important, although obvious, connection between porous sets and Fréchet differentiability.

Remark 3.4.1. If E is a porous set in a Banach space X, then the *distance function* $f(x) := \mathrm{dist}(x, E)$ is nowhere Fréchet differentiable on E.

Proof. Let E be a porous with constant c and let $x \in E$. Then $f(x) = 0$, and since f attains its minimum at x the only possible derivative of f at x is 0. For $\varepsilon > 0$, let $z \in X$ satisfy $\|z\| \leq \varepsilon$ and

$$B(x + z, c\|z\|) \cap E = \emptyset.$$

Then

$$f(x + z) - f(x) \geq c\|z\|,$$

which makes $f'(x) = 0$ impossible. $\square$

A slightly more involved variant of this remark is that even a σ-porous set is contained in the set of points of Fréchet nondifferentiability of some real-valued Lipschitz function. This was proved in [40] (see also [4, Theorem 6.48]) for countable unions of closed porous sets, and Kirchheim observed that "closed" is in fact not needed.

Lemma 3.4.2. *Suppose that X has a Fréchet smooth norm and $E \subset X$ is porous with constant c. Then there is $f : X \longrightarrow [0, 1]$ with $\mathrm{Lip}(f) \leq 2$ such that*

(i) $\displaystyle\liminf_{u \to 0} \frac{f(x + u) + f(x - u) - 2f(x)}{\|u\|} \geq 0$ *for every $x \in X$, and*

(ii) $\displaystyle\limsup_{u \to 0} \frac{f(x + u) + f(x - u) - 2f(x)}{\|u\|} \geq \frac{c}{180}$ *for every $x \in E$.*

Proof. For every $x \in X \setminus \overline{E}$ we define $r(x) = \min\{1, \frac{1}{6}\,\mathrm{dist}(x, E)\}$. By the $5r$-covering theorem (see, e.g., [33, Theorem 2.1]), there are $x_j \in X \setminus \overline{E}$ such that the balls $B(x_j, r_j)$, where $r_j = r(x_j)$, are disjoint, and for every $x \in X \setminus \overline{E}$ there is j such that $B(x, r(x)) \subset B(x_j, 5r_j)$. Notice also that the balls $B(x_j, 6r_j)$ still do not meet E. Define

$$f(x) = \begin{cases} r_j - \|x - x_j\|^2 / r_j & \text{if } x \in B(x_j, r_j), \\ 0 & \text{if } x \text{ belongs to no } B(x_j, r_j). \end{cases}$$

Clearly, $0 \leq f(x) \leq 1$ and f has Lipschitz constant at most two on the closure of each ball $B(x_j, r_j)$. Since on the boundary of these balls both formulas defining f coincide, this implies that $\mathrm{Lip}(f) \leq 2$.

The assertion (i) is obvious at the points where $f(x) = 0$. When $f(x) > 0$, x belongs to one of the balls $B(x_j, r_j)$. In that case f is Fréchet differentiable at x and (i) follows.

To prove (ii), let $x \in E$ and $0 < \varepsilon < 1$. It suffices to find u with $\|u\| < \varepsilon$ such that $f(x + u) \geq \frac{1}{180}c\|u\|$. Indeed, since $f(x) = 0$ and $f(x - u) \geq 0$, this will imply that

$$f(x + u) + f(x - u) - 2f(x) \geq f(x + u) \geq \frac{c}{180}\|u\|$$

and so the statement.

By the porosity assumption on E, there is $y \in B(x, \frac{1}{6}\varepsilon)$ such that

$$B(y, c\|y - x\|) \cap E = \emptyset.$$

Then $y \notin \overline{E}$ and $\operatorname{dist}(y, E) \leq \|y - x\| < 1$, and so $r(y) \geq \frac{1}{6}c\|y - x\|$. Hence there is j such that $B(y, \frac{1}{6}c\|y - x\|) \subset B(x_j, 5r_j)$. In particular,

$$r_j \geq \frac{c}{30}\|y - x\|.$$

Since $y \in B(x_j, 5r_j)$, $\|x_j - x\| \leq \|x - y\| + 5r_j$. Using also that $x \notin B(x_j, 6r_j)$, we get $\|x_j - x\| \geq 6r_j$ and so $r_j \leq \|x - y\|$. Hence $u := x_j - x$ satisfies $\|u\| \leq \|x - y\| + 5r_j \leq 6\|x - y\|$ and

$$f(x + u) = r_j \geq \frac{c}{30}\|y - x\| \geq \frac{c}{180}\|u\|.$$

Since $\|u\| \leq 6\|x - y\| < \varepsilon$, this shows exactly what we needed. □

Theorem 3.4.3. *Let E be a σ-porous subset of a separable Banach space X. Then there is a Lipschitz function from X to $\mathbb{R}$ which is not Fréchet differentiable at any point of E.*

Proof. If X^* is nonseparable, the norm constructed in Corollary 3.2.4 provides the required example. Hence we assume that X^* is separable and so, by Theorem 3.2.1 that its norm is Fréchet smooth.

Let $E = \bigcup_{k=1}^{\infty} E_k$, where E_k is porous with constant c_k, $0 < c_k < 1$. Find the corresponding function f_k according to Lemma 3.4.2 and choose $\alpha_k > 0$ such that

$$\sum_{j>k}^{\infty} \alpha_j \leq \frac{\alpha_k c_k}{360}.$$

We show that the function

$$f(x) = \sum_{k=1}^{\infty} \alpha_k f_k(x)$$

has the required property.

Obviously, f is well defined and Lipschitz. Suppose that $x \in E$ and find k such that $x \in E_k$. By Lemma 3.4.2 (i)

$$\liminf_{u \to 0} \frac{\sum_{j<k} \alpha_j (f_j(x + u) + f_j(x - u) - 2f_j(x))}{\|u\|} \geq 0$$

and a simple estimate by the Lipschitz constant says that

$$\liminf_{u\to 0} \frac{\sum_{j>k} \alpha_j (f_j(x+u) + f_j(x-u) - 2f_j(x))}{\|u\|} \geq -4\sum_{j>k}\alpha_j \geq -\frac{\alpha_k c_k}{360}.$$

Hence, using for f_k the inequality 3.4.2 (ii), we have

$$\limsup_{u\to 0} \frac{f(x+u) + f(x-u) - 2f(x)}{\|u\|} \geq \frac{\alpha_k c_k}{180} - \frac{\alpha_k c_k}{360} \geq \frac{\alpha_k c_k}{360},$$

which shows that f is not Fréchet differentiable at x. □

Remark 3.4.4. If E is porous in the direction of a vector u, the distance from E is at any point of E nondifferentiable in the direction of u. If X is separable, we may use the construction from the previous theorem with a Gâteaux smooth norm on X and so find, for any given σ-directionally porous set $E \subset X$, a real-valued Lipschitz function f such that for every $x \in E$ there is a direction in which f is not differentiable. In particular, f fails to be Gâteaux differentiable at any point of E.

The closure of a porous set is obviously nowhere dense, and thus σ-porous sets are of the first category. In a finite dimensional space porous sets are, by Lebesgue's density theorem, sets of measure zero. We mention here in passing that even in the real line the σ-ideal of σ-porous sets is much smaller that the σ-ideal of sets which are of the first category and measure zero; see [49] or [51, p. 526] where it is pointed out that this result essentially goes back to Beurling and Ahlfors [5].

On the other hand, in infinite dimensional spaces porous sets need not be small in the sense of Gauss measure. In fact, Matoušek and Matoušková [34] (see also [4, Example 6.46]) proved that there is an equivalent norm on ℓ_2 which is Fréchet differentiable only on a Gauss null set. This example was extended by Matoušková [35] to every separable uniformly convex space. (With similar reasoning, we will reuse this example in Chapter 5 to quickly see the difference between Gauss null sets and another σ-ideal of negligible sets, so-called Γ-null sets, which will be introduced there.) An earlier example in [40] (see also [4, Theorem 6.39]), which is, however, not related to differentiability, shows that porous sets are not small in the sense of Gauss measure in any infinite dimensional separable Banach space: any such space can be decomposed into two sets, one σ-porous and the other Gauss null.

3.5   SETS OF FRÉCHET DIFFERENTIABILITY POINTS

In this short section we show that for an arbitrary map between Banach spaces the set of its points of Fréchet differentiability is Borel; in fact it has type $F_{\sigma\delta}$. Such results are classical for real-valued functions of one real variable and have been extended to more dimensions by various authors. The particular extension we treat here is due to Zajíček [50]. Notice that in general the type of these sets cannot be improved even if the map is supposed to be Lipschitz. Indeed, by the result of Zahorski mentioned in

Chapter 2, for any $F_{\sigma\delta}$ subset E of $\mathbb{R}$ whose complement has measure zero there is a Lipschitz function $f\colon \mathbb{R} \longrightarrow \mathbb{R}$ whose set of points of differentiability is precisely E.

In this connection notice that for sets of points of Gâteaux differentiability the situation is quite different. For Lipschitz maps on separable spaces it is easy to check that these sets are Borel (see, for example, the beginning of the proof of [4, Theorem 6.42]), and an inspection of the argument reveals that they are $F_{\sigma\delta}$. But in general they may well be non-Borel. For continuous convex functions on separable spaces they are G_δ, but unlike the sets of their points of Fréchet differentiability, which are G_δ in all Banach spaces, they may fail to be Borel in nonseparable spaces [44] and even in nonseparable Hilbert spaces [21].

The main difficulty in proving that the set of points of differentiability is Borel is that, a priori, this set is obtained as a union over all possible values of the derivative, which is far from being a countable union. The usual way of overcoming this difficulty is via a criterion of Fréchet differentiability that does not need the value of the derivative. Here we use a different approach, which so far as we know is new in this context, via the canonical embedding of the target space into the second dual and compactness of balls in the w^*-topology.

For a map $f\colon X \longrightarrow Y$, $x \in X$, and $r > 0$ denote

$$\varepsilon(f,x,r,X) = \inf_{L\in L(X,Y)} \ \sup_{u\in X,\, 0<\|u\|<r} \frac{\|f(x+u)-f(x)-Lu\|}{\|u\|}.$$

The reason for keeping X as a parameter in $\varepsilon(f,x,r,X)$ will appear in the next section, where for x belonging to a subspace U of X we will look at the relation between differentiability of $f\colon X \longrightarrow Y$ and of its restriction to U, hence at the relation between the quantities $\varepsilon(f,x,r,X)$ and $\varepsilon(f,x,r,U)$.

As the following simple observation says, the limit

$$\varepsilon(f,x,X) := \lim_{r\searrow 0} \varepsilon(f,x,r,X)$$

measures how far f is from being Fréchet differentiable at x.

Observation 3.5.1. *A map f of a Banach space X to a Banach space Y is Fréchet differentiable at $x \in X$ if and only if $\varepsilon(f,x,X) = 0$.*

Proof. Suppose $\varepsilon(f,x,X) = 0$. Then there are $r_k > 0$ and $L_k \in L(X,Y)$ such that $\|f(x+u)-f(x)-L_k u\| \le 2^{-k}\|u\|$ for every $u \in X$ with $\|u\| < r_k$. This implies that

$$\|L_j u - L_k u\| \le \|f(x+u)-f(x)-L_j u\| + \|f(x+u)-f(x)-L_k u\|$$
$$\le (2^{-j}+2^{-k})\|u\|$$

for $\|u\| < \min\{r_j, r_k\}$. Hence $\|L_j - L_k\| \le 2^{-j}+2^{-k}$, implying that the sequence L_k converges to some $L \in L(X,Y)$ and $\|L - L_k\| \le 2^{-k}$. Consequently,

$$\|f(x+u)-f(x)-Lu\| \le \|f(x+u)-f(x)-L_k u\| + \|L-L_k\|\|u\| \le 2^{-k+1}\|u\|,$$

showing that L is the Fréchet derivative of f at x.

The opposite implication is obvious. $\qquad\square$

Notice that the notion of ε-Fréchet differentiability, which we have already mentioned in Remark 3.2.5, is, since we are normally interested in small $\varepsilon > 0$, for all practical purposes equivalent to the requirement that $\varepsilon(f, x, X) \leq \varepsilon$. More precisely, $\varepsilon(f, x, X) < \varepsilon$ implies that f is ε-Fréchet differentiable at x, and this implies that $\varepsilon(f, x, X) \leq \varepsilon$. In particular, the ε-Fréchet differentiability of f at x for every $\varepsilon > 0$ is equivalent to the fact that $\varepsilon(f, x, X) = 0$, and so Observation 3.5.1 translates to the often quoted

Observation 3.5.2. *Any map $f \colon X \longrightarrow Y$ which is ε-Fréchet differentiable at a point $x \in X$ for every $\varepsilon > 0$ is Fréchet differentiable at x.*

In general, the infimum defining $\varepsilon(f, x, r, X)$ is not attained, but it is attained in the presence of compactness. This also allows precise description of ε-differentiability using the quantities $\varepsilon(f, x, r, X)$.

Observation 3.5.3. *Suppose that $f \colon X \longrightarrow Y$, where Y is a dual Banach space and $\varepsilon(f, x, r, X) < \infty$. Then there is $L \in L(X, Y)$ such that for every u with $\|u\| < r$.*

$$\|f(x + u) - f(x) - Ly\| \leq \varepsilon(f, x, r, X)\|u\|.$$

Consequently, f is ε-Fréchet differentiable at x if and only if $\varepsilon(f, x, r, X) \leq \varepsilon$ for some $r > 0$.

Proof. For every $i \geq 1$ pick $L_i \in L(X, Y)$ such that

$$\|f(x + u) - f(x) - L_i u\| \leq \left(\varepsilon(f, x, r, X) + 2^{-i}\right)\|u\|$$

for every u with $\|u\| < r$. Since then $\|L_i(u) - L_1(u)\| \leq 2(\varepsilon(f, x, r, X) + 1)\|u\|$ for $\|u\| < r$, the operators L_i form a bounded sequence in $L(X, Y)$. Let $K > 0$ be such that $\|L_i\| \leq K$. Choose a free ultrafilter $\mathfrak{U}$ on $\mathbb{N}$. Assuming that $Y = Z^*$, we consider for every $u \in X$ and $z \in Z$ the limit

$$\lim_{\mathfrak{U}}(L_i u)z.$$

Since the absolute value of this limit is dominated by $K\|u\|\|z\|$, it defines for every fixed u an element of $Z^* = Y$ and, consequently, it also defines a bounded linear operator $L \in L(X, Y)$ satisfying

$$(Lu)z = \lim_{\mathfrak{U}}(L_i u)z.$$

It follows that for every $z \in Z$, $u \in X$, and $\eta > 0$ there are arbitrarily large $i \in \mathbb{N}$ such that $|(L_i u)z - (Lu)z| \leq \eta\|z\|$. Then for every $u \in X$ with $\|u\| < r$ we find a norm one $z \in Z$ such that

$$\|f(x + u) - f(x) - Lu\| < \eta + (f(x + u) - f(x) - Lu)z$$

and infer that

$$\|f(x + u) - f(x) - Lu\| < \eta + (f(x + u) - f(x) - L_i u)z + |(L_i u)z - (Lu)z|$$
$$\leq \left(\varepsilon(f, x, r, X) + 2^{-i}\right)\|u\| + 2\eta.$$

Given $\eta > 0$, i can be arbitrarily large, and then letting $\eta \to 0$ we obtain the statement.

$\square$

After this digression we return to the real theme of this section.

Proposition 3.5.4. *Suppose that* $f \colon X \longrightarrow Y$, *where* Y *is a dual Banach space. Then for every* $c > 0$ *the set* $\{x \in X \mid \varepsilon(f, x, X) < c\}$ *is* F_σ.

Proof. Let

$$E_{j,k} = \{x \in X \mid \varepsilon(f, x, \tfrac{1}{j}, X) < c - \tfrac{3}{k}\}.$$

We show that $\varepsilon(f, x, \tfrac{1}{j}, X) < c$ for every $x \in \overline{E_{j,k}}$. This will establish that the set in question is $\bigcup_{j=1}^\infty \bigcup_{k=1}^\infty \overline{E_{j,k}}$, hence F_σ.

Let $x \in \overline{E_{j,k}}$ and find $x_i \in E_{j,k}$ such that $x_i \to x$. (Note that this implicitly means that $c > 3/k$.) For each i we choose $L_i \in L(X, Y)$ such that

$$\|f(x_i + u) - f(x_i) - L_i u\| \le \left(c - \frac{3}{k}\right) \|u\| \quad \text{for } \|u\| < \frac{1}{j}.$$

If i_0 is such that $\|x_i - x_{i_0}\| < \frac{1}{2j}$ for $i \ge i_0$ then for every $\|u\| < \frac{1}{2j}$,

$$
\begin{aligned}
\|L_i u - L_{i_0} u\| &\le \|f(x_i + u) - f(x_i) - L_i u\| \\
&\quad + \|f(x_i + u) - f(x_{i_0}) - L_{i_0}(x_i + u - x_{i_0})\| \\
&\quad + \|f(x_i) - f(x_{i_0}) - L_{i_0}(x_i - x_{i_0})\| \\
&\le c(\|u\| + \|x_i + u - x_{i_0}\| + \|x_i - x_{i_0}\|) \le \frac{2c}{j}.
\end{aligned}
$$

Hence $\|L_i\| \le \|L_{i_0}\| + 2c$ and so (L_i) is a bounded sequence. Arguing in the same way as in the proof of Observation 3.5.2, we may find $L \in L(X, Y)$ such that for every $z \in Z$, $u \in X$, and $\eta > 0$ there is $i \in \mathbb{N}$ such that $\|(L_i u)z - (Lu)z\| \le \eta \|z\|$.

Suppose that $0 < \|u\| < 1/k$. We find a unit vector $z \in Z$ such that

$$\|f(x + u) - f(x) - Lu\| < \big(f(x + u) - f(x) - Lu\big)(z) + \frac{1}{k}\|u\|.$$

Then we find $i \in \mathbb{N}$ such that $\|(L_i u)z - (Lu)z\| \le \frac{1}{k}\|u\|$ and conclude that

$$
\begin{aligned}
\|f(x + u) - f(x) - Lu\| &< \big(f(x + u) - f(x) - L_i u\big)(z) + \frac{1}{k}\|u\| \\
&\quad + (L_i u)z - (Lu)z \\
&\le \left(c - \frac{3}{k}\right)\|u\| + \frac{1}{k}\|u\| + \frac{1}{k}\|u\| = \left(c - \frac{1}{k}\right)\|u\|.
\end{aligned}
$$

Hence $\varepsilon(f, x, \tfrac{1}{j}, X) \le c - \frac{1}{k} < c$. $\square$

Corollary 3.5.5. *The set of points of* X *at which a map* f *from* X *to a Banach space* Y *is Fréchet differentiable is* $F_{\sigma\delta}$.

Proof. We may consider f as a mapping into Y^{**}, as this does not change the notion of Fréchet differentiability. Then we see from Observation 3.5.1 that its set of points of Fréchet differentiability is $\bigcap_{k=1}^{\infty}\{x \in X : \varepsilon(f,x,X) < \frac{1}{k}\}$, which is $F_{\sigma\delta}$ by Lemma 3.5.4. $\square$

3.6   SEPARABLE DETERMINATION

The key idea behind separable determination (or separable reduction) is that some notions with which we wish to work in nonseparable Banach spaces may in fact use countability strongly enough so that statements about them hold in a nonseparable space provided they hold in its separable subspaces. There are several approaches to this. Here we follow the approach that started in [38], as modified by various authors. Although in all applications one makes the final deduction using just one separable subspace, it is convenient to know that the family of subspaces that can be used is so large that one easily join countably many arguments together. We will therefore use the concept of rich families of subspaces introduced in [6] by Borwein, Moors, and an unnamed mathematician (about whom the authors say "whose incisive comments formed the genesis of this paper").

Definition 3.6.1. Let X be a Banach space. A family $\mathcal{R}$ of separable subspaces of X is called *rich* if

(i) for every increasing sequence R_i in $\mathcal{R}$, $\overline{\bigcup_{i=0}^{\infty} R_i}$ belongs to $\mathcal{R}$, and

(ii) each separable subspace of X is contained in an element of $\mathcal{R}$.

Recall that for us "a subspace" means "a closed subspace," so this is indeed the definition from [6], from which also comes the following statement.

Proposition 3.6.2. *The intersection of countably many rich families is a rich family.*

Proof. The requirement (i) of the definition is obvious. To show (ii), let $(\mathcal{R}_n)$ be a sequence of rich families and Y a separable subspace of X. Let (n_i) be a sequence of natural numbers in which each number occurs infinitely often. Denote $R_0 = Y$ and for $k = 1, 2, \ldots$ use the property (i) recursively to choose $R_k \in \mathcal{R}_{n_k}$ such that $R_k \supset R_{k-1}$. Observing that the subspace $R := \overline{\bigcup_{k=1}^{\infty} R_k}$ satisfies, for each n,

$$R = \overline{\bigcup\{R_k \mid n_k = n, \ k \in \mathbb{N}\}},$$

we infer from the property (i) of $\mathcal{R}_n$ that $R \in \mathcal{R}_n$. Hence R belongs to the intersection of the $\mathcal{R}_n$, and it obviously contains Y. $\square$

The notion of separable determination could be defined formally in the following way. A property $\mathcal{P}$ of pairs (x, U), where $x \in X$ and U is a subspace of X, is *separably determined* if there is a rich family $\mathcal{R}$ of separable subspaces of X such that for every $R \in \mathcal{R}$ and $x \in R$, the pair (x, X) has property $\mathcal{P}$ if and only if (x, R) does. Usually, however, one does not feel that a given property is of this form, and so we state and prove the separable reduction statements without using this notion. To illustrate this point, separable determination of Fréchet differentiability would, according to this definition, be expressed as: for any $f: X \longrightarrow Y$, the property "f is Fréchet differentiable at the point x in direction U" of pairs (x, U) is separably determined. In our opinion, it is more revealing (although slightly longer) to say that there is a rich family $\mathcal{R}$ of separable subspaces of X such that for every $R \in \mathcal{R}$ and $x \in R$, the function f is Fréchet differentiable at x if and only if its restriction to R is Fréchet differentiable at x.

We will now present a new general setup for a number of separable determination questions. For simplicity, let us concentrate only on Fréchet differentiability. This is a problem of linear approximation, so one may imagine that the real statement is that even the error of the best linear approximation of our function on balls around points of R is the same for the original problem and for its restriction to R. We may measure the error of this approximation inside a subspace U by

$$\inf_{L \in L(U,Y)} \sup_{u \in U, 0 < \|u\| < r} \frac{\|f(x + u) - f(x) - L(u)\|}{\|u\|},$$

but a technically more convenient expression is

$$\inf_{L \in L(U,Y)} \sup_{\substack{u,v \in U \\ 0 < \|u\| + \|v\| < r}} \frac{\|f(x + u) - f(x + v) - L(u - v)\|}{\|u\| + \|v\|}. \tag{3.6}$$

To allow more applications, we will write the fraction as

$$F(u, v, f(x + u), f(x + v)),$$

where F is a function on $U^2 \times Y^2$. We also allow an arbitrary number of variables instead of just two. A more delicate point is that we have not a single function, but a collection of them, and this collection depends on the subspace U of the ambient space X. The last point is important, as it leads to the idea of measuring the error as a supremum not over the whole space but over its separable subspaces, and thereby it leads to the question of separable determination of a somewhat strange looking minimax type problem.

The general setup we arrived at is as follows. For every separable subspace U of X, let $\mathfrak{F}(U)$ be a collection of non-negative functions on $U^p \times Y^p$ such that the restriction of functions from $\mathfrak{F}(U)$ to $V \subset U$ belongs to $\mathfrak{F}(V)$. By $\mathfrak{F}$ we denote the collection of all $\mathfrak{F}(U)$. (These collections depend also on p and Y, but since p and Y are always clear from the context, we do not indicate this dependence.)

To simplify the notation, we denote for $x \in X$ and $u \in X^p$,

$$x + u = (x + u_1, \ldots, x + u_p) \text{ and } f(x + u) = (f(x + u_1), \ldots, f(x + u_p)).$$

Also, for $u \in X^p$ we put

$$\|u\| = \|u_1\| + \|x_2\| + \cdots + \|u_p\|.$$

The quantities measuring the approximability of f are now defined in the natural way. First, for $F \in \mathfrak{F}(U)$ we let

$$\beta(f, x, U, F) = \sup_{u \in U^p} F(u, f(x + u))$$

and then define

$$\beta(f, x, W, \mathfrak{F}) = \sup_{U \subset W \text{ separable}} \inf_{F \in \mathfrak{F}(U)} \beta(f, x, U, F).$$

Notice that for separable W we have a simpler formula,

$$\beta(f, x, W, \mathfrak{F}) = \inf_{F \in \mathfrak{F}(W)} \beta(f, x, W, F) = \inf_{F \in \mathfrak{F}(W)} \sup_{u \in W^p} F(u, f(x + u)).$$

This formula holds for any subspace of X provided we have a family $\mathfrak{F}$ of functions on $X^p \times Y^p$ and each $\mathfrak{F}(U)$ is the collection of their restrictions to $U^p \times Y^p$. This simpler formula will be true in some but not all of our applications. In the opposite direction, it would be natural to restrict the domain of $\mathfrak{F}(U)$ only to finite dimensional U, but we do not have any application of this more general approach.

For the validity of separable determination, we still need another assumption. For that recall that in (3.6) the linear operators L are (as long as we bound their norms) uniformly equicontinuous. So, up to a countable decomposition (which we can handle using Proposition 3.6.2), we assume that our collections are uniformly equicontinuous. This is enough for our purposes, but let us note that we will actually use only the following consequence of uniform equicontinuity. For every $\varepsilon > 0$ there is $\delta > 0$ such that whenever $F \in \mathfrak{F}(U)$, $u, v \in U^p$, $y \in Y^p$, and $\|u - v\| < \delta$, then $|F(u, y) - F(v, y)| < \varepsilon$.

Observation 3.6.3. *If ε and δ are as above and W is a subspace of X, then for every $x, \tilde{x} \in W$ with $\|x - \tilde{x}\| < \delta$,*

$$\beta(f, x, W, \mathfrak{F}) \leq \beta(f, \tilde{x}, W, \mathfrak{F}) + \varepsilon.$$

Proof. Fix any $0 < c < \beta(f, x, W, \mathfrak{F})$. By definition of $\beta(f, x, W, \mathfrak{F})$ there is a separable subspace $U \subset W$ such that $\beta(f, x, U, F) > c$ for every $F \in \mathfrak{F}(U)$. Let V be the linear span of $U \cup \{x - \tilde{x}\}$. Then, given any $F \in \mathfrak{F}(V)$, the restriction of F to U belongs to $\mathfrak{F}(U)$, and so we may find $u \in U^p$ such that $F(u, f(x + u)) > c$. It follows that, with $\tilde{u} = u + x - \tilde{x} \in V^p$,

$$F(\tilde{u}, f(\tilde{x} + \tilde{u})) = F(\tilde{u}, f(x + u)) \geq F(u, f(x + u)) - \varepsilon > c - \varepsilon.$$

Hence $\beta(f, \tilde{x}, W, \mathfrak{F}) \geq c - \varepsilon$, proving the statement. $\square$

Lemma 3.6.4. *Suppose that $\mathfrak{F}$ is uniformly equicontinuous. Then for every function $f \colon X \longrightarrow Y$ the family $\mathcal{R}$ of separable subspaces R of X such that for every $x \in R$,*

$$\beta(f, x, R, \mathfrak{F}) = \beta(f, x, X, \mathfrak{F}),$$

is rich on X.

Proof. We first show a slightly stronger version of the first requirement from the definition of richness. If (W_i) is an increasing sequence of separable subspaces of X such that for each i the set of $\tilde{x} \in W_i$ with

$$\beta(f, \tilde{x}, W_{i+1}, \mathfrak{F}) \geq \beta(f, \tilde{x}, X, \mathfrak{F}), \tag{3.7}$$

is dense in W_i, then $W := \overline{\bigcup_{i=0}^{\infty} W_i}$ belongs to $\mathcal{R}$. To this aim we observe that the inequality $\beta(f, x, W, \mathfrak{F}) \leq \beta(f, x, X, \mathfrak{F})$ is obvious. To prove the opposite, suppose that $x \in W$, $\eta > 0$, and $\beta(f, x, W, \mathfrak{F}) < \infty$. By uniform equicontinuity of $\mathfrak{F}$ there is $\delta > 0$ such that $|F(u, y) - F(v, y)| < \eta$ whenever $F \in \mathfrak{F}(W)$, $u, v \in W$, and $\|u - v\| < \delta$. Choose i large enough that $B(x, \delta) \cap W_i \neq \emptyset$. By assumption, there is $\tilde{x} \in B(x, \delta) \cap W_i$ for which (3.7) holds. Hence, applying Observation 3.6.3 twice, we get

$$\beta(f, x, W, \mathfrak{F}) \geq \beta(f, \tilde{x}, W, \mathfrak{F}) - \eta \geq \beta(f, \tilde{x}, W_{i+1}, \mathfrak{F}) - \eta$$
$$\geq \beta(f, \tilde{x}, X, \mathfrak{F}) - \eta \geq \beta(f, x, X, \mathfrak{F}) - 2\eta.$$

To prove the second requirement from the definition of richness we show that there is a way to assign to every separable subspace V of X a separable subspace $\widetilde{V} \supset V$ of X such that $\beta(f, x, \widetilde{V}, \mathfrak{F}) \geq \beta(f, x, X, \mathfrak{F})$ for every x from a dense subset of V. This will finish the proof since, defining $W_1 = \widetilde{V}$ and $W_{k+1} = \widetilde{W}_k$, we infer from what we have proved above that $\overline{\bigcup_{i=0}^{\infty} W_i}$ belongs to $\mathcal{R}$.

To define $\widetilde{V}$, let S be a countable dense subset of V. For every $x \in S$ and rational $c < \beta(f, x, X, \mathfrak{F})$ find a separable subspace $U_{x,c}$ of X such that $\beta(f, x, U_{x,c}, F) > c$ for every $F \in \mathfrak{F}(U_{x,c})$. Then the space $\widetilde{V}$ given as the closed linear span of V together with all these $U_{x,c}$ has the required property.   □

In the following simple applications, we have a family $\mathfrak{F}$ of functions on $X^p \times Y^p$, and $\mathfrak{F}(U)$ is the collection of their restrictions to a subspace U. Hence the simpler formula for $\beta(f, x, W, \mathfrak{F})$ applies to all subspaces of X.

Corollary 3.6.5. *Let $f : X \longrightarrow Y$. Then there is a rich family $\mathcal{R}$ on X such that for every $R \in \mathcal{R}$, $x \in R$, and $r > 0$,*

$$\sup_{u \in R, \|u\| < r} \|f(x + u)\| = \sup_{u \in X, \|u\| < r} \|f(x + u)\|.$$

Proof. For rational $s, C > 0$ let $\mathfrak{F}_{s,C}(U)$ consist of the single function

$$F_{s,C}(u, y) = \min\{C, \|y\|\} \min\{1, \max\{C(s - \|u\|), 0\}\}.$$

By Lemma 3.6.4 there is a rich family $\mathcal{R}_{s,C}$ such that for every $R \in \mathcal{R}_{s,C}$ and $x \in R$,

$$\sup_{u \in R} F_{s,C}(u, f(x + u)) = \sup_{u \in X} F_{s,C}(u, f(x + u)). \tag{3.8}$$

Let $r > 0$. By Proposition 3.6.2 the family

$$\mathcal{R} = \bigcap \{ \mathcal{R}_{s,C} \mid 0 < s < r, \ C > 0 \text{ rational} \}$$

is rich on X. Let $R \in \mathcal{R}$ and $x \in R$. By taking the supremum in (3.8) over rational $0 < s < r$ and $C > 0$ we get

$$\sup_{u \in R, \|u\| < r} \|f(x + u)\| = \sup_{u \in X, \|u\| < r} \|f(x + u)\|. \qquad \square$$

This simple result is all that we need to show that porosity is separably determined. To give a detailed statement, we define for $E \subset X$ the porosity of E at a point $x \in X$ as

$$p(E, x) = \lim_{r \searrow 0} \sup\{c > 0 \mid (\exists y \in B(x, r)) \, B(y, c\|y - x\|) \cap E = \emptyset\}.$$

If for some $r > 0$ there is no such c, we let $p(E, x) = 0$. If Y is a subspace of X, we define the porosity of E at x in the direction of Y by the same formula in which y is restricted to belong to $x + Y$.

Corollary 3.6.6. *Let $E \subset X$. Then there is a rich family $\mathcal{R}$ on X such that for every $R \in \mathcal{R}$ and every $x \in R$, the porosity of E at x in the direction of R is equal to its porosity (in the direction of X).*

Proof. Observe that

$$p(E, x) = \limsup_{r \searrow 0} \; \sup_{\|u\| < r} \frac{\mathrm{dist}(x + u, E)}{r}$$

and use the previous result for $f(x) = \mathrm{dist}(x, E)$. $\qquad \square$

By Proposition 3.6.2 we have an immediate corollary.

Corollary 3.6.7. *Let $E \subset X$ be a σ-porous subset of X. Then there is a rich family $\mathcal{R}$ on X such that for every $R \in \mathcal{R}$, the set $E \cap R$ is σ-porous in R.*

Before discussing differentiability, it is natural to have a brief look at continuity.

Corollary 3.6.8. *For any function $f \colon X \longrightarrow Y$ there is a rich family $\mathcal{R}$ such that for every $R \in \mathcal{R}$ and every $x \in R$, the function f is continuous at x if and only if its restriction to R is.*

Proof. Use Lemma 3.6.4 with, for example, $\mathfrak{F}(U)$ consisting of functions

$$F_r(u, v, y, z) = \min\{1, \|y - z\|\} \frac{\max\{0, r - \|u\| - \|v\|\}}{r}, \quad r > 0.$$

It suffices to observe that f is continuous at x if and only if $\beta(f, x, X, \mathfrak{F}) = 0$. $\qquad \square$

We now turn our attention to linear approximations related to Fréchet differentiability. For that, we recall the quantities $\varepsilon(f, x, r, U)$ from the previous section. As we could have already seen there, constructing a Y-valued linear operator is a nontrivial task, and so more delicate separable determination results need a compactness assumption in the target space Y. We will therefore consider functions with values in dual spaces (weaker assumptions may be also used). Incidentally, one may notice that in this situation the infimum defining $\varepsilon(f, x, r, U)$ is actually attained.

Theorem 3.6.9. *For every function* $f \colon X \longrightarrow Y$, *where* Y *is a dual Banach space, there is a rich family* $\mathcal{R}$ *on* X *such that* $\varepsilon(f, x, r, R) = \varepsilon(f, x, r, X)$ *whenever* $R \in \mathcal{R}$, $x \in R$, *and* $r > 0$.

Proof. Since the inequality $\varepsilon(f, x, r, R) \leq \varepsilon(f, x, r, X)$ is obvious, we show the converse.

Let $\psi_k \colon [0, \infty)^4 \longrightarrow [0, \infty)$, $k \geq 1$, be continuous functions with compact support having the following property. For every $r > 0$ there is a sequence $(k_j) \subset \mathbb{N}$ such that

$$
\begin{aligned}
\psi_{k_j}(s_0, s_1, t_0, t_1) &\nearrow_{j \to \infty} \frac{1}{s_0 + s_1} \quad &&\text{if } s_0, s_1 < r \text{ and } s_0 + s_1 > 0, \\
\psi_{k_j}(s_0, s_1, t_0, t_1) &= 0 \quad &&\text{otherwise.}
\end{aligned}
\tag{3.9}
$$

For $k, l \in \mathbb{N}$ and a separable subspace U of X, define $\mathfrak{F}_{k,l}(U)$ as the collection of functions

$$
F_{k,L}(u, v, y, z) = \psi_k\big(\|u\|, \|v\|, \|y\|, \|z\|\big)\|y - z - L(u - v)\|,
$$

where $L \in L(U, Y)$ and $\|L\| \leq l$. It is easy to see that each $\mathfrak{F}_{k,l}$ satisfies the assumptions of Lemma 3.6.4. Using also Proposition 3.6.2, we find a rich family $\mathcal{R}$ on X such that

$$
\beta(f, x, R, \mathfrak{F}_{k,l}) = \beta(f, x, X, \mathfrak{F}_{k,l})
$$

for every $x \in R$ and $k, l \in \mathbb{N}$.

Suppose now that $R \in \mathcal{R}$, $x \in R$, and $r, \varepsilon > 0$. Find a sequence $(k_j) \subset \mathbb{N}$ such that (3.9) holds. By definition of $\varepsilon(f, x, r, R)$ there is $T \in L(R, Y)$ such that

$$
\sup_{u \in R, \, \|u\| < r} \|f(x + u) - f(x) - Tu\| < (\varepsilon(f, x, r, R) + \varepsilon)\|u\|.
$$

Hence

$$
\sup_{\substack{u, v \in R \\ \|u\| + \|v\| < r}} \|f(x + u) - f(x + v) - T(u - v)\| < (\varepsilon(f, x, r, R) + \varepsilon)(\|u\| + \|v\|),
$$

which shows that

$$
\sup_{\substack{u, v \in R \\ \|u\| + \|v\| < r}} F_{k_j, T}(u, v, f(x + u), f(x + v)) \leq \varepsilon(f, x, r, R) + \varepsilon
$$

for every $j \in \mathbb{N}$. Thus, fixing $l \in \mathbb{N}$ such that $l > \|T\|$, we see

$$
\beta(f, x, X, \mathfrak{F}_{k_j, l}) = \beta(f, x, R, \mathfrak{F}_{k_j, l}) < \varepsilon(f, x, r, R) + \varepsilon
$$

for each j. This means that for every $j \in \mathbb{N}$ and every separable subspace U of X, there is $L_{(j,U)} \in L(U, Y)$ with $\|L_{(j,U)}\| \leq l$ and

$$
F_{k_j, L_{(j,U)}}(u, v, f(x + u), f(x + v)) \leq \varepsilon(f, x, r, R) + \varepsilon \tag{3.10}
$$

for every $u, v \in U$.

Consider on the set $\mathfrak{I} = \{(j, U) \mid j \in \mathbb{N}, U \subset X \text{ separable subspace}\}$ a partial order defined as $(j, U) \preccurlyeq (k, V)$ if and only if $j \leq k$ and $U \subset V$. Then the sets $\{(k, V) \mid (j, U) \preccurlyeq (k, V)\}$, $(j, U) \in \mathfrak{I}$, form a filter base on $\mathfrak{I}$. Let $\mathfrak{U}$ be an ultrafilter extending this filter base. Assuming that $Y = Z^*$ we put for any $u \in X$ and $z \in Z$,

$$\lim_{\mathfrak{U}}(L_{(j,U)}u)z.$$

Since the absolute value of this limit is bounded by $l\|u\|\,\|z\|$, it defines for every fixed $u \in X$ an element of $Y = Z^*$ and, consequently, it also defines a bounded linear operator $L \in L(X, Y)$ satisfying

$$(Lu)z = \lim_{\mathfrak{U}}(L_{(j,U)}u)z.$$

It follows that for every separable subspace U_0, any $u \in U_0$, every unit vector $z \in Z$, any $\eta > 0$, and $j_0 \in \mathbb{N}$, there are $j > j_0$ and a subspace $U \supset U_0$ such that $|(L_{(j,U)}u)z - (Lu)z| < \eta$.

Suppose now that $u \in U_0$ and $0 < \|u\| < r$. Find a unit vector $z \in Z$ such that

$$\|f(x+u) - f(x) - Lu\| < (f(x+u) - f(x) - Lu)(z) + \varepsilon\|u\|.$$

Also, find $j_0 \in \mathbb{N}$ such that for every $j > j_0$,

$$\psi_{k_j}(u, 0, f(x+u), f(x)) \geq \frac{1}{\|u\|} - \frac{\varepsilon}{\|f(x+u) - f(x)\| + lr}.$$

Finally, we find $j > j_0$ and $U \supset U_0$ such that $|(L_{(j,U)}u)z - (Lu)z| < \varepsilon\|u\|$, and estimate

$$
\begin{aligned}
\frac{\|f(x+u) - f(x) - Lu\|}{\|u\|} &< \frac{(f(x+u) - f(x) - Lu)(z)}{\|u\|} + \varepsilon \\
&< \frac{(f(x+u) - f(x) - L_{(j,U)}u)(z)}{\|u\|} + 2\varepsilon \\
&< F_{k_j, L_{(j,U)}}(u, 0, f(x+u), f(x)) + 3\varepsilon \\
&\leq \varepsilon(f, x, r, R) + 4\varepsilon,
\end{aligned}
$$

where the last step follows from (3.10). Hence $\varepsilon(f, x, r, X) \leq \varepsilon(f, x, r, R) + 4\varepsilon$, and so, since $\varepsilon > 0$ is arbitrary, $\varepsilon(f, x, r, X) \leq \varepsilon(f, x, r, R)$. $\qquad\square$

It is now immediate to deduce the separable determination statement for Fréchet differentiability. It is a variant of Zajíček's [50] strengthening of the separable reduction statement originating in [38] (see also [27]).

Theorem 3.6.10. *For every $f \colon X \longrightarrow Y$ there is a rich family $\mathcal{R}$ on X such that for every $R \in \mathcal{R}$, f is Fréchet differentiable (as a function on X) at every $x \in R$ at which its restriction to R is Fréchet differentiable (as a function on R).*

Proof. We may consider f as a mapping into Y^{**}; this does not change the notion of Fréchet differentiability. Hence the statement follows from Theorem 3.6.9 and Observation 3.5.1. ☐

Remark 3.6.11. The proof of the previous theorem cannot be repeated verbatim for ε-Fréchet differentiability, where $\varepsilon > 0$ is fixed. But if the target space Y is a dual space, we may combine Theorem 3.6.9 and Observation 3.5.3 to get that for every $f\colon X \longrightarrow Y$ there is a rich family $\mathcal{R}$ on X such that for every $R \in \mathcal{R}$, f is ε-Fréchet differentiable (as a function on X) at every $x \in R$ at which its restriction to R is ε-Fréchet differentiable (as a function on R).

Notice that Corollary 3.6.5 applied to the distance from E shows that a set E in X is nowhere dense if and only if there is a rich family in each element of which E is nowhere dense. Hence first category sets are separably determined in the same way as σ-porous sets are in Corollary 3.6.7. More interestingly, this statement may be "lifted" to the Borel Γ_n- and Γ-null sets that will be introduced in Definition 5.1.1. In Corollary 5.6.2 we will prove that Γ_n- and Γ-nullness of Borel sets are separably determined.

By a straightforward use of separable reduction, a number of results from this book may be easily extended to a nonseparable setting. Since this would be rather mechanical, we will not state the results that may be obtained, but just give several examples illustrating the technique of proving nonseparable versions of some of the theorems that we prove later.

(1) *Real-valued Lipschitz, and even cone-monotone functions on (nonseparable) Asplund spaces (i.e., spaces such that for every separable subspace $Z \subset X$ the space Z^* is separable) have points of Fréchet differentiability.* The Lipschitz statement is an immediate consequence of Theorems 12.1.1 and 3.6.10. (Notice that this is how the Fréchet differentiability result for Lipschitz functions was extended to nonseparable Asplund spaces already in [39], but that the slicing technique of [27] can prove it without the use of separable reduction.) In the case of functions monotone with respect to a cone C we find the rich family $\mathcal{R}$ from Theorem 3.6.10 and choose $x_0 \in C$. Further we observe that the family of $R \in \mathcal{R}$ that contain x_0 is rich in X and the restriction of f to such R is cone-monotone. Hence we conclude the argument by a reference to Theorem 12.1.3. We also notice that this argument shows that the additional (mean value) statements of Theorems 12.1.1 and 12.1.3 also hold in nonseparable Asplund spaces.

(2) *σ-porous sets in (nonseparable) Asplund spaces are Γ_1-null.* For this, just combine Theorem 10.4.1 with Corollaries 3.6.7 and 5.6.2.

(3) *For any (possibly uncountable) set Δ, every Lipschitz map of $\ell_p(\Delta)$ to $\mathbb{R}^n$, where $2 \le n \le p < \infty$, has points of Fréchet differentiability.* To see this, notice first that the family of sets $\{x \in \ell_p(\Delta) \mid x(s) = 0 \text{ for } s \notin C\}$, where C runs through countable subsets of Δ, is rich in $\ell_p(\Delta)$. Suppose now that $f\colon \ell_p(\Delta) \longrightarrow \mathbb{R}^n$, where $2 \le n \le p < \infty$, is Lipschitz. By Theorem 3.6.10 combined with Proposition 3.6.2, we can find a rich subfamily $\mathcal{R}$ of the above family such that for every

$R \in \mathcal{R}$, $x \in R$, the function f is Fréchet differentiable at x iff its restriction to R is. By Theorem 13.1.1 the latter happens at some points of R, and so f is Fréchet differentiable at these points.

(4) *For any (uncountable) set Δ, every Lipschitz map of $c_0(\Delta)$ to a Banach space with the RNP is Fréchet differentiable Γ-almost everywhere.* Arguing as in the previous point, we find a rich subfamily $\mathcal{R}$ consisting of spaces isomorphic to c_0 such that for every $R \in \mathcal{R}$, $x \in R$, the given Lipschitz map $f \colon c_0(\Delta) \longrightarrow Y$ is Fréchet differentiable at x iff its restriction to R is. By Theorem 6.4.3 the latter occurs Γ-almost everywhere in R. Since, by Corollary 3.5.5, the set of points of Fréchet nondifferentiability of f is a Borel subset of $c_0(\Delta)$, we conclude from Lemma 5.6.1 that f is Fréchet differentiable at Γ-almost every $x \in c_0(\Delta)$.

Chapter Four

ε-Fréchet differentiability

In the context of all Fréchet differentiability results, or even of almost Fréchet differentiability ones, the results presented here are highly exceptional: they prove almost Fréchet differentiability in some situations when we know that the closed convex hull of all (even almost) Fréchet derivatives may be strictly smaller than the closed convex hull of the Gâteaux derivatives (see Chapter 14). Because of the possible future importance of this, so far only, foray into the otherwise impenetrable fortress of the problem of existence of derivatives in such situations, and because they have never appeared in book form before, we discuss the concepts and arguments leading to these results in some detail. One of these concepts, the idea of asymptotic uniform smoothness, will play a major role in our investigations in the following chapters.

4.1   ε-DIFFERENTIABILITY AND UNIFORM SMOOTHNESS

Although in this text we will prove, among other results, that Lipschitz maps of a Hilbert space into $\mathbb{R}^2$ have points of Fréchet differentiability, it is still unknown whether the same holds for maps of a Hilbert space into $\mathbb{R}^3$. In fact, in Chapter 14 we will learn that there is a qualitative difference between the two situations: while in the former the closed convex hull of all Fréchet derivatives is equal to the closed convex hull of the Gâteaux derivatives, this surely fails in the latter. Of course, this would happen for any nowhere Fréchet differentiable Lipschitz map of a Hilbert space into $\mathbb{R}^3$ (provided such a map exists). The above described phenomenon could be and, in fact, has been considered as evidence for existence of such maps. We show here that this evidence is rather shaky: we present a seemingly only slightly weakened notion of Fréchet differentiability for which the results of Chapter 14 still imply the above phenomenon, but for which the existence result can be proved.

The notion of Fréchet differentiability will be weakened in a natural way. Instead of approximating the increment of the function f at the point x_0 with error $o(\|x\|)$, we will approximate it with error $\varepsilon\|x\|$, where $\varepsilon > 0$ may be as small as we wish but x_0 may depend on ε.

Definition 4.1.1. A map $f\colon X \longrightarrow Y$ between Banach spaces is said to be *ε-Fréchet differentiable at x_0*, for some $\varepsilon > 0$ and $x_0 \in X$, if there are a $\delta > 0$ and a bounded linear operator $T\colon X \longrightarrow Y$ such that

$$\|f(x_0 + x) - f(x_0) - Tx\| \le \varepsilon \|x\|$$

whenever $\|x\| \le \delta$.

Somewhat loosely, one sometimes says that f is ε-Fréchet differentiable (without any specification of x_0 or ε) meaning that for every $\varepsilon > 0$ there is a point at which f is ε-Fréchet differentiable. We will avoid using such terminology, but in comments or remarks we may speak about such functions as almost Fréchet differentiable.

By Observation 3.5.2, if a map f is ε-Fréchet differentiable at some x_0 for every $\varepsilon > 0$, then it is Fréchet differentiable at x_0. However, if we only know that for every $\varepsilon > 0$ there is a point x_ε at which f is ε-Fréchet differentiable, then we cannot or do not know how to deduce Fréchet differentiability at any point. The point x_ε may change with ε in a way that we are not able to control. It is trivial to construct examples of nowhere differentiable continuous functions on $\mathbb{R}$ that are, for every $\varepsilon > 0$, ε-differentiable at some point, and it is similarly easy to construct such examples of real-valued Lipschitz functions on separable Banach spaces with nonseparable dual.

If at a point x, the function f is both Gâteaux differentiable and ε-Fréchet differentiable, then clearly $\|T - f'(x)\| \le \varepsilon$, where T is the operator from Definition 4.1.1. It follows that at points of Gâteaux differentiability we could have defined ε-Fréchet differentiability (up to an unimportant rescaling of ε) just with $T = f'(x)$. We have chosen the definition because it does not require any a priori differentiability assumption.

The fact that finding, even for any $\varepsilon > 0$, a point of ε-Fréchet differentiability is easier than finding a point of Fréchet differentiability is hardly surprising. In some sense, it is the starting point (and only the starting point) of any known proof of existence of points of Fréchet differentiability of real-valued Lipschitz functions in Asplund spaces. This can be traced back to [39]. These proofs are still not easy, even though easier proofs than the original one have been given in [27] and [29]. The proof we present in Chapter 12 is also not easy, but here some of the difficulty arises because we are proving a new, more general result. We will shortly see in Proposition 4.1.3 how simple the proof of just almost Fréchet differentiability of real-valued Lipschitz functions may be. For vector-valued maps, however, this is much harder. The first result [26] proved almost Fréchet differentiability of Lipschitz maps of superreflexive spaces into finite dimensional spaces. The main idea of their (very complicated) proof was to invent an infinite dimensional analogy of approximate continuity and prove the analogy of the (easy) finite dimensional result that a Lipschitz map is differentiable at every point at which its derivative has an approximate limit. A considerably easier argument, which we will follow here, was found in [22], where also the class of spaces to which the method applies was extended.

It should be also mentioned that most known results on the nonexistence of Fréchet derivatives actually show the nonexistence of ε-Fréchet derivatives for ε small enough. Corollary 3.2.4 may serve as a typical example: while the often quoted statement is that separable spaces with nonseparable dual admit a nowhere Fréchet differentiable norm, we have purposely stated it as producing an equivalent norm not ε-Fréchet differentiable at any point, with some fixed $\varepsilon > 0$. However, this is not the case for results claiming nonexistence of points of differentiability in certain subsets such as Theorem 3.4.3. This may be easily seen by inspecting the proof of Proposition 4.1.3. As we explain in Remark 4.3.4, a variant of its proof also shows that certain σ-porous sets (already in $\mathbb{R}^2$) contain, for every real-valued Lipschitz function f and every $\varepsilon > 0$, points of ε-differentiability of f. Nevertheless, Theorem 3.4.3 shows that some such functions are not differentiable at any point of the σ-porous set. (Of course, in this case such a function may be easily found directly.)

To show the key idea, we shall present a simple proof of an almost differentiability result for Lipschitz functions in uniformly smooth spaces. Since such spaces are Asplund, we know (and will see in Chapter 12) that such functions actually have points of Fréchet differentiability. However, the simplicity of the argument proving this special statement is what allowed generalization to the vector-valued case we will see later. To formulate the statement we recall the definition of uniform smoothness and uniform convexity of a space X. Later we will revisit these notions in Definition 4.2.1 in a different setting.

Definition 4.1.2. A Banach space X is said to be *uniformly smooth* if its norm satisfies

$$\sup_{\|x\|=1} \left(\|x + y\| + \|x - y\| - 2 \right) = o(\|y\|), \; y \to 0. \tag{4.1}$$

A Banach space is said to be *uniformly convex* if

$$\lim_{n \to \infty} \|x_n - y_n\| = 0$$

for any sequences (x_n) and (y_n) of unit vectors satisfying

$$\lim_{n \to \infty} \|x_n + y_n\| = 2.$$

Note that writing (4.1) for a fixed x, we recover the criterion of Fréchet differentiability of the norm at x from Proposition 3.1.3. Here we require that the condition for Fréchet differentiability holds uniformly in x from the unit sphere. It is well known that a uniformly smooth space is reflexive. Also, X is uniformly smooth if and only if X^* is uniformly convex (see, e.g., [31, Proposition 1.e.2]).

The result given in the following proposition is contained in the main results of this chapter. As we said before, the purpose of giving it separately is to show how easy it may be to find points of ε-Fréchet differentiability in simple situations.

Proposition 4.1.3. *Assume that X is a separable, uniformly smooth Banach space and let $f\colon X \longrightarrow \mathbb{R}$ be a Lipschitz function. Then for every $\varepsilon > 0$ the function f has a point of ε-Fréchet differentiability.*

Before we pass to the proof of the proposition we make two simple observations. The first is the following. Assume that X is uniformly smooth, $e \in X$, and $x^* \in X^*$ with $\|e\| = x^*(e) = \|x^*\| = 1$. Then for every $\eta > 0$ there is a $\delta = \delta(\eta) > 0$ such that

$$\|e + v\| \leq 1 + \eta\|v\|$$

whenever $\|v\| \leq \delta$ and $x^*(v) = 0$. This follows from (4.1) and from the fact that $\|e - v\| \geq x^*(e - v) = 1$.

The second observation is that if $\|e\| = x^*(e) = \|x^*\| = 1$ and $\varepsilon > 0$, then there is $\eta > 0$ such that any functional $z^* \in X^*$ with $\|z^*\| = 1$ and $z^*(e) \geq 1 - \eta$ satisfies

$$\|z^* - x^*\| < \varepsilon.$$

Indeed, if $z_n^*(e) \to 1$ with $\|z_n^*\| = 1$, then $\|z_n^* + x^*\| \to 2$ and, by the uniform convexity of X^*, $\|z_n^* - x^*\| \to 0$.

Proof of Proposition 4.1.3. Let $f \colon X \longrightarrow \mathbb{R}$ have $\mathrm{Lip}(f) = 1$ and let $\varepsilon > 0$. By Proposition 2.4.1 we choose $x \in X$ such that f is Gâteaux differentiable at x and $\|f'(x)\| > 1 - \delta\eta$, where the parameter $\eta > 0$ will be determined later to be small enough and $0 < \delta = \delta(\eta) \leq 1$ is defined in the first observation above. Find $e \in X$ with $\|e\| = 1$ such that

$$f'(x)(e) \geq 1 - \delta\eta.$$

Put $z^* = f'(x)$. Since f is Gâteaux differentiable at x there is a $\sigma > 0$ such that

$$|f(x + se) - f(x) - z^*(se)| < \delta\eta\,|s| \quad \text{whenever } |s| < \sigma. \tag{4.2}$$

Fix $x^* \in X^*$ with $\|x^*\| = x^*(e) = 1$, and consider any $y \in X$ such that $\|y\| < \sigma\delta/3$. Our plan is to show that for all such y we have

$$|f(x + y) - f(x) - z^*(y)| \leq \varepsilon\|y\|.$$

Write y in the form $y = u + v$, where $u = x^*(y)e$. Then $\|u\| \leq \|y\|$, $\|v\| \leq 2\|y\|$, and $x^*(v) = 0$. Put $r = \|v\|/\delta$. Notice that the condition $\|y\| < \sigma\delta/3$ implies

$$\|u + re\| \leq \|u\| + |r| \leq \|y\| + \frac{2\|y\|}{\delta} < \sigma,$$

and so by (4.2)

$$\begin{aligned}
|f(x + u + re) - f(x) - z^*(u + re)| &\leq \delta\eta\,|x^*(y) + r| \leq 3\eta\|y\|, \\
|f(x + u - re) - f(x) - z^*(u - re)| &\leq \delta\eta\,|x^*(y) - r| \leq 3\eta\|y\|.
\end{aligned} \tag{4.3}$$

By subtracting we obtain

$$\begin{aligned}
|f(x + u + re) - f(x + u - re)| &\geq 2rz^*(e) - 6\eta\|y\| \\
&\geq 2r(1 - \delta\eta) - 6\eta\|y\| \\
&= 2r - 2\eta\|v\| - 6\eta\|y\| \\
&\geq 2r - 10\eta\|y\|.
\end{aligned} \tag{4.4}$$

The fact that $\mathrm{Lip}(f) = 1$ and the first observation before the proof imply

$$\begin{aligned}
|f(x + y) - f(x + u + re)| &\leq \|re - v\| = r\|e - v/r\| \\
&\leq r(1 + \eta\|v\|/r) \leq r + 2\eta\|y\|.
\end{aligned} \tag{4.5}$$

Similarly

$$|f(x+y) - f(x+u-re)| \le r + 2\eta\|y\|. \tag{4.6}$$

We now use the "midpoint inequality" (which is trivial but basic, as we will see later in this chapter)

$$|a| + b \le \max\{|a+b|, |a-b|\}, \quad a, b \in \mathbb{R},$$

with the choice

$$a = f(x+y) - \tfrac{1}{2}\big(f(x+u+re) + f(x+u-re)\big) \text{ and }$$
$$b = \tfrac{1}{2}\big(f(x+u+re) - f(x+u-re)\big).$$

It allows us to deduce from (4.4), (4.5), and (4.6) that

$$\big|f(x+y) - \tfrac{1}{2}\big(f(x+u+re) + f(x+u-re)\big)\big|$$
$$\le r + 2\eta\|y\| - (r - 5\eta\|y\|) = 7\eta\|y\|. \tag{4.7}$$

If η was chosen small enough, the second observation before the proof implies that $\|x^* - z^*\| < \tfrac{1}{8}\varepsilon$. Hence

$$|z^*(v)| = |(x^* - z^*)(v)| \le 2\|x^* - z^*\| \, \|v\| \le \frac{\varepsilon}{2}\|y\|. \tag{4.8}$$

We are now ready for the final estimation. From (4.3), (4.7), and (4.8) we deduce that

$$|f(x+y) - f(x) - z^*(y)|$$
$$\le \big|f(x+y) - \tfrac{1}{2}\big(f(x+u+re) + f(x+u-re)\big)\big| + |z^*(v)|$$
$$+ \tfrac{1}{2}\big|f(x+u+re) - f(x) - z^*(u+re)\big|$$
$$+ \tfrac{1}{2}\big|f(x+u-re) - f(x) - z^*(u-re)\big|$$
$$\le 7\eta\|y\| + \frac{\varepsilon}{2}\|y\| + \frac{3\eta}{2}\|y\| + \frac{3\eta}{2}\|y\|$$
$$\le 10\eta\|y\| + \frac{\varepsilon}{2}\|y\| \le \varepsilon\|y\|,$$

provided $\eta > 0$ is chosen such that $\eta < \varepsilon/20$ and satisfies the requirements imposed on it in order to use the second observation. $\qquad\square$

Remark 4.1.4. If we could find x such that $\|f'(x)\| = 1$, we would use the reflexivity of X to find e with $\|e\| = 1$ such that $f'(x)(e) = 1$, and the proof above would show that f is Fréchet differentiable at x. However, in general such x fails to exist: consider, for example, $f \colon \mathbb{R} \longrightarrow \mathbb{R}$, $f(x) = e^{-|x|}$.

In case we are given two Lipschitz functions $f, g \colon X \to \mathbb{R}$ and we are interested in finding a common point of ε-Fréchet differentiability of f and g, the proof above can be used only if we are able to find a point x such that $\|f'(x)\|$ and $\|g'(x)\|$ are both close to $\mathrm{Lip}(f)$ and $\mathrm{Lip}(g)$, respectively. Such x clearly fails to exist in general. In [26] this difficulty is overcome in a rather complicated way by using a kind of density point argument that enables us to find a point x where $\|f'(x)\|$ and $\|g'(x)\|$ are both close to their respective approximate local maxima at x. The idea of local maximum was used in [22] in a simpler way (without density point considerations) and in a more general setting. We shall start an exposition of this result by presenting this setting in the next section.

Notice that the use of uniform convexity in the proof of Proposition 4.1.3 was essential. The value $\delta\eta$ had to be known before $f'(x)$ was chosen, so even local uniform convexity of the dual norm would not suffice. Nevertheless, a very similar argument is used in the known proofs of existence of points of Fréchet differentiability of real-valued Lipschitz functions in Asplund spaces. In a different context it was recognized in [17] that the basis of the argument lies not in the smoothness of the norm but in existence of small w^*-slices of bounded sets in the dual. In [27], this approach was combined with further (considerably harder) ideas to give the currently simplest proof of the Fréchet differentiability result we just mentioned. We will not follow this approach here, because so far, unlike the methods we develop here, this method has not led to a proof of more general results.

4.2 ASYMPTOTIC UNIFORM SMOOTHNESS

Before coming to the key notion of asymptotic uniform smoothness, let us recall briefly the well-known moduli of uniform convexity and smoothness (we refer to [31, Section 1.e], for more details and references to the original literature). Notice that these moduli are defined slightly differently (but equivalently) by different authors. The definition we have chosen is reasonably standard and fits well with what will follow.

Definition 4.2.1. The *modulus of (uniform) convexity* $\delta_X(t)$, $0 < t \leq 2$, of a Banach space X is

$$\delta_X(t) = \inf_{\substack{\|u\|,\|v\|\leq 1 \\ \|u-v\|\geq t}} 1 - \frac{\|u+v\|}{2},$$

and its *modulus of (uniform) smoothness* $\rho_X(t)$, $t > 0$ is

$$\rho_X(t) = \sup_{\|x\|=1,\|y\|\leq t} \frac{\|x+y\|+\|x-y\|}{2} - 1.$$

X is said to be *uniformly convex* if $\delta_X(t) > 0$ for every $0 < t \leq 2$, and *uniformly smooth* if $\lim_{\tau\to 0}\rho_X(t)/t = 0$. These notions of uniformly convex (or smooth) spaces coincide with those already introduced in Definition 4.1.2 without explicit use of the modulus of convexity or smoothness.

There is a duality relation between the moduli of uniform convexity and smoothness, which implies, in particular, that X is uniformly convex if and only if X^* is uniformly smooth. For a Hilbert space H the moduli are easy to compute. If $\dim H \geq 2$,

$$\delta_H(t) = 1 - \left(1 - \tfrac{1}{4}t^2\right)^{1/2}, \ 0 < t \leq 2, \quad \rho_H(t) = (1+t^2)^{1/2} - 1, \ t > 0.$$

The Hilbert space has the best moduli of convexity or smoothness in the sense that

$$\delta_X(t) \le \delta_H(t),\ 0 < t \le 2,\ \ \rho_X(\tau) \ge \rho_H(\tau),\ \tau > 0$$

for every Banach space X with $\dim X \ge 2$. In the spaces $L_p([0,1])$ the moduli are also well known but their computation is less straightforward:

$$\delta_{L_p}(\varepsilon) = \begin{cases} \frac{1}{8}(p-1)\varepsilon^2 + o(\varepsilon^2), & 1 < p \le 2, \\ \varepsilon^p/(p2^p) + o(\varepsilon^p), & 2 < p < \infty. \end{cases}$$

$$\rho_{L_p}(\tau) = \begin{cases} \tau^p/p + o(\tau^p), & 1 < p \le 2, \\ (p-1)\tau^2/2 + o(\tau^2), & 2 < p < \infty. \end{cases}$$

Much of the development presented in this book centers on the interplay between the quality of asymptotic uniform smoothness (of norms or bump-like functions), measured by its modulus, and differentiability. Here we introduce this modulus and the related modulus of asymptotic uniform convexity for norms. These moduli measure how smooth or convex the space is "at infinity," and the key point is that they may behave much better than the moduli introduced in Definition 4.2.1; in particular, the Hilbert space is no longer the space with the best modulus. They were first considered by Milman [36], and their relation to differentiability was first recognized in [22].

Definition 4.2.2. Let X be an infinite dimensional Banach space.

(i) The *modulus of asymptotic uniform convexity* of X is defined by

$$\bar{\delta}_X(t) = \inf_{\|x\|=1} \sup_{\dim(X/Y)<\infty} \inf_{\substack{y \in Y \\ \|y\| \ge t}} \|x+y\| - 1,\ t > 0.$$

(ii) The *modulus of asymptotic uniform smoothness* of X is defined by

$$\bar{\rho}_X(t) = \sup_{\|x\|=1} \inf_{\dim(X/Y)<\infty} \sup_{\substack{y \in Y \\ \|y\| \le t}} \|x+y\| - 1,\ t > 0.$$

The space is called *asymptotically uniformly convex* if $\bar{\delta}_X(t) > 0$ for every $t > 0$ and *asymptotically uniformly smooth* if $\lim_{t \to 0} \bar{\rho}_X(t)/t = 0$.

We make a comment on the notation of asymptotic moduli. Strictly speaking, we should write $\bar{\rho}_{(X,\|\cdot\|)}$ and $\bar{\delta}_{(X,\|\cdot\|)}$ instead of $\bar{\rho}_X$ and $\bar{\delta}_X$, respectively, since the change of norms clearly changes the corresponding moduli. However, the latter notation is widely accepted in the literature and so we will use it as well. Only in a few exceptional cases of two norms on the same space will we distinguish the corresponding moduli by writing, for example, $\bar{\rho}_{\|\cdot\|}$ and $\bar{\rho}_{\|\cdot\|_1}$.

The definition of $\bar{\rho}$ makes sense even if X is finite dimensional and gives $\bar{\rho}_X(t) = 0$ for every t. However, in this situation the definition of $\bar{\delta}$ does not makes sense: the supremum in the middle involves (and should in fact be realized when) $Y = \{0\}$, in which case there is no y to take the infimum over. This leads to a rather unpleasant exception, which we overcome by letting $\bar{\delta}_X(t) = t$ if X is finite dimensional. The motivation behind this choice is that $\bar{\delta}(t) \leq t$ for every space, and so we say that X is as convex as it can be (as well as as smooth as it can be) in the direction of $\{0\}$. Of course, if $\dim(X) < \infty$ the asymptotic moduli play no role, but in some results this allows us not to distinguish unnecessarily between finite and infinite dimensional spaces.

It is easy to compute the asymptotic moduli for ℓ_p, $1 \leq p < \infty$, and c_0.

$$\bar{\delta}_{\ell_p}(t) = \bar{\rho}_{\ell_p}(t) = (1 + t^p)^{1/p} - 1, \quad 1 \leq p < \infty, \quad t > 0,$$
$$\bar{\delta}_{c_0}(t) = \bar{\rho}_{c_0}(t) = \max\{t - 1, 0\}, \, t > 0.$$

The same results hold if we replace ℓ_p (resp. c_0) by any subspace X of the space $\left(\sum_n \oplus X_n\right)_{\ell_p}$ (resp. $\left(\sum_n \oplus X_n\right)_{c_0}$), where (X_n) is any sequence of finite dimensional spaces.

Some of the intuition behind the asymptotic versions of moduli may be seen from the following. Assume $x \in X$ with $\|x\| = 1$ in an asymptotically uniformly smooth space X. Then for a small perturbation y from a suitable finite codimensional subspace Y we obtain that

$$\|x + y\| = \|x\| + o(\|y\|).$$

In case of a sequence space the equation above suggests that the norm behaves like a sup-norm for almost disjointly supported vectors x and y (especially if we imagine the $o(\|y\|)$ going to zero very fast). Thus c_0 is the most "smooth" among asymptotically uniformly smooth Banach spaces. The use of finite codimensional subspaces causes a similar effect also for asymptotic uniform convexity. Here the norm of the perturbed element $x + y$ is (if we imagine that the modulus is "large")

$$\|x + y\| \approx \|x\| + \kappa\|y\| \quad \text{for some } \kappa > 0.$$

This mimics the norm in ℓ_1 for vectors that have almost disjoint support. The space ℓ_1 is the most "convex" among the asymptotically uniformly convex Banach spaces. Notice that in contrast to the situation with uniformly convex or smooth spaces the spaces c_0 and ℓ_1 are not reflexive.

It is known (see [22] for references) that for reflexive spaces there is a complete duality between asymptotic uniform convexity and asymptotic uniform smoothness. For nonreflexive spaces this duality has been investigated in the literature, but it is still not completely understood.

Because of its intimate connection to the midpoint inequality whose importance we have already seen in the proof of Proposition 4.1.3, we now introduce a slight modification of the modulus of convexity. We could call it an (asymptotic) midpoint modulus, but we will not use it often enough to warrant a name. One can notice other reasonable replacements for the term $\|x + y\| - 1$ when defining asymptotic convexity or smoothness, but none of them appear to be better suited to the problems we study. For asymptotic smoothness, $\frac{1}{2}(\|x + y\| + \|x - y\|) - 1$ seems to be better adjusted for more general functions than norms, and so we will adopt it when defining asymptotic smoothness of "bump functions." For norms, the modulus defined in this way would be smaller than $\bar\rho$, but by no more than a factor $\frac{1}{2}$, and so for nearly all our purposes (but not for some finer results in this chapter) they would be equivalent.

Definition 4.2.3. For a subspace Y of X, $x \in X$, and $t > 0$ we denote by $\bar\mu(t, x, Y)$ (or $\bar\mu_X(t, x, Y)$ if X needs to be specified) the greatest number $\mu \leq t$ such that

$$\max\{\|x + y\|, \|x - y\|\} - \|x\| \geq \mu\|x\| \text{ for } y \in Y, \|y\| \geq t\|x\|,$$

and we let

$$\bar\mu_X(t) = \inf_{x \in X} \sup_{\dim(X/Y) < \infty} \bar\mu_X(t, x, Y).$$

Clearly, $\bar\mu(t, x, Y) \geq 0$ and notice that $\bar\mu(t, cx, Y) = \bar\mu(t, x, Y)$ for $c \neq 0$. So the infimum in $\bar\mu_X$ may as well be taken over $\|x\| = 1$. The artificial looking requirement that $\mu \leq t$ is automatically true when $x \neq 0$, but it was added to handle the special case $x = 0$ in a natural way. In particular, it gives that $\bar\mu_X(t) = t$ if $\dim X < \infty$. Notice also that the nonasymptotic moduli $\bar\mu_X(t, x, Y)$ are of huge interest. To see why, we reformulate their definition in the following way.

Midpoint inequality. *If $x \in X$, $y \in Y$, $t > 0$, and*

$$\max\{\|x + y\|, \|x - y\|\} \leq (1 + \bar\mu_X(t, x, Y))\|x\|$$

then $\|y\| \leq t\|x\|$.

For example, one immediately sees that $\bar\mu_{\mathbb{R}}(t, x, \mathbb{R}) = t$, and the above inequality is equivalent to the midpoint inequality we have used in the proof of Proposition 4.1.3.

In analogy with the definition of $\bar\mu(t, x, Y)$, it is sometimes convenient to have a similar notation for the innermost infimum (resp. supremum) in the definition of asymptotic moduli. So for Y a subspace of X, $x \in X \setminus \{0\}$ and $t > 0$ we let

$$\bar\delta_X(t, x, Y) = \inf_{\substack{y \in Y \\ \|y\| \geq t\|x\|}} \frac{\|x + y\| - \|x\|}{\|x\|}, \quad \bar\rho_X(t, x, Y) = \sup_{\substack{y \in Y \\ \|y\| \leq t\|x\|}} \frac{\|x + y\| - \|x\|}{\|x\|}.$$

These quantities are used only for finding asymptotic moduli, and so we may always imagine that Y is a subspace of the kernel of a norm one functional $x^* \in X^*$ such that $x^*(x) = \|x\|$. In that case, both $\bar\delta_X(t, x, Y)$ and $\bar\rho_X(t, x, Y)$ are non-negative.

Here are some simple facts concerning asymptotic moduli. First notice that the triangle inequality yields trivial estimates for all moduli

$$t - 2 \leq \bar\delta_X(t), \ \bar\mu_X(t), \ \bar\rho_X(t) \leq t.$$

Further simple observations are contained in the following.

Proposition 4.2.4. *For any Banach space X:*

(i) *If $X_0 \subset X$ then $\bar\delta_{X_0}(t) \geq \bar\delta_X(t)$, $\bar\rho_{X_0}(t) \leq \bar\rho_X(t)$, and $\bar\mu_{X_0}(t) \geq \bar\mu_X(t)$.*

(ii) *The functions $\bar{\delta}_X(t)$, $\bar{\rho}_X(t)$ and $\bar{\mu}_X(t)$ are increasing, and have Lipschitz constant at most 1, $\bar{\delta}_X(t) \le \bar{\mu}_X(t) \le \bar{\rho}_X(t)$, and $\bar{\rho}_X(t)$ is also convex.*

(iii) $\delta_X(t) \le \bar{\delta}_X(t)$ and $2\rho_X(t) \ge \bar{\rho}_X(t)$ for $0 < t \le 1$.

Proof. Parts (i) and (ii) are direct consequences of the definitions. Only the convexity of $\bar{\rho}_X(t)$ may need a word of explanation. As a function of t, $\bar{\rho}_X(t, x, Y)$ is easily seen to be convex. The final supremum over x preserves convexity, but the infimum over Y may look as if it could destroy it. But, since the family of finite codimensional subspaces is closed under finite intersection, the family of functions whose infimum we are taking is directed downward by (i), and so convexity is preserved.

To check part (iii), let $x \in X$ and $x^* \in X^*$ be such that $x^*(x) = \|x\| = \|x^*\| = 1$. Let Y be the kernel of x^*. Then

$$\|x - y\| \ge x^*(x + y) = 1$$

for every $y \in Y$ and

$$\sup_{\substack{y \in Y \\ \|y\| \le t}} \frac{\|x + y\| + \|x - y\|}{2} - 1 \ge \frac{1}{2} \sup_{\substack{y \in Y \\ \|y\| \le t}} \left(\|x + y\| - 1 \right).$$

Hence $\bar{\rho}_X(t) \le 2\rho_X(t)$. To prove the first part of (iii), let $0 < t < 1$ and $\varepsilon > 0$ be given. By definition of $\bar{\delta}_X(t)$ choose $x \in X$ with $\|x\| = 1$ such that for every Y of finite codimension

$$\bar{\delta}_X(t) > \bar{\delta}_X(t, x, Y) - \varepsilon.$$

Let again $x^* \in X^*$ be such that $x^*(x) = \|x^*\| = 1$ and consider $Y \subset \ker x^*$. There is $y \in Y$ with $t \le \|y\| \le 1$ such that

$$\bar{\delta}_X(t, x, Y) > \|x + y\| - 1 - \varepsilon,$$

and so

$$\bar{\delta}_X(t) > \|x + y\| - 1 - 2\varepsilon.$$

Notice that $\|x + y\| \ge 1$. We put

$$u = \frac{x + y}{\|x + y\|}, \quad v = u - y = \frac{x}{\|x + y\|} + \left(1 - \frac{1}{\|x + y\|} \right)(-y).$$

The vectors u and v are contained in the unit ball of X and $\|u - v\| \ge t$. Consequently,

$$\delta_X(t) \le 1 - \tfrac{1}{2}\|u + v\|$$

$$\le 1 - \tfrac{1}{2}x^*(u + v) = 1 - \frac{1}{\|x + y\|}$$

$$\le 1 - \frac{1}{\bar{\delta}_X(t) + 1 + 2\varepsilon} \le \bar{\delta}_X(t) + 2\varepsilon.$$

Since $\varepsilon > 0$ was arbitrary we get the desired result. $\qquad\square$

In particular, a uniformly convex (resp. smooth) space is asymptotically uniformly convex (resp. smooth). The converse is obviously false as is evident by considering ℓ_1 or c_0.

A norm equivalent to an asymptotically uniformly convex or smooth norm may easily fail to have the same property. For example, in ℓ_2 the norm $\max\{\|x\|_{\ell_2}, 2\|x\|_\infty\}$ is an equivalent norm, but it is not asymptotically uniformly convex, and the norm $\|x\|_{\ell_2} + \max\{|x_i| + |x_j|, i \neq j\}$ is not asymptotically uniformly smooth. In both cases one can check the claimed properties by considering $x = e_1$ and $y = te_n$ for large n.

However, in some tasks it is convenient to modify the original norm and preserve the asymptotic behavior. For simplicity, when considering spaces with several norms, we will index $\bar\mu$ and $\bar\delta$ by $\|\cdot\|$ instead of the more precise $(X, \|\cdot\|)$. The following simple observation suffices for the application to proving the mean value estimate in Theorem 4.3.3. The modulus of smoothness of a pseudonorm is defined analogically to that of a norm, or equivalently by passing to the quotient. In the application we will in fact only need to know the obvious fact that if $e \in X$ and $e^* \in X^*$ are such that $\|e^*\| = \|e\| = e^*(e)$, then for $\|x\| = \|x\| + |e^*(x)|$ we have $\bar\rho_{\|\cdot\|}(t) = \bar\rho_{\|\cdot\|}(t)$.

Observation 4.2.5. *For every continuous pseudonorm ν on $(X, \|\cdot\|)$ there are constants $c_0, C_0 > 0$ such that $\|x\|_1 = \|x\| + \nu(x)$ and $\|x\|_\infty = \max\{\|x\|, \nu(x)\}$ satisfy*

$$\bar\mu_{\|\cdot\|_1}(t) \geq c_0 \bar\mu_{\|\cdot\|}(c_0 t), \quad \bar\delta_{\|\cdot\|_1}(t) \geq c_0 \bar\delta_{\|\cdot\|}(c_0 t)$$

and

$$\bar\rho_{\|\cdot\|_\infty}(t) \leq \max\{\bar\rho_{\|\cdot\|}(C_0 t), \bar\rho_\nu(C_0 t)\}$$

for all $t > 0$.

Proof. Choose $c_0 > 0$ such that $\|\cdot\| \geq c_0\|\cdot\|_1$. Let $x \in X \setminus \{0\}$ and $x^* \in X^*$, with $\nu^*(x^*) = 1$ and $x^*(x) = \nu(x)$. Assuming that $y \in Y$ and Y is contained in the kernel of x^*, we have $\nu(x \pm y) \geq \nu(x)$, and so

$$\frac{\max\{\|x+y\|_1, \|x-y\|_1\} - \|x\|_1}{\|x\|_1} \geq c_0 \frac{\max\{\|x+y\|, \|x-y\|\} - \|x\|}{\|x\|}$$

and

$$\frac{\|x+y\|_1 - \|x\|_1}{\|x\|_1} \geq c_0 \frac{\|x+y\| - \|x\|}{\|x\|}.$$

Since $\|y\|_1 \geq t\|x\|_1$ implies that $\|y\| \geq c_0\|y\|_1 \geq c_0 t\|x\|_1 \geq c_0 t\|x\|$, which shows the first two inequalities.

For the last statement, choose $C_0 \geq 9$ such that $\|\cdot\|_\infty \leq C_0\|\cdot\|$. The condition $C_0 \geq 9$ was chosen because then

$$\bar\rho_{\|\cdot\|_\infty}(t) \leq t \leq C_0 t - 2 \leq \bar\rho_{\|\cdot\|}(C_0 t) \quad \text{for } t \geq \tfrac{1}{4}.$$

Hence we consider only $t < \frac{1}{4}$ and, similarly to the previous part, we intend to use the inequality

$$\frac{\|x+y\|_\infty - \|x\|_\infty}{\|x\|_\infty} \leq \max\left\{\frac{\|x+y\| - \|x\|}{\|x\|}, \frac{\|x+y\| - \|x\|}{\|x\|_\infty}\right\}.$$

If the above inequality holds with the first term in the maximum, we just notice that $\|y\|_\infty \leq t\|x\|_\infty$ implies $\|y\| \leq C_0 t\|x\|$, and so we get the desired statement. In the opposite case we have

$$\|x + y\| - \|x\| < \|x + y\|_\infty - \|x\|_\infty. \tag{4.9}$$

If it were $\nu(x + y) < \|x + y\|_\infty$, then necessarily $\|x + y\| = \|x + y\|_\infty$ which contradicts (4.9). Therefore $\nu(x + y) = \|x + y\|_\infty$ and

$$\|x\|_\infty \geq \nu(x) \geq \nu(x + y) - \nu(y)$$
$$\geq \|x + y\|_\infty - \|y\|_\infty$$
$$\geq \|x\|_\infty - 2\|y\|_\infty \geq \tfrac{1}{2}\|x\|_\infty.$$

Hence $\nu(x) > 0$ and $\nu(y) \leq t\|x\|_\infty \leq 2t\nu(x) \leq C_0 t\nu(x)$, which implies that

$$\frac{\|x + y\|_\infty - \|x\|_\infty}{\|x\|_\infty} \leq \frac{\nu(x + y) - \nu(x)}{\nu(x)}. \qquad \square$$

A brief survey of the literature on this subject is presented in [22]. The next statement was proved in [22] in a slightly weaker form with $\bar{\delta}_Y(t)$ instead of $\bar{\mu}_Y(t)$.

Proposition 4.2.6. *Let X and Y be two Banach spaces such that, for some $t > 0$, $\bar{\rho}_X(t) < \bar{\mu}_Y(t)$. Then every bounded linear operator from X to Y is compact.*

Proof. Let $T: X \longrightarrow Y$ be a norm one operator. By assumption and by part (ii) of Proposition 4.2.4 there are $\rho, \eta > 0$ and $0 < s < 1$ such that $\bar{\rho}_X(t) < \rho$ and $1 + \rho < (1 + \bar{\mu}_Y(st) - \eta)(1 - \eta)$.

It is enough to find a subspace $X_0 \subset X$ of finite codimension such that

$$\|T|_{X_0}\| \leq s.$$

Indeed, by part (i) of Proposition 4.2.4 we can iterate this procedure to get, for every n, a subspace X_n of X of finite codimension such that

$$\|T|_{X_n}\| \leq s^n.$$

This property clearly implies that the image of the unit ball is totally bounded; hence T is compact.

Let $x \in X$ be a unit vector such that $\|Tx\| > 1 - \eta$. Choose a finite codimensional subspace Y_0 of Y such that

$$\bar{\mu}_Y(st, Tx, Y_0) > \bar{\mu}_Y(st) - \eta.$$

Since $\bar{\rho}_X(t) < \rho$ there is a finite codimensional space X_0 of X such that

$$\|x + u\| - 1 \leq \rho$$

for every $u \in X_0$ with $\|u\| \leq t$. In addition one can require that $TX_0 \subset Y_0$. We show that this X_0 is the desired subspace. Let $u \in X_0$ with $\|u\| = t$. Then

$$\|Tx \pm Tu\| \leq \|x \pm u\| \leq 1 + \rho < (1 + \bar{\mu}_Y(st, Tx, Y_0))\|Tx\|,$$

hence by the midpoint inequality $\|Tu\| < st\|Tx\| \leq s\|u\|$, as required. $\square$

The next proposition from [22] has been proved before in [20] in a slightly weaker form. Notice that, because of the convexity of $\bar{\rho}$, the assumption that $\bar{\rho}_X(t) < t$ for some $t > 0$ is equivalent to requiring that $\bar{\rho}_X(t) < ct$ for some $c < 1$ and all small enough t. Also notice that, although we prove the statement for separable spaces only, we do not assume separability of the space in the statement, since in this case the transfer to the nonseparable situation is trivial.

Proposition 4.2.7. *Let X be a Banach space such that $\bar{\rho}_X(t) < t$ for some $t > 0$. Then X is an Asplund space. In particular, an asymptotically uniformly smooth space is Asplund.*

Proof. Assume that X is separable but X^* is nonseparable. Choose $0 < \tilde{c} < 1$ and $t > 0$ such that $\bar{\rho}_X(t) < \tilde{c}t$. Let $\tilde{c} < c < 1$ be such that $ct - 1 + c = \tilde{c}t$ and find by Theorem 3.2.3 an equivalent norm $\|\cdot\|$ such that $\|\cdot\| \leq \|\cdot\|$ and for every $z \in X$, $r > 0$, and every subspace Y of X with $(X/Y)^*$ separable we have

$$\sup_{y \in Y, \|y\| \leq r} \left(\|z + y\| - \|z\| \right) > cr. \tag{4.10}$$

Using this for $z = 0$, $r = 1$, and $Y = X$, we see that there is $x \in X$ with $\|x\| \leq 1$ such that $\|x\| > c$. Replacing x by $x/\|x\|$ if necessary, we may assume that $\|x\| = 1$.

Since $\bar{\rho}_X(t) < \tilde{c}t$, we can find a finite codimensional subspace Y of X such that $\|x + y\| - 1 < \tilde{c}t$ for $y \in Y$, $\|y\| \leq t$. Choosing by (4.10) a $y \in Y$ with $\|y\| \leq t$ such that $\|x + y\| - \|x\| > ct$, we get the required contradiction

$$\tilde{c}t > \|x + y\| - 1 \geq \|x + y\| - \|x\| - 1 + c > ct - 1 + c = \tilde{c}t. \qquad \square$$

When working with asymptotic notions, we are often faced with the situation when a finite codimensional subspace Y of X has uncontrollable codimension, and so we have no useful estimate of the norm of a projection of X onto Y. Possibly somewhat surprisingly, the following simple lemma often provides a remedy. We will use it extensively throughout this book.

Lemma 4.2.8. *Let Y be a finite codimensional subspace of a Banach space X. Then for every $0 < \varepsilon < 1$ there is a finite dimensional subspace U of X such that*

$$(1 - \varepsilon)B_X \subset B_U + 2B_Y.$$

Proof. Since X/Y is finite dimensional there is a finite $\frac{1}{2}\varepsilon$-net $(y_i)_{i \in I}$ in $B_{X/Y}$ with $\|y_i\| < 1$ for all $i \in I$. Denote by

$$Q \colon X \longrightarrow X/Y$$

the corresponding quotient map. For each $i \in I$ there is an element $x_i \in B_X$ such that $Qx_i = y_i$. Put $U := \text{span}\{x_i \mid i \in I\}$.

Let $x \in B_X$. We can find a point y_i such that $\|Qx - y_i\| < \frac{1}{2}\varepsilon$. Noticing that $Qx - y_i = Q(x - x_i)$ we can further find a vector $y \in Y$ such that

$$\|x - x_i - y\| < \frac{\varepsilon}{2}.$$

It follows that $\|y\| \leq \|x_i\| + \|x\| + \frac{1}{2}\varepsilon \leq 2 + \frac{1}{2}\varepsilon$, and since clearly

$$x = x_i + y + (x - x_i - y)$$

we obtain

$$B_X \subset B_U + (2 + \tfrac{1}{2}\varepsilon)B_Y + \frac{\varepsilon}{2}B_X.$$

After n iterations of the inclusion above we get

$$B_X \subset \left(1 + \frac{\varepsilon}{2} + \cdots + \left(\frac{\varepsilon}{2}\right)^{n-1}\right)\left(B_U + (2 + \tfrac{1}{2}\varepsilon)B_Y\right) + \left(\frac{\varepsilon}{2}\right)^n B_X,$$

and thus, letting $n \to \infty$, we have

$$B_X \subset \frac{1}{1 - \frac{1}{2}\varepsilon}\left(B_U + (2 + \tfrac{1}{2}\varepsilon)B_Y\right),$$

from which the conclusion of the lemma follows.   □

We will often use the following reformulation of Lemma 4.2.8 using inequality between norms instead of inclusions of balls and with the particular choice $\varepsilon = \frac{1}{2}$.

Corollary 4.2.9. *Let Y be a finite codimensional subspace of a Banach space X. Then there is a finite dimensional subspace U of X such that every $x \in X$ can be written as $x = u + y$, where $u \in U$, $y \in Y$, and*

$$\|u\| \leq 3\|x\|, \quad \|y\| \leq 3\|x\|.$$

4.3 ε-FRÉCHET DIFFERENTIABILITY OF FUNCTIONS ON ASYMPTOTICALLY UNIFORMLY SMOOTH SPACES

The main purpose of this section is to prove that in an asymptotically smooth Banach space X any finite set of real-valued Lipschitz functions on X has, for every $\varepsilon > 0$, a common point of ε-Fréchet differentiability. More generally, we show that every Lipschitz mapping from X to an RNP space Y with $\bar{\rho}_X(t) = o(\bar{\mu}_Y(ct))$ has, for every $\varepsilon > 0$, a point of both ε-Fréchet and Gâteaux differentiability, and that even the one-dimensional mean value estimate holds.

We begin with a simple lemma, a variation of the standard estimates of the increment of a function with the help of the value of its derivative at another point.

Lemma 4.3.1. *Let X and Y be Banach spaces, let C be an open convex set in X, and let $f: C \longrightarrow Y$ be a Lipschitz map. Assume that $x \in C$ is a point at which f is both Gâteaux and ε-Fréchet differentiable. Then there is $\delta > 0$ such that*

$$\|f(x + u + v) - f(x + u) - f'(x; v)\| \leq 4\varepsilon(\|u\| + \|v\|)$$

whenever $\|v\| < \delta$ and $u \in X$ with $x + u \in C$.

Proof. Assume that $\mathrm{Lip}(f) \neq 0$; otherwise the statement is trivial. Since f is ε-Fréchet differentiable at x, there are $\delta_1 > 0$ and a bounded linear operator $T \colon X \longrightarrow Y$ such that

$$\|f(x + v) - f(x) - Tv\| \leq \varepsilon\|v\|$$

for $\|v\| < \delta_1$. It also follows that for such v we have $\|f'(x; v) - Tv\| \leq \varepsilon\|v\|$, and so

$$\|f(x + v) - f(x) - f'(x; v)\| \leq 2\varepsilon\|v\|$$

for $\|v\| < \delta_1$.

Let $\delta = 2\varepsilon\delta_1 / \mathrm{Lip}(f)$ and suppose that $\|v\| < \delta$. If

$$\|u\| + \|v\| \geq \frac{\|v\| \, \mathrm{Lip}(f)}{2\varepsilon}$$

then the required inequality is obvious. In the opposite case we have $\|u\| + \|v\| < \delta_1$, and so

$$
\begin{aligned}
\|f(x + u + v) &- f(x + u) - f'(x; v)\| \\
&\leq \|f(x + u + v) - f(x) - f'(x; u + v)\| + \|f(x + u) - f(x) - f'(x; u)\| \\
&\leq 2\varepsilon(\|u\| + \|v\|) + 2\varepsilon\|v\| \leq 4\varepsilon(\|u\| + \|v\|).
\end{aligned}
$$
$\qquad\square$

We turn now to the first main result of this chapter. It will be contained, and its mean value part slightly strengthened, in the next theorem for whose proof it is the main ingredient.

Theorem 4.3.2. *Suppose that X is a separable asymptotically smooth Banach space and $f_1, \ldots, f_n$ are real-valued Lipschitz functions defined on a nonempty open subset D of X. Then for every $\varepsilon > 0$, the functions $f_1, \ldots, f_n$ have a common point of ε-Fréchet differentiability.*

Moreover, whenever $S \subset D$ has Haar null complement in D and $\eta > 0$, we may further require that the point x of ε-Fréchet differentiability belongs to S and

$$\|f_1'(x)\| > \mathrm{Lip}(f_1) - \eta.$$

Proof. We introduce some notation which will be used in the proof. It will be convenient to denote by $\mathrm{diff}^{\varepsilon F}(f)$ the set of points at which f is ε-Fréchet differentiable. In addition to $\varepsilon > 0$ and the functions $f_1, \ldots, f_n$ (in fact, the reader may imagine that we have an infinite sequence $f_1, f_2, \ldots$, but we still find common points of ε-Fréchet differentiability for finite subsequences only) we assume that $\eta > 0$ and $S \subset D$ is a subset with Haar null complement in D at which all f_i are Gâteaux differentiable. We denote

$$
\begin{aligned}
s_1 &= \sup\{\|f_1'(x)\| \mid x \in S\}, \\
S_1(t) &= \{x \in S \mid \|f_1'(x)\| > s_1 - t\},
\end{aligned}
$$

and for $k \geq 2$ we define recursively

$$s_k = \lim_{t \searrow 0} \sup\{\|f_k'(x)\| \mid x \in S_{k-1}(t)\},$$
$$S_k(t) = \{x \in S_{k-1}(t) \mid \|f_k'(x)\| > s_k - t\}.$$

For unit vectors $u_1, \ldots, u_k$ in X we also denote

$$S_k(t; u_1, \ldots, u_k) = \{x \in S \mid f_i'(x; u_i) > s_i - t, \ i = 1, \ldots, k\}.$$

Clearly, $\emptyset \neq S_{k+1}(t) \subset S_k(t)$ for any $k \in \mathbb{N}$ and $t > 0$. Notice that for every choice $u_1, \ldots, u_k$ of unit vectors and $t > 0$ we have

$$S_k(t; u_1, \ldots, u_k) \subset S_k(t).$$

Although proving the theorem by induction with respect to the number of functions is a natural idea, it is not clear how one could deduce the ε-differentiability of the next function assuming just the ε-differentiability of the previous ones. We will therefore prove by induction the following assertions:

(A_k) For every $\varepsilon > 0$ there is a $t > 0$ such that for every $x \in S_k(t)$ there is $\delta > 0$ with the property that for every $\|u\|, \|h\| < \delta$,

$$|f_k(x + h + u) - f_k(x + h)| \leq s_k\|u\| + \varepsilon(\|h\| + \|u\|).$$

(B_k) For every $\varepsilon > 0$ there is an $s > 0$ such that $S_k(s) \subset \operatorname{diff}^{\varepsilon F}(f_k)$.

(C_k) For every $\varepsilon, t > 0$ there is an $s > 0$ such that for all unit vectors $u_1, \ldots, u_k$ and every $x \in S_k(s; u_1, \ldots, u_k)$ there is a $\delta > 0$ such that

$$\mathscr{L}^k\left\{\sigma \in [0, q]^k \ \Big| \ x + u + \sum_{i=1}^{k} \sigma_i u_i \notin S_k(t; u_1, \ldots, u_k)\right\} < \varepsilon(q + \|u\|)\, q^{k-1}$$

whenever $0 < q < \delta$ and $\|u\| < \delta$ are such that $x + u + \sum_{i=1}^{k} \sigma_i u_i \in S$ for $\mathscr{L}^k$-almost all $\sigma \in [0, q]^k$.

We show that (A_k), (B_k), and (C_k) all hold for $k = 1, \ldots, n$ (or, if the reader wishes, for all $k = 1, 2, \ldots$). The statement of the theorem is a consequence of (B_k). Indeed, for each $k = 1, \ldots, n$ we have that $S_k(t_k) \subset \operatorname{diff}^{\varepsilon F}(f_k)$ for some $t_k > 0$. Then all functions $f_1, \ldots, f_n$ are ε-Fréchet differentiable and satisfy $\|f_1'(x)\| > s_1 - \eta = \operatorname{Lip}(f_1) - \eta$ at all points $x \in S_n(t)$ where $t = \min\{\eta, t_1, \ldots, t_n\}$.

Since (A_1) follows from the fact that $s_1 = \operatorname{Lip}(f_1)$, the theorem will be proved once we show the following schema of implications.

$$(A_k) \Rightarrow (B_k), \quad (B_1) \Rightarrow (C_1), \quad (C_k) \Rightarrow (A_{k+1}), \quad (B_{k+1}) \wedge (C_k) \Rightarrow (C_{k+1}).$$

$(A_k) \Rightarrow (B_k)$: If $s_k = 0$, it suffices to take the t from (A_k) and observe that (A_k) already says that f is ε-Fréchet differentiable at every point of $S_k(t)$. So assume $s_k > 0$. Let $\varepsilon > 0$ be given and choose $\eta > 0$ to satisfy

$$\eta s_k \leq \tfrac{1}{6}\varepsilon.$$

Find $0 < t < 1$ such that $\bar{\rho}_X(t) < \tfrac{1}{2}\eta t$ and then choose $0 < \zeta \leq \varepsilon/6$ with

$$s_k(1 + \tfrac{1}{2}\eta t) + \zeta(2t + 3) < (1 + \eta t)(s_k - \zeta).$$

Finally we use (A_k) to find $0 < s < \min\{\zeta, s_k\}$ such that for every $x \in S_k(s)$ there is a $0 < \delta(x) < 1$ with $B(x, \delta(x)) \subset D$ and

$$\left| f_k(x + h + u) - f_k(x + h) \right| \leq s_k \|u\| + \zeta(\|h\| + \|u\|) \tag{4.11}$$

whenever $\|u\|, \|h\| < \delta(x)$.

Take a point $x \in S_k(s)$. By definition, there are unit vectors $u_1, \dots, u_k$ such that $x \in S_k(s; u_1, \dots, u_k)$. Since $\bar{\rho}_X(t) < \tfrac{1}{2}\eta t$, there is a finite codimensional subspace Y of X contained in the kernel of $f_k'(x)$ such that

$$\|u_k + v\| \leq 1 + \tfrac{1}{2}\eta t \text{ for every } v \in Y, \|v\| \leq t. \tag{4.12}$$

By Corollary 4.2.9 there is a finite dimensional subspace U of X containing u_k such that every $y \in X$ can be written $y = u + v$, $u \in U$, $v \in Y$, and $\|u\|, \|v\| \leq 3\|y\|$. Since U is finite dimensional and f_k is Gâteaux differentiable at x, it is Fréchet differentiable at x in the direction of U. Hence by Lemma 4.3.1 there is a $0 < \delta < \tfrac{1}{6}t\delta(x)$ such that

$$\left| f_k(x + \widetilde{u} + u) - f_k(x + \widetilde{u}) - f_k'(x; u) \right| \leq \zeta(\|\widetilde{u}\| + \|u\|) \tag{4.13}$$

whenever $u, \widetilde{u} \in U$ and $\|u\|, \|\widetilde{u}\| < 6\delta/t$.

To prove that f_k is ε-Fréchet differentiable at x, suppose that $0 < \|y\| < \delta$ is given and write $y = u + v$, $u \in U$, $v \in Y$, and $\|u\|, \|v\| \leq 3\|y\|$. Denoting $r = 3\|y\|/t$, we use (A_k), (4.13), (4.11), and (4.12) with $\pm v/r$ (which is allowed since $\|v/r\| \leq t$) to estimate

$$\begin{aligned}
\Big| f_k(x + y) &- f_k(x + u) \pm f_k'(x; ru_k) \Big| \\
&\leq \left| f_k(x + u + v) - f_k(x + u \mp ru_k) \right| \\
&\quad + \left| f_k(x + u \mp ru_k) - f_k(x + u) \pm f_k'(x; ru_k) \right| \\
&\leq s_k\|v \pm ru_k\| + \zeta(\|v \pm ru_k\| + \|u \pm ru_k\| + r) + \zeta(\|u\| + r) \\
&\leq s_k(1 + \tfrac{1}{2}\eta t)r + \zeta(2\|u\| + \|v\| + 3r) \\
&\leq s_k(1 + \tfrac{1}{2}\eta t)r + \zeta(2t + 3)r.
\end{aligned}$$

The choice of ζ allows us to continue

$$< (1 + \eta t)(s_k - s)\, r < (1 + \eta t)\left| f_k'(x; ru_k) \right|.$$

Hence by the midpoint inequality,

$$\left| f_k(x+y) - f_k(x+u) \right| \le \eta t \left| f_k'(x; ru_k) \right| \le \eta t r s_k.$$

Recalling that $y - u = v$ belongs to the kernel of $f_k'(x)$, we conclude that

$$\left| f_k(x+y) - f_k(x) - f_k'(x;y) \right| \le \eta t r s_k + \left| f_k(x+u) - f_k(x) - f_k'(x;u) \right|$$
$$\le \tfrac{1}{2}\varepsilon\|y\| + \zeta\|u\| \le \varepsilon\|y\|.$$

$(B_1) \Rightarrow (C_1)$: Denote $\eta = \varepsilon t/5$ and find $0 < s < \eta$ such that $S_1(s) \subset \operatorname{diff}^{\eta F}(f_1)$. Suppose that $x \in S_1(s; u_1)$. By Lemma 4.3.1 there is a $\delta > 0$ such that

$$\left| f_1(x+u+v) - f_1(x+u) - f_1'(x;v) \right| \le 4\eta(\|u\| + \|v\|)$$

whenever $\|u\|, \|v\| < \delta$.

Fix $0 < q < \delta$ and $\|u\| < \delta$ such that $x + u + \sigma u_1 \in S$ for almost all $\sigma \in [0, q]$, and denote

$$m = \mathscr{L}^1\{\sigma \in [0,d] \mid x + u + \sigma u_1 \notin S_1(t; u_1)\}.$$

Since $s_1 = \operatorname{Lip}(f_1)$ we have

$$f_1(x+u+qu_1) - f_1(x+u) = \int_0^q f_1'(x+u+\tau u_1; u_1)\, d\tau$$
$$\le s_1(q - m) + (s_1 - t)m$$
$$= s_1 q - tm.$$

On the other hand, the choice of δ implies that

$$f_1(x+u+qu_1) - f_1(x+u) \ge f_1'(x; qu_1) - 4\eta(\|u\| + q)$$
$$\ge (s_1 - s)q - 4\eta(\|u\| + q)$$
$$\ge s_1 q - \eta q - 4\eta(\|u\| + q).$$

Hence we conclude that

$$m \le \frac{5\eta(\|u\| + q)}{t} = \varepsilon(\|u\| + q)$$

as required.

$(C_k) \Rightarrow (A_{k+1})$: Let $\varepsilon > 0$ be given. By the definition of s_{k+1} we find $t > 0$ such that

$$\|f_{k+1}'(z)\| \le s_{k+1} + \frac{\varepsilon}{4} \tag{4.14}$$

for every $z \in S_k(t)$. We assume that $\operatorname{Lip}(f_{k+1}) > 0$; otherwise A_{k+1} is trivially satisfied. Denote

$$\eta = \frac{\varepsilon^2}{4\operatorname{Lip}(f_{k+1})\,(\varepsilon + 4k\operatorname{Lip}(f_{k+1}))}. \tag{4.15}$$

By (C_k) we can find $0 < s < t$ such that for all unit vectors $u_1, \ldots, u_k$ and for every $x \in S_k(s; u_1, \ldots, u_k)$ there is $\delta = \delta(x; u_1, \ldots, u_k) > 0$ such that

$$\mathscr{L}^k \left\{ \sigma \in [0, q]^k \, \Big| \, x + u + \sum_{i=1}^k \sigma_i u_i \notin S_k(t; u_1, \ldots, u_k) \right\} < \eta(q + \|u\|) \, q^{k-1},$$

whenever $0 < q < \delta$ and $\|u\| < 2\delta$ are such that $x + u + \sum_{i=1}^k \sigma_i u_i \in S$ for almost all $\sigma \in [0, q]^k$.

We show that for all $x \in S_{k+1}(s; u_1, \ldots, u_{k+1})$ the inequality

$$\left| f_{k+1}(x + h + u) - f_{k+1}(x + h) \right| < s_{k+1} \|u\| + \varepsilon(\|h\| + \|u\|)$$

holds whenever $0 < \|h\|, \|u\| < \delta$. It clearly suffices to consider only the case when $\varepsilon(\|h\| + \|u\|) < \mathrm{Lip}(f_{k+1}) \, \|u\|$.

Let U be the linear span of $u_1, \ldots, u_k, u$. If necessary we use Corollary 2.2.4 to move h slightly such that almost every point of $x + h + U$ belongs to S. Let

$$q = \frac{\varepsilon(\|h\| + \|u\|)}{4k \, \mathrm{Lip}(f_{k+1})}. \tag{4.16}$$

Then $q < \|u\|/(4k) < \delta$ and so for almost every $\tau \in [0, 1]$, since $\|h + \tau u\| < 2\delta$,

$$\mathscr{L}^k \left\{ \sigma \in [0, q]^k \, \Big| \, x + h + \tau u + \sum_{i=1}^k \sigma_i u_i \notin S_k(t; u_1, \ldots, u_k) \right\}$$
$$< \eta(q + \|h + \tau u\|) q^{k-1}.$$

Thus

$$\mathscr{L}^{k+1} \left\{ (\tau, \sigma) \in [0, 1] \times [0, q]^k \, \Big| \, x + h + \tau u + \sum_{i=1}^k \sigma_i u_i \notin S_k(t; u_1, \ldots, u_k) \right\}$$
$$< \eta(q + \|h\| + \|u\|) \, q^{k-1}.$$

Hence there is a $\sigma \in [0, q]^k$ such that, denoting $y := x + h + \sum_{i=1}^k \sigma_i u_i$, we have by (4.15) and (4.16) that

$$\mathscr{L}^1 \left\{ \tau \in [0, 1] \mid y + \tau u \notin S_k(t; u_1, \ldots, u_k) \right\} < \frac{\eta(q + \|h\| + \|u\|)}{q}$$
$$= \frac{\varepsilon}{4 \, \mathrm{Lip}(f_{k+1})}.$$

Recall that by (4.14)
$$|f'_{k+1}(z; u)| \leq \left(s_{k+1} + \tfrac{1}{4}\varepsilon \right) \|u\|$$
for every $z \in S_k(t; u_1, \ldots, u_k) \subset S_k(t)$. Hence we conclude that

$$\mathscr{L}^1 \left\{ \tau \in [0, 1] \mid |f'_{k+1}(y + \tau u; u)| > (s_{k+1} + \tfrac{1}{4}\varepsilon) \|u\| \right\} < \frac{\varepsilon}{4 \, \mathrm{Lip}(f_{k+1})}.$$

It follows that

$$
|f_{k+1}(y+u) - f_{k+1}(y)| \le \int_0^1 |f'_{k+1}(y + \tau u; u)|\, d\tau
$$

$$
\le \frac{\varepsilon \operatorname{Lip}(f_{k+1}) \|u\|}{4 \operatorname{Lip}(f_{k+1})} + \left(s_{k+1} + \tfrac{1}{4}\varepsilon\right)\|u\|
$$

$$
= s_{k+1}\|u\| + \tfrac{1}{2}\varepsilon\|u\|.
$$

Thus using that $y = x + h + \sum_{i=1}^{k} \sigma_i u_i$ and (4.16), we obtain

$$
\begin{aligned}
\big|f_{k+1}(x + h + u) &- f_{k+1}(x + h)\big| \\
&\le \big|f_{k+1}(x + h + u) - f_{k+1}(y + u)\big| \\
&\quad + \big|f_{k+1}(y + u) - f_{k+1}(y)\big| + \big|f_{k+1}(y) - f_{k+1}(x + h)\big| \\
&\le s_{k+1}\|u\| + \tfrac{1}{2}\varepsilon\|u\| + 2\operatorname{Lip}(f_{k+1})\left\|\sum_{i=1}^{k} \sigma_i u_i\right\| \\
&\le s_{k+1}\|u\| + \tfrac{1}{2}\varepsilon\|u\| + 2kq\operatorname{Lip}(f_{k+1}) \\
&\le s_{k+1}\|u\| + \varepsilon(\|h\| + \|u\|)
\end{aligned}
$$

as required.

$(C_k) \wedge (B_{k+1}) \Rightarrow (C_{k+1})$: Let $\varepsilon > 0$ and $0 < t < 1$ be given. We first choose $0 < \eta < \varepsilon/(16t)$ such that

$$
\eta < \frac{\varepsilon t}{4(\operatorname{Lip}(f_{k+1}) + 4k + 6)},
$$

and find $0 < t_0 < t$ such that $\|f'_{k+1}(z)\| \le s_{k+1} + \eta$ for all $z \in S_k(t_0)$. Then we use (B_{k+1}) and (C_k) to choose $0 < s < \eta$ with

$$
S_{k+1}(s) \subset \operatorname{diff}^{\eta F}(f_{k+1}) \tag{4.17}
$$

such that for all unit vectors $u_1, \dots, u_k$ and every $x \in S_k(s; u_1, \dots, u_k)$ there is a $\delta(x; u_1, \dots, u_k) > 0$ such that

$$
\mathcal{L}^k \left\{ \sigma \in [0,q]^k \;\middle|\; x + u + \sum_{i=1}^{k} \sigma_i u_i \notin S_k(t_0; u_1, \dots, u_k) \right\} \\
< \eta^2 (q + \|u\|)\, q^{k-1}, \tag{4.18}
$$

whenever $0 < q < \delta(x; u_1, \dots, u_k)$ and $\|u\| < \delta(x; u_1, \dots, u_k)$ are such that $x + u + \sum_{i=1}^{k} \sigma_i u_i \in S$ for $\mathcal{L}^k$-almost all $\sigma \in [0,q]^k$.

Fix from now on unit vectors $u_1, \dots, u_{k+1}$ and $x \in S_{k+1}(s, u_1, \dots, u_{k+1})$. Since $x \in \operatorname{diff}^{\eta F}(f_{k+1})$ by (4.17), and since f_{k+1} is Gâteaux differentiable at x, we may use Lemma 4.3.1 to find

$$
0 < \delta < \frac{\delta(x; u_1, \dots, u_k)}{k + 1}
$$

such that

$$\left| f_{k+1}(x+u+v) - f_{k+1}(x+u) - f'_{k+1}(x;v) \right| \le 4\eta(\|u\| + \|v\|) \tag{4.19}$$

whenever $\|u\|, \|v\| < (k+1)\delta$.

The rest of the proof is devoted to showing that the inequality required by (C_{k+1}) holds for all $0 < q < \delta$ and $\|u\| < \delta$ such that $x + u + \sum_{i=1}^{k+1} \sigma_i u_i \in S$ for almost all $\sigma \in [0, q]^{k+1}$. To shorten the formulas, we will drop the $u_1, u_2, \ldots$ in the notation, but use square brackets to indicate that this has been done. So, for example, we denote $S_k[s] = S_k(s; u_1, \ldots, u_k)$ and $S_{k+1}[s] = S_{k+1}(s; u_1, \ldots, u_k, u_{k+1})$.

Since for almost all $\sigma_{k+1} \in [0, q]$ we have that

$$x + h + \sigma_{k+1} u_{k+1} + \sum_{i=1}^{k} \sigma_i u_i \in S$$

holds for almost all $\sigma \in [0, q]^k$, we infer from (4.18) that for such $\sigma_{k+1} \in [0, q]$,

$$\mathcal{L}^k \left\{ \sigma \in [0, q]^k \;\middle|\; x + u + \sum_{i=1}^{k+1} \sigma_i u_i \notin S_k[t_0] \right\}$$
$$< \eta^2(q + \|u + \sigma_{k+1} u_{k+1}\|) q^{k-1} \le \eta^2(2q + \|u\|) q^{k-1}.$$

Hence

$$\mathcal{L}^{k+1} \left\{ \sigma \in [0, q]^{k+1} \;\middle|\; x + u + \sum_{i=1}^{k+1} \sigma_i u_i \notin S_k[t_0] \right\} < \eta^2(2q + \|u\|) q^k.$$

Consequently, denoting by M the set of those $\sigma \in [0, q]^k$ for which

$$\mathcal{L}^1 \left\{ \tau \in [0, q] \;\middle|\; x + u + \sum_{i=1}^{k} \sigma_i u_i + \tau u_{k+1} \notin S_k[t_0] \right\} < \eta q, \tag{4.20}$$

we get

$$\eta^2(2q + \|u\|) q^k$$
$$> \int_{[0,q]^k} \mathcal{L}^1 \left\{ \tau \in [0, q] \;\middle|\; x + h + \sum_{i=1}^{k} \sigma_i u_i + \tau u_{k+1} \notin S_k[t_0] \right\} d\sigma$$
$$\ge \mathcal{L}^k \left([0, 1]^k \setminus M \right) \eta q.$$

So

$$\mathcal{L}^k \left([0, 1]^k \setminus M \right) < \eta(2q + \|u\|) q^{k-1}. \tag{4.21}$$

Fix, for a while, $\sigma \in M$ and denote

$$y = x + u + \sum_{i=1}^{k} \sigma_i u_i,$$

$$m = \mathscr{L}^1 \left\{ \tau \in [0, q] \;\middle|\; y + \tau u_{k+1} \in S_k[t_0] \setminus S_{k+1}[t] \right\},$$

$$m_1 = \mathscr{L}^1 \left\{ \tau \in [0, q] \;\middle|\; y + \tau u_{k+1} \notin S_k[t_0] \right\},$$

$$m_2 = \mathscr{L}^1 \left\{ \tau \in [0, q] \;\middle|\; y + \tau u_{k+1} \in S_k[t_0] \cap S_{k+1}[t] \right\}.$$

Observe that $t_0 < t$ implies that

$$f'_{k+1}(z; u_{k+1}) \leq \begin{cases} s_{k+1} + \eta & \text{for } z \in S_k[t_0], \\ s_{k+1} - t & \text{for } z \in S_k[t_0] \setminus S_{k+1}[t]. \end{cases}$$

Also, (4.20) written in the above notation yields $m_1 < \eta q$. It implies that

$$\begin{aligned}
f_{k+1}(y + q u_{k+1}) - f_{k+1}(y) &= \int_0^q f'_{k+1}(y + \tau u_{k+1}; u_{k+1}) \, d\tau \\
&\leq \mathrm{Lip}(f_{k+1}) m_1 + (s_{k+1} - t) m + (s_{k+1} + \eta) m_2 \\
&\leq \mathrm{Lip}(f_{k+1}) \eta q + q s_{k+1} - t m + \eta q.
\end{aligned}$$

On the other hand, since $q < \delta$ and $\|y - x\| \leq (k+1)\delta$, we get from (4.19),

$$f_{k+1}(y + q u_{k+1}) - f_{k+1}(y) \geq f'_{k+1}(x; q u_{k+1}) - 4\eta(\|y - x\| + q).$$

Recalling that $x \in S_{k+1}[s]$ and $s < \eta$ we continue in estimating

$$\begin{aligned}
&\geq (s_{k+1} - s) q - 4\eta(\|u\| + (k+1)q) \\
&\geq s_{k+1} q - \eta(4\|u\| + (4k+5)q).
\end{aligned}$$

Combining the last two estimates we conclude that

$$m \leq \frac{1}{t} \left(\left(\mathrm{Lip}(f_{k+1}) + 4k + 6 \right) \eta q + 4\eta \|h\| \right) \leq \frac{\varepsilon}{4} \left(\|u\| + q \right)$$

because of the choice of η. This implies that for $\sigma \in M$ we have

$$\mathscr{L}^1 \left\{ \sigma_{k+1} \in [0, q] \;\middle|\; x + u + \sum_{i=1}^{k+1} \sigma_i u_i \notin S_{k+1}[t] \right\}$$

$$\leq m + m_1 \leq \frac{\varepsilon}{4} \left(\|u\| + q \right) + \eta q < \frac{\varepsilon}{2} \left(\|u\| + q \right).$$

Finally, using (4.21) we get

$$\mathscr{L}^{k+1}\Big\{\sigma \in [0,q]^{k+1} \ \Big|\ x + u + \sum_{i=1}^{k+1} \sigma_i u_i \notin S_{k+1}[t]\Big\}$$

$$\leq \int_M \mathscr{L}^1\Big\{\sigma_{k+1} \in [0,q] \ \Big|\ x + u + \sum_{i=1}^{k+1} \sigma_i u_i \notin S_{k+1}[t]\Big\}\, d\sigma$$

$$+ q\mathscr{L}^k([0,1]^k \setminus M)$$

$$\leq \frac{\varepsilon}{2}\,(\|u\| + q)q^k + \eta(2q + \|u\|)q^k \leq \varepsilon(\|u\| + q)q^k.$$

This concludes the proof of Theorem 4.3.2.   □

The following statement was proved in a slightly weaker form in [22]. They prove it with $\bar{\delta}$ instead of $\bar{\mu}$, and in the mean value estimate they estimate only the norm.

Theorem 4.3.3. *Suppose that X and Y are Banach spaces, X is separable, and Y is asymptotically uniformly convex and has the RNP. Suppose further that for every $c > 0$ there is $0 < t < 1$ such that $\bar{\rho}_X(t) < \bar{\mu}_Y(ct)$. Then every Lipschitz map f from a nonempty open set $D \subset X$ to Y has, for every $\varepsilon > 0$, a point of ε-Fréchet differentiability.*

Moreover, whenever $S \subset D$ has Haar null complement in D, $\eta > 0$, $y^* \in Y^*$, and $a, b \in D$ are such that $[a, b] \subset D$, we may further require that the point x of ε-Fréchet differentiability of f belongs to S and satisfies the mean value estimate

$$(y^* \circ f)'(x)(b - a) > (y^* \circ f)(b) - (y^* \circ f)(a) - \eta.$$

Proof. We may assume that f is Gâteaux differentiable at all points of S. Our plan is to use, under the assumptions of the additional part, Observation 2.4.2, which will yield the mean value estimate. To this aim we first notice that the modification of norms required in this observation preserves assumptions of the Theorem. Indeed, consider the norms

$$\|x\| = \max\{\|x\|,\ C\|x - e^*(x)e\|\} \text{ on } X \text{ and}$$
$$\|y\| = \|y\| + |y^*(y)| \text{ on } Y$$

for some $C > 0$, $e \in X$, $e^* \in X^*$ with $\|e\| = \|e^*\| = e^*(e) = 1$ and $y^* \in Y^*$. Since $\bar{\delta}_{\|\cdot\|}(t) \geq \bar{\delta}_{\|\cdot\|}(t)$, the space $(Y, \|\cdot\|)$ is asymptotically uniformly convex. The RNP being invariant under equivalent renorming is preserved as well. It remains to verify that for every $c > 0$ there is $0 < t < 1$ such that

$$\bar{\rho}_{\|\cdot\|}(t) < \bar{\mu}_{\|\cdot\|}(ct). \tag{4.22}$$

Let $0 < c_0 \leq 1 \leq C_0$ be the constants from Observation 4.2.5. Since clearly $\bar{\rho}_{\|\cdot\|}(t) = \bar{\rho}_{\|\cdot\|}(t)$, we infer from Observation 4.2.5 that

$$\bar{\mu}_{\|\cdot\|}(t) \geq c_0\bar{\mu}_{\|\cdot\|}(c_0 t) \quad \text{and} \quad \bar{\rho}_{\|\cdot\|}(t) \geq \bar{\rho}_{\|\cdot\|}(t/C_0)$$

for all $t > 0$. For given $0 < c \le c_0^2/C_0$, we find $0 < t_0 < 1$ such that

$$\bar{\rho}_{\|\cdot\|}(t_0) < \bar{\mu}_{\|\cdot\|}(c^2 t_0)$$

and we put $t = ct_0/c_0$. Then

$$\begin{aligned}
\bar{\mu}_{\|\cdot\|}(ct) &\ge c_0 \bar{\mu}_{\|\cdot\|}(c_0 ct) \\
&= c_0 \bar{\mu}_{\|\cdot\|}(c^2 t_0) > c_0 \bar{\rho}_{\|\cdot\|}(t_0) \\
&\ge \bar{\rho}_{\|\cdot\|}(c_0 t_0) \ge \bar{\rho}_{\|\cdot\|}(c_0 t_0/C_0) \ge \bar{\rho}_{\|\cdot\|}(t).
\end{aligned}$$

For sake of simplicity we denote the norms $\|\cdot\|$ on X and Y by the standard symbol $\|\cdot\|$, moduli $\bar{\rho}_{\|\cdot\|}$ and $\bar{\mu}_{\|\cdot\|}$ by $\bar{\rho}_X$ and $\bar{\mu}_Y$, and $f + T$ from Observation 4.2.5 by f only.

All we have to find now is a point $x \in S$ of ε-Fréchet differentiability of f such that $\|f'(x)\| > \mathrm{Lip}(f) - \eta$. Observation 2.4.2, with D being the set of points of S at which f is ε-Fréchet differentiable, will imply the mean value estimate.

For finding the point required in the previous paragraph, we may assume that $\mathrm{Lip}(f) = 1$. Let $0 < \varepsilon, \eta < 1$ and find $0 < t < 1$ and $\rho > 0$ such that

$$\bar{\rho}_X(t) < \rho < \bar{\mu}_Y(\tfrac{1}{12}\varepsilon t).$$

Also choose $0 < \tau < \tfrac{1}{4}\eta$ such that $\rho + 53\tau < \bar{\mu}_Y(\tfrac{1}{12}\varepsilon t)$. In particular, we will use that the last condition also implies $53\tau < \tfrac{1}{2}\varepsilon$. Put

$$A = \{f'(x; e) \in Y \mid x \in S, \|e\| = 1\}.$$

Since the supremum of the norm of elements of A is $\mathrm{Lip}(f) = 1$ by Proposition 2.4.1, we may choose $w \in A$ and $w^* \in Y^*$ such that $\|w^*\| = 1$ and $w^*(w) = \|w\| > 1 - \tfrac{1}{2}\eta$. Recalling that Y has the RNP, we find $y_0^* \in Y^*$ and $\sigma > 0$ such that $\|y_0^* - w^*\| < \tau$ and the slice

$$M = \{y \in A \mid y_0^*(y) > \sigma\}$$

is nonempty and has diameter less than τ (see [37, Theorem 5.20]). Let $y_0 \in M$. By the definition of $\bar{\mu}_Y$ there is a finite codimensional subspace Y_0 of Y such that

$$\bar{\mu}_Y(ct, y_0, Y_0) > \rho + 53\tau.$$

Choose norm one functionals $y_1^*, \ldots, y_n^* \in Y^*$ such that for every $y \in Y$,

$$\mathrm{dist}(y, Y_0) \le 2\max\{y_i^*(y) \mid i = 1, \ldots, n\}.$$

By Theorem 4.3.2 we can find $x \in S$ at which all $y_j^* f$, $j = 0, \ldots, n$, are τ-Fréchet differentiable and $\|(y_0^* f)'(x)\| > \sigma$. Denote $L = f'(x)$ and let $e \in X$ be a unit vector such that $\|y_0^*(Le)\| > \sigma$. Since both y_0 and Le belong to the slice M we infer that $\|y_0 - Le\| < \tau$.

The condition $\bar{\rho}_X(t) < \rho$ allows us to find a subspace X_0 of the intersection of the kernels of all $y_i^* L$, $i = 0, \ldots, n$, such that $\dim(X/X_0) < \infty$ and $\bar{\rho}_X(t, e, X_0) < \rho$. By

Corollary 4.2.9 one can choose a finite dimensional subspace $E \subset X$ satisfying $e \in E$ such that every $u \in X$ can be written as $u = x_0 + v$, where $x_0 \in X_0$, $v \in E$, and $\|x_0\|, \|v\| \leq 3\|u\|$.

According to Lemma 4.3.1 there is $\delta > 0$ such that $B(x, 6\delta/t) \subset D$, and for every $0 \leq j \leq n$ and $u, v \in X$ with $\|u\|, \|v\| < 3\delta/t$,

$$|y_i^* f(x + u + v) - y_i^* f(x + u) - y_i^* Lv| \leq 4\tau(\|u\| + \|v\|). \qquad (4.23)$$

Since E is finite dimensional, the restriction of f to E is Fréchet differentiable and so we may choose this δ small enough that

$$\|f(x + v) - f(x) - Lv\| \leq \tau\|v\|$$

for every $v \in E$, $\|v\| < 3\delta/t$.

Suppose that $u \in X$ and $0 < \|u\| < \delta$. Write $u = x_0 + v$, where $x_0 \in X_0$, $v \in E$, and $\|x_0\|, \|v\| \leq 3\|u\|$. Letting $\gamma = 3\|u\|/t$, we see from $\|x_0\| \leq t\|\gamma e\|$ that $\|x_0 \pm \gamma e\| \leq (1 + \rho)\gamma$.

Since $y_i^* Lx_0 = 0$ for $i = 1, \ldots, n$ we conclude that $Lx_0 \in Y_0$, and so

$$\begin{aligned} \|Lx_0 \pm \gamma y_0\| &\leq \|Lx_0 \pm \gamma Le\| + \gamma\|y_0 - Le\| \\ &\leq \|L\|\|x_0 \pm \gamma e\| + \tau\gamma \\ &\leq (1 + \rho)\gamma + \tau\gamma < \left(1 + \bar{\mu}_Y\left(\tfrac{1}{12}\varepsilon t, y_0, Y_0\right)\right)\gamma. \end{aligned}$$

The midpoint inequality shows that $\|Lx_0\| \leq \frac{1}{12}\varepsilon t\gamma = \frac{1}{4}\varepsilon\|u\|$.

For $1 \leq i \leq n$ we again use that $y_i^*(Lu_0) = 0$ and (4.23) to infer

$$y_j^*\big(f(x + u) - f(x + v)\big) = y_j^*\big(f(x + v + x_0) - f(x + v) - Lx_0\big) \leq 24\tau\|u\|.$$

Hence the distance of $f(x + u) - f(x + v)$ from Y_0 is at most $48\tau\|u\|$. There is a point $z \in Y_0$ such that $\|f(x + u) - f(x + v) - z\| < 49\tau\|u\|$. Then

$$\begin{aligned} \|z \mp \gamma y_0\| &\leq \|f(x + v \pm \gamma e) - f(x + u)\| + \|f(x + u) - f(x + v) - z\| \\ &\quad + \|f(x + v) - f(x) - Lv\| + \gamma\|y_0 - Le\| \\ &\quad + \|f(x + v \pm \gamma e) - f(x) - L(v \pm \gamma e)\| \\ &\leq \|x_0 \pm \gamma e\| + 49\tau\|u\| + 3\tau\|u\| + \tau\gamma + 3\tau\|u\| + \tau\gamma \\ &\leq (1 + \rho)\gamma + 53\tau\gamma < \left(1 + \mu_Y\left(\tfrac{1}{12}\varepsilon t, y_0, Y_0\right)\right)\gamma. \end{aligned}$$

The midpoint inequality now yields that $\|z\| \leq \frac{1}{12}\varepsilon t\gamma = \frac{1}{4}\varepsilon\|u\|$. To finish the proof of ε-Fréchet differentiability of f at x we estimate

$$\begin{aligned} \|f(x + u) &- f(x) - Lu\| \\ &\leq \|f(x + u) - f(x + v) - z\| + \|z\| + \|Lx_0\| + \|f(x + v) - f(x) - Lv\| \\ &\leq 50\tau\|u\| + \tfrac{1}{4}\varepsilon\|u\| + \tfrac{1}{4}\varepsilon\|u\| + 3\tau\|u\| < \varepsilon\|u\|. \end{aligned}$$

It only remains to notice that, since $w \in A$, $y_0^*(w) \leq y_0^*(y_0) + \tau$. Hence

$$1 = w^*(w) \leq y_0^*(w) + \|y_0^* - w^*\| < y_0^*(y_0) + 2\tau \leq (1 + \tau)\|y_0\| + 2\tau,$$

which gives $\|y_0\| > 1 - 3\tau$. So we have, as required,

$$\|f'(x)\| \geq \|f'(x;e)\| = \|Le\| \geq \|y_0\| - \|y_0 - Le\| > 1 - 4\tau > 1 - \eta. \qquad \Box$$

Remark 4.3.4. It should be noted that the difference between the proof of the warm up Proposition 4.1.3 and the results of this section is not only technical. To see it, recall the proof of Proposition 4.1.3 and notice that it in fact did not need to use Gâteaux derivatives. Instead one can observe that a Lipschitz function f on a uniformly smooth space is ε-Fréchet differentiable at every point x for which one can find a unit vector e such that the directional derivative $f'(x;e)$ exists and has value close to the Lipschitz constant of f. This approach removes the assumption of separability, but, more interestingly, shows that points of ε-Fréchet differentiability may be found in very small sets. For example, if X is separable such points may be found in the union of lines passing through two different points of a fixed countable dense subset of X, which is an (even directionally) σ-porous subset of X as long as dim $X \geq 2$.

The above remark applies also to the proof of Theorem 4.3.2 provided X is uniformly smooth and not only asymptotically uniformly smooth. In that case we also did not need the full strength of Gâteaux differentiability; we only needed that the set $S_k(t; u_1, \ldots, u_k)$ consists of points at which f is differentiable in the direction of the linear span of $u_1, \ldots, u_k$. This implies, for example, the result of [12] that every collection of $n - 1$ real-valued Lipschitz functions on $\mathbb{R}^n$ has, for every $\varepsilon > 0$, points of ε-differentiability in the countable union of rational hyperplanes.

However, when X is not uniformly smooth, the use of Gâteaux differentiability in our arguments was essential. It permitted us to treat, via Corollary 4.2.9 to Lemma 4.2.8, the directions not belonging to the finite codimensional subspace inside which we had full control from asymptotic uniform smoothness.

As has been already mentioned above, Theorems 4.3.2 and 4.3.3 and their proofs are only minor modifications of those in [22]. The same paper also contains much other highly interesting information on asymptotically uniformly convex and smooth spaces as well as applications of the ε-differentiability results. In connection with Theorem 4.3.3 we should at least recall that [22] raises the possibility that points of ε-Fréchet differentiability, or even of Fréchet differentiability, of Lipschitz maps from separable X to Y, where Y has the RNP, exist provided every bounded linear operator from X to Y is compact. This question, unless it has an easy counterexample, is very far from being solved. It is not even clear whether Theorem 4.3.3 can be proved only under the assumptions of Theorem 4.2.6 (plus RNP of Y, of course), and yet this would be only a minor step in attempts to give a positive answer to this question.

Chapter Five

Γ-null and Γ_n-null sets

We define the notions of Γ- and Γ_n-null sets that will play a major role in our investigations of the interplay between differentiability, porosity, and smallness on curves or surfaces. Here we relate these notions to Gâteaux differentiability and investigate their basic properties. Somewhat unexpectedly, we discover an interesting relation between Γ- and Γ_n-null $G_{\delta\sigma}$ sets, and this will turn out to be very useful in finding a new class of spaces for which the strong Fréchet differentiability result holds in Theorem 10.6.2.

5.1 INTRODUCTION

In this chapter we introduce σ-ideals of subsets of a Banach space X called Γ-null sets or Γ_n-null sets, respectively. The Γ-null sets first appeared in [28], while Γ_n-null sets are new. Like Haar null sets and Gauss null (or equivalently Aronszajn null) sets, they play a role of exceptional sets in questions dealing with differentiability of functions. Definitions of both Γ- and Γ_n-null sets link topological and measure theoretical notions in a sophisticated way that distinguishes them from the two above mentioned purely measure theoretic approaches to the negligible sets. While the idea behind the purely measure theoretic approach is to define a set to be null if it is null for *all* measures from a certain class, here we define a set to be null if it is null for *typical* measures from a certain class, where "typical" is understood in the sense of Baire category. Of course, these notions depend on the choice of the class of measures (and on its topology), and so it is obvious that we should use a class adjusted to the intended applications to differentiability. For example, consider the space of Borel probability measures on $\mathbb{R}$ with the vague topology. Since the measures with finite support are dense, typical measures are concentrated on any given dense G_δ subset of $\mathbb{R}$. Hence the real line may be decomposed into a Lebesgue null set and a set which is null for typical Borel probability measures on $\mathbb{R}$. Such a situation is clearly unsuitable for the study of differentiability. Hence for our purposes the most natural classes are formed by measures associated with n-dimensional C^1 surfaces (which we use to define Γ_n-null sets) or with their infinite dimensional analogues (which we use to define Γ-null sets). We also show that C^1 may be equivalently replaced by suitable Sobolev spaces.

The Γ_n-null sets, Γ-null sets, and Gauss null sets are not comparable in general. A simple example of the difference between the latter two is given by any compact convex set C that is not Gauss null: since the distance from C is a continuous convex function, which is not Fréchet differentiable at any point of C, the set C is Γ-null by Corollary 6.3.10 to be proved in Chapter 6. In a Hilbert space, a much stronger example showing the difference between Γ-null and Gauss null sets may be obtained from the example of Matoušek and Matoušková [34], mentioned at the end of Chapter 3, of a continuous convex function on a Hilbert space whose set of points of Fréchet differentiability is Gauss null. Since by Corollary 6.3.10 such functions are necessarily Fréchet differentiable except for a Γ-null set, this decomposes the Hilbert space into a Γ-null set and a Gauss null set; that is, there is a set which is Γ-null whose complement is Gauss null. This example was extended to all superreflexive spaces in [35]. In fact, as we will see in Example 5.4.11, every separable infinite dimensional Banach space admits such a decomposition, but our construction is not related to differentiability. Other relations between these null sets will be discussed in Section 5.4.

Let X be a Banach space, and let $T := [0, 1]^{\mathbb{N}}$ be endowed with the product topology and product Lebesgue measure $\mathscr{L}^{\mathbb{N}}$. We denote by $\Gamma(X)$ the space of continuous mappings

$$\gamma \colon T \longrightarrow X$$

having continuous partial derivatives $D_j\gamma$ (we consider one-sided derivatives at points where the jth coordinate is 0 or 1). The elements of $\Gamma(X)$ will be called *surfaces*. We equip $\Gamma(X)$ with the topology generated by the pseudonorms

$$\|\gamma\|_\infty = \sup_{t \in T} \|\gamma(t)\| \text{ and } \|\gamma\|_k = \sup_{t \in T} \|D_k\gamma(t)\|, \ k \geq 1.$$

Equivalently, this topology may be defined by the pseudonorms

$$\|\gamma\|_{\leq k} = \max\{\|\gamma\|_\infty, \|\gamma\|_1, \dots, \|\gamma\|_k\}.$$

The space $\Gamma(X)$ with this topology is a Fréchet space; in particular, it is a Polish space.

We also define $\Gamma_n(X) = C^1([0, 1]^n, X)$. It will be technically convenient to consider on $\Gamma_n(X)$ the norm $\|\cdot\|_{\leq n}$, which is equivalent to more standard norms such as

$$\|\gamma\|_{C^1} = \max\{\|\gamma\|_\infty, \|\gamma'\|_\infty\}.$$

Clearly, $\|\gamma\|_{\leq n} \leq \|\gamma\|_{C^1} \leq n\|\gamma\|_{\leq n}$ for every $\gamma \in \Gamma_n(X)$. Notice also that $\Gamma_n(X)$ is a subspace of $\Gamma(X)$ in the sense that the surfaces $\gamma \in \Gamma(X)$ depending on the first n coordinates only are naturally identified with the surfaces from $\Gamma_n(X)$.

We are now able to define the notions of Γ-null and Γ_n-null sets.

Definition 5.1.1. A Borel set $A \subset X$ will be called Γ-null if

$$\mathscr{L}^{\mathbb{N}}\{t \in T \mid \gamma(t) \in A\} = 0$$

for residually many $\gamma \in \Gamma(X)$. In other words, A is Γ-null if the set

$$\{\gamma \in \Gamma(X) \mid \mathscr{L}^{\mathbb{N}}\{t \in T \mid \gamma(t) \in A\} > 0\}$$

is a first category (or, in more recent terminology, meager) subset of $\Gamma(X)$. Analogically, a Borel set $A \subset X$ will be called Γ_n-null if

$$\mathscr{L}^n\{t \in [0, 1]^n \mid \gamma(t) \in A\} = 0$$

for residually many $\gamma \in \Gamma_n(X)$. A possibly non-Borel set $A \subset X$ will be called Γ-null (resp. Γ_n-null) if it is contained in a Borel set which is Γ-null (resp. Γ_n-null).

In a sense, Γ-null sets are Γ_n-null set with $n = \infty$. We could in fact call them Γ_∞-null sets and use this to unify some of the treatment of Γ-null and Γ_n-null sets. But we feel that this would obscure the point that the freedom in the construction of infinite dimensional surfaces is much bigger than for finite dimensional surfaces. As an example, the reader may compare the statement of Corollary 5.2.2 and the arguments leading to it with its finite dimensional counterpart Lemma 5.3.5.

In this chapter we will often use the following notation for a suitable finite dimensional modification of a given surface $\gamma \in \Gamma(X)$. If $s \in T$ and $n \in \mathbb{N}$, we denote

$$\gamma^{n,s}(t) := \gamma(t_1, \ldots, t_n, s) = \gamma(t_1, \ldots, t_n, s_1, s_2, \ldots).$$

It is easy to observe that for every $\gamma \in \Gamma(X)$, $m \in \mathbb{N}$, and $\varepsilon > 0$ there is $n \in \mathbb{N}$ such that $\|\gamma^{n,s} - \gamma\|_{\leq m} < \varepsilon$ for every $s \in T$. This follows immediately from the uniform continuity of γ and its partial derivatives. This observation says that $\bigcup_{n=1}^{\infty} \Gamma_n(X)$ is dense in the space $\Gamma(X)$. Further information about dense subsets of $\Gamma_n(X)$, and so also of $\Gamma(X)$, will be proved when it is needed, in Section 10.2.

We also recall that the tangent space $\mathrm{Tan}(\gamma, t)$ of a surface γ at a point $t \in T$ is the closed linear span in X of the vectors $D_k\gamma(t)$, $k \geq 1$,

$$\mathrm{Tan}(\gamma, t) = \overline{\mathrm{span}}\{D_k\gamma(t) \mid k \geq 1\}.$$

5.2 Γ-NULL SETS AND GÂTEAUX DIFFERENTIABILITY

Here we show that the standard results on Gâteaux differentiability such as Theorem 2.3.1 remain valid if we interpret the notion of exceptional sets as Γ-null sets. The argument is again straightforward; the only additional information we need to know is that derivatives of typical surfaces span the whole space.

The following lemma shows how easy it is to use the infinite dimensionality of surfaces from $\Gamma(X)$ to create any tangent vectors we wish. Its immediate corollary is that in separable spaces typical surfaces have the property that $\mathrm{Tan}(\gamma, t) = X$ for all t.

Lemma 5.2.1. Let $u_1, \ldots, u_k \in X$ and let $\varepsilon > 0$. Then the set of surfaces $\gamma \in \Gamma(X)$ for which one can find $m \in \mathbb{N}$ and $c > 0$ with the property

$$\sup_{t \in T} \|cD_{m+j}\gamma(t) - u_j\| < \varepsilon \text{ for all } j = 1, \ldots, k$$

is open and dense in $\Gamma(X)$.

Proof. By the definition of the topology of $\Gamma(X)$ it is clear that this set is open. To see that it is dense it suffices to show that its closure contains $\Gamma_n(X)$ for every $n \in \mathbb{N}$. Let $\gamma_0 \in \Gamma_n(X)$, $\eta > 0$, $m \geq n$, and consider the surface

$$\gamma(t) = \gamma_0(t) + \eta \sum_{j=1}^{k} t_{m+j} u_j.$$

Then

$$\|\gamma - \gamma_0\|_\infty \leq k\eta \max_{1 \leq j \leq k} \|u_j\|.$$

If $1 \leq j \leq m$ or $j > m + k$, we have $\|\gamma - \gamma_0\|_j = 0$. For the remaining values $m < j \leq m + k$ the norm is $\|\gamma - \gamma_0\|_j = \eta\|u_{j-m}\|$. Finally, $D_{m+j}\gamma(t) = \eta u_j$ for $1 \leq j \leq m$, and so γ has the property of the lemma with $c = 1/\eta$. $\qquad \square$

Corollary 5.2.2. *If X is separable, then residually many $\gamma \in \Gamma(X)$ have the property* $\mathrm{Tan}(\gamma, t) = X$ *for every $t \in T$.*

Proof. Using Lemma 5.2.1 with $k = 1$ we get that for a given $u \in X$ and $\varepsilon > 0$ the surfaces $\gamma \in \Gamma(X)$ satisfying $\mathrm{dist}(u, \mathrm{Tan}(\gamma, t)) < \varepsilon$ for every $t \in T$ form an open and dense set in $\Gamma(X)$. Intersection of such open dense sets for $\varepsilon \searrow 0$ rational and u from a countable dense set in X gives the desired result. $\qquad \square$

We are now ready to prove the announced Gâteaux differentiability result for Γ-null sets.

Theorem 5.2.3. *Let X be separable and let Y have the RNP. Then every Lipschitz function from an open set G in X into Y is Gâteaux differentiable outside a Γ-null set.*

Proof. We recall first that the set of points at which f fails to be Gâteaux differentiable is a Borel set. We recall next that Rademacher's theorem holds also for Lipschitz maps from $\mathbb{R}^k$ into a space Y having the RNP (see, e.g., [4, Proposition 6.41]). Consider now an arbitrary surface γ. Using Fubini's theorem, we get that for almost every $t \in \gamma^{-1}(G)$ the mapping

$$(s_1, \ldots, s_k) \rightarrow f\big(\gamma(s_1, \ldots, s_k, t_{k+1}, \ldots)\big)$$

is differentiable at $(t_1, \ldots, t_k)$. Since f is Lipschitz, it follows that for almost all t, the function f has directional derivatives in all directions

$$v \in \mathrm{span}\{D_j\gamma(t) \mid j \geq 1\}$$

at $x = \gamma(t)$. Since these directional derivatives depend linearly on v, the function f is Gâteaux differentiable at x with respect to the subspace

$$\overline{\mathrm{span}}\{D_j\gamma(t) \mid j \geq 1\} = \mathrm{Tan}(\gamma, t);$$

see, for example, [4, Lemma 6.40]. In particular, for every surface γ from the residual set obtained in Corollary 5.2.2, f is Gâteaux differentiable at $x = \gamma(t)$ for almost all $t \in \gamma^{-1}(G)$. This proves the theorem. $\qquad \square$

Remark 5.2.4. An example of a simple but useful consequence of Theorem 5.2.3 is that any set $E \subset X$ which is porous in the direction of a finite dimensional subspace U of X is Γ-null (see Section 1.1 for definitions). It suffices to notice that the compactness of the unit sphere in U implies that E is actually σ-directionally porous. Writing $E = \bigcup_n E_n$, where each E_n is directionally porous, we observe that for every n the function $x \mapsto \mathrm{dist}(x, E_n)$ is not Gâteaux differentiable at any $x \in E_n$, and so E_n is Γ-null.

5.3 SPACES OF SURFACES, AND Γ- AND Γ_n-NULL SETS

In this section we give a useful criterion for a set to be Γ- or Γ_n-null, investigate tangential properties of typical surfaces from $\Gamma_n(X)$, and in finite dimensional spaces relate negligibility in the Γ- and Γ_n- sense to negligibility in the sense of the Lebesgue measure.

We start by giving more precise information about dense subsets of $\Gamma(X)$, which is useful in several questions. For that we need a simple extension result. Although it is a special case of the Whitney extension theorem (for Banach space–valued functions), to make the treatment complete we give a simple proof.

Since in some statements we will use functions $\gamma \colon S \subset \mathbb{R}^n \longrightarrow X$ that are only piecewise C^1, we should extend to them the norms defined in $\Gamma_n(X)$ by, for example,

$$\max\{\sup_{s \in S} \|\gamma(s)\|, \ \mathrm{Lip}_S(\gamma)\}.$$

However, we believe that, in spite of not being formally correct, using standard notation $\|\cdot\|_{C^1}$ or $\|\cdot\|_{\leq n}$ also in these cases will not cause any confusion.

A cube in $\mathbb{R}^n$ will be a closed axis-parallel cube, that is, any product $\prod_{k=1}^n I_k$, where I_k are closed intervals of equal length.

Lemma 5.3.1. *Let $S_0 \subset S_1$ be two concentric cubes with $\mathscr{L}^n S_0 = c^n \mathscr{L}^n S_1$, where $0 < c < 1$. Then for every $\gamma_0 \in C^1(S_0, X)$ there is $\gamma \in C^1(\mathbb{R}^n, X)$ such that $\gamma = \gamma_0$ on S_0, $\gamma = 0$ outside S_1, and*

$$\|\gamma\|_{C^1(\mathbb{R}^n)} \leq \frac{K_n}{1-c} \|\gamma_0\|_{C^1(S_0)},$$

where the constant K_n depends only on n.

Proof. We may suppose that $S_1 = [-1,1]^n$ and $S_0 = [-c,c]^n$. It suffices to find a surface $\tilde{\gamma}_0 \in C^1(\mathbb{R}^n, X)$ satisfying all the requirements of the lemma but with the first one replaced by $\|\tilde{\gamma}_0 - \gamma_0\|_{C^1(S_0)} \leq \frac{1}{2}\|\gamma_0\|_{C^1(S_0)}$. Indeed, we denote recursively $\gamma_k = \gamma_{k-1} - \tilde{\gamma}_{k-1}$, and find the corresponding $\tilde{\gamma}_k$ satisfying $\tilde{\gamma}_k = 0$ outside S_1, $\|\gamma_k - \tilde{\gamma}_k\|_{C^1(S_0)} \leq \frac{1}{2}\|\gamma_k\|_{C^1(S_0)}$, and

$$\|\tilde{\gamma}_k\|_{C^1(\mathbb{R}^n)} \leq \frac{K}{1-c} \|\gamma_k\|_{C^1(S_0)} \tag{5.1}$$

for some constant K depending only on n. Then

$$\|\gamma_k\|_{C^1(S_0)} = \|\gamma_{k-1} - \tilde{\gamma}_{k-1}\|_{C^1(S_0)} \leq \frac{1}{2} \|\gamma_{k-1}\|_{C^1(S_0)},$$

that is, $\|\gamma_k\|_{C^1(S_0)} \leq 2^{-k}\|\gamma_0\|_{C^1(S_0)}$. This condition together with (5.1) guarantees that the series $\gamma = \sum_{k=0}^{\infty} \tilde{\gamma}_k$ converges in $C^1(\mathbb{R}^n, X)$. Moreover, for $t \in S_0$

$$\sum_{k=0}^{\infty} \tilde{\gamma}_k(t) = \sum_{k=1}^{\infty} (\gamma_{k-1}(t) - \gamma_k(t)) = \gamma(t),$$

and obviously $\gamma = 0$ outside S_1. Finally,

$$\|\gamma\|_{C^1(\mathbb{R}^n)} \leq \sum_{k=0}^{\infty} \|\widetilde{\gamma}_k\|_{C^1(\mathbb{R}^n)}$$

$$\leq \frac{K}{1-c} \sum_{k=0}^{\infty} \|\gamma_k\|_{C^1(S_0)}$$

$$\leq \frac{2K}{1-c} \|\gamma_0\|_{C^1(S_0)}$$

and we obtain the statement with $K_n = 2K$.

Choose $c < a < b < 1$ and define

$$\widetilde{\gamma}(t) = \begin{cases} \gamma_0(ct/a) & \text{if } t \in [-a, a]^n, \\ 0 & \text{if } t \notin [-b, b]^n. \end{cases}$$

We can extend $\widetilde{\gamma}$ to a map of $\mathbb{R}^n$ to X bounded by the same constant and such that

$$\mathrm{Lip}_{\mathbb{R}^n}(\widetilde{\gamma}) \leq \frac{K}{b-a} \|\gamma_0\|_{C^1(S_0)}.$$

(This can be done easily directly, or with the help of a Lipschitz extension result we will see in Lemma 5.5.3.) It follows that

$$\|\widetilde{\gamma}\|_{C^1(\mathbb{R}^n)} \leq \|\widetilde{\gamma}\|_{\infty} + \mathrm{Lip}_{\mathbb{R}^n}(\widetilde{\gamma}) \leq \frac{K+1}{b-a} \|\gamma_0\|_{C^1(S_0)}.$$

Assuming that a is close enough to c and b is close enough to 1, we convolve $\widetilde{\gamma}$ with a smooth mollifier supported in a small enough neighborhood of the origin to obtain $\widetilde{\gamma}_0 \in C^1(\mathbb{R}^n, X)$ such that $\widetilde{\gamma}_0 = \gamma_0$ on S_0, $\widetilde{\gamma} = 0$ outside S_1 and

$$\|\widetilde{\gamma}_0\|_{C^1(\mathbb{R}^n)} \leq \frac{K+2}{1-c} \|\gamma_0\|_{C^1(S_0)}. \qquad \square$$

A special case of Lemma 5.3.1 is that we may always assume that a surface from $\Gamma_n(X)$ actually belongs to $C^1(\mathbb{R}^n, X)$. We will often use this fact without any reference.

We have already noted that $\bigcup_{n=1}^{\infty} \Gamma_n(X)$ is dense in $\Gamma(X)$. The following simple statement slightly improves it.

Lemma 5.3.2. *Whenever (X_k) is an increasing sequence of subspaces of X whose union is dense in X, then the union $\bigcup_{k=1}^{\infty} \Gamma_n(X_k)$ is dense in $\Gamma_n(X)$. Consequently, $\bigcup_{n=1}^{\infty} \Gamma_n(X_n)$ is dense in $\Gamma(X)$.*

Proof. Let $\gamma \in C^1(\mathbb{R}^n, X)$ and $\varepsilon > 0$. Take a triangulation of $[-1, 2]^n$ such that the piecewise affine approximation γ_1 to γ that agrees with γ on the vertices of the triangulation satisfies

$$\|\gamma_1 - \gamma\|_{C^1([-1,2]^n)} < \tfrac{1}{3}\varepsilon.$$

For sufficiently large k, the piecewise affine approximation γ_2 to γ_1 obtained by replacing the values of γ at each vertex by a close value from X_k satisfies

$$\|\gamma_2 - \gamma_1\|_{C^1([-1,2]^n)} < \tfrac{1}{3}\varepsilon.$$

Convolving γ_2 with a smooth mollifier supported in a small neighborhood of the origin gives the desired approximation of γ on $[0,1]^n$. □

Recall that one of many equivalent ways of defining that a set E in a metric space M is nowhere dense is that for every element x of a dense subset D of M and for every $\varepsilon > 0$, there is a nonempty open set G contained in $B(x,\varepsilon) \setminus E$. The next results show that for certain sets that occur in the definition of Γ-nullness, the set $D \subset \Gamma(X)$ may be taken rather far from being dense: it may consist only of affine functions depending on finitely many variables. To state a general version, let us call a property $P(t,\gamma)$ on $[0,1]^n \times \Gamma_n(X)$ local if, once $P(t,\gamma)$ holds and $\widetilde{\gamma} = \gamma$ on a neighborhood of t, then $P(t,\widetilde{\gamma})$ holds as well.

Proposition 5.3.3. *Let $P(t,\gamma)$ be a local property on $[0,1]^n \times \Gamma_n(X)$ such that for every affine map $\gamma_0 \colon \mathbb{R}^n \longrightarrow X$ and every $\varepsilon > 0$ there are $\widetilde{\gamma} \in \Gamma_n(X)$ and $\delta > 0$ such that $\|\widetilde{\gamma} - \gamma_0\|_{\leq n} < \varepsilon$ and*

$$\mathscr{L}^n\{t \in [0,1]^n \mid P(t,\gamma)\,\text{fails}\} < \varepsilon$$

for every $\gamma \in \Gamma_n(X)$ with $\|\gamma - \widetilde{\gamma}\|_{\leq n} < \delta$.
Then for every $\varepsilon > 0$ the set

$$\big\{\gamma \in \Gamma_n(X) \mid \mathscr{L}^n\{t \in [0,1]^n \mid P(t,\gamma)\,\text{fails}\} < \varepsilon\big\}$$

contains an open and dense subset of $\Gamma_n(X)$. Consequently, residually many surfaces $\gamma \in \Gamma_n(X)$ have the property that $P(t,\gamma)$ holds for almost every t.

Proof. Let $\gamma_0 \in C^1(\mathbb{R}^n, X)$ and $\varepsilon > 0$. Take a triangulation of $[0,1]^n$ such that the piecewise affine approximation γ_1 to γ_0 that agrees with γ_0 on the vertices of the triangulation satisfies $\|\gamma_1 - \gamma_0\|_{\leq n} < \tfrac{1}{2}\varepsilon$. Choose $k \geq 1$ large enough that when we divide $[0,1]^n$ into nonoverlapping cubes of side-length $1/k$, the union of those cubes on which γ_1 is affine covers all of $[0,1]^n$ up to a set of measure less than $\tfrac{1}{2}\varepsilon$. Let $\mathcal{S}$ be the family of all these cubes.
Choose $0 < \eta < 1$ small enough to satisfy

$$1 - (1-\eta)^n + k^n \leq \frac{\varepsilon}{2}, \quad 2K_n n\eta \leq \varepsilon,$$

where K_n is the constant from Lemma 5.3.1. For each cube $S \in \mathcal{S}$ we denote by γ_S the extension of γ_1 from S to an affine map from $\mathbb{R}^n$ to X. By the assumption, one can find $\widetilde{\gamma}_S \in \Gamma_n(X)$ and $\delta_S > 0$ such that $\|\widetilde{\gamma}_S - \gamma_S\|_{\leq n} < \eta^2$ and

$$\mathscr{L}^n\{t \in [0,1]^n \mid P(t,\gamma)\,\text{fails}\} < \eta \tag{5.2}$$

for every $\gamma \in \Gamma_n(X)$ with $\|\widetilde{\gamma}_S - \gamma\|_{\leq n} < \delta_S$.

For each $S \in \mathcal{S}$ let S_0 denote the cube concentric with S with volume $\mathscr{L}^n S_0 = (1-\eta)^n \mathscr{L}^n S$. Using Lemma 5.3.1, we find $\beta_S \in C^1(\mathbb{R}^n, X)$ such that $\beta_S = \widetilde{\gamma}_S - \gamma_S$ on S_0, $\beta_S = 0$ outside S and

$$\|\beta_S\|_{C^1(\mathbb{R}^n)} \leq \frac{K_n}{\eta} \|\widetilde{\gamma}_S - \gamma_S\|_{C^1(S_0)} < K_n n\eta \leq \frac{\varepsilon}{2}.$$

Put $\widetilde{\gamma}(t) = \gamma_1(t) + \sum_{S \in \mathcal{S}} \beta_S(t)$. Then

$$\|\widetilde{\gamma} - \gamma_0\|_{\leq n} \leq \|\gamma_1 - \gamma_0\|_{\leq n} + \|\widetilde{\gamma} - \gamma_1\|_{\leq n}$$
$$\leq \frac{\varepsilon}{2} + \left\|\sum_{S \in \mathcal{S}} \beta_S\right\|_{\leq n}$$
$$= \frac{\varepsilon}{2} + \max_{S \in \mathcal{S}} \|\beta_S\|_{\leq n} < \frac{\varepsilon}{2} + \frac{\varepsilon}{2} = \varepsilon.$$

Let $\delta = \min\{\eta\delta_S/K_n \mid S \in \mathcal{S}\}$ and consider any $\gamma \in \Gamma(X)$ such that $\|\gamma - \widetilde{\gamma}\|_{\leq n} < \delta$. Our plan is to prove that

$$\mathscr{L}^n\{t \in [0,1]^n \mid P(t,\gamma) \text{ fails}\} < \varepsilon. \tag{5.3}$$

By showing it we complete the argument that (5.3) holds for γ belonging to a dense open subset of $\Gamma_n(X)$.

To this aim take any $S \in \mathcal{S}$. We apply Lemma 5.3.1 to the cubes $S_0 \subset S$ and to the function $\gamma - \widetilde{\gamma}_S$ to get an extension, whose $\|\cdot\|_{\leq n}$-norm is strictly less than $K_n\delta/\eta \leq \delta_S$. Then we add this extension back to $\widetilde{\gamma}_S$, obtaining in this way a function $\widehat{\gamma}_S \in \Gamma_n(X)$ which agrees with γ in S_0 and satisfies the same control

$$\|\widehat{\gamma}_S - \widetilde{\gamma}_S\|_{\leq n} < \delta_S.$$

Hence by (5.2) we have

$$\mathscr{L}^n\{t \in S_0 \mid P(t,\gamma) \text{ fails}\} \leq \mathscr{L}^n\{t \in [0,1]^n \mid P(t,\widehat{\gamma}_S) \text{ fails}\} < \eta.$$

Finally,

$$\mathscr{L}^n\{t \in [0,1]^n \mid P(t,\gamma) \text{ fails}\}$$
$$\leq \frac{\varepsilon}{2} + (1-(1-\eta)^n) + \sum_{S \in \mathcal{S}} \mathscr{L}^n\{t \in S_0 \mid P(t,\gamma) \text{ fails}\}$$
$$< \frac{\varepsilon}{2} + (1-(1-\eta)^n) + k^n\eta \leq \varepsilon. \qquad \square$$

Corollary 5.3.4. *Suppose that $A \subset X$ is a Borel set such that for every affine map $\gamma_0 \colon \mathbb{R}^n \longrightarrow X$ and $\varepsilon > 0$ there are $\widetilde{\gamma} \in \Gamma_n(X)$ and $\delta > 0$ such that $\|\widetilde{\gamma} - \gamma_0\|_{\leq n} < \varepsilon$ and $\mathscr{L}^n\gamma^{-1}(A) < \varepsilon$ for every $\gamma \in \Gamma_n(X)$ with $\|\gamma - \widetilde{\gamma}\|_{\leq n} < \delta$. Then A is Γ_n-null.*
If the above criterion holds for all sufficiently large $n \in \mathbb{N}$, then A is Γ-null.

Proof. The first statement follows immediately from Proposition 5.3.3 since the property $P(t, \gamma) = \{(t, \gamma) \mid \gamma(t) \notin A\}$ is local on $[0, 1]^n \times \Gamma_n(X)$.

As for the second statement, let $\gamma_0 \colon \mathbb{R}^n \longrightarrow X$ be affine, where n is large enough, and $\varepsilon > 0$. Using Proposition 5.3.3 with the same property as in the previous part, find $\gamma_1 \in \Gamma_n(X)$ and $\delta > 0$ such that $\|\gamma_1 - \gamma_0\|_{\leq n} < \varepsilon$ and $\mathscr{L}^n \gamma^{-1}(A) < \varepsilon$ for every $\gamma \in \Gamma_n(X)$ with $\|\gamma - \gamma_1\|_{\leq n} < \delta$.

Let now $\gamma \in \Gamma(X)$ and $\|\gamma - \gamma_1\|_{\leq n} < \delta$. Then for every $s \in T$, $\gamma^{n,s} \in \Gamma_n(X)$ and $\|\gamma^{n,s} - \gamma_1\|_{\leq n} < \delta$. Hence $\mathscr{L}^n(\gamma^{n,s})^{-1}(A) < \varepsilon$ for every s, and Fubini's theorem implies that $\mathscr{L}^n \gamma^{-1}(A) < \varepsilon$.

Since the increasing union $\bigcup_{n=1}^{\infty} \Gamma_n(X)$ is dense in $\Gamma(X)$, it follows that for every $\varepsilon > 0$ the set of $\gamma \in \Gamma(X)$ such that $\mathscr{L}^n \gamma^{-1}(A) < \varepsilon$ contains a dense open subset of $\Gamma(X)$. Letting $\varepsilon \searrow 0$ rational we conclude that A is Γ-null. $\qquad\square$

We now turn our attention to tangential properties of residually many surfaces from $\Gamma_n(X)$, or, in other words, to analogues of Corollary 5.2.2 for these spaces. It is easy to see that there is a nonempty open set of $\gamma \in \Gamma_1(\mathbb{R})$ each of which has a point with $\gamma'(t) = 0$. Hence the exact analogue of Corollary 5.2.2 fails. However, it remains true when we restrict ourselves to almost every t.

Lemma 5.3.5. *The set of those surfaces $\gamma \in \Gamma_n(X)$ that satisfy*

$$\operatorname{rank} \gamma'(t) = \min\{n, \dim X\} \text{ for almost all } t \in [0, 1]^n,$$

is residual in $\Gamma_n(X)$.

Proof. The property

$$P(t, \gamma) = \{(t, \gamma) \mid \operatorname{rank} \gamma'(t) = \min\{n, \dim X\}\}$$

is clearly local, and for any affine map $\gamma_0 \in \Gamma_n(X)$ with this property the same holds for every (t, γ) provided γ is close enough to γ_0. Now it suffices to apply Proposition 5.3.3. $\qquad\square$

The first of the following corollaries is just a reformulation of Lemma 5.3.5 and the second is a special case of the first.

Corollary 5.3.6. *Residually many $\gamma \in \Gamma_n(X)$ have the property that*

$$\dim \operatorname{Tan}(\gamma, t) = \min\{n, \dim X\}$$

for almost every $t \in [0, 1]^n$.

Corollary 5.3.7. *If $\dim X \leq n$, then residually many $\gamma \in \Gamma_n(X)$ have the property that $\operatorname{Tan}(\gamma, t) = X$ for almost every $t \in [0, 1]^n$.*

We show next that in a finite dimensional space X the classes of Γ-null sets and Γ_n-null sets for $n \geq \dim X$ coincide with the class of sets of Lebesgue measure zero (just as for Gauss and Haar null sets).

Theorem 5.3.8. *In a finite dimensional space X, Γ-null sets and Lebesgue null sets coincide. If $n \geq \dim X$ then they also coincide with Γ_n-null sets.*

Proof. Let $A \subset X$, where $\dim X = k$, be a Borel set with $\mathscr{L}^k A > 0$. Find an affine map γ_0 of $\mathbb{R}^k$ onto X such that $\mathscr{L}^k(A \cap \gamma_0([0,1]^k)) > 0$. If $\gamma \in \Gamma(X)$ and $\|\gamma - \gamma_0\|_{\leq k}$ is sufficiently small, then for every $s \in T$ the mapping $\gamma^{k,s}$ is a diffeomorphism of $[0,1]^k$ onto a subset of X. Making the norm $\|\gamma - \gamma_0\|_{\leq k}$ even smaller, we may assume that all $\gamma^{k,s}([0,1]^k)$ meet A in a set of positive measure. Hence for every $s \in T$, we have $\mathscr{L}^k(\gamma^{k,s})^{-1}(A) > 0$. Consequently, by Fubini's theorem,

$$\mathscr{L}^{\mathbb{N}} \gamma^{-1}(A) = \int_T \mathscr{L}^k(\gamma^{k,s})^{-1}(A) \, d\mathscr{L}^{\mathbb{N}}(s) > 0$$

and we infer that A is not Γ-null. A similar argument works also for Γ_n-null sets, $n \geq k$.

If $\mathscr{L}^k A = 0$, we use that for every $n \geq k$ the set of affine maps γ_0 of $\mathbb{R}^n$ to X whose restriction to $\mathbb{R}^k$ has rank k is dense in the space of affine maps of $[0,1]^n$ to X. Let γ_0 be such affine mapping. If $\gamma \in \Gamma_n(X)$ and $\|\gamma - \gamma_0\|_{\leq k}$ is sufficiently small, then for every $s \in [0,1]^{n-k}$ the mapping $\gamma^{k,s}$ is a diffeomorphism of $[0,1]^k$ onto a subset of X. Hence for every $s \in [0,1]^{n-k}$, we have $\mathscr{L}^k(\gamma^{k,s})^{-1}(A) > 0$ and Fubini's theorem shows that

$$\mathscr{L}^n \gamma^{-1}(A) = \int_{[0,1]^n} \mathscr{L}^k(\gamma^{k,s})^{-1}(A) \, d\mathscr{L}^{n-k}(s) = 0.$$

Corollary 5.3.4 now yields that A is Γ_n-null as well as Γ-null. □

Finally, we observe that under direct products, Γ-nullness behaves in the expected way.

Remark 5.3.9. If π is a projection of X onto its subspace Y then a set $A \subset Y$ is Γ-null in Y if and only if $\pi^{-1}(A)$ is Γ-null in X. This follows by observing that the map $\varphi \colon \Gamma(X) \longrightarrow \Gamma(Y)$ defined by

$$\varphi(\gamma) = \pi \circ \gamma$$

is onto, hence open (by the open mapping theorem), and so it maps residual sets onto residual sets. In particular, if Y is finite dimensional and $A \subset Y$ has Lebesgue measure zero, then $\pi^{-1}(A)$ is Γ-null. A similar remark applies to Γ_n-null sets: $A \subset Y$ is Γ_n-null if and only if $\pi^{-1}(A)$ is Γ_n-null. In particular, if Y is finite dimensional and $A \subset Y$ has Lebesgue measure zero then $\pi^{-1}(A)$ is Γ_n-null for all $n \geq \dim Y$.

5.4 Γ- AND Γ_n-NULL SETS OF LOW BOREL CLASSES

We will see that for F_σ sets there are several simple relations between our classes of null sets. Unfortunately, in differentiability problems F_σ sets occur rather rarely. In our context, the only examples are the set of Fréchet nondifferentiability points of continuous convex functions or the set of Gâteaux nondifferentiability points of continuous convex functions on separable spaces. For neither of these two differentiability problems, the results obtained here are particularly useful. However, our main result will show a relation between Γ_n-null sets and Γ-null sets that is valid for $G_{\delta\sigma}$ sets. This is very interesting from our point of view, since sets of Fréchet nondifferentiability points are $G_{\delta\sigma}$ (this was proved in Corollary 3.5.5). Indeed, our result will be the key ingredient in the proof of Theorem 10.6.2, which gives a new result on Fréchet differentiability of Lipschitz functions Γ-almost everywhere on spaces with norms asymptotically smooth with modulus better than any power.

We first point out that it is easy to find the Borel type of sets in $\Gamma_n(X)$ and $\Gamma(X)$ that are defined by an estimate of the measure of the intersection with a Borel set in X.

Lemma 5.4.1. *Let A be a Borel subset of X, $n \in \mathbb{N}$, and $\alpha \geq 0$. Then the sets*

$$U = \{\gamma \in \Gamma_n(X) \mid \mathscr{L}^n \gamma^{-1}(A) \geq \alpha\} \text{ and } V = \{\gamma \in \Gamma(X) \mid \mathscr{L}^{\mathbb{N}} \gamma^{-1}(A) \geq \alpha\}$$

are Borel. Moreover, if A is closed then U and V are both closed in the $\|\cdot\|_\infty$ norm, and if A is G_δ then they are G_δ in the $\|\cdot\|_\infty$ norm.

Proof. We give the proof for U only; a proof for V may be obtained by a simple modification of the notation. We first prove the additional statements.

Let $A \subset X$ be closed, $\gamma_0 \in \Gamma_n(X)$ with $\mathscr{L}^n \gamma_0^{-1}(A) < \alpha$. Then there is a compact set $C \subset [0,1]^n \setminus \gamma_0^{-1}(A)$ with measure $\mathscr{L}^n C > 1 - \alpha$. Since $\gamma_0(C)$ is a compact subset of the open set $X \setminus A$, there is an ε-neighborhood of $\gamma_0(C)$ still contained in the complement of A. Any surface γ with $\|\gamma - \gamma_0\|_\infty < \varepsilon$ then satisfies that $\gamma(t) \notin A$ for $t \in C$. Hence $\gamma^{-1}(A) \subset [0,1]^n \setminus C$, and so $\mathscr{L}^n \gamma^{-1}(A) < \alpha$.

If A is G_δ, write $A = \bigcap_{k=1}^\infty A_k$, where A_k is a decreasing sequence of open sets. Then $\mathscr{L}^n \gamma^{-1}(A_k) \searrow \mathscr{L}^n \gamma^{-1}(A)$ for each γ, implying that

$$\{\gamma \in \Gamma_n(X) \mid \mathscr{L}^n \gamma^{-1}(A) \geq \alpha\} = \bigcap_{k=1}^\infty \left\{\gamma \in \Gamma_n(X) \,\Big|\, \mathscr{L}^n \gamma^{-1}(A_k) > \alpha - \frac{1}{k}\right\}$$

is a countable intersection of sets that are open by the previous part.

For Borel A the statement may be proved by transfinite induction, which would actually show that U, V belong to the same multiplicative Borel class to which A does, or by observing that the family of all Borel sets A for which the statement holds contains all closed sets and is closed under countable monotone unions and intersections, and so it contains all Borel sets. $\square$

We are now ready to prove the main result of this section.

Theorem 5.4.2. *If $A \subset X$ is a $G_{\delta\sigma}$ set which is Γ_n-null for infinitely many values of $n \in \mathbb{N}$, then it is Γ-null.*

Proof. Since countable unions of Γ-null sets are again Γ-null, it suffices to prove the statement for G_δ sets only.

Let A be a G_δ set which is Γ_n-null for infinitely many n. We show that the set

$$Q = \{\gamma \in \Gamma(X) \mid \mathscr{L}^{\mathbb{N}} \gamma^{-1}(A) \geq \alpha\}$$

is nowhere dense in $\Gamma(X)$ for every $\alpha > 0$. To this aim, let $G \subset \Gamma(X)$ be a nonempty open set. One can find $n \in \mathbb{N}$ such that the intersection $G \cap \Gamma_n(X)$ is nonempty and, at the same time, A is Γ_n-null. Fix this n and put

$$Q_n := \{\gamma \in \Gamma_n(X) \mid \mathscr{L}^n \gamma^{-1}(A) \geq \alpha\}.$$

Lemma 5.4.1 gives that Q_n is a G_δ set. By assumption, Q_n is also of the first category in $\Gamma_n(X)$ in the norm $\|\cdot\|_\infty$, and so also in the norm $\|.\|_{\leq n}$. Any G_δ set of the first category in a complete metric space is necessarily nowhere dense. Hence there is a nonempty open set $H \subset G \cap \Gamma_n(X)$ which does not meet Q_n,

$$H \cap Q_n = \emptyset.$$

Then the set

$$G_0 = \{\gamma \in G \mid \gamma^{n,s} \in H \text{ for all } s \in T\}$$

is a nonempty open subset of G. Indeed, any $\gamma \in H$ (considered as an element of $\Gamma(X)$) belongs to G_0, since in this case $\gamma^{n,s} = \gamma$ for all s. So $G_0 \neq \emptyset$. Let now $\gamma \in G_0$ be arbitrary. Notice that the family of surfaces

$$K = \{\gamma^{n,s} \mid s \in T\} \subset \Gamma_n(X)$$

forms a compact set. Since this set is contained in H, one can find $\varepsilon > 0$ such that the ε-neighborhood of K is still a subset of H. Thus any surface $\tilde{\gamma} \in \Gamma(X)$ satisfying $\|\gamma - \tilde{\gamma}\|_{\leq n} < \varepsilon$ belongs to G_0. Hence G_0 is open in $\Gamma(X)$.

We show that $G_0 \cap Q = \emptyset$. Assuming that $\gamma \in Q$, we have $\mathscr{L}^\mathbb{N} \gamma^{-1}(A) \geq \alpha$. Since by Fubini's theorem

$$\mathscr{L}^\mathbb{N} \gamma^{-1}(A) = \int_T \mathscr{L}^n (\gamma^{n,s})^{-1}(A) \, d\mathscr{L}^\mathbb{N}(s),$$

there exists $s \in T$ such that $\mathscr{L}^n (\gamma^{n,s})^{-1}(A) \geq \alpha$. For such s the corresponding $\gamma^{n,s} \notin H$, and the surface γ itself cannot belong to G_0. Consequently, the intersection $G_0 \cap Q = \emptyset$, which finishes the proof. $\qquad\square$

We now consider the less interesting case of F_σ sets.

Proposition 5.4.3. *If $E \subset X$ is a Γ_k-null, F_σ set, then E is Γ_n-null for every $n < k$. If E is a Γ-null, F_σ set, then it is Γ_n-null for every n.*

Proof. Since countable unions of Γ_n-null sets are again Γ_n-null, it suffices to prove the statement a closed set $E \subset X$. By Lemma 5.4.1, for every $\alpha > 0$ the set U_α of those $\gamma \in \Gamma_n(X)$ such that $\mathscr{L}^n \gamma^{-1}(E) < \alpha$ is open. To see that it is dense, let $\tilde{\gamma} \in \Gamma_n(X)$ and $\varepsilon > 0$. Use that E is Γ_k-null to find $\gamma \in \Gamma_k(X)$ such that $\|\gamma - \tilde{\gamma}\|_{\leq n} < \varepsilon$ and $\mathscr{L}^k \gamma^{-1}(E) = 0$. Since

$$\mathscr{L}^k \gamma^{-1}(E) = \int_{[0,1]^{k-n}} \mathscr{L}^n (\gamma^{n,s})^{-1}(E) \, d\mathscr{L}^{k-n}(s),$$

by Fubini's theorem $\mathscr{L}^n(\gamma^{n,s})^{-1}(E) = 0$ for almost every s. Since all such $\gamma^{n,s}$ are within ε of $\widetilde{\gamma}$ and belong to U_α, we infer that U_α is indeed an open dense subset of $\Gamma_n(X)$. Intersecting the U_α over (say rational) α, we get the residual subset of $\Gamma_n(X)$ required to show that E is Γ_n-null.

The same proof in which Γ_k is replaced by Γ and $\mathscr{L}^k$ by $\mathscr{L}^{\mathbb{N}}$ gives the second statement. $\qquad\square$

Proposition 5.4.4. *If an F_σ set is null on a dense set of n-dimensional affine surfaces, then it is Γ_n-null.*

Proof. It suffices to consider the case when $E \subset X$ is closed. For any $\alpha > 0$ the set of $\gamma \in \Gamma_n(X)$ that meet A in measure at least α is, by Lemma 5.4.1, closed. Hence for every n-dimensional affine surface $\gamma_0 \in \Gamma_n(X)$ that meets E in a null set there is $\varepsilon > 0$ such that $\mathscr{L}^n \gamma^{-1}(A) < \alpha$ for every $\gamma \in \Gamma_n(X)$ with $\|\gamma - \gamma_0\| < \varepsilon$. By Corollary 5.3.4, A is Γ_n-null. $\qquad\square$

Combining this with Corollary 2.2.4, we get

Corollary 5.4.5. *Gauss null, and even Haar null, sets of type F_σ are Γ-null.*

We now show that no further relations between these sets hold even for sets of low Borel classes. Since in the examples below we obtain decomposition of the space into two sets both of which are null in some sense, by considering the set or the complement one can see that the Borel class in the above results cannot be improved.

Example 5.4.6. Suppose that $n < \dim X < \infty$. Then X contains a set which is F_σ and Γ_n-null and the complement has Lebesgue measure zero.

Proof. Let $(\gamma_i)_{i=1}^\infty$ be a countable dense subset of $\Gamma_n(X)$. For each $i \geq 1$ the image $\gamma_i([0,1]^n)$ has Lebesgue measure zero. Choose open sets $G_i \subset X$ of measure at most 2^{-i} containing $\gamma_i([0,1]^n)$. The required set A is defined as the complement of

$$B = \bigcap_{k=1}^\infty \bigcup_{i=k}^\infty G_i.$$

Clearly, B is a G_δ set and since

$$B \subset \bigcup_{i \geq k} G_i \quad \text{for all } k \geq 1,$$

it has Lebesgue measure zero. It remains to show that A is Γ_n-null. Let

$$H_k := \left\{ \gamma \in \Gamma_n(X) \,\Big|\, \gamma([0,1]^n) \subset \bigcup_{i \geq k} G_i \right\}, \quad k \geq 1.$$

Each H_k is open and contains the dense set $\{\gamma_i \in \Gamma_n(X) \mid i \geq k\}$. Consequently, $H = \bigcap_{k=1}^\infty H_k$ is residual in $\Gamma_n(X)$. Let $\gamma \in H$. Then

$$\gamma([0,1]^n) \subset \bigcap_{k=1}^\infty \bigcup_{i \geq k} G_i = B,$$

which means that $A = X \setminus B$ is Γ_n-null. $\qquad\square$

Example 5.4.7. For every $n < \min\{k, \dim X\}$ there is an F_σ subset $E \subset X$ which is Γ_n-null and has Γ_k-null complement. In particular, E is not Γ_k-null.

Proof. Let π be a projection of X onto its k-dimensional subspace Y. By the previous example find a Γ_n-null, F_σ subset A of Y whose complement in Y has Lebesgue measure zero. Then $\pi^{-1}(A)$ is an F_σ subset of X that is Γ_n-null by Remark 5.3.9. Its complement is $\pi^{-1}(Y \setminus A)$, which is Γ_k-null by the same remark because $Y \setminus A$ has Lebesgue measure zero. $\square$

Example 5.4.8. For every $n < \dim X$ there is an F_σ subset $E \subset X$ which is Γ_n-null and has Γ-null complement.

Proof. If $\dim X < \infty$, this is immediate from the case $k = \dim X$ of the previous example and Theorem 5.3.8. If $\dim X = \infty$, we use the previous example to find for each $k > n$ an F_σ subset $E_k \subset X$ which is Γ_n-null and has Γ_k-null complement. By Theorem 5.4.2, the complement of $E := \bigcup_{k=n+1}^{\infty} E_k$ is Γ-null. Clearly, E is an F_σ set and we are done. $\square$

Example 5.4.9. Any separable infinite dimensional space X contains an $F_{\sigma\delta}$ subset $E \subset X$ which is Γ_n-null for all n and has Γ-null complement. In particular, E is not Γ-null.

Proof. For each $n \in \mathbb{N}$ take E_n from the previous example and let $E = \bigcap_{n=1}^{\infty} E_n$. $\square$

We finish this section by the promised example of the difference between Γ-null and Gauss null sets.

Lemma 5.4.10. *Suppose that $\gamma \in \Gamma(X)$ has range contained in a finite dimensional subspace of an infinite dimensional space X. Then for every $n \in \mathbb{N}$ and $\varepsilon > 0$ there is $\tilde{\gamma} \in \Gamma_n(X)$ with $\|\tilde{\gamma} - \gamma\|_{\leq n} < \varepsilon$ whose range is contained in an open set $U \subset X$ that intersects every line in a set of measure less than ε.*

Proof. Let V be a finite dimensional subspace of X containing the range of γ. Choose any $w \in X \setminus V$ and let π be a projection of X onto the span W of $V \cup \{w\}$. Since the statement does not change if we replace the original norm by an equivalent one, we may assume that on W the norm is Hilbertian, w is orthogonal to V, $\|w\| = 1$, and $\|\pi\| = 1$. Choose first a large $R > 0$ and then a small $0 < \tau < \frac{1}{2}R$ (precise estimates will be clear later), and let

$$U = \{x \in X \mid \|x - \pi x\| < \tau, \, \big|\|\pi x - Rw\| - R\big| < \tau\}.$$

Clearly, U is an open subset of X. To estimate the measure of its intersection with lines, suppose that $x, y \in U$ are such that $\|x - y\| > \varepsilon$ and let

$$\tilde{x} = Rw + R\frac{\pi x - Rw}{\|\pi x - Rw\|} \quad \text{and} \quad \tilde{y} = Rw + R\frac{\pi y - Rw}{\|\pi y - Rw\|}.$$

Then $\|\tilde{x} - \pi x\| < \tau$, and similarly $\|\tilde{y} - \pi y\| < \tau$. Write any $z \in U \cap [x, y]$ as $z = \alpha x + \beta y$, where $\alpha, \beta \geq 0$ and $\alpha + \beta = 1$. Then

$$\|z - (\alpha\tilde{x} + \beta\tilde{y})\| \leq \alpha\|x - \tilde{x}\| + \beta\|y - \tilde{y}\|$$
$$\leq \alpha\big(\|x - \pi x\| + \|\pi x - \tilde{x}\|\big) + \beta\big(\|y - \pi y\| + \|\pi y - \tilde{y}\|\big)$$
$$< 2\tau,$$

and $\|\pi z - (\alpha\tilde{x} + \beta\tilde{y})\| \leq \alpha\|\pi x - \tilde{x}\| + \beta\|\pi y - \tilde{y}\| < \tau$. Assuming that $\tau < \frac{1}{8}\varepsilon$, we obtain

$$\|\tilde{x} - \tilde{y}\| \geq \|x - y\| - \|\tilde{x} - \pi x\| - \|\tilde{y} - \pi y\| - \|\pi x - x\| - \|\pi y - y\|$$
$$> \|x - y\| - 4\tau > \varepsilon - 4\tau \geq \frac{1}{2}\varepsilon.$$

Since W is equipped with the Hilbertian norm some elementary calculation yields

$$\big|\|\alpha\tilde{x} + \beta\tilde{y} - Rw\| - R\big| = \frac{\|\alpha\tilde{x} + \beta\tilde{y} - Rw\|^2 - R^2}{\|\alpha\tilde{x} + \beta\tilde{y} - Rw\| + R}$$
$$\geq \frac{\|\alpha\tilde{x} + \beta\tilde{y} - Rw\|^2 - R^2}{3R} = \frac{\alpha\beta\|\tilde{x} - \tilde{y}\|^2}{3R}$$
$$\geq \frac{\alpha\beta\varepsilon^2}{12R}.$$

On the other hand,

$$\big|\|\alpha\tilde{x} + \beta\tilde{y} - Rw\| - R\big| \leq \|\alpha\tilde{x} + \beta\tilde{y} - \pi z\| + \big|\|\pi z - Rw\| - R\big| < 2\tau$$

since $z \in U$. Combining the last two estimates we get that $2\tau > \alpha\beta\varepsilon^2/(12R)$. Assuming, as we may, that τ has been chosen such that $\tau < 2^{-9}\varepsilon^4/(3R^3)$, we infer that either $\alpha < \varepsilon/8R$ or $\beta < \varepsilon/8R$. Noting that

$$\|x - y\| \leq \|\pi x - \pi y\| + \|x - \pi x\| + \|y - \pi y\| < 2(R + \tau) + 2\tau \leq 4R,$$

we see that either

$$\alpha < \frac{\varepsilon}{2\|x - y\|} \quad \text{or} \quad \beta < \frac{\varepsilon}{2\|x - y\|}.$$

It follows that any $z \in U \cap [x, y]$ is $\frac{1}{2}\varepsilon$ close to one of the endpoints of $[x, y]$. Hence $[x, y] \cap U$ is covered by two intervals of length less than $\frac{1}{2}\varepsilon$, showing that every line meets U in a set of measure less than ε.

It remains to observe that if R is big enough, the function $\tilde{\gamma} \in \Gamma_n(X)$,

$$\tilde{\gamma}(t) = \gamma(t) + (R - \sqrt{R^2 - \|\gamma(t)\|^2})w,$$

is well defined and satisfies $\|\tilde{\gamma} - \gamma\|_{\leq n} < \varepsilon$. $\qquad\square$

In the last example below notice that, just writing an F_σ set as a union of closed sets gives an example of a closed Γ-null set that is not Gauss null.

Example 5.4.11. Every infinite dimensional separable Banach space X contains an F_σ, Γ-null set E whose complement is null on all straight lines. In particular, the complement of E is Gauss null, and so E is not Gauss null.

Proof. Let X_n be an increasing sequence of finite dimensional subspaces of X whose union is dense in X. For each $n \in \mathbb{N}$ choose a sequence $(\gamma_{n,j})_{j=1}^\infty$ dense in $\Gamma_n(X_n)$. By Lemma 5.4.10 find $\gamma_{n,j,k} \in \Gamma_n(X)$ with

$$\|\gamma_{n,j,k} - \gamma_{n,j}\|_{\leq n} < 2^{-n-j-k}.$$

whose image is contained in an open set $U_{n,j,k} \subset X$ that intersects every line in a set of measure less than 2^{-n-j-k}. Then $U = \bigcap_{i=1}^\infty \bigcup_{n,j,k=i}^\infty U_{n,j,k}$ is a G_δ set meeting every line in a set of measure zero. The sets

$$\left\{ \gamma \in \Gamma(X) \,\Big|\, \gamma(T) \subset \bigcup_{n,j,k \geq i} U_{n,j,k} \right\}$$

are open and dense for every $i \geq 1$. Since any γ from the intersection of the sets above does not meet $E := X \setminus U$ we may conclude that E is the required F_σ and Γ-null set whose complement is null on every line. $\qquad\square$

5.5   EQUIVALENT DEFINITIONS OF Γ$_n$-NULL SETS

In this section we spend some time on a question which though not directly connected to our main theme nevertheless is quite natural and of some interest. The definition of Γ$_n$-null sets uses mappings γ of class C^1. One may ask whether the requirement concerning the existence of the derivative γ' may be weakened. Although we cannot completely remove the requirement of the existence of derivative we can still expect that reasonable spaces of functions having only a distributional derivative may lead to reasonable notions of negligible sets. We will show that this is indeed true, but, at least if we consider as "reasonable spaces" the usual Sobolev spaces, it does not bring anything new: Theorem 5.5.8 shows that null sets so defined would coincide with Γ$_n$-null sets.

   We first recall some basic definitions and facts about Sobolev spaces of mappings with values in a Banach space X.

Definition 5.5.1. Let $g: D \longrightarrow X$ be a locally integrable map of an open subset D of $\mathbb{R}^n$ into a Banach space X. The map $g': D \longrightarrow L(\mathbb{R}^n, X)$ is called the weak (or distributional) derivative of g if g' belongs to the space $L^1_{loc}(D, L(\mathbb{R}^n, X))$ and for all $\varphi \in C_0^\infty(D)$

$$\int_D g' \varphi \, d\mathscr{L}^n = - \int_D g \varphi' \, d\mathscr{L}^n.$$

The weak partial derivative $D_j g$ of g is a locally integrable map of D into X satisfying

$$\int_D D_j g \, \varphi \, d\mathscr{L}^n = - \int_D g D_j \varphi \, d\mathscr{L}^n.$$

Both the weak derivative and the weak partial derivative are defined almost everywhere in D. Clearly, $g'(x)(e_j) = D_j g(x)$ for a.a. $x \in D$.

Definition 5.5.2. For $1 \leq p < \infty$ we define $\Gamma_n^p(X)$ as the space of all measurable mappings $\gamma \colon [0,1]^n \longrightarrow X$ such that γ belongs to $L^p([0,1]^n, X)$ and its weak derivative γ' belongs to $L^p([0,1]^n, L(\mathbb{R}^n, X))$. The space $\Gamma_n^p(X)$ will be equipped with the norm

$$\|\gamma\|_{\Gamma_n^p} = \left(\int_{[0,1]^n} \|\gamma(t)\|^p d\mathscr{L}^n(t) + \int_{[0,1]^n} \|\gamma'(t)\|^p d\mathscr{L}^n(t) \right)^{1/p}.$$

Of course, in this definition we really have in mind the open set D to be $(0,1)^n$, but adding the boundary of measure zero does not change anything and makes the comparison with $\Gamma_n(X)$ slightly more pleasant.

We should also point out that a perhaps geometrically more natural definition of surfaces may require that γ be continuous on $[0,1]^n$ and not only measurable as in Definition 5.5.2. The reader may easily see that, when the space of such continuous surfaces is equipped with an appropriate norm, namely, $\|\gamma\|_\infty + \|\gamma'\|_{L^p([0,1]^n)}$, the arguments below will also show that the null sets defined with the help of such spaces again coincide with Γ_n-null sets.

The key to the validity of the result alluded to above is a Lusin type theorem for Sobolev spaces. These are well known and easily accessible in the literature for Sobolev spaces of real-valued functions (see, for example, [52]). We could have said that the standard proofs carry over to the Banach space–valued case, but since they include a variant of the Whitney extension theorem and some (simple) estimates that are not so readily available, we prefer to give a proof. However, we derive the Lusin property in Proposition 5.5.6 in a rather weak form, so that readers who prefer to provide their own arguments may skip directly to this lemma and prove it themselves.

The argument that we present here is a small addition to the Lipschitz truncation result of Acerbi and Fusco [1]. They construct a Lipschitz truncation using the weak type inequality for maximal operator

$$\mathscr{L}^n \{ u \in [0,1]^n \mid \mathbf{M}\varphi(u) > \lambda \} \leq K_{n,p} \lambda^{-p} \|\varphi\|_{L^p}^p,$$

where

$$\mathbf{M}\varphi(u) = \sup_{r>0} \fint_{B(u,r)} \|\varphi(u)\| \, du.$$

Lemmas 5.5.4 and 5.5.5 come from [1]; a small addition is contained in Proposition 5.5.6. Since their work may not be easily available, we also give the proofs. The transfer to the Banach space–valued functions is handled via an important result of [23] (which for us is just the technical Lemma 5.5.3).

In addition to the statements that we prove or quote below, we also need to know the standard properties of smoothing (by convolution with a mollifier with small support) and the fact that C^1 functions are dense in $\Gamma_n^p(X)$. This may be easily seen by smoothing.

The following result is well known. Its proof may be found in the original reference [23] (with linear dependence of K_n on n) or elsewhere. Recent developments of such extension problems may be found in [25]. Notice that the statement about the bound on the supremum is added only for our convenience; were the lemma stated without it, we would have to replace the extended function by $\kappa \circ \widetilde{\gamma}$, where $\kappa(x) = rx/\|x\|$ for $\|x\| \geq r$, $\kappa(x) = x$ for $\|x\| \leq r$, and $r = \sup_{t \in E} \|\gamma\|$.

Lemma 5.5.3. *Any Lipschitz map γ from a subset E of $\mathbb{R}^n$ to a Banach space may be extended to a Lipschitz $\widetilde{\gamma} \colon \mathbb{R}^n \longrightarrow X$ such that*

$$\sup_{t \in \mathbb{R}^n} \|\widetilde{\gamma}\| \leq \sup_{t \in E} \|\gamma\| \ and \ \mathrm{Lip}(\widetilde{\gamma}) \leq K \, \mathrm{Lip}(\gamma),$$

where K may depend only on the dimension n.

Lemma 5.5.4. *For any $\gamma \in \Gamma_n(X)$ and $\lambda > 0$, the restriction of γ to the set $H_\lambda = \{u \in [0,1]^n \mid \mathbf{M}\gamma'(u) < \lambda\}$ has Lipschitz constant at most $K\lambda$, where K depends only on dimension n.*

Proof. For $u \in [0,1]^n$ let $M_r(u) = [0,1]^n \cap B(u,r)$. Observe that there is a constant $c > 0$ (depending on n only) such that for every $u \in [0,1]^n$ and $r > 0$ we have

$$\mathscr{L}^n M_r(u) \geq c\mathscr{L}^n B(u,r),$$

and, if $v \in [0,1]^n$ and $r = |u - v|$,

$$\mathscr{L}^n(M_r(u) \cap M_r(v)) \geq c\mathscr{L}^n(M_r(u) \cup M_r(v)).$$

We show that the statement holds with $K = 4/c^2$. For this, fix $\lambda > 0$, denote

$$S_{k,r}(u) = \left\{v \in M_r(u) \mid \|\gamma(u) - \gamma(v)\| > \tfrac{1}{2}K\lambda|u - v|\right\}$$

and for $u \in H_\lambda$ estimate

$$\frac{\lambda K}{2} \frac{\mathscr{L}^n S_{k,r}(u)}{\mathscr{L}^n M_r(u)} \leq \fint_{M_r(u)} \frac{\|\gamma(v) - \gamma(u)\|}{|v - u|} \, dv$$

$$\leq \fint_{M_r(u)} \frac{1}{|v - u|} \int_0^1 \|\gamma'(u + t(v - u); v - u)\| \, dt dv$$

$$\leq \fint_{M_r(u)} \int_0^1 \|\gamma'(u + t(v - u))\| \, dt dv$$

$$= \int_0^1 \fint_{M_r(u)} \|\gamma'(u + t(v - u))\| \, dv dt.$$

Substituting $u + t(v - u)$ by w we continue

$$
\begin{aligned}
&= \frac{1}{\mathscr{L}^n M_r(u)} \int_0^1 \frac{1}{t^n} \int_{M_{tr}(u)} \|\gamma'(w)\| \, dw dt \\
&\le \frac{1}{c \mathscr{L}^n B(u,r)} \int_0^1 \frac{1}{t^n} \int_{M_{tr}(u)} \|\gamma'(w)\| \, dw dt \\
&\le \frac{1}{c} \int_0^1 \mathbf{M}\gamma'(u) \, dt < \frac{\lambda}{c}.
\end{aligned}
$$

Hence $\mathscr{L}^n S_{k,r}(u) < \frac{1}{2} c \mathscr{L}^n M_r(u)$, showing that for any $u, v \in H_\lambda$ we can find, letting $r = |u - v|$, a point $w \in M_r(u) \cap M_r(v) \setminus (S_{k,r}(u) \cup S_{k,r}(v))$. Then

$$
\begin{aligned}
\|\gamma(u) - \gamma(v)\| &\le \|\gamma(u) - \gamma(w)\| + \|\gamma(w) - \gamma(v)\| \\
&\le \tfrac{1}{2} K\lambda|u - w| + \tfrac{1}{2} K\lambda|w - v| \le K\lambda|u - v|. \qquad \square
\end{aligned}
$$

Lemma 5.5.5. *For any $\gamma \in \Gamma_n^p(X)$ and $\lambda > 0$, there is a Lipschitz map $\widetilde{\gamma} \colon \mathbb{R}^n \longrightarrow X$ such that $\max\{\|\widetilde{\gamma}\|_\infty, \mathrm{Lip}(\widetilde{\gamma})\} \le \lambda$ and*

$$
\mathscr{L}^n\{t \in [0,1]^n \mid \gamma(t) \ne \widetilde{\gamma}(t)\} < K_{n,p}\lambda^{-p}\|\gamma\|_{\Gamma_n^p}^p,
$$

where $K_{n,p}$ depends only on n and p.

Proof. We denote by K a constant with which the weak type inequality for maximal operator, Lemma 5.5.3 and Lemma 5.5.4 all hold. Then we put $K_{n,p} = 1 + 2^p K^{2p+1}$.

Let $\gamma_j \in \Gamma_n(X)$ be such that $\|\gamma_j - \gamma\|_{\Gamma_n^p}^p \le 2^{-j}$ and at the same time $\gamma_j(u)$ converges to $\gamma(u)$ for almost every $u \in [0,1]^n$. The maximal operator inequality together with $|\mathbf{M}\varphi - \mathbf{M}\psi| \le \mathbf{M}(\varphi - \psi)$ shows that for every $r > 0$,

$$
\begin{aligned}
\mathscr{L}^n\{u \in [0,1]^n \mid |\mathbf{M}\gamma'(u) - \mathbf{M}\gamma_j'(u)| > r\} &\le K r^{-p}\|\gamma' - \gamma_j'\|_{L^p}^p \\
&\le K r^{-p}\|\gamma - \gamma_j\|_{\Gamma_n^p}^p \le K 2^{-j} r^{-p}.
\end{aligned}
$$

Given $\lambda > 0$, let E be the set of all $u \in [0,1]^n$ such that

$$
\|\gamma(u)\| < \lambda, \quad \mathbf{M}\gamma'(u) < \lambda/K^2, \text{ and } \mathbf{M}\gamma_j'(u) \to \mathbf{M}\gamma'(u).
$$

Since the latter condition holds, by the above estimate, almost everywhere, we infer from the maximal operator inequality that

$$
\begin{aligned}
\mathscr{L}^n([0,1]^n \setminus E) &\le \mathscr{L}^n\{u \in [0,1]^n \mid \|\gamma(u)\| \ge \lambda\} \\
&\quad + \mathscr{L}^n\{u \in [0,1]^n \mid \mathbf{M}\gamma'(u) > \lambda/(2K^2)\} \\
&\le \lambda^{-p}\|\gamma\|_{L^p}^p + 2^p K^{2p+1}\lambda^{-p}\|\gamma'\|_{L^p}^p \\
&\le (1 + 2^p K^{2p+1})\lambda^{-p}\|\gamma\|_{\Gamma_n^p}^p \\
&= K_{n,p}\lambda^{-p}\|\gamma\|_{\Gamma_n^p}^p.
\end{aligned}
$$

Moreover, any $u, v \in E$ belong to

$$H_{\lambda/K^2} := \left\{ u \in [0,1]^n \mid \mathbf{M}\gamma_j' < \lambda/K^2 \right\}$$

for large enough j, so by Lemma 5.5.4

$$\|\gamma(u) - \gamma(v)\| = \lim_{j \to \infty} \|\gamma_j(u) - \gamma_j(v)\| \leq \lambda \|u - v\|/K.$$

Hence the Lipschitz constant of the restriction of γ to E is at most λ/K, and the statement follows by an application of Lemma 5.5.3. $\qquad\square$

Proposition 5.5.6. *For any $\varepsilon > 0$ there is $\delta > 0$ with the following property: If $\gamma \in \Gamma_n^p(X)$ with $\|\gamma\|_{\Gamma_n^p} < \delta$, then there is $\widetilde{\gamma} \in \Gamma_n(X)$ such that $\|\widetilde{\gamma}\|_{\leq n} < \varepsilon$ and $\mathscr{L}^n\{t \in [0,1]^n \mid \gamma(t) \neq \widetilde{\gamma}(t)\} < \varepsilon$.*

Proof. We denote $\varepsilon_k = 2^{-k-2}\varepsilon$ and $\delta_k = \varepsilon_k^{1+1/p}/K_{n,p}^{1/p}$, where $K_{n,p}$ is the constant from Lemma 5.5.5, and show that the statement holds with $\delta = \delta_0$.

Suppose that $\|\gamma\|_{\Gamma_n^p} < \delta$. By Lemma 5.5.5 we find Lipschitz $\gamma_0 : \mathbb{R}^n \longrightarrow X$ such that $\|\gamma_0\|_\infty \leq \varepsilon_0$, $\mathrm{Lip}(\gamma_0) \leq \varepsilon_0$ and $\mathscr{L}^n E_0 < \varepsilon_0$, where

$$E_0 := \{t \in [0,1]^n \mid \gamma(t) \neq \gamma_0(t)\}.$$

By smoothing γ_0, we find a function $\widetilde{\gamma}_0 \in C^1(\mathbb{R}^n, X)$ such that

$$\|\widetilde{\gamma}_0\|_\infty \leq \varepsilon_0, \ \mathrm{Lip}(\widetilde{\gamma}_0) \leq \varepsilon_0 \ \text{and} \ \|\widetilde{\gamma}_0 - \gamma_0\|_{\Gamma_n^p} < \delta_1.$$

We continue recursively. First, with the help of Lemma 5.5.5 choose for each $k = 1, 2, \dots$ a Lipschitz map $\gamma_k : \mathbb{R}^n \longrightarrow X$ such that $\|\gamma_k\|_\infty \leq \varepsilon_k$, $\mathrm{Lip}(\gamma_k) \leq \varepsilon_k$, and the set

$$E_k := \{t \in [0,1]^n \mid \gamma_{k-1} - \widetilde{\gamma}_{k-1} \neq \gamma_k(t)\}$$

satisfies $\mathscr{L}^n E_k < \varepsilon_k$. Then we smooth γ_k to find $\widetilde{\gamma}_k \in C^1(\mathbb{R}^n, X)$ such that

$$\|\widetilde{\gamma}_k\|_\infty \leq \varepsilon_k, \ \mathrm{Lip}(\widetilde{\gamma}_k) \leq \varepsilon_k, \ \text{and} \ \|\gamma_k - \widetilde{\gamma}_k\|_{\Gamma_n^p} < \delta_{k+1}.$$

Let $\widetilde{\gamma} = \sum_{k=0}^\infty \widetilde{\gamma}_k$. Since $\|\widetilde{\gamma}_k\|_{\leq n} \leq \varepsilon_k$, the series converges uniformly in the norm $\|\cdot\|_{\leq n}$. So $\widetilde{\gamma} \in C^1(\mathbb{R}^n, X)$. Moreover,

$$\|\widetilde{\gamma}\|_{\leq n} \leq \sum_{k=0}^\infty \varepsilon_k = 2\varepsilon_0 < \varepsilon.$$

Finally, the set $E = \bigcup_{k=0}^\infty E_k$ has measure less than $\sum_{k=0}^\infty \varepsilon_k < \varepsilon$, and for almost every $t \in [0,1]^n \setminus E$,

$$\widetilde{\gamma}(t) = \sum_{k=1}^\infty \left(\gamma_{k-1}(t) - \gamma_k(t)\right) = \gamma_0(t) = \gamma(t). \qquad\square$$

Lemma 5.5.7. *Let* $1 \leq p < \infty$. *For any dense open set* $U \subset \Gamma_n(X)$ *and any* $\varepsilon > 0$ *the set*

$$V = \left\{ \gamma \in \Gamma_n^p(X) \mid (\exists \tilde{\gamma} \in U) \ \mathscr{L}^n\{t \in [0,1]^n \mid \gamma(t) \neq \tilde{\gamma}(t)\} < \varepsilon \right\} \qquad (5.4)$$

contains a subset that is open and dense in $\Gamma_n^p(X)$.

Conversely, for any dense open set $V \subset \Gamma_n^p(X)$ *and any* $\varepsilon > 0$ *the set*

$$U = \left\{ \tilde{\gamma} \in \Gamma_n(X) \mid (\exists \gamma \in V) \ \mathscr{L}^n\{t \in [0,1]^n \mid \gamma(t) \neq \tilde{\gamma}(t)\} < \varepsilon \right\} \qquad (5.5)$$

contains a subset that is open and dense in $\Gamma_n(X)$.

Proof. For the first statement, assume that $\gamma_0 \in \Gamma_n^p(X)$ and $\eta > 0$. Since the space $\Gamma_n(X)$ is dense in $\Gamma_n^p(X)$ and the norm $\|\cdot\|_{\Gamma_n^p}$ is weaker than the norm $\|\cdot\|_{\leq n}$, the set U is dense in $\Gamma_n^p(X)$. So one can find $\gamma_1 \in U$ and $\varepsilon > 0$ with the property that $\|\gamma_0 - \gamma_1\|_{\Gamma_n^p} < \eta$ and, at the same time, every $\gamma \in \Gamma_n(X)$ with $\|\gamma - \gamma_1\|_{\leq n} < \varepsilon$ belongs to U. Finally, let $\delta > 0$ be as in Proposition 5.5.6. We show that the open ball

$$B := \{\gamma \in \Gamma_n^p \mid \|\gamma - \gamma_1\|_{\Gamma_n^p} < \delta\}$$

is contained in V. Indeed, for any $\gamma \in B$ there is by Proposition 5.5.6 a surface $\tilde{\gamma} \in \Gamma_n(X)$ such that $\|\tilde{\gamma}\|_{\leq n} < \varepsilon$ and the measure of the points $t \in [0,1]^n$, where $\gamma(t) - \gamma_1(t) \neq \tilde{\gamma}(t)$ is less than ε. Since then $\gamma_1 + \tilde{\gamma} \in U$, we are done.

Similarly, for the second statement, assume that $\tilde{\gamma}_0 \in \Gamma_n(X)$ and $\varepsilon > 0$. Find $\delta > 0$ according to Proposition 5.5.6. Since V is dense in $\Gamma_n^p(X)$, there is $\gamma_1 \in V$ such that $\|\gamma_1 - \tilde{\gamma}_0\|_{\Gamma_n^p} < \delta$ and, by Proposition 5.5.6, there is also $\tilde{\gamma}_1 \in \Gamma_n(X)$ such that

$$\|\tilde{\gamma}_1\|_{\leq n} < \varepsilon \quad \text{and} \quad \mathscr{L}^n\{t \in [0,1]^n \mid \tilde{\gamma}_1(t) \neq \gamma_1(t) - \tilde{\gamma}_0(t)\} < \varepsilon.$$

Finally, let $\delta_0 > 0$ be small enough to guarantee that

$$\{\gamma \in \Gamma_n^p(X) \mid \|\gamma - \gamma_1\|_{\Gamma_n^p} < 2n\delta_0\} \subset V.$$

We show that the open ball

$$B := \{\tilde{\gamma} \in \Gamma_n(X) \mid \|\tilde{\gamma} - (\tilde{\gamma}_0 + \tilde{\gamma}_1)\|_{\leq n} < \delta_0\}$$

is contained in U, which will complete the proof of the second statement. Let $\tilde{\gamma} \in B$ and put $\gamma = \tilde{\gamma} - (\tilde{\gamma}_0 + \tilde{\gamma}_1) + \gamma_1$. Clearly, the surfaces γ and $\tilde{\gamma}$ differ on a set of measure less than ε. Also

$$\|\gamma - \gamma_1\|_{\Gamma_n^p} = \|\tilde{\gamma} - (\tilde{\gamma}_0 + \tilde{\gamma}_1)\|_{\Gamma_n^p} \leq 2n\|\tilde{\gamma} - (\tilde{\gamma}_0 + \tilde{\gamma}_1)\|_{\leq n} < 2n\delta_0,$$

which implies that $\gamma \in V$. $\qquad \square$

The main result of this section is the already announced theorem showing that Γ_n-null sets may be equivalently defined using surfaces from $\Gamma_n^p(X)$ (for any p) instead of those from $\Gamma_n(X)$.

Theorem 5.5.8. *Suppose that X is a Banach space, $n \in \mathbb{N}$, and $1 \le p < \infty$. A Borel set $E \subset X$ is Γ_n-null if and only if the set*

$$\left\{ \gamma \in \Gamma_n^p(X) \mid \mathscr{L}^n \gamma^{-1}(E) = 0 \right\}$$

is residual in $\Gamma_n^p(X)$.

Proof. Suppose first that $E \subset X$ is a Borel set which is null in the Γ_n^p sense, that is,

$$V := \left\{ \gamma \in \Gamma_n^p(X) \mid \mathscr{L}^n \gamma^{-1}(E) = 0 \right\}$$

is residual in $\Gamma_n^p(X)$. There are dense open subsets V_k of $\Gamma_n^p(X)$ with $\bigcap_{k=1}^\infty V_k \subset V$. Define $U_k \subset \Gamma_n(X)$ by (5.5) with $V = V_k$ and $\varepsilon = 2^{-k}$. By Lemma 5.5.7, each U_k contains a dense open subset of $\Gamma_n(X)$, so their intersection $U := \bigcap_{k=1}^\infty U_k$ is residual in $\Gamma_n(X)$.

For any $\tilde{\gamma} \in \bigcap_{k=1}^\infty U_k$ and $k \ge 1$ we may find $\gamma_k \in V_k$ such that

$$\mathscr{L}^n \{ t \in [0,1]^n \mid \gamma_k(t) \ne \tilde{\gamma}(t) \} < 2^{-k}.$$

Hence $\mathscr{L}^n \tilde{\gamma}^{-1}(E) \le 2^{-k} + \mathscr{L}^n \gamma_k^{-1}(E) = 2^{-k}$. Since $k \ge 1$ is arbitrary, we have $\mathscr{L}^n \tilde{\gamma}^{-1}(E) = 0$, as required.

As for the reverse implication we prove the statement for any E which is a G_δ set. Let $E \subset X$ be a Γ_n-null and G_δ. The set of those surfaces $\gamma \in \Gamma_n(X)$ with $\mathscr{L}^n \gamma^{-1}(E) < 2^{-k}$ is residual, and by Lemma 5.4.1 it is an F_σ set. Such a set necessarily contains an open dense set $U_k \subset \Gamma_n(X)$. Define $V_k \subset \Gamma_n^p(X)$ by (5.4) with $U = U_k$ and $\varepsilon = 2^{-k}$. By Lemma 5.5.7, each V_k contains a dense open subset of $\Gamma_n^p(X)$. So $V := \bigcap_{k=1}^\infty V_k$ is residual in $\Gamma_n^p(X)$.

Consider any $\gamma \in \bigcap_{k=1}^\infty V_k$. Then for each $k \ge 1$ we may find $\tilde{\gamma}_k \in U_k$ such that

$$\mathscr{L}^n \{ t \in [0,1]^n \mid \gamma(t) \ne \tilde{\gamma}_k(t) \} < 2^{-k}.$$

Hence we infer that $\mathscr{L}^n \gamma^{-1}(E) \le 2^{-k} + \mathscr{L}^n \tilde{\gamma}_k^{-1}(E) = 2^{-k+1}$. Since $k \ge 1$ is arbitrary, we have $\mathscr{L}^n \tilde{\gamma}^{-1}(E) = 0$.

To complete the argument we notice that the family of Borel sets in X for which the just proved reverse implication holds true is closed under countable monotone unions and intersections. Since it holds also for all G_δ subsets E, in particular for all open and closed subsets, we established the validity for all Borel sets. $\qquad \square$

5.6 SEPARABLE DETERMINATION

As promised in Section 3.6 (where the reader should look for definitions), we show that for Borel sets Γ-nullness is separably determined. The same statement, with the same proof, holds also for Γ_n-nullness: just replace Γ by Γ_n and work with the fixed n (and with T replaced by $[0,1]^n$) throughout.

Lemma 5.6.1. *Let $M \subset \Gamma(X)$ be nowhere dense in $\Gamma(X)$. Then there is a rich family $\mathcal{R}$ on X such that for every $R \in \mathcal{R}$, the set $M \cap \Gamma(R)$ is nowhere dense in $\Gamma(R)$.*

Proof. For $\gamma \in \Gamma(X)$, $r > 0$, and $n \in \mathbb{N}$ denote

$$B(\gamma, r, n) = \{\widetilde{\gamma} \in \Gamma(X) : \|\widetilde{\gamma} - \gamma\|_{\leq n} < r\}.$$

Define $\mathcal{R}$ as the family of separable subspaces R of X having the following property:

- For any surface $\gamma \in \Gamma(R)$, rational $\varepsilon > 0$, and $n \in \mathbb{N}$ there are $\widetilde{\gamma} \in \Gamma(R)$, $\widetilde{\varepsilon} > 0$, and $\widetilde{n} \in \mathbb{N}$ satisfying

$$\|\widetilde{\gamma} - \gamma\|_{\leq n} < \varepsilon \text{ and } B(\widetilde{\gamma}, \widetilde{\varepsilon}, \widetilde{n}) \cap M = \emptyset.$$

Obviously, such a property of spaces $R \in \mathcal{R}$ guarantees that $M \cap \Gamma(R)$ is nowhere dense in R.

The proof will be complete once we verify that $\mathcal{R}$ is a rich family; see Definition 3.6.1. As for the first condition, suppose that (R_k) is an increasing sequence of subspaces belonging to $\mathcal{R}$ and let $R = \bigcup_{k=0}^{\infty} R_k$.

If $\gamma \in \Gamma(R)$, rational $\varepsilon > 0$, and $n \in \mathbb{N}$, one can find with the help of Lemma 5.3.2 an index $k \in \mathbb{N}$ and $\gamma_k \in \Gamma(R_k)$ such that $\|\gamma - \gamma_k\|_{\leq n} < \frac{1}{2}\varepsilon$. Since $R_k \in \mathcal{R}$, there are $\widetilde{\gamma} \in \Gamma(R_k)$, $\widetilde{\varepsilon} > 0$, and $\widetilde{n}$ such that $\|\widetilde{\gamma} - \gamma_k\|_{\leq n} < \frac{1}{2}\varepsilon$ and $B(\widetilde{\gamma}, \widetilde{r}, \widetilde{n}) \cap M = \emptyset$. Consequently, $\widetilde{\gamma} \in \Gamma(R)$,

$$\|\widetilde{\gamma} - \gamma\|_{\leq n} \leq \|\gamma - \gamma_k\|_{\leq n} + \|\widetilde{\gamma} - \gamma_k\|_{\leq n} < \varepsilon,$$

and also $B(\widetilde{\gamma}, \widetilde{r}, \widetilde{n}) \cap M = \emptyset$, as required.

To show the second property of a rich family let Y be a separable subspace of X. We choose a countable set S dense in $\Gamma(Y)$ and enlarge Y as follows. For each $\gamma \in S$, rational $r > 0$, and $n \in \mathbb{N}$ we use that M is nowhere dense in X to find $\widetilde{\gamma} \in \Gamma(X)$ such that, for some $\widetilde{r} > 0$ and $\widetilde{n} \in \mathbb{N}$,

$$\|\widetilde{\gamma} - \gamma\|_{\leq n} < r \text{ and } B(\widetilde{\gamma}, \widetilde{r}, \widetilde{n}) \cap M = \emptyset.$$

The set of all so chosen $\widetilde{\gamma}$ is countable, so the union of their images is separable. Hence the closed linear span $\widetilde{Y}$ of Y together with this union is also separable.

Define now $Y_0 = Y$, $Y_{k+1} = \widetilde{Y_k}$ and $R = \bigcup_{k=0}^{\infty} Y_k$. Clearly, R is separable and contains Y. We show that it belongs to $\mathcal{R}$. For this, let $\gamma \in \Gamma(R)$, rational $\varepsilon > 0$, and $n \in \mathbb{N}$ be given. By Lemma 5.3.2 we find $k \in \mathbb{N}$ and $\gamma_k \in \Gamma(Y_k)$ such that

$$\|\gamma - \gamma_k\|_{\leq n} < \frac{\varepsilon}{2}.$$

Choose rational $0 < r < \varepsilon/2$. The definition of Y_{k+1} provides $\widetilde{\gamma} \in \Gamma(Y_{k+1})$, $\widetilde{r} > 0$, and $\widetilde{n}$ such that

$$\|\widetilde{\gamma} - \gamma_k\|_{\leq n} < r \text{ and } B(\widetilde{\gamma}, \widetilde{r}, \widetilde{n}) \cap M = \emptyset.$$

Since then

$$\|\widetilde{\gamma} - \gamma\|_{\leq n} \leq \|\widetilde{\gamma} - \gamma_k\|_{\leq n} + \|\gamma_k - \gamma\|_{\leq n} < \frac{\varepsilon}{2} + \frac{\varepsilon}{2} = \varepsilon,$$

we conclude that $R \in \mathcal{R}$. $\square$

Corollary 5.6.2. *Let $E \subset X$ be a Borel set. Then E is Γ-null in X if and only if there is a rich family $\mathcal{R}$ on X such that for every $R \in \mathcal{R}$, $E \cap R$ is Γ-null in R.*

Proof. If E is Γ-null, we write the set of $\gamma \in \Gamma(X)$ such that $\mathscr{L}^{\mathbb{N}} \gamma^{-1}(E) > 0$ as a countable union $\bigcup_n M_n$ of nowhere dense sets M_n. Using Lemma 5.6.1 for each of these sets, we obtain countably many rich families $\mathcal{R}_n$. The intersection of all such families is, in view of Proposition 3.6.2, again a rich family. Let $R \in \bigcap_n \mathcal{R}_n$. Then $M_n \cap \Gamma(R)$ is nowhere dense in $\Gamma(R)$ for all n and so $\bigcup_n M_n \cap \Gamma(R)$ is the first category set in $\Gamma(R)$. Consequently, E is Γ-null in R.

If E is not Γ-null, we use that E is Borel to infer from Lemma 5.4.1 that the set M of $\gamma \in \Gamma(X)$ such that $\mathscr{L}^{\mathbb{N}} \gamma^{-1}(E) = 0$ is Borel. In particular, it has the Baire property. Since it is not residual, there is a nonempty open set $G \subset \Gamma(X)$ such that $M \cap G$ is of the first category. Arguing as in the previous paragraph, we find a rich family $\mathcal{R}_0$ on X such that for every $R \in \mathcal{R}_0$, the set $M \cap G \cap \Gamma(R)$ is of the first category in $\Gamma(R)$.

Let $\mathcal{R}$ be any rich family on X. Pick arbitrary $\gamma_0 \in G$. The image of γ_0 is clearly separable. If $R \in \mathcal{R} \cap \mathcal{R}_0$ which, in addition, contains the image of γ_0 then, first, $G \cap \Gamma(R) \neq \emptyset$ and second, $M \cap \Gamma(R)$ is not residual in $\Gamma(R)$. Hence $E \cap R$ is not Γ-null in R. $\square$

Chapter Six

Fréchet differentiability except for Γ-null sets

We give an account of the known genuinely infinite dimensional results proving Fréchet differentiability almost everywhere. This is where Γ-null sets, porous sets, and special features of the geometry of the space enter the picture. Γ-null sets provide the only notion of negligible sets with which a Fréchet differentiability result is known. Porous sets appear as sets at which Gâteaux derivatives can behave exceptionally badly (we call this behavior irregular), and they turn out to be the only obstacle to validity of a Fréchet differentiability result Γ-almost everywhere. Finally, geometry of the space may (or may not) guarantee that porous sets are Γ-null.

6.1 INTRODUCTION

In this chapter we show, following the arguments of [28], that on some infinite dimensional Banach spaces countable collections of real-valued Lipschitz functions, and even of fairly general Lipschitz maps to infinite dimensional spaces, have a common point of Fréchet differentiability.

The key result is proved in Section 6.3. It states that a Lipschitz map between separable Banach spaces is Fréchet differentiable Γ-almost everywhere provided it is regularly Gâteaux differentiable Γ-almost everywhere and the Gâteaux derivatives stay within a norm separable space of operators. One of the consequences of this statement is that if every σ-porous set in X is Γ-null, then any Lipschitz $f: X \longrightarrow Y$ with Y having the RNP whose set of Gâteaux derivatives is norm separable is Fréchet differentiable Γ-almost everywhere. It also follows that continuous convex functions on any space X with X^* separable are Fréchet differentiable Γ-almost everywhere. In particular, if X^* is separable, $f: X \longrightarrow \mathbb{R}$ is convex and continuous, and $g: X \longrightarrow Y$ is Lipschitz with Y having the RNP, then there is a point x (actually Γ-almost any point) at which f is Fréchet differentiable and g is Gâteaux differentiable. Existence of such points cannot be deduced using previous concepts of null sets.

In Section 6.4 we prove that for $X = c_0$ or more generally $X = C(K)$ with K countable compact, and for some closely related spaces, every σ-porous set in them is indeed Γ-null. Thus combined with the main result of Section 6.3 we get a general result on existence of points of Fréchet differentiability for Lipschitz maps $f: X \longrightarrow Y$, where X is as above and Y has the RNP. This is the first, and so far only, result on existence of points of Fréchet differentiability for general Lipschitz mappings for certain pairs of infinite dimensional spaces.

6.2 REGULAR POINTS

Various notions of "regular" behavior of functions at points at which they are not Fréchet differentiable or where their derivative is not continuous have been studied, in recent times mainly in connection with the investigation of generalized subdifferentials. Our definition below of regular points of Lipschitz maps is tailored to the applications we have in mind. For real-valued functions, the reader familiar with subdifferentials may recognize a reformulation via a variant of the Michel-Penot type subdifferential

$$\left\{ x^* \in X^* \mid (\forall v \in X) \; x^*(v) \le \sup_{c>0} \limsup_{t \to 0} \; \sup_{\|u\| \le ct} \frac{f(x + u + tv) - f(x + u)}{t} \right\}.$$

Definition 6.2.1. Suppose that f is a map from an open set G in X to Y. We say that a point $x \in G$ is a *regular point of f* if for every $v \in X$ for which the directional derivative $f'(x; v)$ exists,

$$\lim_{t \to 0} \frac{f(x + tu + tv) - f(x + tu)}{t} = f'(x; v)$$

uniformly for $\|u\| \le 1$. A point which is not regular is called *irregular*.

Note that in the definition it is enough to take the limit for $t \searrow 0$ only, since we may replace u by $-u$. It is easy to see that a point at which f is Fréchet differentiable is a regular point of f. In fact, we have already seen a more precise form of this fact in Lemma 4.3.1. Also, a point at which f is not differentiable in any direction is, by our definition, a regular point of f.

The following simple fact, in addition to being an important source in the proof in Corollary 6.3.10 of Fréchet differentiability of convex functions Γ-almost everywhere, illustrates one of the ways in which regular points may occur.

Proposition 6.2.2. *For a continuous convex function $f \colon X \longrightarrow \mathbb{R}$ every point of X is a regular point of f.*

Proof. Given $x \in X$, $v \in X$, and $\varepsilon > 0$, we find $r > 0$ such that

$$|f(x + tv) - f(x) - f'(x; tv)| < \varepsilon |t| \quad \text{for } 0 < |t| \le r,$$

and such that f is Lipschitz on the ball $B(x, 2r(1 + \|v\|))$ with constant $K > 0$. If $\|u\| \le 1$ and $0 < t \le \min\{r, \varepsilon r/2K\}$, then by convexity

$$\frac{f(x + tu + tv) - f(x + tu)}{t} \le \frac{f(x + tu + rv) - f(x + tu)}{r}$$
$$\le \frac{f(x + rv) - f(x)}{r} + \frac{2Kt\|u\|}{r}$$
$$< f'(x; v) + 2\varepsilon,$$

and similarly,

$$\frac{f(x + tu + tv) - f(x + tu)}{t} \geq \frac{f(x + tu) - f(x + tu - rv)}{r}$$

$$\geq \frac{f(x) - f(x - rv)}{r} - \frac{2Kt\|u\|}{r}$$

$$> f'(x; u) - 2\varepsilon. \qquad \square$$

Remark 6.2.3. It is well known and as easy to prove as the statement above that convex functions satisfy a stronger condition of regularity (sometimes called Clarke regularity), namely, that

$$\lim_{\substack{z \to x \\ t \to 0}} \frac{f(z + tv) - f(z)}{t} = f'(x; v)$$

whenever $f'(x; v)$ exists. We do not use here this stronger regularity concept since, while every point of Fréchet differentiability of f is a point of regularity of f in our sense, this is no longer true for the stronger regularity notion. Indeed, it suffices to consider an indefinite integral of the characteristic function of a set $E \subset \mathbb{R}$ such that both E and its complement have positive measure in every interval. For such f the above limit does not exist at any point x. Therefore the stronger form of regularity is not useful in proving existence of points of differentiability for Lipschitz maps.

We will now introduce particular sets, whose complements will turn out to be porous, and which cover the irregularity points of a given Lipschitz function. Here we just need the consequence that the irregularity points form a σ-porous set. The more detailed information will be utilized later, mainly in Chapter 11.

Definition 6.2.4. Let $G \subset X$ be an open set, $f: G \longrightarrow Y$, $u \in X$, $y \in Y$, and $\zeta, \rho > 0$. Denote by $R(f; u, y; \zeta, \rho)$ the set of all $x \in G$ such that at least one of the following statements holds:

(i) $B(x, \rho) \not\subset G$; or

(ii) there is $r \in \mathbb{R}$ such that $\|ru\| < \rho$ and $\|f(x + ru) - f(x) - ry\| > \zeta\|ru\|$; or

(iii) there is $\delta > 0$ such that $\|f(x + v + ru) - f(x + v) - ry\| \leq \zeta(\|v\| + \|ru\|)$ whenever $\|v\| + \|ru\| \leq \delta$.

The set $R(f; u, y; \zeta, \rho)$ represents, in the given approximation, the regular points of f. This is clearly seen from condition (iii), which expresses what may be called "regular differentiability" of f in the direction u, with "derivative" y and error ζ. Requirement (ii) takes care of the situation when f is nondifferentiable in the direction u. Recall that nondifferentiability implies, by our definition, regular behavior. The remaining condition (i) is to avoid consideration of points outside G in the main conditions (ii) and (iii).

If a point x does not belong to $R(f; u, y; \zeta, \rho)$ then the failure of all (i), (ii), and (iii) guarantees in a simple way that the complement of $R(f; u, y; \zeta, \rho)$ is porous at x.

Lemma 6.2.5. *If f is Lipschitz on G, each of the sets $G \setminus R(f; u, y; \zeta, \rho)$ is porous in X.*

Proof. We assume that $\mathrm{Lip}(f) > 0$ and show that the set $G \setminus R(f; u, y; \zeta, \rho)$ is porous with constant $c = \zeta/2\,\mathrm{Lip}(f)$. For this, let $x \in G \setminus R(f; u, y; \zeta, \rho)$ and $0 < \delta < \rho$. Since $x \notin R(f; u, y; \zeta, \rho)$, we have $B(x, \rho) \subset G$, and the failure of the requirement (iii) provides us with $r \in \mathbb{R}$ and $v \in X$ such that $\|v\| + |r|\,\|u\| < \rho$ and

$$\|f(x + v + ru) - f(x + v) - ry\| > \zeta(\|v\| + |r|\,\|u\|). \tag{6.1}$$

Moreover, the failure of (ii) at x implies that $v \neq 0$.

Consider now the ball $B(x + v, c\|v\|)$ and let $z \in B(x + v, c\|v\|)$ be arbitrary. We show that $z \in R(f; u, y; \zeta, \rho)$. This is obvious if $B(z, \rho) \not\subset G$, so suppose that $B(z, \rho) \subset G$ and use (6.1) to estimate

$$
\begin{aligned}
\|f(z + ru) &- f(z) - ry\| \\
&\geq \|f(x + v + ru) - f(x + v) - ry\| - 2\,\mathrm{Lip}(f)\,\|z - (x + v)\| \\
&> \zeta(\|v\| + |r|\,\|u\|) - 2c\,\mathrm{Lip}(f)\,\|v\| = \zeta |r|\,\|u\|.
\end{aligned}
$$

Hence the condition (ii) holds for z, implying that $z \in R(f; u, y; \zeta, \rho)$ and, consequently, that the ball $B(x + v, c\|v\|)$ is contained in $R(f; u, y; \zeta, \rho)$. □

An immediate consequence of Lemma 6.2.5 is the next proposition, showing that for a Lipschitz map the set of regular points is big.

Proposition 6.2.6. *Let f be a Lipschitz map from an open subset G of a separable Banach space X to a Banach space Y. Then the set of irregular points of f is σ-porous.*

Proof. Since $f(X)$ is separable, we may assume that Y is separable as well. For $p, q \in \mathbb{N}$, u from a countable dense subset of X, and y from a countable dense subset of Y, we consider the sets $R(f; u, y; \frac{1}{p}, \frac{1}{q})$.

Let $x \in G$ be an irregular point of f. Then, by definition there are a unit vector u_0 and $p \in \mathbb{N}$ such that $f'(x; u_0)$ exists and

$$\limsup_{t \to 0}\ \sup_{\|v\| \leq t}\ \left\| \frac{f(x + v + tu_0) - f(x + v)}{t} - f'(x; u_0) \right\| > \frac{5}{p}. \tag{6.2}$$

One can further find $q \in \mathbb{N}$ such that $B(x, 1/q) \subset G$ and

$$\|f(x + tu_0) - f(x) - tf'(x; u_0)\| \leq \frac{|t|}{3p} \quad \text{for all } |t| < \frac{1}{q}.$$

Let u and y be from the given countable dense sets such that

$$1 \leq \|u\| \leq 2, \quad \mathrm{Lip}(f)\,\|u - u_0\| < \frac{1}{3p}, \quad \|f'(x; u_0) - y\| < \frac{1}{3p}.$$

With the help of Lemma 6.2.5 the proof will be completed by showing that x does not belong to the set $R(f; u, y; \frac{1}{p}, \frac{1}{q})$. Since $B(x, 1/q) \subset G$, requirement (i) in Definition 6.2.4 fails to hold. Also,

$$\|f(x + tu) - f(x) - ty\| \leq \|f(x + tu_0) - f(x) - tf'(x; u_0)\| + \frac{|t|}{3p} + \frac{|t|}{3p}$$

$$\leq \frac{|t|}{p} \leq \frac{\|tu\|}{p},$$

which violates (ii). Finally, the choice of u and y yields that for all $\|v\| \leq t$

$$\frac{\|f(x + v + tu) - f(x + v) - ty\|}{\|v\| + \|tu\|}$$

$$\geq \frac{\|f(x + v + tu_0) - f(x + v) - tf'(x; u_0)\|}{\|v\| + \|tu\|}$$

$$- \frac{1}{3p} \frac{|t|}{\|v\| + \|tu\|} - \frac{1}{3p} \frac{|t|}{\|v\| + \|tu\|}$$

$$\geq \left\|\frac{f(x + v + tu_0) - f(x + v)}{t} - f'(x; u_0)\right\| \frac{|t|}{\|v\| + \|tu\|} - \frac{1}{3p} - \frac{1}{3p}$$

$$\geq \left\|\frac{f(x + v + tu_0) - f(x + v)}{t} - f'(x; u_0)\right\| \frac{1}{3} - \frac{2}{3p}.$$

The inequality (6.2) then gives

$$\limsup_{t \to 0} \sup_{\|v\| \leq t} \frac{\|f(x + v + tu) - f(x + v) - ty\|}{\|v\| + \|tu\|} > \frac{1}{p},$$

which contradicts (iii). □

6.3   A CRITERION OF FRÉCHET DIFFERENTIABILITY

In this section we prove the main criterion for Fréchet differentiability of Lipschitz functions in terms of Γ-null sets. The first lemma is a direct consequence of the definition of regularity. It enables us to use the regularity assumption in a more convenient way in subsequent arguments.

Lemma 6.3.1. *Suppose that f is Lipschitz on a neighborhood of x and that, at x, it is regular and differentiable in the direction of a finite dimensional subspace V of X. Then for every C and $\varepsilon > 0$ there is a $\delta > 0$ such that*

$$\|f(x + v + u) - f(x + v)\| \geq \|f(x + u) - f(x)\| - \varepsilon\|u\|$$

whenever $\|u\| \leq \delta$, $v \in V$ and $\|v\| \leq C\|u\|$.

Proof. Let $r > 0$ be such that f is Lipschitz on $B(x, r)$ and assume that $\mathrm{Lip}(f) > 0$. Let S be a finite subset of $Q := \{v \in V \mid \|v\| \leq C\}$ such that for every $w \in Q$ there is $v \in S$ satisfying $\|w - v\| < \varepsilon/(6 \, \mathrm{Lip}(f))$. By the definition of regularity, there is $0 < \delta < r/(1 + C)$ such that

$$\|f(x + tu + tv) - f(x + tu) - tf'(x; v)\| \leq \frac{\varepsilon t}{3} \tag{6.3}$$

whenever $0 \leq t \leq \delta$, $\|u\| \leq 1$ and $v \in S$.

Suppose that $0 < \|u_0\| \leq \delta$, $v_0 \in V$, and $\|v_0\| \leq C\|u_0\|$. By using (6.3) with $t = \|u_0\|$, $u = u_0/\|u_0\|$, and $v \in S$ with

$$\left\| v - \frac{v_0}{\|u_0\|} \right\| < \frac{\varepsilon}{6 \, \mathrm{Lip}(f)},$$

we get

$$\|f(x + u_0 + tv) - f(x + u_0) - tf'(x; v)\| \leq \frac{\varepsilon t}{3}. \tag{6.4}$$

Similarly, by using (6.3) with the same t and v but with $u = 0$, we get

$$\|f(x + tv) - f(x) - tf'(x; v)\| \leq \frac{\varepsilon t}{3}. \tag{6.5}$$

Hence the triangle inequality, (6.4), and (6.5) imply that

$$\left\| (f(x + tv + u_0) - f(x + tv)) - (f(x + u_0) - f(x)) \right\| \leq \frac{2\varepsilon t}{3}.$$

Recalling that $t = \|u_0\|$, we thus get

$$\begin{aligned}
\|f(x &+ v_0 + u_0) - f(x + v_0)\| \\
&\geq \|f(x + tv + u_0) - f(x + tv)\| - 2 \, \mathrm{Lip}(f)\|v_0 - tv\| \\
&\geq \|f(x + u_0) - f(x)\| - \frac{2\varepsilon t}{3} - \frac{\varepsilon t}{3} \\
&= \|f(x + u_0) - f(x)\| - \varepsilon\|u_0\|.
\end{aligned}$$

$\square$

Recall that $T = [0, 1]^{\mathbb{N}}$ as in Chapter 5. In this chapter it will also be convenient to denote for $k \in \mathbb{N}$, $t \in T$, and $r > 0$,

$$Q_k(t, r) = \big\{ s \in T \mid \max_{1 \leq i \leq k} |s_i - t_i| \leq r \big\}.$$

We will call these sets cubes, even though some of the noninteresting ones are mere cuboids. We will need to use the Vitali covering theorem for the cubes $Q_k(t, r)$ with fixed k. Since we are in the infinite dimensional space T, its applicability is not obvious. We will therefore use only a special form of this theorem that easily follows from its standard version.

Lemma 6.3.2. *Suppose that Q is a family of cubes in T, $A \subset T$, $k \in \mathbb{N}$, and for each $t \in A$ there are arbitrarily small $r > 0$ such that $Q_k(t, r) \in Q$. Then for each $\varepsilon > 0$ there is a finite disjoint subfamily $\{Q_k(t_i, r_i) \mid i = 1, \dots, j\}$ of Q such that $\mathscr{L}^{\mathbb{N}}(A \setminus \bigcup_{i=1}^{j} Q_k(t_i, r_i)) < \varepsilon$.*

Proof. Let π_k be the projection of T onto the first k coordinates. Then $\pi_k Q_k(t, r)$, where $Q_k(t, r) \in Q$ are cubes in $\mathbb{R}^k$ covering $\pi_k(A)$ in the sense of Vitali. Hence there is a finite disjoint subfamily $\{\pi_k Q_k(t_i, r_i) \mid i = 1, \dots, j\}$ covering the set $\pi_k(A)$ up to a set of $\mathscr{L}^k$-measure less than ε. It follows that the disjoint cubes $Q_k(t_i, r_i) = \pi_k^{-1}\pi_k Q_k(t_i, r_i)$ cover $\pi_k^{-1}\pi_k(A)$, and so also A, up to a set of $\mathscr{L}^{\mathbb{N}}$-measure less than ε. $\qquad\square$

Our next lemma examines the consequence of having a Lipschitz function which is Gâteaux but not Fréchet differentiable at a regular point: It shows that if a certain surface contains such a point x, then, after a suitable small perturbation of the surface, the derivative near x, in the mean, is not close to its value at x. Moreover, this property persists for all surfaces close enough to the perturbed surface. Notice that we will make an essential use of the fact that we work with infinite dimensional surfaces $\gamma \in \Gamma(X)$. These surfaces can be well approximated by k-dimensional surfaces in $\Gamma_k(X)$. The surfaces in $\Gamma_k(X)$ can in turn be approximated by surfaces in $\Gamma_{k+1}(X)$, and we are quite free to do appropriate constructions on the $(k+1)$th coordinate in order to get an approximation with desired properties.

Lemma 6.3.3. *Let $f \colon G \longrightarrow Y$ be a Lipschitz function with G an open set in a separable Banach space X. Let E be a Borel subset of G consisting of points where f is Gâteaux differentiable and regular. Let $\eta > 0$, $\gamma_0 \in \Gamma_k(X)$, $t \in T$ such that $x = \gamma_0(t) \in E$, and assume that*

$$\alpha := \limsup_{u \to 0} \frac{1}{\|u\|} \|f(x + u) - f(x) - f'(x; u)\| > 0.$$

Then there are $0 < r < \eta$, $\delta > 0$, and $\widetilde{\gamma} \in \Gamma_{k+1}(X)$ such that

- *$\|\widetilde{\gamma} - \gamma_0\|_{\leq k} < \eta$;*

- *$\widetilde{\gamma}(s) = \gamma_0(s)$ for $s \in T \setminus Q_k(t, r)$;*

- *whenever $\gamma \in \Gamma(X)$ has the property that*

$$\|\gamma(s) - \widetilde{\gamma}(s)\| + \|D_{k+1}(\gamma(s) - \widetilde{\gamma}(s))\| < \delta, \quad s \in Q_k(t, r),$$

then either

$$\mathscr{L}^{\mathbb{N}}\big(Q_k(t, r) \setminus \gamma^{-1}(E)\big) \geq \frac{\alpha}{8(1 + \mathrm{Lip}(f))} \, \mathscr{L}^{\mathbb{N}} Q_k(t, r)$$

or

$$\int_{Q_k(t,r) \cap \gamma^{-1}(E)} \|f'(\gamma(s)) - f'(x)\| \, d\mathscr{L}^{\mathbb{N}}(s) \geq \frac{\alpha}{2} \, \mathscr{L}^{\mathbb{N}} Q_k(t, r).$$

Proof. We shall assume that $\alpha > \eta > 0$ and will define the values of r and δ later in the proof.

Let $0 < \zeta < \min\{1, \alpha/5\}$ be such that

$$(\alpha - 5\zeta)\,\mathscr{L}^{\mathbb{N}}Q_k(t, r(1-\zeta)) \geq \frac{3\alpha}{4}\,\mathscr{L}^{\mathbb{N}}Q_k(t, r) \tag{6.6}$$

whenever $t \in T$ and $r < 1$. This can evidently be accomplished even though for some t, $Q_k(t, r)$ is not an entire cube.

Choose a continuously differentiable function $\psi\colon \ell_\infty \longrightarrow [0, 1]$ depending on the first k coordinates only such that

$$\psi(s) = \begin{cases} 1 & \text{if } \|\pi_k s\| \leq 1 - \zeta, \\ 0 & \text{if } \|\pi_k s\| \geq 1, \end{cases}$$

where π_k denotes the projection of ℓ_∞ on its first k coordinates. We choose the constant $K \geq \alpha$ such that both γ_0 and ψ are K-Lipschitz and the function

$$g(z) := f(z) - f(x) - f'(x; z - x)$$

is also K-Lipschitz on the ball $B(x, 2\delta_1(1 + K^2/\eta)) \subset G$ for some $0 < \delta_1 < \eta$.

We let $C = 4K^2/\eta$ and $V = \mathrm{Tan}(\gamma_0, t)$. Clearly, V is of dimension at most k since $\gamma_0 \in \Gamma_k(X)$ (see end of Section 5.1 for definition). We use Lemma 6.3.1 to find a $0 < \delta_2 < \delta_1 < \eta$ such that

$$\|g(x + v + u) - g(x + v)\| \geq \|g(x + u)\| - \zeta\|u\| \tag{6.7}$$

whenever $\|u\| \leq \delta_2$, $v \in V$, and $\|v\| \leq C\|u\|$. Also, let $\delta_3 > 0$ be such that

$$\|\gamma_0(s) - x - \gamma_0'(t; s - t)\| \leq \frac{\zeta}{C}\,\|\pi_k(s - t)\| \tag{6.8}$$

whenever $s \in T$ and $\|\pi_k(s - t)\| \leq \delta_3$. By the definition of α we can find an element $u \in X$ such that

$$\|g(x + u)\| \geq (\alpha - \zeta)\|u\| \quad \text{and} \quad 0 < \|u\| < \min\left\{\delta_2, \frac{2K\delta_3}{C}, \frac{\eta^2}{2K}\right\}.$$

Let $s \in T$ be such that $\|\pi_k(s - t)\| \leq C\|u\|/2K < \delta_3$. To estimate the norm of difference $\|g(\gamma_0(s) + u) - g(\gamma_0(s))\|$ we use both (6.7) with $v = \gamma_0'(t; s - t)$, which is applicable since $\|\gamma_0'(t; s - t)\| \leq k\|\pi_k(s - t)\| < C\|u\|$, and (6.8). So we infer that

$$\begin{aligned}
\|g(\gamma_0(s) + u) &- g(\gamma_0(s))\| \\
&\geq \|g(x + v + u) - g(x + v)\| - 2K\|\gamma_0(s) - (x + v)\| \\
&\geq \|g(x + v + u) - g(x + v)\| - \frac{2K\zeta}{C}\,\|\pi_k(s - t)\| \\
&\geq \|g(x + u)\| - \zeta\|u\| - \zeta\|u\| \\
&\geq (\alpha - 3\zeta)\|u\|. \tag{6.9}
\end{aligned}$$

We are now ready to define $r := 2K\|u\|/\eta$ and put

$$\widetilde{\gamma}(s) := \gamma_0(s) + s_{k+1}\psi((s-t)/r)\,u.$$

By the choice of $\|u\|$ we have $0 < r < \eta$. Also, $\|\widetilde{\gamma} - \gamma_0\|_\infty \le \|u\| < \eta$ and $\|D_j(\widetilde{\gamma} - \gamma_0)\|_\infty \le K\|u\|/r < \eta$ for $1 \le j \le k$, and thus $\|\widetilde{\gamma} - \gamma_0\|_{\le k} < \eta$ as required.

We show that the statement of the lemma holds with $\delta = \zeta\|u\|/2K$. To this aim suppose that $\gamma \in \Gamma(X)$ satisfies

$$\|\gamma(s) - \widetilde{\gamma}(s)\| + \|D_{k+1}(\gamma(s) - \widetilde{\gamma}(s))\| < \delta \text{ for } s \in Q_k(t,r),$$

and that

$$\mathscr{L}^{\mathbb{N}}\big(Q_k(t,r) \setminus \gamma^{-1}(E)\big) < \frac{\alpha}{8(1 + \mathrm{Lip}(f))}\,\mathscr{L}^{\mathbb{N}}Q_k(t,r).$$

Consider any $s \in Q_k(t, r(1 - \zeta))$, and put

$$\overline{s} := s + (1 - s_{k+1})e_{k+1} \quad \text{and} \quad \underline{s} := s - s_{k+1}e_{k+1},$$

where e_{k+1} denotes the $(k+1)$th unit vector in ℓ_∞. Notice also that the $(k+1)$-coordinate of $\overline{s}$ (resp. $\underline{s}$) is 1 (resp. 0). Further, we denote $I_s = [-s_{k+1}, 1 - s_{k+1}]$ and

$$J_s = \big\{\sigma \in I_s \mid \gamma(s + \sigma e_{k+1}) \in E\big\}.$$

Let $u^* \in Y^*$ be a linear functional with $\|u^*\| = 1$ which realizes the norm of $g(\gamma(\overline{s})) - g(\gamma(\underline{s}))$, that is,

$$u^*\big(g(\gamma(\overline{s})) - g(\gamma(\underline{s}))\big) = \|g(\gamma(\overline{s})) - g(\gamma(\underline{s}))\|.$$

In order to continue our argument we have to verify that all four values $\widetilde{\gamma}(\overline{s})$, $\widetilde{\gamma}(\underline{s})$, $\gamma(\overline{s})$, and $\gamma(\underline{s})$ belong to the ball $B(x, 2\delta_1(1 + K^2/\eta))$, on which the function g is K-Lipschitz.

$$\begin{aligned}
\|\widetilde{\gamma}(\overline{s}) - x\| &= \|\widetilde{\gamma}(\overline{s}) - \gamma_0(t)\| \\
&\le \|\widetilde{\gamma} - \gamma_0\|_\infty + \|\gamma_0(\overline{s}) - \gamma_0(t)\| \\
&\le \|u\| + K\|\pi_k(\overline{s} - t)\| \le \|u\| + Kr \\
&= \|u\|\Big(1 + \frac{2K^2}{\eta}\Big) < \delta_1\Big(1 + \frac{2K^2}{\eta}\Big).
\end{aligned}$$

Since $\|\gamma(\overline{s}) - \widetilde{\gamma}(\overline{s})\| < \delta < \delta_1$ we conclude that also

$$\|\widetilde{\gamma}(\overline{s}) - x\| < 2\delta_1\Big(1 + \frac{K^2}{\eta}\Big).$$

Similarly for $\underline{s}$.

Since we know that $\|\pi_k(s-t)\| < r = 2K\|u\|/\eta = C\|u\|/2K$, we may use (6.9) together with the inequalities $\|\gamma(\bar{s}) - \tilde{\gamma}(\bar{s})\| < \delta$, $\|\gamma(\underline{s}) - \tilde{\gamma}(\underline{s})\| < \delta$ to get

$$
\begin{aligned}
\int_{I_s} \frac{\partial}{\partial\sigma} u^* g(\gamma(s + \sigma e_{k+1}))\, d\sigma &= \|g(\gamma(\bar{s})) - g(\gamma(\underline{s}))\| \\
&\geq \|g(\tilde{\gamma}(\bar{s})) - g(\tilde{\gamma}(\underline{s}))\| - 2K\delta \\
&= \|g(\gamma_0(s) + u) - g(\gamma_0(s))\| - \zeta\|u\| \\
&\geq (\alpha - 4\zeta)\|u\|.
\end{aligned}
$$

Since $s \in Q_k(t, r(1-\zeta))$ we have $D_{k+1}\tilde{\gamma}(s + \sigma e_{k+1}) = u$ for $\sigma \in I_s$, and so $\|D_{k+1}\gamma(s + \sigma e_{k+1}) - u\| < \delta$. Using also that $\mathrm{Lip}(g) \leq 2\,\mathrm{Lip}(f)$ and that $K \geq \alpha$, we infer that

$$
\begin{aligned}
\int_{J_s} &\|g'(\gamma(s + \sigma e_{k+1}))\|\, d\sigma \\
&\geq \frac{1}{\|D_{k+1}\gamma(s + \sigma e_{k+1})\|} \int_{J_s} |u^* g'(\gamma(s + \sigma e_{k+1}); D_{k+1}\gamma(s + \sigma e_{k+1})|\, d\sigma \\
&= \frac{1}{\|D_{k+1}\gamma(s + \sigma e_{k+1})\|} \int_{J_s} \left|\frac{\partial}{\partial\sigma} u^* g(\gamma(s + \sigma e_{k+1}))\right|\, d\sigma \\
&\geq \frac{1}{\|u\| + \delta} \int_{I_s} \left|\frac{\partial}{\partial\sigma} u^* g(\gamma(s + \sigma e_{k+1}))\right|\, d\sigma - \mathscr{L}^1(I_s \setminus J_s)\,\mathrm{Lip}(g) \\
&\geq \frac{\alpha - 4\zeta}{1 + \zeta/2K} - 2\mathscr{L}^1(I_s \setminus J_s)\,\mathrm{Lip}(f) \\
&\geq (\alpha - 4\zeta)\left(1 - \frac{\zeta}{2K}\right) - 2\mathscr{L}^1(I_s \setminus J_s)\,\mathrm{Lip}(f) \\
&\geq \alpha - 5\zeta - 2\mathscr{L}^1(I_s \setminus J_s)\,\mathrm{Lip}(f).
\end{aligned}
$$

Integrating over $s \in Q_k(t, r(1-\zeta))$ and recalling the choice of ζ in (6.6), we finally get the desired result.

$$
\begin{aligned}
\int_{Q_k(t,r)\cap\gamma^{-1}(E)} &\|f'(\gamma(s)) - f'(x)\|\, d\mathscr{L}^{\mathbb{N}}(s) \\
&= \int_{Q_k(t,r)\cap\gamma^{-1}(E)} \|g'(\gamma(s))\|\, d\mathscr{L}^{\mathbb{N}}(s) \\
&\geq \int_{Q_k(t,r(1-\zeta))\cap\gamma^{-1}(E)} \|g'(\gamma(s))\|\, d\mathscr{L}^{\mathbb{N}}(s) \\
&= \int_{Q_k(t,r(1-\zeta))} \int_{J_s} \|g'(s + \sigma e_{k+1})\|\, d\sigma\, d\mathscr{L}^{\mathbb{N}}(s) \\
&\geq (\alpha - 5\zeta)\,\mathscr{L}^{\mathbb{N}} Q_k(t, r(1-\zeta)) - 2\,\mathrm{Lip}(f)\,\mathscr{L}^{\mathbb{N}}\big(Q_k(t,r) \setminus \gamma^{-1}(E)\big) \\
&\geq \frac{3\alpha}{4}\,\mathscr{L}^{\mathbb{N}} Q_k(t,r) - \frac{2\,\mathrm{Lip}(f)}{8(1 + \mathrm{Lip}(f))}\,\alpha\,\mathscr{L}^{\mathbb{N}} Q_k(t,r) \\
&\geq \frac{\alpha}{2}\,\mathscr{L}^{\mathbb{N}} Q_k(t,r). \qquad\qquad \square
\end{aligned}
$$

We now extend the notation introduced in Chapter 5 from surfaces to mappings in $L_1(T, X)$. For $k \in \mathbb{N}$, $s \in T$, and $g \in L_1(T, X)$ we put

$$g^{k,s}(t) = g(t_1, \ldots, t_k, s_1, s_2, \ldots).$$

Notice that, for every $k \geq 1$ and $s \in T$,

$$\int_{[0,1]^k} g^{k,s}(t) \, d\mathscr{L}^k(t) = \int_T g^{k,s}(t) \, d\mathscr{L}^{\mathbb{N}}(t).$$

Then, using that $[0,1]^k \times T \approx T$, Fubini's theorem says that for every k we have

$$\int_T g(t) \, d\mathscr{L}^{\mathbb{N}}(t) = \int_T \int_T g^{k,s}(t) \, d\mathscr{L}^{\mathbb{N}}(t) \, d\mathscr{L}^{\mathbb{N}}(s)$$

$$= \int_T \int_T g^{k,s}(t) \, d\mathscr{L}^{\mathbb{N}}(s) \, d\mathscr{L}^{\mathbb{N}}(t).$$

As in the case of continuous functions, the functions $g^{k,s}$ approximate g for large enough k. The precise formulation of this statement is given in

Lemma 6.3.4. *Let $g \in L_1(T, X)$. Then for any $\eta > 0$ there is $k_0 \in \mathbb{N}$ such that*

$$\mathscr{L}^{\mathbb{N}}\{s \in T \mid \|g^{k,s} - g\|_{L_1} > \eta\} < \eta$$

for all $k \geq k_0$.

Proof. Let $h \colon T \longrightarrow X$ be a continuous function depending on the first k_0 variables only such that $\|g - h\|_{L_1} < \eta^2/(1 + \eta)$, and let $k \geq k_0$. By Fubini's theorem,

$$\int_T \|g^{k,s} - h\|_{L_1} \, d\mathscr{L}^{\mathbb{N}}(s) = \|g - h\|_{L_1} < \frac{\eta^2}{1 + \eta},$$

so by Chebyshev's inequality

$$\mathscr{L}^{\mathbb{N}}\{s \in T \mid \|g^{k,s} - g\|_{L_1} > \eta\}$$
$$\leq \mathscr{L}^{\mathbb{N}}\{s \in T \mid \|g^{k,s} - h\|_{L_1} > \eta - \|g - h\|_{L^1}\}$$
$$\leq \mathscr{L}^{\mathbb{N}}\{s \in T \mid \|g^{k,s} - h\|_{L_1} > \eta/(1 + \eta)\} < \eta. \qquad \square$$

The next lemma is a version of Lebesgue's differentiation theorem for functions defined on the infinite torus T. Notice, however, the order of the quantifiers. Since A depends on k, the statement does not say that the averages of the function converge to the function value in the perhaps expected sense. Recall that the symbol $\fint_A f \, d\mu$ defines the average integral

$$\fint_A f \, d\mu = \frac{1}{\mu A} \int_A f \, d\mu, \quad 0 < \mu A < \infty.$$

Lemma 6.3.5. *Let X be a Banach space and suppose that $g \in L_1(T, X)$. Then for every $\varepsilon > 0$ there is a $k_0 \in \mathbb{N}$ such that for all $k \geq k_0$ there are $\delta > 0$ and $A \subset T$ with $\mathscr{L}^{\mathbb{N}} A < \varepsilon$ such that*

$$\fint_{Q_k(t,r)} \|g(s) - g(t)\| \, d\mathscr{L}^{\mathbb{N}}(s) < \varepsilon$$

for every $t \in T \setminus A$ and $0 < r < \delta$.

Proof. Let $h: T \longrightarrow X$ be a continuous function depending on the first k_0 coordinates such that $\|g - h\|_{L_1} < \varepsilon^2/9$. If we put

$$N = \{t \in T \mid \|g(t) - h(t)\| \geq \tfrac{1}{3}\varepsilon\},$$

we get by Chebyshev's inequality that $\mathscr{L}^{\mathbb{N}} N < \tfrac{1}{3}\varepsilon$.

Fix an arbitrary $k \geq k_0$ and put

$$\zeta(s) = \int_T \|g^{k,\sigma}(s) - h(s)\| \, d\mathscr{L}^{\mathbb{N}}(\sigma) \quad \text{and} \quad S = \{s \in T \mid \zeta(s) \geq \tfrac{1}{3}\varepsilon\}.$$

Then as above, we get from Fubini's theorem that $\mathscr{L}^{\mathbb{N}} S < \tfrac{1}{3}\varepsilon$.

Let

$$A_n = \left\{ t \notin S \ \middle| \ \fint_{Q_k(t,r)} \zeta(s) \, d\mathscr{L}^{\mathbb{N}}(s) \geq \frac{\varepsilon}{3} \ \text{for some } 0 < r < \frac{1}{n} \right\}.$$

Since ζ depends only on the first k coordinates, by Lebesgue's differentiation theorem,

$$\lim_{r \to 0} \fint_{Q_k(t,r)} \zeta(s) \, d\mathscr{L}^{\mathbb{N}}(s) = \zeta(t)$$

for $\mathscr{L}^{\mathbb{N}}$-almost every $t \in T$. Hence A_n is a decreasing sequence of measurable sets whose intersection has measure zero. Let n_0 be such that $\mathscr{L}^{\mathbb{N}} A_{n_0} < \tfrac{1}{3}\varepsilon$ and put

$$A = N \cup S \cup A_{n_0}.$$

Then $\mathscr{L}^{\mathbb{N}} A < \varepsilon$. For every $t \in T$ and $r > 0$ we have, by Fubini's theorem and the fact that h depends on the first $k_0 \leq k$ coordinates, that

$$\int_{Q_k(t,r)} \|g(s) - h(s)\| \, d\mathscr{L}^{\mathbb{N}}(s)$$
$$= \int_{Q_k(t,r)} \int_T \|g^{k,\sigma}(s) - h(s)\| \, d\mathscr{L}^{\mathbb{N}}(\sigma) \, d\mathscr{L}^{\mathbb{N}}(s)$$
$$= \int_{Q_k(t,r)} \zeta(s) \, d\mathscr{L}^{\mathbb{N}}(s).$$

Hence, choosing $0 < \delta < 1/n_0$ such that $\|h(s) - h(t)\| < \frac{1}{3}\varepsilon$ whenever $s \in Q_k(t, \delta)$, we obtain for every $t \in T \setminus A$ and $0 < r < \delta$ that

$$\fint_{Q_k(t,r)} \|g(s) - g(t)\|\, d\mathscr{L}^{\mathbb{N}}(s)$$

$$\leq \fint_{Q_k(t,r)} \|g(s) - h(s)\|\, d\mathscr{L}^{\mathbb{N}}(s) + \fint_{Q_k(t,r)} \|h(s) - h(t)\|\, d\mathscr{L}^{\mathbb{N}}(s)$$

$$+ \|g(t) - h(t)\|$$

$$< \frac{\varepsilon}{3} + \frac{\varepsilon}{3} + \frac{\varepsilon}{3} = \varepsilon.$$

$\square$

Our next lemma makes clear why the separability assumption on $\mathcal{L}$ in the statement of the main theorem (Theorem 6.3.9 below) is needed. Before stating it, we list some assumptions and definitions that enter into its statement. We assume that X and Y are separable Banach spaces and the function f is a Lipschitz function from an open set $G \subset X$ into Y. We further assume that $\mathcal{L}$ is a norm separable subspace of the space $L(X, Y)$ of bounded linear operators from X to Y and

$$E := \big\{ x \in G \mid f \text{ is regular at } x,\ f'(x) \text{ exists},\ f'(x) \in \mathcal{L} \big\}.$$

We also put

$$\psi_E(x) = \begin{cases} f'(x) & \text{for } x \in E, \\ 0 & \text{if } x \notin E. \end{cases}$$

Finally, let $\Phi(\gamma) = \mathbf{1}_E \circ \gamma$ and $\Psi(\gamma) = \psi_E \circ \gamma$.

Lemma 6.3.6. *With the above notation, the set E is Borel and the maps*

$$\psi_E \colon X \longrightarrow \mathcal{L}, \quad \Phi \colon \Gamma(X) \longrightarrow L_1(T) \quad \text{and} \quad \Psi \colon \Gamma(X) \longrightarrow L_1(T, \mathcal{L})$$

are all Borel measurable. In particular there is a residual subset H of $\Gamma(X)$ such that the restrictions of Φ and Ψ to H are continuous.

Proof. For $L \in \mathcal{L}$, $u, v \in X$, and $\sigma, \tau > 0$ denote by $M(L, u, v, \sigma, \tau)$ the set of $x \in G$ such that

- $\operatorname{dist}(x, X \setminus G) \geq \tau(\|u\| + \|v\|)$;

- $\|f(x + su + sv) - f(x + sv) - sL(u)\| \leq |s|\sigma\|u\|$ whenever $|s| < \tau$.

Clearly each $M(L, u, v, \sigma, \tau)$ is a closed subset of X.

Let $\mathcal{S}$ be a countable dense subset of $\mathcal{L}$, D a dense countable subset of X, and $\mathbb{Q}^+$ the set of positive rational numbers. Then

$$E = \bigcap_{\sigma \in \mathbb{Q}^+} \bigcup_{L \in \mathcal{S}} \bigcap_{u \in D} \bigcup_{\tau \in \mathbb{Q}^+} \bigcap_{\substack{v \in D \\ \|v\| \leq 1}} M(L, u, v, \sigma, \tau),$$

and hence E is a Borel set.

For every $L \in \mathcal{L}$ and $\varrho > 0$ we have

$$\{x \in E \mid \|\psi_E(x) - L\| \leq \varrho\} = E \cap \bigcap_{u \in D} \bigcap_{\substack{\sigma \in \mathbb{Q}^+ \\ \sigma > \varrho}} \bigcup_{\tau \in \mathbb{Q}^+} M(L, u, 0, \sigma, \tau).$$

Since E is Borel, $\mathcal{L}$ is separable, and $\psi_E(x) = 0$ outside E, it follows that also ψ_E is Borel measurable.

Since ψ_E is bounded and Borel measurable, the Borel measurability of Ψ will be established once we show that for every bounded, Borel measurable $h \colon X \longrightarrow \mathcal{L}$ the mapping $\Psi_h \colon \Gamma(X) \longrightarrow L_1(T, \mathcal{L})$ defined by

$$\Psi_h(\gamma) = h \circ \gamma$$

is Borel measurable. The proof is by transfinite induction with respect to the Baire class of the mapping h:

- If h is continuous, then so is Ψ_h.

- If $(h_n)_{n=1}^\infty$ are uniformly bounded in norm and $h_n \to h$ pointwise, and if all Ψ_{h_n} are Borel measurable, then Ψ_{h_n} converge pointwise to Ψ_h, so Ψ_h is Borel measurable.

The same argument shows the Borel measurability of Φ.

The last statement of the lemma follows from the general fact that a Borel measurable (or even Baire measurable) mapping between complete separable metrizable spaces has a continuous restriction to a suitable residual subset (see, e.g., [24, Theorem 8.38]). $\qquad\square$

Remark 6.3.7. Without assuming the separability of $\mathcal{L}$ not only does the proof not work but the statement is actually false. Indeed, if the set

$$\{f'(x) \mid x \in E\} \subset L(X, Y)$$

were not separable, then the image of $\Gamma(X)$ under Ψ would be nonseparable as well. It would follow that Ψ is not Borel since the Borel image of a complete separable metric space is again separable.

The final lemma before the proof of the main theorem combines much of what have we done in this section so far. According to Lemma 6.3.3, if f is Gâteaux but not Fréchet differentiable at a regular point on a surface, then a suitable small local deformation of the surface causes a nonsmall perturbation of the function $t \to f'(\gamma(t))$. Here we perform a finite number of such local perturbations (on suitable disjoint neighborhoods chosen via the Vitali covering theorem) and get the same effect globally.

Lemma 6.3.8. *Suppose that G is an open subset of a separable Banach space X, $f \colon G \longrightarrow Y$ is a Lipschitz function, and $\mathcal{L}$ is a norm separable subspace of the space $L(X, Y)$. Let*

$$E := \{x \in G \mid f \text{ is regular at } x, \ f'(x) \text{ exists}, \ f'(x) \in \mathcal{L}\}.$$

Suppose that $\varepsilon, \eta > 0$, $k_0 \in \mathbb{N}$, and $\gamma_0 \in \Gamma_{k_0}(X)$ are such that the set

$$S = \left\{ t \in \gamma_0^{-1}(E) \; \middle| \; \limsup_{u \to 0} \frac{1}{\|u\|} \left\| f(\gamma_0(t) + u) - f(\gamma_0(t)) - f'(\gamma_0(t); u) \right\| > \varepsilon \right\}$$

has $\mathcal{L}^{k_0}$-measure greater than ε. Then one can find $k \geq k_0$, $\delta > 0$, and $\widetilde{\gamma} \in \Gamma_{k+1}(X)$ such that $\|\widetilde{\gamma} - \gamma_0\|_{\leq k} < \eta$ and

$$\int_T \left(\left| \mathbf{1}_E(\gamma(t)) - \mathbf{1}_E(\gamma_0(t)) \right| + \left\| \psi_E(\gamma(t)) - \psi_E(\gamma_0(t)) \right\| \right) d\mathcal{L}^{\mathbb{N}}(t)$$

$$> \frac{\varepsilon^2}{32(1 + \mathrm{Lip}(f))}$$

whenever $\gamma \in \Gamma(X)$ and $\|\gamma - \widetilde{\gamma}\|_{\leq k+1} < \delta$.

Proof. By Lemma 6.3.5 find $k \geq k_0$, $\delta_0 > 0$, and a set $A \subset T$ with $\mathcal{L}^{\mathbb{N}} A < \varepsilon/4$ such that for every $t \in T \setminus A$ and $0 < r < \delta_0$,

$$\fint_{Q_k(t,r)} \left| \mathbf{1}_E(\gamma_0(s)) - \mathbf{1}_E(\gamma_0(t)) \right| d\mathcal{L}^{\mathbb{N}}(s) < \frac{\varepsilon}{16(1 + \mathrm{Lip}(f))} \qquad (6.10)$$

and

$$\fint_{Q_k(t,r)} \left\| \psi_E(\gamma_0(s)) - \psi_E(\gamma_0(t)) \right\| d\mathcal{L}^{\mathbb{N}}(s) < \frac{\varepsilon}{4}. \qquad (6.11)$$

For every $t \in S \setminus A$ and $n \in \mathbb{N}$ we use Lemma 6.3.3 with

$$\eta_n = \frac{1}{n+1} \min\{\eta, \delta_0\}$$

to find $0 < r_{t,n} < \eta_n$, $\delta_{t,n} > 0$, and $\widetilde{\gamma}_{t,n} \in \Gamma_{k+1}(X)$ such that $\|\widetilde{\gamma}_{t,n} - \gamma_0\|_{\leq k} < \eta_n$, $\widetilde{\gamma}_{t,n}(s) = \gamma_0(s)$ for $s \in T \setminus Q_k(t, r_{t,n})$ and either

$$\mathcal{L}^{\mathbb{N}}\left(Q_k(t, r_{t,n}) \setminus \gamma^{-1}(E) \right) \geq \frac{\varepsilon}{8(1 + \mathrm{Lip}(f))} \, \mathcal{L}^{\mathbb{N}} Q_k(t, r_{t,n}) \qquad (6.12)$$

or

$$\int_{Q_k(t,r_{t,n}) \cap \gamma^{-1}(E)} \left\| f'(\gamma(s)) - f'(\gamma_0(t)) \right\| d\mathcal{L}^{\mathbb{N}}(s) \geq \frac{\varepsilon}{2} \mathcal{L}^{\mathbb{N}} Q_k(t, r_{t,n}), \qquad (6.13)$$

whenever $\gamma \in \Gamma(X)$ and $\|\gamma(s) - \widetilde{\gamma}_{t,n}(s)\| + \|D_{k+1}(\gamma(s) - \widetilde{\gamma}_{t,n}(s))\| < \delta_{t,n}$ for all $s \in Q_k(t, r_{t,n})$.

The reason to introduce the extra parameter n here is to enable us to use the Vitali covering theorem later on.

If (6.12) holds, it follows using (6.10) and the fact that $1_E(\gamma_0(t)) = 1$ that

$$\int_{Q_k(t,r_{t,n})} \left|1_E(\gamma(s)) - 1_E(\gamma_0(s))\right| d\mathscr{L}^{\mathbb{N}}(s)$$

$$\geq \int_{Q_k(t,r_{t,n})} \left|1_E(\gamma(s)) - 1_E(\gamma_0(t))\right| d\mathscr{L}^{\mathbb{N}}(s)$$

$$- \int_{Q_k(t,r_{t,n})} \left|1_E(\gamma_0(s)) - 1_E(\gamma_0(t))\right| d\mathscr{L}^{\mathbb{N}}(s)$$

$$> \frac{\varepsilon}{8(1 + \mathrm{Lip}(f))} \mathscr{L}^{\mathbb{N}} Q_k(t, r_{t,n}) - \frac{\varepsilon}{16(1 + \mathrm{Lip}(f))} \mathscr{L}^{\mathbb{N}} Q_k(t, r_{t,n})$$

$$= \frac{\varepsilon}{16(1 + \mathrm{Lip}(f))} \mathscr{L}^{\mathbb{N}} Q_k(t, r_{t,n}).$$

On the other hand, if (6.13) holds, we get using (6.11) that

$$\int_{Q_k(t,r_{t,n})} \|\psi_E(\gamma(s)) - \psi_E(\gamma_0(s))\| d\mathscr{L}^{\mathbb{N}}(s)$$

$$\geq \int_{Q_k(t,r_{t,n})} \|\psi_E(\gamma(s)) - \psi_E(\gamma_0(t))\| d\mathscr{L}^{\mathbb{N}}(s)$$

$$- \int_{Q_k(t,r_{t,n})} \|\psi_E(\gamma_0(s)) - \psi_E(\gamma_0(t))\| d\mathscr{L}^{\mathbb{N}}(s)$$

$$> \frac{\varepsilon}{2} \mathscr{L}^{\mathbb{N}} Q_k(t, r_{t,n}) - \frac{\varepsilon}{4} \mathscr{L}^{\mathbb{N}} Q_k(t, r_{t,n})$$

$$= \frac{\varepsilon}{4} \mathscr{L}^{\mathbb{N}} Q_k(t, r_{t,n}).$$

Hence, in any case,

$$\int_{Q_k(t,r_{t,n})} \left(\left|1_E(\gamma(s)) - 1_E(\gamma_0(s))\right| + \|\psi_E(\gamma(s)) - \psi_E(\gamma_0(s))\|\right) d\mathscr{L}^{\mathbb{N}}(s)$$

$$> \frac{\varepsilon}{16(1 + \mathrm{Lip}(f))} \mathscr{L}^{\mathbb{N}} Q_k(t, r_{t,n}) \qquad (6.14)$$

for all $t \in S \setminus A$, $n \geq 1$ and for all surfaces $\gamma \in \Gamma(X)$ satisfying that

$$\|\gamma(s) - \tilde{\gamma}_{t,n}(s)\| + \|D_{k+1}(\gamma(s) - \tilde{\gamma}_{t,n}(s))\| < \delta_{t,n} \text{ for all } s \in Q_k(t, r_{t,n}).$$

By the Vitali covering theorem (in the form of Lemma 6.3.2) applied to the system

$$\{Q_k(t_i, r_{t,n}) \mid t \in S \setminus A, \ n \geq 1\}$$

there is a finite disjoint family $\{Q_k(t_i, r_{t_i,n_i}) \mid i = 1, \ldots, j\}$ covering the set $S \setminus A$ up to a set of $\mathscr{L}^{\mathbb{N}}$-measure less than $\varepsilon/4$. Consequently,

$$\sum_{i=1}^{j} \mathscr{L}^{\mathbb{N}} Q_k(t_i, r_{t_i,n_i}) \geq \mathscr{L}^{\mathbb{N}}(S \setminus A) - \frac{\varepsilon}{4} \geq \frac{\varepsilon}{2}.$$

Define now

$$\widetilde{\gamma}(s) = \begin{cases} \widetilde{\gamma}_{t_i,n_i}(s) & \text{if } s \in Q_k(t_i, r_{t_i,n_i}), \, i = 1, \dots, j, \\ \gamma_0(s) & \text{otherwise.} \end{cases}$$

Clearly $\|\widetilde{\gamma} - \gamma_0\|_{\leq k} < \eta$. Let $\delta = \min\{\delta_{t_i,n_i} \mid i = 1, \dots, j\}$; then for every $\gamma \in \Gamma(X)$ with $\|\gamma - \widetilde{\gamma}\|_{\leq k+1} < \delta$ we get by adding (6.14) over all the j cubes $Q_k(t_i, r_{t_i,n_i})$ that

$$\int_T \left(\left| \mathbf{1}_E(\gamma(s)) - \mathbf{1}_E(\gamma_0(s)) \right| + \|\psi_E(\gamma(s)) - \psi_E(\gamma_0(s))\| \right) d\mathscr{L}^{\mathbb{N}}(s)$$

$$> \frac{\varepsilon}{16(1 + \mathrm{Lip}(f))} \sum_{i=1}^{j} \mathscr{L}^{\mathbb{N}} Q_k(t_i, r_{t_i,u_i})$$

$$\geq \frac{\varepsilon^2}{32(1 + \mathrm{Lip}(f))}. \qquad \square$$

We are now ready to prove the main theorem.

Theorem 6.3.9. *Suppose that G is an open subset of a separable Banach space X, $\mathcal{L}$ a norm separable subspace of $L(X,Y)$, and $f: G \longrightarrow Y$ a Lipschitz function. Then f is Fréchet differentiable at Γ-almost every point $x \in X$ at which it is regular, Gâteaux differentiable, and $f'(x) \in \mathcal{L}$.*

Proof. We may assume that Y is separable and continue to use the notation of the previous lemmas. By Lemma 6.3.6 there is a residual subset H of $\Gamma(X)$ such that the restrictions of Φ and Ψ to H are continuous. Recall that $\Phi(\gamma) = \mathbf{1}_E \circ \gamma$ and $\Psi(\gamma) = \psi_E \circ \gamma$, where

$$E = \{x \in G \mid f \text{ is regular in } x, \, f'(x) \text{ exists}, \, f'(x) \in \mathcal{L}\}.$$

Fix $\varepsilon > 0$ and put

$$E_0 = \left\{ x \in E \, \middle| \, \limsup_{u \to 0} \frac{1}{\|u\|} \|f(x+u) - f(x) - f'(x; u)\| > \varepsilon \right\}.$$

To prove the theorem it suffices to show that the set

$$M = \left\{ \gamma \in H \mid \mathscr{L}^{\mathbb{N}} \gamma^{-1}(E_0) > \varepsilon \right\}$$

is nowhere dense in $\Gamma(X)$. Assuming that this is not the case, we can find a nonempty open set U in the closure of M. Let $\widehat{\gamma} \in M \cap U$. Making U smaller if necessary we achieve that

$$\|\Phi(\gamma) - \Phi(\widehat{\gamma})\|_{L_1} + \|\Psi(\gamma) - \Psi(\widehat{\gamma})\|_{L_1} < \frac{\varepsilon^2}{64(1 + \mathrm{Lip}(f))} \tag{6.15}$$

for every $\gamma \in U \cap H$.

Let $n \in \mathbb{N}$ and $\eta > 0$ be such that every γ with $\|\gamma - \widehat{\gamma}\|_{\leq n} < 3\eta$ belongs to U. Noticing that $\Phi(\widehat{\gamma}^{k,s}) = (\Phi(\widehat{\gamma}))^{k,s}$, and similarly for Ψ, we use Lemma 6.3.4 to find $k_0 \geq n$ such that for all $k \geq k_0$

- $\|\widehat{\gamma}^{k,s} - \widehat{\gamma}\|_{\leq n} < \eta$ for every $s \in T$;

- $\mathscr{L}^{\mathbb{N}}\left\{ s \in T \mid \|\Phi(\widehat{\gamma}^{k,s}) - \Phi(\widehat{\gamma})\|_{L_1} \geq \dfrac{\varepsilon^2}{128(1 + \mathrm{Lip}(f))} \right\} < \dfrac{\varepsilon}{2};$ and

- $\mathscr{L}^{\mathbb{N}}\left\{ s \in T \mid \|\Psi(\widehat{\gamma}^{k,s}) - \Psi(\widehat{\gamma})\|_{L_1} \geq \dfrac{\varepsilon^2}{128(1 + \mathrm{Lip}(f))} \right\} < \dfrac{\varepsilon}{2}.$

Fubini's theorem shows that there is $s \in T$ such that $\gamma_0 := \widehat{\gamma}^{k_0,s}$ has the properties that $\mathscr{L}^{\mathbb{N}}\gamma_0^{-1}(E_0) > \varepsilon$ and at the same time

$$\|\Phi(\gamma_0) - \Phi(\widehat{\gamma})\|_{L_1} + \|\Psi(\gamma_0) - \Psi(\widehat{\gamma})\|_{L_1} < \dfrac{\varepsilon^2}{64(1 + \mathrm{Lip}(f))}. \tag{6.16}$$

By Lemma 6.3.8 with $S = \gamma_0^{-1}(E_0)$ we can find $k \geq k_0$, $\delta > 0$, and a surface $\widetilde{\gamma} \in \Gamma_{k+1}(X)$ such that $\|\widetilde{\gamma} - \gamma_0\|_{\leq k} < \eta$ and

$$\begin{aligned}
\|\Phi(\gamma) - \Phi(\gamma_0)\|_{L_1} &+ \|\Psi(\gamma) - \Psi(\gamma_0)\|_{L_1} \\
&= \int \left(\left| \mathbf{1}_E(\gamma(t)) - \mathbf{1}_E(\gamma_0(t)) \right| + \|\psi_E(\gamma(t)) - \psi_E(\gamma_0(t))\| \right) d\mathscr{L}^{\mathbb{N}}(t) \\
&> \dfrac{\varepsilon^2}{32(1 + \mathrm{Lip}(f))}
\end{aligned} \tag{6.17}$$

whenever $\gamma \in \Gamma(X)$ and $\|\gamma - \widetilde{\gamma}\|_{\leq k+1} < \delta$. Since $\widetilde{\gamma} \in U$ and H is dense in U, we can choose γ in H satisfying $\|\gamma - \widetilde{\gamma}\|_{\leq k+1} < \min\{\delta, \eta\}$. Then

$$\|\gamma - \widetilde{\gamma}\|_{\leq n} \leq \|\gamma - \widetilde{\gamma}\|_{\leq n} + \|\widetilde{\gamma} - \gamma_0\|_{\leq n} + \|\gamma_0 - \widetilde{\gamma}\|_{\leq n} < 3\eta,$$

and so $\gamma \in H \cap U$. For such γ, on the one hand, the estimate (6.17) holds and hence, using (6.16) we obtain

$$\begin{aligned}
\|\Phi(\gamma) - \Phi(\widehat{\gamma})\|_{L_1} &+ \|\Psi(\gamma) - \Psi(\widehat{\gamma})\|_{L_1} \\
&\geq \|\Phi(\gamma) - \Phi(\gamma_0)\|_{L_1} + \|\Psi(\gamma) - \Psi(\gamma_0)\|_{L_1} \\
&\quad - \|\Phi(\gamma_0) - \Phi(\widehat{\gamma})\|_{L_1} - \|\Psi(\gamma_0) - \Psi(\widehat{\gamma})\|_{L_1} \\
&> \dfrac{\varepsilon^2}{64(1 + \mathrm{Lip}(f))}.
\end{aligned}$$

On the other hand, (6.15) must hold as well, which is the contradiction we wanted to complete the proof. $\qquad \square$

Corollary 6.3.10. *Assume that $X^\star$ is separable. Then any continuous convex function on an open subset of X is Γ-almost everywhere Fréchet differentiable.*

Proof. This follows immediately from Theorem 5.2.3, Proposition 6.2.2, and Theorem 6.3.9. $\qquad \square$

Corollary 6.3.11. *Assume that $X^\star$ is separable. Then any Lipschitz function f from an open subset G of X into $\mathbb{R}$ is Γ-almost everywhere Fréchet differentiable if and only if every σ-porous set in X is Γ-null.*

Proof. The "if" part is an immediate consequence of Theorem 5.2.3, Proposition 6.2.6, and Theorem 6.3.9.

The "only if" part is trivial: as already noticed in Remark 3.4.1, Chapter 3, if A is porous, the Lipschitz function $f(x) = \text{dist}(x, A)$ is nowhere Fréchet differentiable on A.   □

6.4 FRÉCHET DIFFERENTIABILITY EXCEPT FOR Γ-NULL SETS

In this section we apply the porosity criteria from the previous sections together with the results of Chapter 10 where we will identify a class of Banach spaces in which σ-porous sets are Γ-null. These spaces are considered in detail in Section 10.6. For example, spaces whose modulus of asymptotic smoothness $\bar{\rho}_{\|\cdot\|}(x; t)$ satisfies

$$\sup_{\|x\|=1} \bar{\rho}_{\|\cdot\|}(x; t) = o(t^n), \ t \searrow 0$$

for each $n \in \mathbb{N}$ have the property.

Here we will just discuss the following spaces about which we will prove in Theorem 10.6.8 that they have the property that all their σ-porous subsets are Γ-null: c_0, $C(K)$ with K compact countable, the Tsirelson space, subspaces of c_0. All these spaces have separable dual. Combining this result with Corollary 6.3.11, we get

Theorem 6.4.1. *The following spaces have the property that every real-valued Lipschitz mapping on them is Fréchet differentiable Γ-almost everywhere: c_0, $C(K)$ with K compact countable, the Tsirelson space, subspaces of c_0.*

Stronger results follow by application of Theorem 6.3.9. For optimal results in this direction it is desirable to have information about when $L(X, Y)$ is separable for every space Y with the RNP.

Lemma 6.4.2. *Consider the following property of a Banach space X:*

(**O**) *$L(X, Y)$ is separable for every separable Y with the RNP.*

*The class of spaces X with the property (**O**) is closed under finite direct sums and under infinite direct sums in the sense of c_0.*

Proof. Stability of the property (**O**) for finite direct sums is obvious. Assume that $X = (\sum_{i=1}^{\infty} \oplus X_i)_{c_0}$ with each X_i having (**O**). Denote by π_k the natural projection of X onto $(\sum_{i=1}^{k} \oplus X_i)_{c_0}$ and put

$$V_k = (\text{Id} - \pi_k)X, \ \ k \geq 1.$$

Let $L \in L(X, Y)$ with Y separable and having the RNP. For $\varepsilon > 0$ we choose $0 < \eta < 1$ such that $\eta/(1 - \eta) < \varepsilon$.

Since Y has the RNP, the set $\{Lx \mid \|x\| \leq 1\} \subset Y$ has slices with arbitrarily small diameter. Thus there are $y^* \in Y^*$ and a real number c such that the set

$$S = \{Lx \mid \|x\| \leq 1,\ y^*(Lx) > c\}$$

is nonempty and of diameter $< \eta$. Let $x \in X$ with $\|x\| \leq 1$ be such that $Lx \in S$, and let $k \geq 1$ be such that $\|x - \pi_k x\| < \eta$. Whenever $v \in V_k$ with $\|v\| \leq 1$ then

$$\|x \pm (1 - \eta)v\| = \max\{\|\pi_k x\|,\ \|(x - \pi_k x) \pm (1 - \eta)v\|\} \leq 1.$$

It follows that at least one of the vectors $L(x \pm (1 - \eta)v)$ also belongs to S. Using that $\operatorname{diam} S < \eta$ we obtain

$$\eta > \|Lx - L(x \pm (1 - \eta)v)\| = (1 - \eta)\|Lv\|.$$

Since $v \in V_k$ is an arbitrary vector with $\|v\| \leq 1$ we conclude that

$$\|L - L \circ \pi_k\| = \|L(\operatorname{Id} - \pi_k)\| < \frac{\eta}{1 - \eta} < \varepsilon.$$

Finally, the spaces $L(\sum_{i=1}^{k} \oplus X_i)_{c_0}$ are separable for every k, which implies that $L(X, Y)$ is separable as well.   □

Since $\mathbb{R}$ clearly has (**O**), it follows from Lemma 6.4.2 that c_0 has (**O**) and by countable transfinite induction that $C(K)$ has (**O**) for every compact countable K.

We show next that every subspace Z of c_0 has property (**O**). Let

$$V_k = \{x \in Z \mid x_1 = \cdots = x_k = 0\}.$$

By Corollary 4.2.9 there is for every k a finite dimensional subspace U_k of Z such that every $z \in Z$ can be written as $z = u + v$ with $u \in U_k$, $v \in V_k$ and $\|u\|, \|v\| \leq 3\|z\|$. For every $L \in L(Z, Y)$ and every $\varepsilon > 0$ there is a k such that $\|L|_{V_k}\| \leq \varepsilon$. Indeed, if this were false, we would get a sequence of unit vectors $v_k \in V_k$ such that

$$Z_0 := \overline{\operatorname{span}}\{v_k \mid k \geq 1\}$$

is isomorphic to c_0 and L is an isomorphism of Z_0 onto $L(Z_0) \subset Y$. This contradicts the assumption that Y has the RNP.

Let $\varepsilon > 0$ and $L \in L(Z, Y)$ be given. Find k such that $\|L|_{V_k}\| \leq \varepsilon$. Since U_k is finite dimensional, there is a countable set $\mathcal{L}_k \subset L(Z, Y)$ which is dense in the subspace of operators vanishing on V_k. It follows that there is an $\tilde{L} \in \mathcal{L}_k$ such that $\|Lv - \tilde{L}v\| = \|Lv\| \leq \varepsilon\|v\|$ for $v \in V_k$ and $\|\tilde{L}u - Lu\| \leq \varepsilon\|u\|$ for $u \in U_k$. By decomposing each $z \in Z$ as above, we deduce that $\|\tilde{L}z - Lz\| \leq 6\varepsilon\|z\|$ for every $z \in Z$. Hence every $L \in L(Z, Y)$ has distance at most 6ε from the countable set $\bigcup_k \mathcal{L}_k$ and this proves the separability of $L(Z, Y)$.

Using these arguments we deduce

Theorem 6.4.3. *The following spaces have the property that every Lipschitz mapping of them into a space with the RNP is Fréchet differentiable Γ-almost everywhere: $C(K)$ for compact countable K, subspaces of c_0.*

Proof. The spaces in the statement possess (**O**). By Proposition 6.2.6 the set of irregular points is σ-porous and so Γ-null in the spaces under consideration by Theorem 10.6.8. In view of Theorem 5.2.3 every Lipschitz map into a space with the RNP is Gâteaux differentiable at Γ-almost all points. Theorem 6.3.9 now finishes the proof. □

Theorem 6.4.3 does not hold for subspaces of $C(K)$ with K countable. Indeed, the Schreier space which is the completion of the space of eventually zero sequences $(a_i)_{i=1}^\infty$ with respect to the norm

$$\left\|(a_i)\right\|_S = \sup\left\{ \sum_{k=1}^n |a_{i_k}| \ \bigg| \ n \in \mathbb{N} \text{ and } n \le i_1 < i_2 < \cdots < i_n \right\}$$

is isomorphic to a subspace of $C(\omega^\omega)$ and, as remarked in [22, Example 7.7], there is a Lipschitz map from this space into ℓ_2 which is nowhere Fréchet differentiable. This is explained in more detail in the following proposition.

Proposition 6.4.4. *The Schreier space S is isomorphic to a subspace of $C(\omega^\omega)$, and there is a Lipschitz map $f\colon S \longrightarrow \ell_2$ which, for any $0 < \varepsilon < 1$, does not have a point of ε-Fréchet differentiability.*

Proof. We assign to every $a = (a_i) \in S$ a function $\varphi_a \in C(\omega^\omega)$ given by the following formula. If $\sigma = (\sigma_0, \sigma_1, \dots) \in \omega^\omega$ then

$$\varphi_a(\sigma) = \begin{cases} \displaystyle\sum_{i=1}^{\sigma_0} a_{\sigma_i} & \text{if } \sigma_0 \le \sigma_1 < \cdots < \sigma_{\sigma_0}, \\[2mm] 0 & \text{otherwise.} \end{cases}$$

Clearly, $\|\varphi_a\|_\infty \le \|a\|_S$. To estimate $\|a\|_S$ by a multiple of $\|\varphi_a\|_\infty$, find subsets of indices $I = \{i_1, \dots, i_n\} \subset \mathbb{N}$, $n \le i_1 < \cdots < i_n$, and $I_0 \subset I$ such that

$$\|a\|_S \le 2 \sum_{i \in I} |a_i| \le 4 \left| \sum_{i \in I_0} a_i \right|.$$

This implies that $\|a\|_S \le 4\|\varphi_a\|_\infty$ and so the mapping $a \mapsto \varphi_a$ is an isomorphism of S onto a subspace of $C(\omega^\omega)$.

Define a map $f\colon S \longrightarrow \ell_2$ by

$$f(a) = \sum_{i=1}^\infty |a_i| e_i, \quad a = (a_i).$$

In order to see that f is well defined and Lipschitz we show that

$$\|a\|_2 \le \sqrt{8}\|a\|_S.$$

Assume that $\|a\|_S = 1$ and let $b = (b_i)$ be a nonincreasing rearrangement of the sequence $(|a_i|)$. Then

$$\|a\|_2^2 = \|b\|_2^2 = \sum_{k=0}^{\infty} \sum_{i=2^k}^{2^{k+1}-1} b_i^2 \le \sum_{k=0}^{\infty} 2^k b_{2^k}^2. \tag{6.18}$$

Since $\|b\|_S \le \|a\|_S = 1$ we have $2^{k-1} b_{2^k} \le 1$ for all $k \ge 1$, which allows us to continue the estimate (6.18)

$$\le \sum_{k=0}^{\infty} 2^k 2^{-2(k-1)} = 8.$$

Let $x = (x_i) \in S$ and $0 < \varepsilon < 1$. We show that f is not ε-Fréchet differentiable at x. Supposing that it is ε-Fréchet differentiable, there are a $\delta > 0$ and an operator $A \in L(S, \ell_2)$ satisfying

$$\|f(x+u) - f(x) - Au\|_2 \le \varepsilon\|u\|_S$$

whenever $\|u\|_S \le \delta$. Choose $\eta > 0$ such that $\varepsilon + \eta < 1$, and let $n \in \mathbb{N}$ be large enough to verify

$$\left\| \sum_{i=n+1}^{\infty} x_i e_i \right\|_S \le \frac{\eta\delta}{2\sqrt{8}}.$$

Then we obtain at the point $y = \sum_{i=1}^{n} x_i e_i$ the following estimate for all $\|u\|_S \le \delta$:

$$\|f(y+u) - f(y) - Au\|_2 \le \|f(x+u) - f(x) - Au\|_2 + 2\sqrt{8}\|x - y\|_S$$
$$\le \varepsilon\delta + \eta\delta = (\varepsilon + \eta)\delta.$$

Hence for any $i > n$

$$2\delta = \|(\delta e_i - \delta Ae_i) + (\delta e_i + \delta Ae_i)\|_2$$
$$= \|f(y + \delta e_i) - f(y) - \delta Ae_i\|_2 + \|f(y - \delta e_i) - f(y) + \delta Ae_i\|_2$$
$$\le 2(\varepsilon + \eta)\delta < 2\delta,$$

a contradiction. □

We conclude this chapter by showing that even the strongest multidimensional form of the mean value estimates (as discussed in Section 2.4) holds for Lipschitz, Γ-almost everywhere Fréchet differentiable maps. As we will see in Chapter 14, this fact shows that in many spaces, for example in infinite dimensional Hilbert spaces, Lipschitz functions are *not* Fréchet differentiable Γ-almost everywhere. The key to this is the following fact.

Proposition 6.4.5. *Suppose that $f : G \longrightarrow Y$ is a Lipschitz mapping which is Gâteaux differentiable at Γ-almost every point of an open subset G of a Banach space X. Then for every $v_1, \ldots, v_n \in X$, $y_1^*, \ldots, y_n^* \in Y^*$ and $c \in \mathbb{R}$, the set E of those $x \in G$ at which f is Gâteaux differentiable and $\sum_{i=1}^n y_i^* f'(x; v_i) > c$ is either empty or not Γ-null.*

Proof. We may assume without loss of generality that $\sum_{i=1}^n \|v_i\|$, $\sum_{i=1}^n \|y_i^*\|$, and $\mathrm{Lip}(f)$ are all bounded by one.

Suppose there is a point $x_0 \in G$ at which the function f is Gâteaux differentiable and $\sum_{i=1}^n y_i^* f'(x_0; v_i) > c$. Choose $0 < \eta < 1$ such that

$$\sum_{i=1}^n y_i^* f'(x_0; v_i) > c + 6\eta,$$

and let $g(x) := f(x) - f(x_0) - f'(x_0; x - x_0)$. Since g is Fréchet differentiable at x_0 in the direction of the linear span V of $\{v_1, \ldots, v_n\}$ with derivative zero, there is $r > 0$ such that

$$\|g(x_0 + v)\| \leq \eta \|v\|$$

for $v \in V$ and $\|v\| \leq r$. Moreover, we assume that $2r < \mathrm{dist}(x_0, X \setminus G)$.

Let $\gamma_0(t) = x_0 + r \sum_{k=1}^n t_k v_k$ and consider any surface $\gamma \in \Gamma(X)$ satisfying $\|\gamma - \gamma_0\|_{\leq n} < \eta r$. Notice that $\gamma(T) \subset G$ and choose this γ such that, in addition, the mapping f is Gâteaux differentiable at $\gamma(t)$ for $\mathscr{L}^{\mathbb{N}}$ almost every t. We introduce for each $s \in T$ the vector field

$$t \in [0,1]^n \to \left(y_1^* g(\gamma^{n,s}(t)), \ldots, y_n^* g(\gamma^{n,s}(t)) \right),$$

whose jth component is bounded by

$$|y_j^* g(\gamma^{n,s}(t))| \leq |y_j^* g(\gamma^{n,s}(t)) - y_j^* g(\gamma_0(t))| + |y_j^* g(\gamma_0(t))|$$

$$\leq \|y_j^*\| \, \mathrm{Lip}(g) \, \|\gamma^{n,s} - \gamma_0\|_\infty + \|y_j^*\| \eta r \sum_{i=1}^n \|v_i\| \leq 2\eta r \|y_j^*\|.$$

The divergence theorem (or Fubini's theorem together with the fundamental theorem of calculus) shows that

$$\left| \int_{[0,1]^n} \sum_{i=1}^n y_i^* g'(\gamma^{n,s}(t); D_i \gamma^{n,s}(t)) \, d\mathscr{L}^n(t) \right|$$

$$= \left| \int_{[0,1]^n} \sum_{i=1}^n \frac{\partial}{\partial t_i} y_i^* \left(g(\gamma^{n,s}(t)) \right) d\mathscr{L}^n(t) \right|$$

$$\leq 4\eta r \sum_{i=1}^n \|y_i^*\| \leq 4\eta r.$$

Hence, observing that

$$\left| y_i^* g'(\gamma(t); rv_i) - y_i^* g'(\gamma(t); D_i\gamma(t)) \right| \leq 2\|y_i^*\| \|rv_i - D_i\gamma(t)\|$$
$$= 2\|y_i^*\| \|D_i\gamma_0(t) - D_i\gamma(t)\|$$
$$\leq 2\eta r \|y_i^*\|,$$

we get with the help of Fubini's theorem

$$\int_T \sum_{i=1}^n y_i^* \left(g'(\gamma(t); rv_i) \right) d\mathscr{L}^{\mathbb{N}}(t)$$

$$\geq \int_T \sum_{i=1}^n y_i^* \left(g'(\gamma(t); D_i\gamma(t)) \right) d\mathscr{L}^{\mathbb{N}}(t) - 2\eta r \sum_{i=1}^n \|y_i^*\|$$

$$\geq \int_T \int_{[0,1]^n} \left(\sum_{i=1}^n \left(y_i^* \left(g'(\gamma(t); D_i\gamma(t)) \right) \right) \right)^{n,s} d\mathscr{L}^n(t) \, d\mathscr{L}^{\mathbb{N}}(s) - 2\eta r$$

$$= \int_T \int_{[0,1]^n} \sum_{i=1}^n y_i^* \left(g'(\gamma^{n,s}(t); D_i\gamma^{n,s}(t)) \right) d\mathscr{L}^n(t) \, d\mathscr{L}^{\mathbb{N}}(s) - 2\eta r$$

$$\geq -6\eta r.$$

for any $\gamma \in \Gamma(X)$ such that f is Gâteaux differentiable at $\gamma(t)$ for $\mathscr{L}^{\mathbb{N}}$-almost every $t \in T$ and $\|\gamma - \gamma_0\|_{\leq n} < \eta r$. Hence for any such surface γ there is a set of positive $\mathscr{L}^{\mathbb{N}}$-measure in which

$$\sum_{i=1}^n y_i^* f'(\gamma(t); v_i) = \sum_{i=1}^n y_i^* f'(x_0; v_i) + \sum_{i=1}^n y_i^* g'(\gamma(t); v_i) > c.$$

Since these surfaces form a set which is not meager in $\Gamma(X)$, the statement follows. □

If f is Γ-almost everywhere Fréchet differentiable, the set E from Proposition 6.4.5 contains a Fréchet derivative. By definition, this means that the multidimensional mean value estimate holds for Fréchet derivative of f. We record this for future reference.

Corollary 6.4.6. *Suppose that $f: G \longrightarrow Y$ is a Lipschitz mapping which is Fréchet differentiable at Γ-almost every point of an open subset G of a Banach space X. Then the multidimensional mean value estimate holds for Fréchet derivatives of f.*

Chapter Seven

Variational principles

Compared to direct recursive constructions such as those used in Chapter 15, the arguments of the following chapters will be significantly better organized and simplified by the use of variational principles (of Ekeland type). Our description of these principles as games and analysis following from it should also help to understand technical peculiarities of our arguments that stem from the careful order of the choice of parameters. In addition to the abstract variational principle, we also deduce some technical variants that will be used for special tasks later.

7.1 INTRODUCTION

The goal of the variational principles that we consider in this short chapter is to modify a given function $f \colon M \longrightarrow \mathbb{R}$ defined on a metric space (M, d) to a function of the form

$$h(x) := f(x) + \sum_{j=0}^{\infty} F_j(x),$$

where F_j are recursively chosen small perturbations such that the perturbed function h attains its minimum. During the recursion, we also define a minimizing sequence $x_j \in M$. For our more delicate results, it is important to know various technical estimates saying, for example, how quickly the sequence x_j converges to the point at which the minimum of h is attained. For this reason, we will follow the approach to the smooth variational principles from [7] rather than the interesting approach from [13], where the whole perturbation function is chosen from a larger class of allowed perturbations by the Baire category theorem.

At first, the recursion defining the minimizing sequence x_j looks extremely simple:

- $f(x_0)$ is ε_0-close to the infimum of f,

- $f(x_1) + F_0(x_1)$ is ε_1-close to the infimum of $f + F_0$,

- etc.

The fine points appear in the order of the choice of the F and ε. The only way in which we may hope to estimate, say, that $d(x_1, x_2) \leq r$ is by establishing that $s_1 := \inf_{d(x_1, x) > r} (F_1(x) - F_1(x_1)) > \varepsilon_1$. If this holds, and $f(x_2) + F_0(x_2) + F_1(x_2)$ is $(s_1 - \varepsilon_1)$-close to the infimum of $f + F_0 + F_1$, then automatically $d(x_1, x_2) \leq r$.

Indeed, if it were $d(x_1, x_2) > r$, then

$$f(x_2) + F_0(x_2) + F_1(x_2) \geq f(x_2) + F_0(x_2) + F_1(x_1) + s_1$$
$$\geq \inf_{y \in M} \left(f(y) + F_0(y)\right) + F_1(x_1) + s_1$$
$$> f(x_1) + F_0(x_1) - \varepsilon_1 + F_1(x_1) + s_1,$$

which would mean that $f(x_2) + F_0(x_2) + F_1(x_2)$ is not $(s_1 - \varepsilon_1)$-close to the infimum of $f + F_0 + F_1$.

The point of the previous discussion is that when we are choosing x_1 as an almost minimizing point of $f + F_0$ we have to know something about the behavior of the function F_1, but this function will be added to the function that we minimize only in the future. This is the reason why usual formulations of such variational principles prescribe the whole "perturbation scheme," which is essentially the collection of functions that may be added as F_j, in advance. For example, the variational principle of [7] uses $F_j(x) = \Phi_j(d(x_j, x))$, where the functions Φ_j are fixed in advance. Even more apparent is the case of Ekeland's variational principle, where $F_j(x) = 2^{-j}d(x_j, x)$, say. Although in our main applications the choice of the perturbation functions F_j will not be so simple, they will still not be as general as Theorem 7.2.1 allows. The need for the advance choice of some data reappears in the form of strangely indexed parameters, and being aware of the reason for this helps to avoid mistakes. We will therefore present the general forms of the variational principles in a way in which the order of choice is easy to see: as an infinite game between two players. However, the principles that we directly apply, Corollary 7.2.4 and Theorem 7.3.5, are formulated in the standard way without game language, and the reader who wishes to may easily modify our approach to provide their proof along more standard lines.

Notice that above we spoke more about the minimizing sequence than about the actual minimizer. This reflects the fact that for us the importance of the sequence approximating the minimum is at least equal to the importance of the minimum itself. In fact, our starting principle, Theorem 7.2.1, on which all our variational principles are based, will just show the existence of a minimizing sequence satisfying certain conditions, not the existence of the minimum.

Apart from the introduction, this chapter has two sections. In the first we describe the game that is used in Theorem 7.2.1 to formulate an abstract version of the variational principle. We then show how to specialize it to more standard formulations. The main aim of the specialized principles in this section is to allow as free choice of the perturbation functions as possible, since we have in mind applications in which delicate technical choices of such functions are needed. On the other hand, in these applications we have standard topological assumptions of completeness and lower semicontinuity, and so we do not attempt to weaken them there.

The situation in the final section is of opposite nature: the perturbation functions are relatively simple, but completeness and lower semicontinuity fail. We will informally describe one such case at the beginning of the section. This will naturally lead us to considering cases when completeness and lower semicontinuity hold only in a bimetric sense. We deduce from the abstract Theorem 7.2.1 a variational principle using such concepts and, in the final result of this section, give its technical variant whose main use is in proving residuality of certain sets. Since this section is geared toward particular applications, we state these principles in the classical way with "perturbation schemes" given in advance.

7.2 VARIATIONAL PRINCIPLES VIA GAMES

The variational principles that we consider here are based on a simple but careful recursive choice of points where certain functions that change during the process have values close to their infima. Like many other recursive constructions, the choice has a natural description using the language of infinite two-player games with perfect information. The advantage of this description is that it reveals that some choices made during the recursion are made "in advance," so that the choice of the next point can already use part of the information that will be fully available only after the choice has been made. In our applications, this aspect is rather hidden because all such information is known even before the game starts, but it still appears as an important technical difficulty that has to be carefully handled.

We now describe what objects the players, say (α) and (β), choose, and in what order they choose them.

The starting setup of the game, which we will call the *perturbation game,* is a metric space (M, d), a function $f \colon M \longrightarrow \mathbb{R}$, and a number $\varepsilon_0 > 0$. The game begins by (α) choosing a positive number η_0; then (β) chooses a function $F_0 \colon M \longrightarrow \mathbb{R}$, a point $x_0 \in M$, and positive numbers r_0, ε_1, then (α) chooses a positive number η_1, then (β) chooses a function $F_1 \colon M \longrightarrow \mathbb{R}$, a point $x_1 \in M$, and positive numbers r_1, ε_2, then (α) chooses a positive number η_2, etc. The constants $r_k, \eta_k, \varepsilon_k$ limit (β)'s choice of the function F_k. For (α), the (ideal) goal of the game is that the sequence (x_k) converges to a point at which $f + \sum_{k=0}^{\infty} F_k$ attains its minimum.

We now give a precise description of the rules for the choices in the perturbation game.

Setup of the perturbation game. The players are given

- a metric space (M, d),
- a function $f \colon M \longrightarrow \mathbb{R}$ bounded from below, and
- a number $0 < \varepsilon_0 < \infty$.

For $k = 0, 1, \ldots$ we require the following rules.

Move k of (α). Player (α) chooses $\eta_k > 0$ and denotes

$$h_k(x) = f(x) + \sum_{0 \le j < k} F_j(x). \tag{7.1}$$

Move k of (β). Player (β) chooses

- numbers $r_k > 0$ and $0 < \varepsilon_{k+1} < \infty$,
- a point $x_k \in M$ and a function $F_k \colon M \longrightarrow [0, \infty]$ such that

$$h_k(x_k) + F_k(x_k) < \eta_k + \inf_{x \in M} h_k(x) \quad \text{and} \quad \inf_{d(x_k, y) > r_k} F_k(y) > \varepsilon_k.$$

We will not define the meaning of win or loss in the perturbation game, because all our results will just say that player (α) has a strategy that achieves a certain goal no matter how (β) plays. Of course, one could define that achieving this goal means a win for (α). However, there is little point in it since (β) can never win. Thus calling it a win could be misleading also from the point of view that the goals are different in different results as the conditions on the choices become stricter. The precise conditions imposed on the objects will be described later. Slight changes to these conditions lead to different versions of the variational principle.

Theorem 7.2.1. *In the perturbation game, player (α) has a strategy with first move $\eta_0 = \varepsilon_0$ that, no matter how (β) plays, guarantees that the sequence (x_j) and the functions h_k have the following properties. For every $0 \le j \le k < \infty$,*

(VP$_1$) $h_k(x_k) < \varepsilon_j + \inf_{x \in M} h_j(x)$,

(VP$_2$) $h_k(x_k) < \varepsilon_j + h_j(x_j)$,

(VP$_3$) $F_j(x_k) < \varepsilon_j$,

(VP$_4$) $d(x_j, x_k) \le r_j$,

(VP$_5$) $h_k(x) \ge h_k(x_k) + \frac{1}{2}\varepsilon_j$ *if $k > j \ge 1$ and $d(x, x_k) > 2r_j$,*

(VP$_6$) *the sequence $h_j(x_j)$ converges and, when h_∞ is defined by (7.1) with $k = \infty$, its limit satisfies $\lim_{j \to \infty} h_j(x_j) \le \inf_{x \in M} h_\infty(x)$.*

Proof. The required strategy of player (α) may be described as follows. If $k = 0$, let $\eta_0 = \varepsilon_0$, $\sigma_0 = h_0(x_0) + F_0(x_0) - \inf_{x \in M} h_0(x)$, and observe that $\sigma_0 < \eta_0$. If $k \ge 1$,

- let $\eta_k = 2^{-k-1} \min\{\eta_0 - \sigma_0, \eta_1 - \sigma_1, \ldots, \eta_{k-1} - \sigma_{k-1}, \varepsilon_k\}$,
- denote $\sigma_k = h_k(x_k) + F_k(x_k) - \inf_{x \in M} h_k(x)$,
- observe that $\sigma_k < \eta_k$.

Suppose that (x_k) is a sequence obtained while (α) was using this strategy. Notice that by the rules of the game, $h_k \ge h_{k-1} \ge \cdots \ge h_0 = f$ and $h_k(x_k)$ and $F_k(x_k)$ are finite. Since $h_k(x) = h_{k-1}(x) + F_{k-1}(x)$,

$$\inf_{x \in M} h_k(x) \le h_{k-1}(x_{k-1}) + F_{k-1}(x_{k-1}) = \sigma_{k-1} + \inf_{x \in M} h_{k-1}(x).$$

Hence for $k \geq j$,

$$h_k(x_k) + F_k(x_k) = \sigma_k + \inf_{x \in M} h_k(x) \leq \sum_{i=j}^{k} \sigma_i + \inf_{x \in M} h_j(x)$$

$$< \eta_j + \inf_{x \in M} h_j(x). \tag{7.2}$$

Since $\eta_j \leq \varepsilon_j$ and $F_k \geq 0$, this inequality gives (VP$_1$) and its corollary (VP$_2$). Also, using (7.2) with $x = x_k$ and expanding the left side, we get

$$h_j(x_k) + F_j(x_k) + F_{j+1}(x_k) + \cdots + F_k(x_k) < \eta_j + h_j(x_k).$$

Since $F_i \geq 0$ and the strict inequality implies that $h_j(x_k)$ is finite, we infer that $F_j(x_k) < \eta_j$, hence (VP$_3$).

Recalling that the functions F_j satisfy $\inf_{d(x_j,y)>r_j} F_j(y) > \varepsilon_j$ by the rule governing the moves of player (β), we infer from (VP$_3$) that

$$F_j(x_k) \leq \varepsilon_j < \inf_{d(x_j,y)>r_j} F_j(y),$$

showing that $d(x_j, x_k) > r_j$ cannot occur.

To prove (VP$_5$), let $k > j \geq 1$ and $d(x, x_k) > 2r_j$. Then (VP$_4$) implies that $d(x, x_j) > r_j$ and, consequently, $F_j(x) \geq \varepsilon_j$. Using also the inequality $h_k(x_k) \leq h_j(x_j) + \eta_j$, which follows from (7.2) with $x = x_j$, we estimate

$$h_k(x) \geq h_{j+1}(x)$$
$$= h_j(x) + F_j(x) \geq h_j(x_j) - \eta_j + \varepsilon_j$$
$$\geq h_k(x_k) - 2\eta_j + \varepsilon_j \geq h_k(x_k) + \tfrac{1}{2}\varepsilon_j.$$

Finally, to see that the sequence $h_k(x_k)$ converges, recall that it is bounded below and $h_k(x_k) \leq h_j(x_j) + \eta_j$ for $k > j$. Since $\eta_j \to 0$, this implies that the sequence $(h_k(x_k))$ is Cauchy, hence convergent. The inequality required in (VP$_6$) follows by noticing that (7.2) with $k = j$ and the inequality $h_j \leq h_\infty$ imply

$$h_j(x_j) \leq \eta_j + \inf_{x \in M} h_j(x) \leq \eta_j + \inf_{x \in M} h_\infty(x)$$

and taking the limit as $j \to \infty$. $\square$

For a variational principle, the statement of Theorem 7.2.1 is somewhat unusual. It does not claim that the minimum of the perturbation of f, that is, of the function h_∞, exists, let alone that the sequence (x_k) is a minimizing sequence for h_∞. In fact, there is no reason why the sequence (x_k) should be even Cauchy. For example, if (β) chooses $r_k = \infty$ for all k, the metric on M becomes irrelevant.

The next simple observation points out very weak sufficient conditions which guarantee that the function h_∞ attains its minimum.

Observation 7.2.2. *If the sequence (x_j) from Theorem 7.2.1 converges to some x_∞ and the lower semicontinuity assumptions*

$$f(x_\infty) \leq \liminf_{k\to\infty} f(x_k) \quad \text{and} \quad F_j(x_\infty) \leq \liminf_{k\to\infty} F_j(x_k) \quad \text{for } j = 0, 1, \ldots$$

hold, then the function h_∞ attains its minimum at x_∞ and the properties (VP$_1$) – (VP$_5$) hold also when $0 \leq j < k = \infty$. Moreover,

$$\begin{aligned} h_\infty(x_\infty) &\leq \lim_{k\to\infty} h_k(x_k) \\ &< \inf\{h_\infty(x) \mid d(x, x_\infty) > 2r_j\} \text{ for every } j \geq 1. \end{aligned} \tag{7.3}$$

Proof. We use that $h_j \leq h_k$ for $j \leq k$ to infer that for $j \geq 0$,

$$h_j(x_\infty) \leq \liminf_{k\to\infty} h_j(x_k) \leq \liminf_{k\to\infty} h_k(x_k) = \lim_{k\to\infty} h_k(x_k).$$

This gives the first inequality in (7.3). Combining it with (VP$_6$) we obtain that h_∞ attains its minimum at x_∞. It also implies that the inequalities (VP$_1$) and (VP$_2$) hold for $0 \leq j < k = \infty$. For these j and k, (VP$_3$) follows from the lower semicontinuity assumption of F_j and (VP$_4$) from the continuity of the metric. Finally, if $j \geq 1$ and $d(x, x_\infty) > 2r_j$, then $d(x, x_k) > 2r_j$ for large enough k, so $h_k(x) \geq h_k(x_k) + \frac{1}{2}\varepsilon_j$ by (VP$_5$), and taking limit as $k \to \infty$ gives both (VP$_5$) for $k = \infty$ and the second inequality in (7.3). $\square$

In the following corollary we record the fact that the more usual form of the smooth variational principle holds under standard completeness and lower semicontinuity assumptions provided the rules of the game require player (β) to make more stringent choices. In the next section we will prove similar results under weaker completeness and lower semicontinuity assumptions.

Corollary 7.2.3. *Suppose that in the setup of the perturbation game, the space (M, d) is complete and the function f is lower semicontinuous. Then, provided (β) plays so that*

(a) *the functions F_j are lower semicontinuous, and*

(b) $\liminf_{j\to\infty} r_j = 0$,

the strategy of (α) from Theorem 7.2.1 guarantees that the sequence (x_j) converges to a point $x_\infty \in M$ at which the function h_∞ attains its minimum and the properties (VP$_1$) – (VP$_5$) hold for $0 \leq j < k \leq \infty$. Moreover,

$$h_\infty(x_\infty) \leq \lim_{k\to\infty} h_k(x_k) < \inf_{d(x,x_\infty)>r} h_\infty(x) \text{ for every } r > 0.$$

Proof. Because of (VP$_4$) and (b), the sequence (x_j) is Cauchy. Since we assume that M is complete, it converges to some point x_∞. The lower semicontinuity of f and all F_j imply the assumptions of Observation 7.2.2 and the statement follows. $\square$

While Corollary 7.2.3 is an abstract version of the usual form of the smooth variational principles, the last assertion of this section specializes it further in several directions: the perturbation functions and constants are prescribed in advance, and (β) always chooses F_k and x_k such that $F_k(x_k) = 0$. With this choice, (β) may also achieve that the sequence $h_k(x_k)$ is decreasing, which is often required in variational principles of this type.

Corollary 7.2.4. *Suppose that $f : M \longrightarrow \mathbb{R}$ is lower bounded and lower semicontinuous on a complete metric space (M, d). Suppose further that $F_j : M \times M \longrightarrow [0, \infty]$, $j \geq 0$, are functions lower semicontinuous in the second variable with $F_j(x, x) = 0$ for all $x \in M$ and that $0 < r_j \leq \infty$ are such that $r_j \to 0$ and*

$$\inf_{d(x,y)>r_j} F_j(x, y) > 0.$$

If $x_0 \in M$ and $(\varepsilon_j)_{j=0}^{\infty}$ is any sequences of positive numbers such that

$$f(x_0) < \varepsilon_0 + \inf_{x \in M} f(x) \quad and \quad \inf_{d(x_0,y)>r_0} F_0(x_0, y) > \varepsilon_0,$$

then one may find a sequence $(x_j)_{j=1}^{\infty}$ of points in M converging to some $x_\infty \in M$ such that the function

$$h(x) := f(x) + \sum_{j=0}^{\infty} F_j(x_j, x)$$

attains its minimum on M at x_∞. Moreover, for each $j \geq 0$,

$$d(x_j, x_\infty) \leq r_j, \quad F_j(x_j, x_\infty) \leq \varepsilon_j,$$

$$h(x_\infty) \leq \varepsilon_j + \inf_{x \in S}\left(f(x) + \sum_{i=0}^{j-1} F_i(x_i, x) \right),$$

and with h_k defined by (7.1), $h_k(x_k) \leq h_j(x_j)$ and $(VP_1) - (VP_5)$ hold whenever $0 \leq j < k + 1 \leq \infty$.

Proof. We note first that making ε_j for $j \geq 1$ smaller makes the statement stronger. So we assume that

$$\inf_{d(x,y)>r_j} F_j(x, y) > \varepsilon_j, \quad j \geq 1.$$

With this assumption, Corollary 7.2.4 becomes a special case of Corollary 7.2.3 when we impose the following additional requirements upon the rules of the kth move of player (β). When $k = 0$, (β) must choose the given starting point x_0. When $k \geq 1$, the point x_k has to satisfy

$$h_k(x_k) \leq \min\{h_k(x_{k-1}), \tfrac{1}{2}\eta_k + \inf_{x \in M} h_k(x)\},$$

further $F_k(x) := F_k(x_k, x)$ and, finally, (β) chooses the given parameters r_k and ε_{k+1}. Notice that this choice still fits in the general rule for (β).

By Corollary 7.2.3 the function h attains its minimum on M at x_∞. The next three statements follow from (VP_4), (VP_3), and (VP_2), respectively. The inequality $h_k(x_k) \leq h_{k-1}(x_{k-1})$ follows from $h_k(x_{k-1}) = h_{k-1}(x_{k-1})$. $\square$

7.3 BIMETRIC VARIATIONAL PRINCIPLES

To explain the bimetric variant of the smooth variational principle, we start by motivating it by its intended applications to porosity, which will be done in detail in Chapter 10. (Further motivation may be found in the variational proof of Fréchet differentiability of everywhere Gâteaux differentiable functions in [29] or in Chapter 12, Section 12.2.)

We recall the space $\Gamma_n(X) = C^1([0,1]^n, X)$ with (one of possible equivalent norms)

$$\|\gamma\| = \max\{\|\gamma\|_\infty, \|\gamma'\|_\infty\}.$$

To show Γ_n-nullness of a subset E of a Banach space X porous in the direction of a family of subspaces (see Definition 10.1.3), we plan to proceed as follows. Assuming, as we may, that E is a G_δ set we wish to find a contradiction from the assumption that, for some $\alpha > 0$, the set

$$M = \left\{\gamma \in \Gamma_n(X) \mid \mathscr{L}^n \gamma^{-1}(E) \geq \alpha\right\}$$

is dense in some ball $B(\gamma_0, r_0) \subset \Gamma_n(X)$. It will be achieved by analyzing the function

$$f(\gamma) = \mathscr{L}^n \gamma^{-1}(\overline{E}).$$

Our key observation is that, under suitable smoothness assumptions on the Banach space X, the function f cannot attain its minimum on $\Gamma_n(X)$ at any point of M. So a reasonable plan could be to try to find an argument yielding that the minimum of f is attained at some point of M. Of course, this has to be just a crude approximation to the real proof, since f may not attain a minimum at all, and even if it did, there would be no reason for the attainment point to belong to M. We will, however, modify this observation. Instead of looking for the minimum of f, we show that f must be $\|\cdot\|_\infty$-lower semicontinuous at some point of M. There are two reasons why this looks like a plausible direction. First, the statement that the function f is $\|\cdot\|_\infty$-lower semicontinuous at γ may be understood as a variational statement. It is equivalent to saying that for some $\|\cdot\|_\infty$-continuous function φ the sum $f + \varphi$ attains its minimum at γ. Second, since f is $\|\cdot\|_\infty$-upper semicontinuous, it belongs to the first class of Baire, and Baire's theorem says that such functions are continuous on a dense G_δ set *provided the space is complete.* Unfortunately, with the norm $\|\cdot\|_\infty$ the space $\Gamma_n(X)$ is incomplete, so Baire's theorem is not directly applicable. (We note that Baire's theorem provides us with many points of $\|\cdot\|$-lower semicontinuity of f; however, our observation requires a much stronger notion of $\|\cdot\|_\infty$-lower semicontinuity points of f.)

A natural attempt that may solve this difficulty is to find a quantitative version of the argument showing that f is $\|\cdot\|_\infty$-lower semicontinuous at some point of M. Based on the smoothness assumption one can introduce a weight function $\Theta \colon \Gamma_n(X) \longrightarrow \mathbb{R}$, called "energy" for short, and show that once $f(\gamma_0) > 0$ there is γ such that

$$f(\gamma) < f(\gamma_0) - \big(\Theta(\gamma) - \Theta(\gamma_0)\big).$$

(A more precise version of this is proved in Lemma 10.2.3.) This approach naturally leads to a sequence (γ_k) of surfaces along which f decreases and the increments of energy are bounded by the decrements of f. One may then try to use suitable properties of the energy Θ to deduce that the sequence (γ_k) actually converges to some γ at which f may not necessarily attain its minimum, but is perhaps at least $\|\cdot\|_\infty$-lower semicontinuous.

Unfortunately, the increment of the energy $\Theta(\gamma) - \Theta(\gamma_0)$ is not easily related to an estimate of $\|\gamma - \gamma_0\|$, and so our program seems to present technical difficulties similar to those encountered when proving full Fréchet differentiability results in Chapter 12. However, the above description is strikingly similar to the rules governing the moves of (β) in the perturbation game. It turns out that, instead of working with approximating sequences (γ_k), it is considerably more convenient to work with a variational principle that actually produces a limiting γ_∞. The main difference between such principle and the more standard versions such as Corollary 7.2.3 is that our function f is not lower semicontinuous and we are looking for a minimum in the space M that is not necessarily complete.

Of course, some forms of completeness and lower semicontinuity assumptions are needed for a variational principle to hold. For that we observe that M is a $\|\cdot\|_\infty$-G_δ subset of $\Gamma_n(X)$. So a $\|\cdot\|$-Cauchy sequence of elements of M converges to an element of M *provided its $\|\cdot\|_\infty$-increments converge to zero fast enough.* A precise definition and statement are given in Definition 7.3.1 and Lemma 7.3.4; here it may suffice to say that "fast enough" means that the distance $\|\gamma_k - \gamma_{k+1}\|_\infty$ is controlled by a quantity depending on $\gamma_0, \dots, \gamma_k$. Similarly, even though our f is not lower semicontinuous, we nevertheless prove that the lower semicontinuity condition holds whenever the surfaces $\gamma_j \in M$ converge in $\|\cdot\|$ to γ and, again, *$\|\cdot\|_\infty$-increments converge to zero fast enough.* This is stated precisely below, in Definition 7.3.2 and Lemma 7.3.3.

We now define the above-mentioned notions of completeness and lower semicontinuity in an abstract form and in Theorem 7.3.5 deduce from Theorem 7.2.1 a general form of the variational principle alluded to above.

To express the requirement that the sequence converges fast enough we could use the language of infinite games again, but the concept is so simple (in particular, there are no advance choices), that we do not see any advantages in doing so.

Definition 7.3.1. Suppose that d_0 a continuous pseudometric on a metric space (M, d). We say that M is (d, d_0)-*complete* if there are functions $\delta_j \colon M^{j+1} \longrightarrow (0, \infty)$ such that every d-Cauchy sequence $(x_j)_{j=0}^\infty$ converges to an element of M provided

$$d_0(x_j, x_{j+1}) \le \delta_j(x_0, \dots, x_j) \text{ for each } j \ge 0.$$

Definition 7.3.2. Suppose that d_0 a continuous pseudometric on a metric space (M, d). We say that a function $f \colon M \longrightarrow \mathbb{R}$ is (d, d_0)-*lower semicontinuous* if there are functions $\delta_j \colon M^{j+1} \longrightarrow (0, \infty)$, $j \ge 0$, such that

$$f(x) \le \liminf_{j \to \infty} f(x_j)$$

whenever $x_j \in M$ converge in metric d to x and

$$d_0(x_j, x_{j+1}) \le \delta_j(x_0, \dots, x_j) \text{ for each } j \ge 0.$$

Notice that the choice $d_0 = 0$ in Definitions 7.3.1 and 7.3.2 yields the standard notions of completeness and lower semicontinuity, respectively. Notice also that in these definitions it is only important that the estimates $d_0(x_j, x_{j+1}) \leq \delta(x_0, \ldots, x_j)$ hold for large j. If they hold for $j \geq k$, we may redefine x_i for $i \leq k$ as $x_i = x_k$ and observe that the new sequence satisfies the estimates for all j and that we have changed neither the value of $\liminf_{j \to \infty} f(x_j)$, nor the convergence or divergence of the sequence.

As usual, if the metrics d and d_0 are induced by norms $\| \cdot \|$ and $\| \cdot \|_0$, we speak about $(\| \cdot \|, \| \cdot \|_0)$-completeness and lower semicontinuity, respectively.

Although in our applications one can verify both (d, d_0)-completeness and (d, d_0)-lower semicontinuity directly, it is often more convenient to use the simple topological criteria given in the following two lemmas.

Lemma 7.3.3. *Let (M, d) be a metric space and d_0 a continuous pseudometric on M. If $f \colon M \longrightarrow \mathbb{R}$ has the property that for each $r \in \mathbb{R}$, the set*

$$\{x \in M \mid f(x) \leq r\}$$

is a G_δ subset of (M, d_0), then f is (d, d_0)-lower semicontinuous.

Proof. For each rational number $q \in \mathbb{Q}$ choose a sequence $(H_i^q)_{i \geq 1}$ of d_0-open sets such that

$$\{x \in M \mid f(x) \leq q\} = \bigcap_{i=1}^{\infty} H_i^q.$$

Order all sets H_i^q, $i \geq 1$, $q \in \mathbb{Q}$ into one sequence $G_0, G_1, \ldots$.

We denote by $B_0[x, r]$ the d_0-closed ball with center x and radius r. Let $\delta_0(x) = 1$ for all $x \in M$. We define the functions δ_j for $j = 1, 2, \ldots$ recursively such that

- $\delta_j(x_0, \ldots, x_j) \leq \frac{1}{2} \delta_{j-1}(x_0, \ldots, x_{j-1})$,

- $B_0[x_j, 2\delta_j(x_0, \ldots, x_j)] \subset G_i$ whenever $0 \leq i \leq j$ and $x_j \in G_i$.

Suppose now that the points x_j converge in (M, d) to some $x_\infty \in M$ and satisfy $d_0(x_j, x_{j+1}) \leq \delta_j(x_0, \ldots, x_j)$ for each $j \geq 0$. Consider any rational number q with $q < f(x_\infty)$. Since the set $\{x \in M \mid f(x) \leq q\}$ is the intersection of a subcollection of the sets G_i, there is i such that

$$\{x \in M \mid f(x) \leq q\} \subset G_i \text{ and } x_\infty \notin G_i.$$

We show that $f(x_j) > q$ for $j > i$. Indeed, $f(x_j) \leq q$ implies that $x_j \in G_i$, and so the ball $B_0[x_j, 2\delta_j(x_0, \ldots, x_j)]$ is contained in G_i. But for all $k \geq j$ we have

$$d_0(x_k, x_{k+1}) \leq \delta_k(x_0, \ldots, x_k) \leq 2^{j-k} \delta_j(x_0, \ldots, x_j),$$

hence

$$d_0(x_j, x_k) \leq d_0(x_j, x_{j+1}) + \cdots + d_0(x_{k-1}, x_k)$$
$$\leq \left(1 + 2^{-1} + \cdots + 2^{j-k+1} \right) \delta_j(x_0, \ldots, x_j) \leq 2\delta_j(x_0, \ldots, x_j).$$

Since the metric d_0 is d-continuous, we can take the limit $k \rightarrow \infty$ to obtain that $d_0(x_j, x_\infty) \leq 2\delta_j(x_0, \ldots, x_j)$. This implies

$$x_\infty \in B_0[x_j, 2\delta_j(x_0, \ldots, x_j)] \subset G_i,$$

which is a contradiction. Hence $f(x_j) > q$ for all $j > i$, and, consequently,

$$f(x_\infty) \leq \liminf_{j \to \infty} f(x_j). \qquad \square$$

Lemma 7.3.4. *Let (X, d) be a complete metric space, d_0 a continuous pseudometric on X, and M a G_δ subset of (X, d_0). Then M is (d, d_0)-complete.*

Proof. Let f be the indicator of the complement of M, $f = 1_{X \setminus M}$. Then f satisfies the assumptions of Lemma 7.3.3. Hence f is (d, d_0)-lower semicontinuous; that is, there are functions $\delta_j \colon X^j \longrightarrow (0, \infty)$ such that

$$f(x) \leq \liminf_{j \to \infty} f(x_j)$$

whenever $x_j \in X$ converge to x in (X, d) and $d_0(x_j, x_{j+1}) \leq \delta_j(x_0, \ldots, x_j)$ for each $j \geq 0$.

Suppose now that a Cauchy sequence $(x_j)_{j=0}^\infty$ in (M, d) satisfies the condition $d_0(x_j, x_{j+1}) \leq \delta_j(x_0, \ldots, x_j)$ for each $j \geq 0$. Since X is complete, the points x_j converge in (X, d) to some $x \in X$. Moreover, by the (d, d_0)-lower semicontinuity of f,

$$f(x) \leq \liminf_{j \to \infty} f(x_j) = 0,$$

implying that $x \in M$, and we are done. $\qquad \square$

We now deduce a variational principle for bimetric spaces from the variational principle of theorem 7.2.1. Since in this section we are guided mostly by particular applications, we do not aim for the same level of generality as in the previous section. In particular, we will not use infinite games in the statement of the theorem: from their point of view we describe a specific strategy of player (β) in advance. If needed, one can use the same arguments to increase the generality.

Theorem 7.3.5. *Let $f \colon M \longrightarrow \mathbb{R}$ be a function bounded from below on a metric space (M, d). Let d_0 be a continuous pseudometric on (M, d) such that*

- *M is (d, d_0)-complete, and*

- *f is (d, d_0)-lower semicontinuous.*

Let $F_j \colon M \times M \longrightarrow [0, \infty]$, $j \geq 0$, be functions (d, d_0)-lower semicontinuous in the second variable with $F_j(x, x) = 0$ for all $x \in M$, and let $r_j \searrow 0$ be such that

$$\inf_{d(x,y) > r_j} F_j(x, y) > 0.$$

If $x_0 \in M$ and $(\varepsilon_j)_{j=0}^{\infty}$ is any sequence of strictly positive numbers such that

$$f(x_0) < \varepsilon_0 + \inf_{x \in M} f(x) \quad and \quad \inf_{d(x_0,x) > r_0} F_0(x_0, x) > \varepsilon_0,$$

then one may find a sequence $(x_j)_{j=1}^{\infty}$ of points in M converging in metric d to some $x_\infty \in M$ and a d_0-continuous function $\varphi \colon M \longrightarrow [0, \infty]$ such that the function

$$h(x) := f(x) + \varphi(x) + \sum_{j=0}^{\infty} F_j(x_j, x)$$

attains its minimum on M at x_∞. Moreover, for all $0 \le j < k + 1 \le \infty$,

$$d(x_j, x_k) \le r_j, \ F_j(x_j, x_k) \le \varepsilon_j, \ f(x_k) \le f(x_0), \ and$$

$$h(x_\infty) \le \varepsilon_j + \inf_{x \in M} \left(f(x) + \varphi(x) + \sum_{i=0}^{j-1} F_i(x_i, x) \right).$$

If, in addition, (M, d_0) is a (metric) subspace of some (N, d_0), then φ may be required to be defined and d_0-continuous on the whole N.

Proof. Let $\delta_j \colon M^j \longrightarrow (0, \infty)$ witness both (d, d_0)-completeness of M and (d, d_0)-lower semicontinuity of f, and let $\delta_j^x \colon M^{j+1} \longrightarrow (0, \infty)$ witness the (d, d_0)-lower semicontinuity of $y \to F_j(x, y)$. Since diminishing ε_j, $j \ge 1$, makes the conclusion of the theorem stronger, we assume that $\sum_{j=1}^{\infty} \varepsilon_j < \infty$ and

$$\inf_{d(x,y) > r_j} F_j(x, y) > \varepsilon_j, \quad j \ge 0.$$

We describe a particular strategy that player (β) will use in the perturbation game. To distinguish between objects introduced in the theorem and those needed in the game, we will add $\tilde{}$ to the latter in cases of possible conflict. In the setup of the game we let $\tilde{\varepsilon}_0 = \varepsilon_0$. Recalling the notation introduced by player (α),

$$h_k(x) = f(x) + \sum_{0 \le j < k} \tilde{F}_j(x),$$

the move $k \ge 0$ of player (β) will consist of the following choices:

- the given x_0 if $k = 0$;

- any x_k with $h_k(x_k) \le \min\{h_k(x_{k-1}), \frac{1}{2}\eta_k + \inf_{x \in M} h_k(x)\}$ when $k \ge 1$;

- $\xi_k = \min\left\{ \delta_k(x_0, \dots, x_k), \min_{j=0,\dots,k} \delta_k^{x_j}(x_0, \dots, x_k) \right\}$;

- $\varphi_k(x) = 2\tilde{\varepsilon}_k \min\left\{ 1, \dfrac{d_0(x_k, x)}{\xi_k} \right\}$;

- $\tilde{F}_k(x) = F_k(x_k, x) + \varphi_k(x)$;

- $\tilde{r}_k = r_k$ and $\tilde{\varepsilon}_{k+1} = \varepsilon_{k+1}$.

With these choices, the rules of the perturbation game are clearly satisfied. Let (x_j) be the sequence obtained when (β) plays according to the described strategy and (α) uses the strategy from Theorem 7.2.1. Then we have that for every $0 \le j \le k < \infty$,

(a) $h_k(x_k) \le \tilde{\varepsilon}_j + \inf_{x \in M} h_j(x)$;

(b) $h_k(x_k) \le h_j(x_j)$;

(c) $\tilde{F}_j(x_k) \le \tilde{\varepsilon}_j$; and

(d) $d(x_j, x_k) \le \tilde{r}_j = r_j$.

Indeed, (a), (c), and (d) are consequences of (VP_1), (VP_3), and (VP_4), respectively, while (b) follows from the fact that $h_k(x_k) \le h_k(x_{k-1}) = h_{k-1}(x_{k-1})$.

We show that even the assumptions of Observation 7.2.2 hold. From (d) and $r_j \searrow 0$ we see that the sequence (x_j) is d-Cauchy. Let now $j \le k$. From (c) we see that $\varphi_j(x_k) \le \tilde{\varepsilon}_j$, which implies

$$d_0(x_j, x_k) \le \xi_j \le \delta_j(x_0, \dots, x_j). \tag{7.4}$$

Hence the sequence (x_j) satisfies the conditions of Definition 7.3.1 and so, in particular, it d-converges to some x_∞. By (7.4) and the (d, d_0)-lower semicontinuity of f we have

$$f(x_\infty) \le \liminf_{j \to \infty} f(x_j).$$

Assume $i \le j \le k$. Since $\xi_j \le \delta_j^{x_i}(x_0, \dots, x_j)$ the estimate (7.4) reveals that $d_0(x_j, x_k) \le \delta_j^{x_i}(x_0, \dots, x_j)$. The functions F_i are (d, d_0)-lower semicontinuous in the second variable, and so

$$F_i(x_i, x_\infty) \le \liminf_{j \to \infty} F_i(x_i, x_j).$$

(We used the observation following Definition 7.3.2 that the estimates of the d_0 distance are needed for large j only.) Noticing that φ_i is d_0-continuous, we see that also

$$\tilde{F}_i(x_\infty) \le \liminf_{j \to \infty} \tilde{F}_i(x_j).$$

In this way we have verified all assumptions of Observation 7.2.2. Consequently, h_∞ attains its minimum at x_∞ and all estimates (a)–(d) hold whenever $0 \le j < k+1 \le \infty$.

The series defining the function $\varphi(x) := \sum_{j=0}^{\infty} \varphi_j(x)$ converges uniformly and so defines a d_0-continuous function on M (or even on N under the additional assumption). Since with this φ, the function h coincides with h_∞, all statements of the theorem follow from (a)–(d), about which we now know that they hold also for $j < k = \infty$. For example, (b) with $k = \infty$ and $j = 0$ gives $f(x_\infty) \le h(x_\infty) \le f(x_0)$, and (a) with $k = \infty$ gives the last inequality. $\qquad \square$

Chapter Eight

Smoothness and asymptotic smoothness

We introduce the smoothness notions that will be used to prove our main results: the modulus of smoothness of a function in the direction of a family of subspaces and the much simpler notion of upper Fréchet differentiability. This leads to the key notion of spaces admitting bump functions smooth in the direction of a family of subspaces with modulus controlled by $\omega(t)$. We show how this notion is related to asymptotic uniform smoothness, and that very smooth bumps, and very asymptotically uniformly smooth norms, exist in all asymptotically c_0 spaces. This allows a new approach to results on Γ-almost everywhere Fréchet differentiability of Lipschitz functions.

8.1 MODULUS OF SMOOTHNESS

In Chapter 4 we have defined the modulus of asymptotic uniform smoothness of the Banach space X by

$$\bar{\rho}_X(t) = \sup_{\|x\|=1} \inf_{\dim X/Y < \infty} \sup_{\substack{y \in Y \\ \|y\| \leq t}} \|x + y\| - 1, \ t > 0,$$

and that this notion turned out to be useful in showing existence of points of almost Fréchet differentiability. Here we define a related notion of asymptotic smoothness for functions that are not necessarily norms, and in subsequent chapters we use this notion substantially to extend known results on existence of points of Fréchet differentiability of Lipschitz maps. The use of functions instead of norms has, in addition to greater generality of the results, the main advantage that a function may be smooth everywhere, which avoids a number of difficulties. (Perhaps these difficulties are only technical, but the arguments are rather involved and so even avoiding some technical difficulties is an advantage.) Also, we wish to treat asymptotic and nonasymptotic moduli together. To see where this leads, recall that *the modulus of smoothness of the function* $\Theta \colon X \longrightarrow \mathbb{R}$ *at a point* $x \in X$ is usually defined for $t \geq 0$ by

$$\rho_\Theta(x; t) = \sup_{y \in X, \|y\| \leq t} \big(\Theta(x + y) + \Theta(x - y) - 2\Theta(x) \big).$$

(The above definition is mainly used in the case of convex functions. For general functions we should perhaps speak about "upper symmetric smoothness.")

We will use the same concept, but make it more flexible by first introducing moduli of smoothness with respect to subspaces and then moduli of (asymptotic) smoothness with respect to families of subspaces of X. Since our moduli will be based on the formula from the above definition of the modulus of smoothness of functions, there will be a slight difference between the moduli of Chapter 4 and those introduced here; this has been already noted in Chapter 4 and will be made more precise shortly.

We will define the notion of smoothness with respect to nonempty families $\mathcal{Y}$ of subspaces of X that are *directed downward,* that is, for every $Y_1, Y_2 \in \mathcal{Y}$ there is $Y \in \mathcal{Y}$ such that $Y \subset Y_1 \cap Y_2$. In fact, our notions of smoothness may be defined also without these assumptions. However, we believe that it is better to agree that saying that "a function is smooth in the direction of $\mathcal{Y}$" implies these properties of $\mathcal{Y}$ instead of cluttering main results with spelling out assumptions that are anyhow needed for all but minor results. (Incidentally, little would be lost if we replaced "directed downward" by "closed under finite intersections.")

Definition 8.1.1. If Y is a subspace of X, *the modulus of smoothness of the function* $\Theta \colon X \longrightarrow \mathbb{R}$ *at a point* $x \in X$ *in the direction of Y is defined for* $t > 0$ *by*

$$\rho_{\Theta,Y}(x;t) = \sup_{y \in Y, \|y\| \le t} \left(\Theta(x+y) + \Theta(x-y) - 2\Theta(x) \right),$$

and if $\mathcal{Y}$ is a nonempty, downward directed family of subspaces of X, *the modulus of smoothness of the function* Θ *at a point* $x \in X$ *in the direction of $\mathcal{Y}$ is defined for* $t > 0$ by

$$\bar{\rho}_{\Theta,\mathcal{Y}}(x;t) = \inf_{Y \in \mathcal{Y}} \rho_{\Theta,Y}(x;t).$$

In the special case when $\mathcal{Y}$ is the family of finite codimensional subspaces of X, we write $\bar{\rho}_{\Theta}(x;t)$ instead of $\bar{\rho}_{\Theta,\mathcal{Y}}(x;t)$ and call it *the modulus of asymptotic smoothness of the function* Θ *at the point* x.

Notice that the choice $y = 0$ shows that $\rho_{\Theta,Y}(x;t) \ge 0$, and so also $\bar{\rho}_{\Theta,\mathcal{Y}}(x;t) \ge 0$, for all x, t, Y and $\mathcal{Y}$. Notice also that the earlier concepts may be considered as simplifications of notation in special cases of the last one since

$$\rho_{\Theta,Y}(x;t) = \bar{\rho}_{\Theta,\{Y\}}(x;t) \quad \text{and} \quad \rho_{\Theta}(x;t) = \bar{\rho}_{\Theta,X}(x;t).$$

In another direction, given $y \in X$, we will simplify the notation for moduli of smoothness by writing $\rho_{\Theta,y}$ instead of $\rho_{\Theta,\text{span}\{y\}}$.

As we have already indicated, the modulus of asymptotic uniform smoothness $\bar{\rho}_X(t)$ of the space is related to the modulus of asymptotic smoothness of its norm $\bar{\rho}_{\|\cdot\|}(x;t)$ by

$$\bar{\rho}_X(t) \le \sup_{\|x\|=1} \bar{\rho}_{\|\cdot\|}(x;t) \le 2\bar{\rho}_X(t).$$

More generally, if the epigraph of Θ has a supporting hyperplane at x, that is, there is $x^* \in X^*$ such that $\Theta(x+y) \ge \Theta(x) + x^*(y)$ for all y, and if $Y \subset \operatorname{Ker} x^*$, then

$$\sup_{y \in Y, \|y\| \le t} \left(\Theta(x+y) - \Theta(x) \right) \le \rho_{\Theta,Y}(x;t) \le 2 \sup_{y \in Y, \|y\| \le t} \left(\Theta(x+y) - \Theta(x) \right).$$

We list the standard examples of the behavior of these moduli on ℓ_p spaces.

Example 8.1.2. Let $X = \ell_p$ and $\Theta(x) = \|x\|^p$.

- If $1 < p < 2$ then $\rho_\Theta(x, t) \leq C_p t^p$, and this estimate cannot be improved.

- If $p = 2$ then $\rho_\Theta(x, t) = 2t^2$.

- If $2 < p < \infty$ then $\rho_\Theta(x, t) \leq C_p t^2$, $0 \leq t \leq 1$, and this estimate cannot be improved.

Example 8.1.3. Let $1 \leq p < \infty$, X be the ℓ_p-sum of finite dimensional spaces, and let $\Theta(x) = \|x\|^p$. Then $\bar\rho_\Theta(x; t) = 2t^p$.

Example 8.1.4. Let $1 \leq p < \infty$, X be the c_0-sum of finite dimensional spaces, and let $\Theta(x) = \|x\|^p$. Then $\bar\rho_\Theta(x; t) = 2\max\{0, t^p - \|x\|^p\}$.

We will use the concept of modulus of smoothness not only in the space X but (more important) in the space $L(\mathbb{R}^n, X)$ of bounded linear operators from $\mathbb{R}^n$ to X. However, we will need it only in the direction of special subspaces. Allowing a slight formal incorrectness, we will use, for $\Theta\colon L(\mathbb{R}^n, X) \longrightarrow \mathbb{R}$, $T \in L(\mathbb{R}^n, X)$, a subspace Y of X, and $t > 0$, the symbol $\rho_{\Theta,Y}(T; t)$ instead of the cumbersome

$$\rho_{\Theta,\{S \in L(\mathbb{R}^n, X) \mid S(\mathbb{R}^n) \subset Y\}}(T; t).$$

Much of the material in the following chapters is based on the idea that $\mathbb{R}^n$-valued Lipschitz functions on spaces on which there is a suitable function Θ with modulus of (asymptotic) smoothness $o(t^n)$ should have points of Fréchet differentiability. (We will prove a better result, namely, that $o(t^n \log^{n-1}(1/t))$ suffices, which is especially important for $n = 2$.) The word "suitable" indicates that further assumptions, including smoothness assumptions of lower order, may be needed. For full differentiability results we will need a stronger version of the assumption that the (nonasymptotic) modulus of smoothness of Θ is $o(t)$, namely, that Θ is everywhere upper Fréchet smooth (or upper Fréchet differentiable).

Definition 8.1.5. We say that $\Theta\colon X \longrightarrow \mathbb{R}$ is *upper Fréchet smooth at* $x \in X$ if there is $x^* \in X^*$ such that

$$\limsup_{y \to 0} \frac{\Theta(x + y) - \Theta(x) - x^*(y)}{\|y\|} \leq 0.$$

The function Θ is said to be *upper Fréchet smooth (on X)* if it is upper Fréchet smooth at every $x \in X$.

One could, if one wished, go farther to the concept of "upper Fréchet differentiability in direction of a family $\mathcal{Y}$ of subspaces of X." Although it seems probable that some of our results could use this concept, we do not see any pertinent examples that would justify the effort.

We give several simple results mainly concerned with estimating moduli of smoothness of functions obtained from functions with known moduli by simple operations. The first result does not need a proof; it is just stated for completeness.

Observation 8.1.6. *Suppose that* $\Theta_1, \Theta_2 \colon X \longrightarrow \mathbb{R}$, $x_0 \in X$, $a \geq 0$, *and* $c \neq 0$, *and let* $\widetilde{\Theta}_1(x) = a\Theta_1(cx - x_0)$ *and* $\widetilde{\Theta}_2(x) = \Theta_1(x) + \Theta_2(x)$. *Then*

(i) $\bar{\rho}_{\widetilde{\Theta}_1, \mathcal{Y}}(x; t) = a\bar{\rho}_{\Theta_1, \mathcal{Y}}(cx - x_0; |c|t)$;

(ii) $\bar{\rho}_{\widetilde{\Theta}_2, \mathcal{Y}}(x; t) \leq \bar{\rho}_{\Theta_1, \mathcal{Y}}(x; t) + \bar{\rho}_{\Theta_2, \mathcal{Y}}(x; t)$;

(iii) *if* Θ_1, Θ_2 *are everywhere upper Fréchet smooth then so are* $\widetilde{\Theta}_1$ *and* $\widetilde{\Theta}_2$.

Lemma 8.1.7. *Suppose that* $\Theta_1, \Theta_2 \colon X \longrightarrow \mathbb{R}$, *and let* $\Theta = \min\{\Theta_1, \Theta_2\}$.

(i) *If* $M = \{x \in X \mid \Theta(x) = \Theta_1(x)\}$, *we have*

$$\bar{\rho}_{\Theta, \mathcal{Y}}(x; t) \leq \bar{\rho}_{\Theta_1, \mathcal{Y}}(x; t)\, \mathbf{1}_M(x) + \left(1 - \mathbf{1}_M(x)\right)\bar{\rho}_{\Theta_2, \mathcal{Y}}(x; t),$$

where $\mathbf{1}_M$ *denotes the indicator of the set* M. *In particular,*

$$\bar{\rho}_{\Theta, \mathcal{Y}}(x; t) \leq \max\{\bar{\rho}_{\Theta_1, \mathcal{Y}}(x; t), \bar{\rho}_{\Theta_2, \mathcal{Y}}(x; t)\}.$$

(ii) *If* Θ_1, Θ_2 *are upper Fréchet smooth at* x, *then so is* Θ.

Proof. The first statement follows on noticing that, if $\Theta(x) = \Theta_1(x)$, then for every $y \in X$,

$$\Theta(x + y) + \Theta(x - y) - 2\Theta(x) \leq \Theta_1(x + y) + \Theta_1(x - y) - 2\Theta_1(x).$$

If moreover $x^* \in X^*$ we obtain

$$\Theta(x + y) - \Theta(x) - x^*(y) \leq \Theta_1(x + y) - \Theta_1(x) - x^*(y),$$

which implies (ii). $\qquad\square$

Lemma 8.1.8. *Let* $\Theta \colon L(\mathbb{R}^n, X) \longrightarrow \mathbb{R}$ *be defined by* $\Theta(T) = \Theta_0(Tu)$, *where* $\Theta_0 \colon X \longrightarrow \mathbb{R}$ *and* u *is a fixed unit vector from* $\mathbb{R}^n$.

(i) $\bar{\rho}_{\Theta, \mathcal{Y}}(T; t) = \bar{\rho}_{\Theta_0, \mathcal{Y}}(Tu; t)$.

(ii) *If* Θ_0 *is everywhere upper Fréchet smooth, then so is* Θ.

Proof. For the first statement, let T, t be given. For any $\varepsilon > 0$ there is $Y \in \mathcal{Y}$ such that $\rho_{\Theta_0, Y}(Tu; t) \leq \bar{\rho}_{\Theta_0, \mathcal{Y}}(Tu; t) + \varepsilon$. Then for every $S \in L(\mathbb{R}^n, X)$ with $S(\mathbb{R}^n) \subset Y$ and $\|S\| \leq t$,

$$\Theta(T + S) + \Theta(T - S) - 2\Theta(T) = \Theta_0(Tu + Su) + \Theta_0(Tu - Su) - 2\Theta_0(Tu)$$
$$\leq \rho_{\Theta_0, Y}(Tu; t) \leq \bar{\rho}_{\Theta_0, \mathcal{Y}}(Tu; t) + \varepsilon.$$

Hence $\bar{\rho}_{\Theta, \mathcal{Y}}(T; t) \leq \bar{\rho}_{\Theta_0, \mathcal{Y}}(Tu; t)$. For the opposite inequality we argue similarly: find $Y \in \mathcal{Y}$ such that $\rho_{\Theta, Y}(T; t) \leq \bar{\rho}_{\Theta, \mathcal{Y}}(T; t) + \varepsilon$. Given any $y \in Y$ with $\|y\| \leq t$, we let $S = y \otimes u$. Since $\|S\| \leq t$ we get

$$\Theta_0(Tu + y) + \Theta_0(Tu - y) - 2\Theta_0(Tu) = \Theta(T + S) + \Theta(T - S) - 2\Theta(T)$$
$$\leq \rho_{\Theta, Y}(T; t) \leq \bar{\rho}_{\Theta, \mathcal{Y}}(Tu; t) + \varepsilon.$$

To prove (ii), let $T \in L(\mathbb{R}^n, X)$ and find $x^* \in X^*$ such that

$$\Theta_0(x) - \Theta_0(Tu) \leq x^*(x - Tu) + o(\|x - Tu\|)$$

for all $x \in X$. The linear functional $x^* \otimes u \in L^*(\mathbb{R}^n, X)$ is the upper Fréchet derivative of Θ at T. Indeed,

$$\begin{aligned} \Theta(S) - \Theta(T) &= \Theta_0(Su) - \Theta_0(Tu) \\ &\leq x^*(Su - Tu) + o(\|Su - Tu\|) \\ &= (x^* \otimes u)(S - T) + o(\|S - T\|). \end{aligned} \qquad \square$$

The following simple result on moduli of smoothness of infinite sums is of great importance. Its statements (ii) and/or (iii), in the form of their direct consequence Lemma 8.2.5, will be used in every proof of a differentiability result in a space admitting a "smooth bump."

In the statement (ii) we meet for the first time the requirement that the modulus is controlled by a given function $\omega \colon (0, \infty) \longrightarrow (0, \infty)$. Throughout, we will assume that ω is increasing.

Lemma 8.1.9. *Suppose that* $x \in X$, $r > 0$, $\Theta_i \colon X \longrightarrow \mathbb{R}$ *are equibounded on* $B(x, r)$ *and* $\sum_{i=1}^{\infty} \lambda_i$ *is a convergent series of non-negative numbers. Let* $\Theta = \sum_{i=1}^{\infty} \lambda_i \Theta_i$. *Then:*

(i) *For every* $0 < t < r$, $\bar{\rho}_{\Theta, \mathcal{Y}}(x; t) \leq \sum_{i=1}^{\infty} \lambda_i \bar{\rho}_{\Theta_i, \mathcal{Y}}(x; t)$.

(ii) *If* $\omega \colon (0, \infty) \longrightarrow (0, \infty)$ *is such that* $\sup_{i \in \mathbb{N}} \bar{\rho}_{\Theta_i, \mathcal{Y}}(x; t) = O(\omega(t))$, *and for each* i, $\bar{\rho}_{\Theta_i, \mathcal{Y}}(x; t) = o(\omega(t))$ *as* $t \searrow 0$, *then*

$$\bar{\rho}_{\Theta, \mathcal{Y}}(x; t) = o(\omega(t)) \text{ as } t \searrow 0.$$

(iii) *If* $\sup_{i \in \mathbb{N}} |\Theta_i(x + y) - \Theta_i(x)| = O(\|y\|)$ *as* $\|y\| \to 0$, *and for each* i, *the function* Θ_i *is upper Fréchet smooth at* x, *then* Θ *is upper Fréchet smooth at* x.

Proof. (i) Let $C < \infty$ be such that $|\Theta_i(z)| \leq C$ for all $i \in \mathbb{N}$ and $z \in B(x, r)$. Let $0 < t < r$. Fixing, for a moment, an $\varepsilon > 0$, we use the convergence of $\sum_{i=1}^{\infty} \lambda_i$ to find $j \in \mathbb{N}$ such that $\sum_{i=j+1}^{\infty} \lambda_i \leq \varepsilon$. For each $i = 1, \dots, j$ find $Y_i \in \mathcal{Y}$ such that $\rho_{\Theta_i, Y_i}(x; t) \leq \varepsilon + \bar{\rho}_{\Theta_i, \mathcal{Y}}(x; t)$. Find $Y \in \mathcal{Y}$, $Y \subset \bigcap_{i=1}^{j} Y_i$. Then for every $y \in Y$ with $\|y\| \leq t$,

$$\sum_{i=1}^{j} \lambda_i \big(\Theta_i(x + y) + \Theta_i(x - y) - 2\Theta_i(x)\big) \leq \varepsilon \sum_{i=1}^{j} \lambda_i + \sum_{i=1}^{j} \lambda_i \bar{\rho}_{\Theta_i, \mathcal{Y}}(x; t).$$

Noticing that also

$$\sum_{i=j+1}^{\infty} \lambda_i \big(\Theta_i(x + y) + \Theta_i(x - y) - 2\Theta_i(x)\big) \leq 4C \sum_{i=j+1}^{\infty} \lambda_i \leq 4C\varepsilon,$$

we infer that

$$\Theta(x+y) + \Theta(x-y) - 2\Theta(x) \le \varepsilon\left(4C + \sum_{i=1}^{\infty}\lambda_i\right) + \sum_{i=1}^{\infty}\lambda_i\bar\rho_{\Theta_i,\mathcal{Y}}(x;t).$$

Since $\varepsilon > 0$ is arbitrary, the first statement follows.

(ii) Let $t_0 > 0$ and $C < \infty$ be such that $\bar\rho_{\Theta_i,\mathcal{Y}}(x;t) \le C\omega(t)$ for every $t \in (0, t_0)$ and $i \in \mathbb{N}$. By (i),

$$\frac{\bar\rho_{\Theta,\mathcal{Y}}(x;t)}{\omega(t)} \le \sum_{i=0}^{\infty}\lambda_i \frac{\bar\rho_{\Theta_i,\mathcal{Y}}(x;t)}{\omega(t)},$$

where the series on the right converges uniformly on $(0, t_0)$ since its terms are majorized by $C\lambda_i$. Hence we may exchange limit and summation, and so infer the statement.

(iii) Let $t_0 > 0$ and $C < \infty$ be such that $|\Theta_i(x+y) - \Theta_i(x)| \le C\|y\|$ for every $\|y\| \le t_0$ and $i \in \mathbb{N}$. For each i we find an upper Fréchet derivative $x_i^* \in X^*$ at x. Then $\|x_i^*\| \le C$, and so the series $\sum_{i=1}^{\infty}\lambda_i x_i^*$ defines an element $x^* \in X^*$. Thus

$$\frac{\Theta(x+y) - \Theta(x) - x^*(y)}{t} \le \sum_{i=1}^{\infty}\lambda_i \sup_{\|z\|\le t}\frac{\Theta_i(x+z) - \Theta_i(x) - x_i^*(z)}{t},$$

where the series on the right converges uniformly on $(0, t_0)$ since its terms are majorized by $2C\lambda_i$. By exchanging limit and summation, the sum on the right has limit zero as $t \searrow 0$, and consequently Θ is upper Fréchet differentiable at x. $\qquad\square$

In the next proposition we show that, in separable spaces and under suitable uniform continuity assumptions, smoothness in direction of a general family of subspaces is reducible to smoothness in direction of its countable subfamily.

Observation 8.1.10. *Let Θ be uniformly continuous on $B(x_0, r)$. Then the function $(x, t) \to \bar\rho_{\Theta,\mathcal{Y}}(x;t)$ is uniformly continuous on the set*

$$\{(x,t) \mid t > 0, \|x - x_0\| + t < r\}.$$

Proof. Given $\varepsilon > 0$, find $\delta > 0$ such that $|\Theta(y) - \Theta(x)| < \frac{1}{5}\varepsilon$ for all $x, y \in B(x_0, r)$ and $\|y - x\| < \delta$.

Suppose that $t, s > 0$, $\|x - x_0\| + t < r$, $\|z - x_0\| + s < r$, and $\|x - z\| + |t - s| < \delta$. Find $Y \in \mathcal{Y}$ such that

$$\rho_{\Theta,Y}(z, s) < \bar\rho_{\Theta,\mathcal{Y}}(z, s) + \frac{\varepsilon}{5}.$$

It follows that for each $u \in Y$ with $\|u\| \le t$ we may find $v \in Y$ with $\|v\| \le s$ such that $\|u - v\| \le |t - s|$. Then all points $x \pm u$ and $z \pm v$ belong to $B(x_0, r)$, and we estimate

$$\Theta(x+u) + \Theta(x-u) - 2\Theta(x)$$

$$\le \Theta(z+v) + \frac{\varepsilon}{5} + \Theta(z-v) + \frac{\varepsilon}{5} - 2\Theta(z) + \frac{2\varepsilon}{5}$$

$$\le \rho_{\Theta,Y}(z, s) + \frac{4\varepsilon}{5} < \bar\rho_{\Theta,\mathcal{Y}}(z; s) + \varepsilon.$$

Hence $\rho_{\Theta,Y}(x;t) \leq \bar{\rho}_{\Theta,\mathcal{Y}}(z;s) + \varepsilon$, implying that $\bar{\rho}_{\Theta,\mathcal{Y}}(x;t) \leq \bar{\rho}_{\Theta,\mathcal{Y}}(z;s) + \varepsilon$. Exchanging (x,t) and (z,s) completes the argument. $\qquad\square$

Proposition 8.1.11. *Let X be a separable Banach space. Suppose that $\Theta\colon X \longrightarrow \mathbb{R}$ is uniformly continuous on bounded sets and $\mathcal{Y}$ is a nonempty family of subspaces of X. Then there is a countable subfamily $\mathcal{Z} \subset \mathcal{Y}$ such that $\bar{\rho}_{\Theta,\mathcal{Z}}(x;t) = \bar{\rho}_{\Theta,\mathcal{Y}}(x;t)$ for every $x \in X$ and $t > 0$.*

Proof. Notice first that $\bar{\rho}_{\Theta,\mathcal{Z}}(x;t) \geq \bar{\rho}_{\Theta,\mathcal{Y}}(x;t)$ holds for any $\mathcal{Z} \subset \mathcal{Y}$. So it suffices only to prove the opposite inequality.

Let $(x_i, t_i, r_i) \in X \times (0,\infty) \times (0,\infty)$, where $i \in \mathbb{N}$, form a dense subset of $X \times (0,\infty) \times (0,\infty)$. For each i we choose a subspace $Y_i \in \mathcal{Y}$ such that

$$\rho_{\Theta,Y_i}(x_i;t_i) < \bar{\rho}_{\Theta,\mathcal{Y}}(x_i;t_i) + r_i.$$

We show that $\mathcal{Z} = \{Y_i \mid i \in \mathbb{N}\}$ has the required property. To this aim suppose that $x \in X$ and $t > 0$ and find a sequence (i_k) such that $x_{i_k} \to x$, $t_{i_k} \searrow t$, and $r_{i_k} \to 0$.

Denote $r = t_{i_1} + \sup_{k\in\mathbb{N}} \|x_{i_k} - x\|$. Let $\varepsilon > 0$. Since Θ is uniformly continuous on $B(x,r)$ we find $\delta > 0$ such that $|\Theta(y) - \Theta(z)| < \varepsilon$ whenever $y, z \in B(x,r)$ and $\|y - z\| < \delta$. Using Observation 8.1.10 we may diminish δ if necessary to achieve that

$$|\bar{\rho}_{\Theta,\mathcal{Y}}(y;\tau) - \bar{\rho}_{\Theta,\mathcal{Y}}(z,s)| < \varepsilon$$

whenever $\|y - z\| + |\tau - s| < \delta$ and $\|y - x\| + |\tau| < r$, $\|z - x\| + |s| < r$.

Taking now k so large that $r_{i_k} < \varepsilon$ and $\|x_{i_k} - x\| + |t_{i_k} - t| < \delta$, we have for every $y \in Y_{i_k}$, $\|y\| \leq t$,

$$\begin{aligned}
\Theta(x + y) + \Theta(x - y) - 2\Theta(x) &\leq \Theta(x_{i_k} + y) + \Theta(x_{i_k} - y) - 2\Theta(x_{i_k}) + 4\varepsilon \\
&\leq \rho_{\Theta,Y_{i_k}}(x_{i_k};t_{i_k}) + 4\varepsilon \\
&< \bar{\rho}_{\Theta,\mathcal{Y}}(x_{i_k};t_{i_k}) + 4\varepsilon + r_{i_k} \\
&\leq \bar{\rho}_{\Theta,\mathcal{Y}}(x;t) + 5\varepsilon.
\end{aligned}$$

Hence $\bar{\rho}_{\Theta,\mathcal{Z}}(x;t) \leq \bar{\rho}_{\Theta,\mathcal{Y}}(x;t) + 5\varepsilon$, and the arbitrariness of $\varepsilon > 0$ shows that $\bar{\rho}_{\Theta,\mathcal{Z}}(x;t) \leq \bar{\rho}_{\Theta,\mathcal{Y}}(x;t)$. $\qquad\square$

Generalizing the qualitative part of Proposition 4.2.7, we now show that even existence of an asymptotically smooth "bump" function implies Asplundness. We go a bit farther: instead of requiring smoothness with respect to the family of subspaces Y for which X/Y is finite dimensional, we require it only for the much larger family of those Y for which $(X/Y)^*$ is separable.

Proposition 8.1.12. *Suppose that X is a separable Banach space for which one can find a family $\mathcal{Y}$ of subspaces of X, a lower semicontinuous function $\Theta\colon X \longrightarrow \mathbb{R}$, and a nonempty bounded set $G \subset X$ such that*

- *$-\infty < \inf_{x\in G} \Theta(x) < \inf_{x\notin G} \Theta(x)$;*

- *X/Y has separable dual for each $Y \in \mathcal{Y}$; and*

- $\bar{\rho}_{\Theta,\mathcal{Y}}(x;t) = o(t)$ as $t \searrow 0$ for every $x \in G$.

Then X is an Asplund space.

Proof. Let $\inf_{x \in G} \Theta(x) < a < b \leq \inf_{x \notin G} \Theta(x)$. We assume that $\bar{\rho}_{\Theta,\mathcal{Y}}(x;t) = o(t)$ as $t \searrow 0$ for every $x \in X$; if necessary we achieve this by replacing Θ by $\min\{b,\Theta\}$. We may also assume that every finite codimensional subspace Z of any $Y \in \mathcal{Y}$ belongs to $\mathcal{Y}$: indeed, for each such Z the dual of X/Z is separable and $\bar{\rho}_{\Theta,\mathcal{Y}}$ becomes smaller when one adds subspaces to $\mathcal{Y}$.

Assume that X is not Asplund. Then by Theorem 3.2.3 we may assume that X is equipped with a norm $\|\cdot\|$ such that, for every $x \in X$, every $Y \in \mathcal{Y}$ with X/Y Asplund (i.e., $(X/Y)^*$ separable), and every $r > 0$, there is $y \in Y$ with $\|y\| \leq r$ and $\|x + y\| - \|x\| > \frac{1}{2}r$. In this new norm the modulus of smoothness of Θ at the point (x,t) is dominated by the value of the original modulus at the point (x, Kt) for some constant K. So our assumptions hold also with this new norm.

Let $r_0 > 0$ be such that $G \subset B(0, r_0)$. The rest of the proof will be a standard application of a perturbational variational principle on the space $M = \overline{B(0, r_0)}$ to the function

$$f(x) = \Theta(x) - \eta\|x\|/r_0,$$

where $\eta = \frac{1}{2}(b - a)$. For definiteness, we will show it using Corollary 7.2.4, although in this case the finer points of this corollary are not needed and so the reader may find other variants of the principle easier to use. We let $F_j(x,y) = 2^{-j-4}\eta\|x - y\|/r_0$, $r_j = 2^{-j}r_0$, $\varepsilon_j = 2^{-j-5}\eta r_j/r_0$, and choose $x_0 \in M$ such that

$$f(x_0) < \varepsilon_0 + \inf_{x \in M} f(x).$$

It follows from Corollary 7.2.4 that there are $x_j \in M$, $j \geq 1$, such that the function

$$h(x) := f(x) + \sum_{j=0}^{\infty} F_j(x_j, x)$$

attains minimum at some point $z \in M$. Then $z \in B(0, r_0)$ since $h(x) - f(x)$ is bounded by η, and so points y on the boundary of $B(0, r_0)$ satisfy

$$h(y) \geq b - \eta > \eta + \inf_{x \in G} \Theta(x) \geq \eta + \inf_{x \in G} f(x) \geq \inf_{x \in G} h(x) \geq \inf_{x \in M} h(x).$$

For $\varepsilon := \eta/r_0$ find $r > 0$ such that $\bar{\rho}_{\Theta,\mathcal{Y}}(z;t) < \frac{1}{4}\varepsilon t$ for every $0 < t \leq r$. Assured by the previous argument that $\|z\| < r_0$, we diminish r to achieve also that $r < r_0 - \|z\|$. Let $Y \in \mathcal{Y}$ be such that

$$\Theta(z + w) + \Theta(z - w) - 2\Theta(z) < \frac{\varepsilon r}{4} \tag{8.1}$$

whenever $w \in Y$ with $\|w\| \leq r$. Find $z^* \in X^*$ with $\|z^*\| = 1$ and $z^*(z) = \|z\|$ and use that $Y \cap \mathrm{Ker}(z^*) \in \mathcal{Y}$ to choose $y \in Y \cap \mathrm{Ker}(z^*)$ such that $\|y\| \leq r$ and $\|z + y\| - \|z\| > \frac{1}{2}r$. Since $\|z - y\| - \|z\| \geq z^*(-y) = 0$,

$$-\varepsilon\|z + y\| - \varepsilon\|z - y\| + 2\varepsilon\|z\| < -\frac{\varepsilon r}{2}. \tag{8.2}$$

The difference $(h - f)$ is Lipschitz with constant at most $\frac{1}{8}\varepsilon$. Thus

$$(h - f)(z + y) + (h - f)(z - y) - 2(h - f)(z) \leq \frac{\varepsilon r}{4}. \tag{8.3}$$

Adding the estimate (8.1) with $w = y$ to the estimates (8.2) and (8.3), we infer that

$$h(z + y) + h(z - y) - 2h(z) < \frac{\varepsilon r}{4} - \frac{\varepsilon r}{2} + \frac{\varepsilon r}{4} = 0,$$

contradicting the fact that h attains its minimum at z.   □

8.2 SMOOTH BUMPS WITH CONTROLLED MODULUS

The key assumption that we will make on a Banach space X to obtain strong Fréchet differentiability results is the existence of a function $\Theta \colon X \longrightarrow \mathbb{R}$ smooth in the direction of a suitable family $\mathcal{Y}$ of subspaces of X and with modulus controlled by some function ω. As usual, we can state this assumption equivalently in several forms, some formally weaker and others stronger. The goal of the next proposition is to show equivalence of these forms, which will allow us to pick in Definition 8.2.3 the weakest one as our key notion.

We will consider smoothness in the direction of a family $\mathcal{Y}$ of subspaces of the given Banach space X. As before, we will always assume that $\mathcal{Y}$ is nonempty and directed downward. The control of smoothness will be established with the help of a function $\omega \colon (0, \infty) \longrightarrow (0, \infty)$, which we will always assume to be increasing and with $\omega(t) \to 0$ as $t \searrow 0$. We will in fact need this control only with functions $\omega = \omega_n$ defined in Section 9.3, in which case (as in the case of all ω satisfying the doubling condition) we may simplify the conditions of the following proposition, and similar conditions later, by letting $c = 1$ instead of saying "for every $c > 0$."

Proposition 8.2.1. *Suppose that for the given function ω and family $\mathcal{Y}$ of subspaces of X we can find a nonempty bounded open set $G \subset X$ and a function $\Theta_0 \colon X \longrightarrow \mathbb{R}$ such that*

(a) $-\infty < \inf_{x \in X} \Theta_0(x) < \inf_{x \notin G} \Theta_0(x)$;

(b) Θ_0 *is lower semicontinuous on* $\overline{G}$;

(c) $\sup_{x \in G} \bar{\rho}_{\Theta_0, \mathcal{Y}}(x; t) = O(\omega(ct))$ *as* $t \searrow 0$ *for every* $c > 0$;

(d) $\bar{\rho}_{\Theta_0, \mathcal{Y}}(x; t) = o(\omega(ct))$ *as* $t \searrow 0$ *for every* $x \in G$ *and* $c > 0$.

Then there is a function $\Theta \colon X \longrightarrow \mathbb{R}$ *such that*

(A) Θ *is bounded,* $\Theta(0) = 0$ *and* $\inf_{\|x\| > s} \Theta(x) > 0$ *for every* $s > 0$;

(B) Θ *is lower semicontinuous on* X;

(C) *for every* $c > 0$, $\bar{\rho}_{\Theta, \mathcal{Y}}(x; t)$ *is bounded by a constant multiple of* $\omega(ct)$;

(D) $\bar{\rho}_{\Theta,y}(x;t) = o(\omega(ct))$ as $t \searrow 0$ for every $x \in X$ and $c > 0$.

If, in addition, Θ_0 is on G continuous, uniformly continuous, Lipschitz, or Lipschitz and upper Fréchet smooth, then Θ has the same property on X.

Proof. First notice that $\inf_{x \in X} \Theta_0(x) = \inf_{x \in G} \Theta_0(x)$ and choose a number κ such that $\inf_{x \in X} \Theta_0(x) < \kappa < \inf_{x \notin G} \Theta_0(x)$. Denote

$$a := \inf_{x \in X} \Theta_0(x) \quad \text{and} \quad \mathcal{Q}(x) := \min\{\kappa, \Theta_0(x)\} - a.$$

Also choose any $0 < \beta_k \nearrow \infty$, let $\tau = \kappa - a$, and find $r > 0$ such that $G \subset B(0,r)$. Observe that $\inf_{x \in X} \mathcal{Q}(x) = 0$, $\mathcal{Q}(x) \leq \tau$ for all $x \in X$, and that $\mathcal{Q}$ satisfies conditions (b)–(d) even with G replaced by X. (For the latter two see Lemma 8.1.7.) As for (c), we actually get a little bit more: since $\mathcal{Q}$ is bounded, $\bar{\rho}_{\mathcal{Q},y}$ is also bounded, and it follows that for every $c > 0$, $\bar{\rho}_{\mathcal{Q},y}(x;t)/\omega(ct)$ is bounded. Hence there are constants C_j such that

$$\bar{\rho}_{\mathcal{Q},y}(x;t) \leq C_j \omega(t/\beta_j^2)$$

for every $x \in X$ and $t > 0$. Finally, we choose

$$0 < \alpha_j \leq \frac{1}{2^j + C_j + \beta_j}.$$

To define Θ, we will use the perturbation game described in Section 7.2 with the following setup: $M = X$, $f = 0$, $\varepsilon_0 = \frac{1}{2}\alpha_0\tau$. We also let $r_k = 2r/\beta_k$ and $\varepsilon_k = \frac{1}{4}\alpha_k\tau$ for $k \geq 1$, so these choices of player (β) have been already determined.

It remains to describe the choice of x_k and F_k. Recall that in the kth move, player (β) is given $\eta_k > 0$ and $h_k(x) = \sum_{0 \leq j < k} F_j(x)$. To make the choice, (β) finds x_k such that

$$h_k(x_k) < \frac{\eta_k}{2} + \inf_{x \in X} h_k(x),$$

chooses $z_k \in X$ such that $\mathcal{Q}(z_k) < \frac{1}{2}\min\{\eta_k, \tau\}$ (so necessarily $\|z_k\| < r$), and defines

$$F_k(x) := \alpha_k \mathcal{Q}(\beta_k(x - x_k) + z_k).$$

If $\|x - x_k\| > r_k$ then $\|\beta_k(x - x_k) + z_k\| > \beta_k r_k - r = r$, which implies that $F_k(x) \geq \alpha_k\tau/2 > \varepsilon_k$. Since also

$$h_k(x_k) + F_k(x_k) < \frac{\eta_k}{2} + \inf_{x \in X} h_k(x) + \frac{\alpha_k}{2}\min\{\eta_k, \tau\} \leq \eta_k + \inf_{x \in X} h_k(x),$$

the move of (β) complies with the rules of the perturbation game. Hence (α), playing according to the strategy from Theorem 7.2.1, achieves, by Corollary 7.2.3, that the sequence (x_k) converges to some $x_\infty \in X$ and the function

$$h(x) := \sum_{j=0}^{\infty} F_j(x)$$

satisfies $h(x_\infty) < \inf\{h(x) \mid s < \|x - x_\infty\|\}$ for every $s > 0$. Consequently, the function

$$\Theta(x) := h(x_\infty + x) - h(x_\infty)$$

satisfies (A).

To prove the remaining statements, we denote $Q_j(x) = Q(\beta_j(x - x_j) + z_j)$. Using Observation 8.1.6 (i) we infer that

$$\bar{\rho}_{Q_j,y}(x;t) = \bar{\rho}_{Q,y}(\beta_j(x - x_j) + z_j; \beta_j t) \leq C_j \omega(t/\beta_j).$$

The functions Q_j are non-negative and bounded from above by τ, so we may apply Lemma 8.1.9 (i) to see that for all $c > 0$,

$$\frac{\bar{\rho}_{\Theta,y}(x;t)}{\omega(ct)} \leq \sum_{j=0}^{\infty} \alpha_j \frac{\bar{\rho}_{Q_j,y}(x;t)}{\omega(ct)}.$$

The series on the right converges uniformly since its terms are majorized by $\alpha_j C_j$ for all j such that $1/\beta_j \leq c$. This implies (D), and recalling that each $\bar{\rho}_{Q_j,y}(x;t)/\omega(ct)$ is a bounded function, it implies (C) as well.

It remains to consider the additional properties. First observe that each function Q_j has the corresponding property on X: this is obvious for all properties except upper Fréchet smoothness, which follows from Lemma 8.1.7 (ii). Since $\sum_{j=0}^{\infty} \alpha_j < \infty$, the series $\sum_{j=0}^{\infty} F_j$ defining h converges uniformly, and so all continuity properties are preserved. If Θ_0 is Lipschitz on G, the function Q is Lipschitz on X and

$$\sum_{j=0}^{\infty} \mathrm{Lip}(F_j) \leq \sum_{j=0}^{\infty} \alpha_j \beta_j \mathrm{Lip}(Q) < \infty.$$

Hence Θ is Lipschitz. If Θ_0 is Lipschitz and everywhere upper Fréchet smooth, we apply Lemma 8.1.9 (iii) to the functions Q_j which yields that Θ is also everywhere upper Fréchet smooth. $\qquad\square$

Remark 8.2.2. In case we need to preserve other properties of Θ_0 than those listed in Proposition 8.2.1, it may be helpful to notice that we have in fact proved more. For any $0 < \beta_k \nearrow \infty$ there are $a_k > 0$ such that whenever $\alpha_k > 0$, $\sum_{k=1}^{\infty} \alpha_k a_k < \infty$ and $\inf_{x \in X} \Theta_0(x) < \kappa < \inf_{x \notin G} \Theta_0(x)$, the required function Θ may be found of the form

$$\Theta(x) = \alpha + \sum_{k=0}^{\infty} \alpha_j \min\{\Theta_0(\beta_k(x - x_k)), \kappa\}, \qquad (8.4)$$

where $\alpha \in \mathbb{R}$ and $x_j \in X$ form a convergent sequence. For example, based on the preceding proof, we can choose $a_k = \max\{C_k, \beta_k\}$ and replace x_k by $x_k - x_\infty + z_k/\beta_k$.

Recall that by our agreement the function $\omega : (0, \infty) \longrightarrow (0, \infty)$ is increasing with $\lim_{t \searrow 0} \omega(t) = 0$.

Definition 8.2.3. We shall say that X *admits a bump smooth in the direction of a family of subspaces* $\mathcal{Y}$ *with modulus controlled by* ω if there is a function $\Theta_0 \colon X \longrightarrow \mathbb{R}$ satisfying the assumptions (a)–(d) of Proposition 8.2.1 with some nonempty bounded open set G.

We will also use the term "bump function smooth in the direction of a family of subspaces $\mathcal{Y}$ with modulus controlled by ω" to describe any function Θ satisfying the conditions of (a)–(d) of Proposition 8.2.1 for some nonempty bounded open set G. The main advantage of this is that natural functions such as $\|x\|^2$ on a Hilbert space become "smooth bumps" without the need for simple but artificial modifications.

To compare our usage of the term "bump" with others, recall that it is often used for (nonzero) functions with bounded support. We could use this concept, since such a function may be obtained from the function Θ in (8.4) by subtracting a constant. Some people also define bumps as non-negative; if we wanted to adopt this we would have to take $C - \Theta$ and so change the notions of smoothness (which, as we have already pointed out, is really "upper symmetric smoothness") and upper differentiability from upper to lower. But we find both these concepts too limiting, and moreover, for the "best" bumps that we will use in our arguments, all conditions (A) – (D) of Proposition 8.2.1 (including that $\Theta(0) = 0$ is the minimum) are rather useful.

Since we will use the existence of "bumps" not on X but on $L(\mathbb{R}^n, X)$, we record the following easy observation that allows the transfer.

Observation 8.2.4. *If X admits a bump smooth in the direction of a family of subspaces $\mathcal{Y}$ of X with modulus controlled by ω, then so does $L(\mathbb{R}^n, X)$. Moreover, each of the properties listed in the additional statement of Proposition 8.2.1 is preserved.*

Proof. Let $\Theta_X \colon X \longrightarrow \mathbb{R}$ be a function satisfying the statements (A) – (D) of Proposition 8.2.1. We define for $T \in L(\mathbb{R}^n, X)$

$$\Theta(T) = \sum_{j=1}^{n} \Theta_X(Te_j).$$

Clearly, Θ is lower semicontinuous. Since

$$\inf\{\Theta(T) \mid \|T\| > s\} \geq \inf\{\Theta_X(x) \mid \|x\| > s/\sqrt{n}\} > 0,$$

the statement (A) is obvious, and in view of Lemma 8.1.8 (i) and Observation 8.1.6 (ii), so are (C) and (D). The additional properties are clearly preserved except possibly the upper Fréchet smooth part, which follows from Lemma 8.1.8 (ii) and Observation 8.1.6 (iii). □

Before coming to construction of smooth bumps in particular situations, we state a simple consequence of Proposition 8.2.1 and Lemma 8.1.9. It spells out the properties of the "best" bumps we will actually use, and is given here partly to enable an immediate reference and partly because the readers wishing to attempt to generalize our results may find it useful to see the minimal set of requirements that we need to prove our results. (Readers who wish to consider the statement of Lemma 8.2.5 as an assumption may further weaken this assumption by requiring the sequence (x_k) to be convergent.)

Lemma 8.2.5. *If X admits a bump smooth in the direction of a family of subspaces $\mathcal{Y}$ of X with modulus controlled by ω, it also admits a lower semicontinuous function $\Theta\colon X \longrightarrow [0,1]$ such that $\inf_{\|x\|>s} \Theta(x) > 0 = \Theta(0)$ for every $s > 0$, and for every convergent series $\sum_{k=0}^{\infty} \lambda_k$ of positive numbers and every sequence $(x_k) \subset X$, the function*

$$\widetilde{\Theta}(x) = \sum_{k=0}^{\infty} \lambda_k \Theta(x - x_k)$$

satisfies for every $x \in X$,

$$\bar{\rho}_{\widetilde{\Theta},\mathcal{Y}}(x;t) = o(\omega(t)) \text{ as } t \searrow 0.$$

If, in addition, the original bump is on the set G associated with it continuous, uniformly continuous, Lipschitz, or Lipschitz and upper Fréchet smooth, then $\widetilde{\Theta}$ may be required to have the same property on X.

Proof. We find the function Θ with properties (A)–(D) of Proposition 8.2.1 and multiply it by a suitable strictly positive constant to achieve that $\Theta \leq 1$. By Observation 8.1.6 (i) and (C), there is $C < \infty$ such that the functions $\Theta_k(x) := \Theta(x - x_k)$ satisfy

$$\sup_{x \in X} \bar{\rho}_{\Theta_k,\mathcal{Y}}(x;t) = \sup_{x \in X} \bar{\rho}_{\Theta,\mathcal{Y}}(x;t) \leq C\omega(t).$$

Hence the conclusion follows from (D) and Lemma 8.1.9 (ii).

The additional statement is obvious except possibly for the upper Fréchet smooth part, which is immediate from Lemma 8.1.9 (iii).   □

The simplest examples of spaces admitting bumps smooth in our sense are for $\mathcal{Y} = \{X\}$ and $\omega_1(t) = t$. If X admits a Fréchet smooth norm, it admits a locally Lipschitz, upper Fréchet differentiable bump smooth in the direction of the family $\mathcal{Y}$ with modulus controlled by ω_1, namely, $\Theta(x) = \|x\|^2$. A more interesting example is provided by a Hilbert space, still with $\mathcal{Y} = \{X\}$ with modulus controlled by $\omega_2(t) = t^2 \log(1/t)$. (The requested bump is again $\Theta(x) = \|x\|^2$.) This fact is behind our result that two real-valued Lipschitz functions on a Hilbert space have a common point of Fréchet differentiability. Notice that the additional logarithmic term (or a similar term) is necessary: the Hilbert space does not admit a bump smooth in the direction of $\mathcal{Y}$ with modulus controlled by t^2. To see this, we may assume that the space in question is $\mathbb{R}$. If $\Theta\colon \mathbb{R} \longrightarrow \mathbb{R}$ is a lower semicontinuous function with $\Theta(0) + \varepsilon < \inf_{|t|>1} \Theta(t)$, there is a point s at which $t \to \Theta(t) - \varepsilon t^2$ attains a local minimum. Then

$$\liminf_{t \to 0} \frac{\Theta(s+t) + \Theta(s-t) - 2\Theta(s)}{t^2} \geq 2\varepsilon,$$

showing that Θ cannot satisfy 8.2.1 (d). Since any Banach space has a subspace linearly isomorphic to $\mathbb{R}$, this argument shows that no Banach space X admits a bump smooth in the direction of $\{X\}$ with modulus controlled by t^2.

As is our custom, when $\mathcal{Y}$ is the family of finite codimensional subspaces of X, we use the term "asymptotically smooth" instead of "smooth in the direction of $\mathcal{Y}$." Classical examples of spaces having this property are provided by ℓ^p spaces with $\Theta(x) = \|x\|^p$. Example 8.1.3 shows that such Θ is a bump asymptotically smooth with modulus controlled by any ω that satisfies $\lim_{t\to 0} \omega(t)/t^p = \infty$. We now show that a remark similar to the one made for Hilbert spaces applies: ℓ^p does not admit a bump asymptotically smooth with the control t^p. It is therefore again important that differentiability results that we will be proving in the following chapters use moduli that have an additional logarithmic term.

Observation 8.2.6. *Suppose $\mathcal{Y}$ is a downward directed family of infinite dimensional subspaces of ℓ^p, and $\Theta \colon \ell^p \longrightarrow \mathbb{R}$ is a lower bounded, lower semicontinuous function such that $\bar{\rho}_{\Theta,\mathcal{Y}}(x,t) = o(t^p)$ for every $x \in \ell^p$. Then $\inf_{x \in \ell^p \setminus B} \Theta(x) = \inf_{x \in \ell^p} \Theta(x)$ for every bounded subset B of ℓ^p.*

Proof. Suppose this is false and modify Θ to achieve $\Theta(0) = 0 < 1 = \inf_{\|x\| \geq 1} \Theta(x)$. Let $\varepsilon, \lambda_j > 0$ be such that $\varepsilon + \sum_{j=0}^{\infty} \lambda_j < 1$ and $\sum_{j=0}^{\infty} \lambda_j < \frac{1}{16}\varepsilon$. We apply the variational principle from Corollary 7.2.4 with $F_j(x,y) = \lambda_j \|x - y\|^p$ to the function

$$f(x) = \Theta(x) - \varepsilon \|x\|^p,$$

and we infer there are $x_j \in \overline{B(0,1)}$ such that the function

$$h(x) := \Theta(x) - \varepsilon \|x\|^p + \sum_{j=0}^{\infty} \lambda_j \|x - x_j\|^p$$

attains its minimum at some $z \in \overline{B(0,1)}$. Notice that necessarily $\|z\| < 1$, since for $\|x\| = 1$, $h(x) \geq 1 - \varepsilon > \sum_{j=0}^{\infty} \lambda_j \geq h(0)$. By assumption, there are an infinite dimensional subspace Y of ℓ^p and $0 < t < \frac{1}{2}(1 - \|z\|)$ such that

$$\Theta(z + y) + \Theta(z - y) - 2\Theta(z) \leq \frac{\varepsilon t^p}{2}$$

for all $y \in Y$, $\|y\| \leq t$. Let now $n \in \mathbb{N}$ be such that $2^{p+1} \sum_{j=n+1}^{\infty} \lambda_j < \frac{1}{4}\varepsilon t^p$, and find $m \in \mathbb{N}$ such that every $y \in \ell^p$ with $\|y\| = t$ and with the first m coordinates zero satisfies

$$\|z + y\|^p \geq \|z\|^p + \frac{t^p}{2} \quad \text{and} \quad \|z - x_i + y\|^p \leq \|z - x_i\|^p + 2t^p \text{ for } i \leq n.$$

Since Y is infinite dimensional, there is $y \in Y$ with $\|y\| = t$ and the first m coordinates zero. Then $z \pm y \in B(0, 1)$ and so a contradiction follows from

$$\frac{\varepsilon t^p}{2} \geq \Theta(z + y) + \Theta(z - y) - 2\Theta(z)$$

$$= h(z + y) + h(z - y) - 2h(z) + \varepsilon(\|z + y\|^p + \|z - y\|^p - 2\|z\|^p)$$

$$- \sum_{j=0}^{\infty} \lambda_j(\|z + y - x_j\|^p + \|z - y - x_j\|^p - 2\|z - x_j\|^p)$$

$$\geq \varepsilon t^p - 4\sum_{j=0}^{n} \lambda_j t^p - 2^{p+1}\sum_{j=n+1}^{\infty} \lambda_j > \frac{\varepsilon t^p}{2}. \qquad \square$$

We now show that further examples of spaces admitting bumps smooth in our sense are provided by spaces with $\bar{\rho}_X(t) = o(\omega(t))$. We do this together with the discussion of the standard situation in which the bump functions are continuous and convex. We show that, similarly to the well-known case of ordinary smoothness, for asymptotic smoothness existence of such bumps is equivalent to the possibility of renorming the space such that $\bar{\rho}_{\|\cdot\|}(x; t)$ has properties (c) and (d) of Proposition 8.2.1 on the unit sphere. In particular, the possibility of renorming the space such that its modulus of asymptotic uniform smoothness introduced in Definition 4.2.2 satisfies $\bar{\rho}_X(t) = o(\omega(ct))$ for every $c > 0$ is equivalent to the existence of a continuous convex bump Θ_0 with the O in (c) replaced by o.

Proposition 8.2.7. *Let X be a Banach space and let $\mathcal{Y}$ be a downward directed family of subspaces of X such that every finite codimensional subspace of an element of $\mathcal{Y}$ contains an element of $\mathcal{Y}$. Suppose further that $t_0 > 0$, and that $\omega: (0, t_0) \longrightarrow (0, \infty)$ is an increasing function such that for every $c > 0$,*

$$\sup_{\|x\|=1} \bar{\rho}_{\|\cdot\|,\mathcal{Y}}(x; t) = O(\omega(ct)), \quad t \searrow 0.$$

Then there is an increasing convex C^1 function $\varphi: [0, \infty) \longrightarrow [0, \infty)$ such that the function $\Theta(x) = \varphi(\|x\|)$ satisfies, for every $c, R > 0$,

$$\sup_{x \in B(0,R)} \bar{\rho}_{\Theta,\mathcal{Y}}(x; t) = O(\omega(ct)), \quad t \searrow 0.$$

Moreover, for any given increasing function $\widetilde{\varphi}: (0, \infty) \longrightarrow (0, \infty)$ we can also require that $\varphi(t) \leq t\widetilde{\varphi}(t)$.

Proof. We may assume that ω is defined on $(0, \infty)$, and that for every $c > 0$ there is $\eta(c) > 0$ such that

$$\bar{\rho}_{\|\cdot\|,\mathcal{Y}}(x; t) < \eta(c)\,\omega(ct) \qquad (8.5)$$

for all $\|x\| = 1$ and $t > 0$. We achieve this, upon observing that the left side is at most $2t$, by keeping the original ω on $(0, \frac{1}{2}t_0]$ and extending it to an increasing function such that $\omega(t) = t^2$ for large t. Since ω is increasing, we may modify η, if necessary, to achieve that it is decreasing on $(0, \infty)$.

Let $\psi\colon [0,\infty) \longrightarrow [0,\infty)$ be a continuous increasing function such that

$$\psi(s) \leq \min\left\{1, s, \frac{1}{\eta(s^2/8)}, \frac{\omega(s^2/4)}{s}, \tilde{\varphi}(s)\right\}, \quad s > 0.$$

We show that

$$\varphi(t) := \int_0^t \psi(s)\, ds$$

has all the desired properties. From the choice of ψ it is obvious that φ is increasing, convex, C^1, and that $\varphi'(0) = 0$. The additional statement is obvious as well. Hence to finish the proof, it suffices to show that for every $c > 0$, $x \in X$, and $0 < t < c$ there is $Y \in \mathcal{Y}$ such that

$$\Theta(x + y) - \Theta(x) \leq \left(1 + \eta(c^2/2)\right)\left(1 + \|x\|\right)\omega(ct)$$

for every $y \in Y$, $\|y\| \leq t$.

To show this inequality, we start by distinguishing two cases depending on how large $\|x\|$ is. If $\|x\| \leq t$, then for any $y \in X$ with $\|y\| \leq t$,

$$\Theta(x + y) - \Theta(x) \leq \Theta(x + y) \leq \varphi(2t) \leq 2t\psi(2t) \leq \omega(t^2) \leq \omega(ct).$$

We will therefore suppose from now on that $\|x\| > t$. If $y \in X$ and $\|y\| \leq \omega(ct)$, we use that φ is Lipschitz with constant one to infer that

$$\Theta(x + y) - \Theta(x) \leq \|y\| \leq \omega(ct).$$

For the remaining case, let $\hat{x} = x/\|x\|$ and find a functional $x^* \in X^*$ such that that $\|x^*\| = x^*(\hat{x}) = 1$. Let $M \subset [\omega(ct), t]$ be a finite set such that for every $\tau \in [\omega(ct), t]$ there is $s \in M$ such that $\frac{1}{2}s \leq \tau \leq s$. By (8.5) with t replaced by $s/\|x\|$ and c by $c\|x\|/2$, there is $Y \in \mathcal{Y}$, which we may assume to be contained in the kernel of x^*, such that

$$\rho_{\|\cdot\|,Y}(\hat{x}; s/\|x\|) < \eta\left(c\|x\|/2\right)\omega(cs/2) \tag{8.6}$$

for every $s \in M$. Let now $y \in Y$, $\|y\| \geq \omega(ct)$. We use that $\|y\| \in [\omega(ct), t]$ to find $s \in M$ such that $\frac{1}{2}s \leq \|y\| \leq s$. Then, denoting $\hat{y} := y/\|x\|$, we infer from (8.6) that

$$\begin{aligned}
0 = x^*(y) &\leq \|x + y\| - \|x\| \\
&= \|x\|(\|\hat{x} + \hat{y}\| - 1) \\
&\leq \|x\|\, \rho_{\|\cdot\|,Y}(\hat{x}; s/\|x\|) \\
&< \|x\|\, \eta\left(c\|x\|/2\right)\omega(cs/2) \leq \|x\|\, \eta\left(c\|x\|/2\right)\omega(ct).
\end{aligned}$$

Since $\psi = \varphi'$ is increasing, we infer that

$$\begin{aligned}
\Theta(x + y) - \Theta(x) &\leq \psi(\|x + y\|)(\|x + y\| - \|x\|) \\
&\leq \psi(2\|x\|)\, \|x\|\, \eta\left(c\|x\|/2\right)\omega(ct).
\end{aligned}$$

So we just have to establish that $\psi(2s)\,\eta(cs/2) \leq 1 + \eta(c^2/2)$ for every $s > 0$. But this can be shown as follows: If $s \geq c$, then

$$\psi(2s)\,\eta(cs/2) \leq \eta(cs/2) \leq \eta(c^2/2).$$

If $s \leq c$, then

$$\psi(2s)\,\eta(cs/2) \leq \frac{1}{\eta(s^2/2)}\,\eta(cs/2) \leq 1. \qquad \square$$

Corollary 8.2.8. *Let X, $\mathcal{Y}$, and ω be as in Proposition 8.2.7 and consider the following four properties of a function $\Theta\colon X \longrightarrow \mathbb{R}$ on a set $S \subset X$.*

- *For every $c > 0$, $\sup_{x \in S} \bar{\rho}_{\Theta,\mathcal{Y}}(x;t) = O(\omega(ct))$ as $t \searrow 0$.*

- *For every $c > 0$, $\sup_{x \in S} \bar{\rho}_{\Theta,\mathcal{Y}}(x;t) = o(\omega(ct))$ as $t \searrow 0$.*

- *For every $c > 0$ and $x \in S$, $\bar{\rho}_{\Theta,\mathcal{Y}}(x;t) = o(\omega(ct))$ as $t \searrow 0$.*

- *At every $x \in S$, the function Θ is Fréchet differentiable.*

If the norm has some or all of these properties with S being the unit sphere, there is a continuous convex function Θ on X having the same properties for every bounded set S and satisfying $\lim_{\|x\| \to \infty} \Theta(x) = \infty$.

Conversely, if there is a continuous convex function Θ on X satisfying

$$\lim_{\|x\| \to \infty} \Theta(x) = \infty$$

and having some or all the above properties for every bounded set S (or just for every set S of the form $\{x \mid \Theta(x) < C\}$), then the space X admits an equivalent norm that has the same properties with S being the unit sphere.

Proof. Assume for a moment that a function $\varphi\colon [0,\infty) \longrightarrow [0,\infty)$ is increasing, C^1, convex, $\mathrm{Lip}(\varphi) \leq 1$, $\varphi'(0) = 0$, and $\varphi(t) = o(\omega(t^2))$ as $t \searrow 0$. We define $\Theta(x) = \varphi(\|x\|)$ and observe the following two facts.

Fact 1. Θ is Fréchet smooth provided $\|\cdot\|$ is Fréchet smooth.
Indeed, at $x \neq 0$ it is obvious, and for $x = 0$ it follows from $\varphi'(0) = 0$.

Fact 2. *If the norm has the penultimate property for S being the unit sphere, then for every $x \in X$ and $c > 0$*

$$\bar{\rho}_{\Theta,\mathcal{Y}}(x;t) = o(\omega(ct)), \ t \searrow 0.$$

Again we distinguish two cases. If $x = 0$, the statement follows from $\varphi(t) = o(\omega(t^2))$, since this is $o(\omega(ct))$ for every $c > 0$. If $x \neq 0$, put $\hat{x} = x/\|x\|$ and consider $Y \in \mathcal{Y}$ which is a subset of the kernel of a norming functional for $\hat{x}$. Since φ is 1-Lipschitz we have $\varphi(\|x \pm y\|) - \varphi(\|x\|) \leq \|x \pm y\| - \|x\|$ for $y \in Y$, and so for $\|y\| \leq t$,

$$\Theta(x+y) + \Theta(x-y) - 2\Theta(x) = \varphi(\|x+y\|) + \varphi(\|x-y\|) - 2\varphi(\|x\|)$$
$$\leq \|x+y\| + \|x-y\| - 2\|x\|$$
$$\leq \|x\|\,\bar{\rho}_{\|\cdot\|,Y}(\hat{x}, t/\|x\|).$$

Hence $\bar{\rho}_{\Theta,\mathcal{Y}}(x,t) \leq \|x\| \, \bar{\rho}_{\|\cdot\|,\mathcal{Y}}(\hat{x}, t/\|x\|)$, proving Fact 2.

If the norm has one or both of the last two properties, the above facts imply that there is Θ having the same properties. It also follows that if the norm satisfies

$$\sup_{\|x\|=1} \bar{\rho}_{\|\cdot\|,\mathcal{Y}}(x;t) = O(\omega(ct)), \ t \searrow 0,$$

for every $c > 0$, then we can apply Proposition 8.2.7 with, for example, $\widetilde{\varphi}(t) = \omega(t^2)$ to get the function Θ with the same properties. Moreover, this Θ will also share any combination of the last two properties that the norm may have.

If the norm has the second property, we first observe that there is an increasing function $\widetilde{\omega} \colon (0,\infty) \longrightarrow (0,\infty)$ such that for every $c > 0$,

$$\sup_{\|x\|=1} \bar{\rho}_{\Theta,\mathcal{Y}}(x;t) = O(\widetilde{\omega}(ct)) \text{ and } \widetilde{\omega}(t) = o(\omega(t)) \text{ as } t \searrow 0.$$

The function Θ obtained from Proposition 8.2.7 used with $\widetilde{\omega}$ (and the same $\widetilde{\varphi}$ as before) will then have the second property as well. If, in addition, the norm is Fréchet smooth, Θ will be Fréchet smooth. Since the second property implies the first and the third, this covers all cases.

To prove the converse, we replace $\Theta(x)$ by $\Theta(x) + \Theta(-x)$ and add a suitable constant to achieve that Θ is even and $\Theta(x) \geq \Theta(0) = 0$. Let $\|\cdot\|$ be the Minkowski functional of $\{x \in X \mid \Theta(x) \leq 1\}$. Notice that, by convexity, $\|z\| \leq \Theta(z)$ whenever $\Theta(z) \geq 1$. Hence, given any x with $\|x\| = 1 = \Theta(x)$ we pick $x^* \in \partial\Theta(x)$ and conclude that for $z \in \operatorname{Ker} x^*$,

$$\Theta(x \pm z) \geq \Theta(x) \pm x^*(z) \geq 1.$$

Hence

$$\|x + z\| + \|x - z\| - 2\|x\| \leq \Theta(x + z) + \Theta(x - z) - 2\Theta(x).$$

This implies that $\|\cdot\|$ is Fréchet smooth if Θ is Fréchet smooth. By considering for every $Y \in \mathcal{Y}$ a $Z \in \mathcal{Y}$ such that $Z \subset Y \cap \operatorname{Ker} x^*$, the remaining parts of the statement follow as well. $\qquad\square$

The following example is an immediate consequence of Corollary 8.2.8. It is pointed out because it covers the situation that occurs in classical spaces.

Example 8.2.9. If the modulus of asymptotic uniform smoothness of the space X satisfies $\bar{\rho}_X(t) = o(\omega(ct))$ as $t \searrow 0$, for all $c > 0$, then X admits a convex Lipschitz bump which is asymptotically smooth with modulus controlled by $\omega(t)$. If the norm of X is, in addition, Fréchet smooth, X admits a Lipschitz, upper Fréchet differentiable bump which is asymptotically smooth with modulus controlled by $\omega(t)$.

It is, of course, clear that not all spaces admitting a convex Lipschitz bump asymptotically smooth with modulus controlled by, say, $\omega_1(t) = t$ are described by this example. Consider any space with separable dual that does not admit an equivalent asymptotically smooth norm. It is, however, not clear to us to what extent or whether at all, separable spaces with (possibly nonconvex) bumps smooth with controlled modulus differ from spaces admitting convex bumps with the same control of the modulus.

We now recall that another notion of "smoothness," existence of an asymptotically c_0 sequence of finite codimensional subspaces, has been successfully used in [28] to show Fréchet differentiability of Lipschitz functions Γ-almost everywhere. We show that spaces satisfying this property have for each strictly positive ω an equivalent norm such that $\bar{\rho}_X(t) = o(\omega(t))$. In fact, we show a similar result for any asymptotically c_0 sequence of subspaces of X. This will allow us to give in Section 10.6 a new, more general approach to the results of [28].

Definition 8.2.10. A nonempty downward directed family $\mathcal{Y}$ of subspaces of a Banach space X is said to be *asymptotically c_0* if there is a constant C such that for every $n \in \mathbb{N}$

$$\left(\exists Y_1 \in \mathcal{Y}\right) \left(\forall y_1 \in Y_1\right) \cdots \left(\exists Y_n \in \mathcal{Y}\right) \left(\forall y_n \in Y_n\right)$$
$$\|y_1 + \cdots + y_n\| \leq C \max\{\|y_1\|, \ldots, \|y_n\|\}. \tag{8.7}$$

The following statement shows that in the situation from this definition we can construct arbitrarily smooth bump functions. Notice that the condition imposed on ω is at infinity, and so has nothing to do with the notion of smoothness: we may redefine any strictly positive ω on an interval away from zero to satisfy it.

Proposition 8.2.11. *Let $\mathcal{Y}$ be an asymptotically c_0 family of subspaces of X and let $\omega \colon (0, \infty) \longrightarrow (0, \infty)$ be increasing and such that $\lim_{t\to\infty} \omega(t)/t = \infty$. Then there is a convex function $\Theta \colon X \longrightarrow \mathbb{R}$ with $\mathrm{Lip}(\Theta) \leq 1$ such that $\|x\| \leq \Theta(x) \leq \|x\| + K$, where K is a suitable constant, and for every $t > 0$ there is $Y \in \mathcal{Y}$ such that*

$$\rho_{\Theta, Y}(x; t) \leq \omega(t)$$

for every $x \in X$.

Proof. Let C be the constant from Definition 8.2.10 and use the limit assumption on ω to find $M > 0$ such that $\omega(t) > 4(Ct + 1)$ for $t \geq M$. Also choose $M_n > 0$ such that $\omega(t) > 4nt$ for $t > M_n$. Extend ω to $[0, \infty)$ by defining $\omega(0) = 0$ and denote

$$\Theta_n(x_0, x_1, \ldots, x_n) = \left\|\sum_{i=0}^{n} x_i\right\| - \frac{1}{4}\sum_{i=1}^{n} \omega(\|x_i\|).$$

Clearly, $\Theta_0(x_0) = \|x_0\|$ and $\Theta_n(x_0, x_1, \ldots, x_{n-1}, 0) = \Theta_{n-1}(x_0, \ldots, x_{n-1})$ for all n. Hence, letting $s = \max\{\|x_1\|, \ldots, \|x_n\|\}$, we see

$$\Theta_n(x_0, x_1, \ldots, x_n) \leq \|x_0\| + ns - \tfrac{1}{4}\omega(s),$$

which shows that $\Theta_n(x_0, x_1, \ldots, x_n) \leq \|x_0\| + nM_n$.

Define recursively for $m = n - 1, n - 2, \ldots, 0$,

$$\Theta_n(x_0, x_1, \ldots, x_m) = \inf_{Y \in \mathcal{Y}} \sup_{y \in Y} \Theta_n(x_0, x_1, \ldots, x_m, y).$$

Using the backward induction over $m = n, n-1, \ldots, 0$ one can verify that

$$\Theta_n(x_0, \ldots, x_m) \geq \Theta_m(x_0, \ldots, x_m).$$

It follows that $\Theta_n(x_0) \geq \Theta_m(x_0)$ whenever $n \geq m$ and, in particular,

$$\Theta_n(x_0) \geq \Theta_0(x_0) = \|x_0\|.$$

The infimum over $Y \in \mathcal{Y}$ defining $\Theta_n(x_0, x_1, \ldots, x_m)$ is actually a limit along $\mathcal{Y}$, since $\sup_{y \in Y} \Theta_n(x_0, x_1, \ldots, x_m, y)$ decreases as Y becomes smaller and $\mathcal{Y}$ is directed downward. It follows that in the first variable $x = x_0$, the functions $\Theta_n(x, x_1, \ldots, x_m)$ are convex and Lipschitz with constant one.

Since $\Theta_n(x)$ is increasing in n we may define the (at this stage possibly infinite) function

$$\Theta(x) = \lim_{n \to \infty} \Theta_n(x) = \sup_{n \in \mathbb{N}} \Theta_n(x).$$

Then clearly $\Theta(x) \geq \|x\|$, and, once we show that Θ is finite, we will also know that it is convex and Lipschitz with constant one.

To get further information about Θ, we write the data contained in (8.7) more formally: we denote the spaces occurring there (for the given n)

$$Y_1^n, \ Y_2^n(y_1), \ Y_3^n(y_1, y_2), \ \ldots, \ Y_n^n(y_1, \ldots, y_{n-1}).$$

Since $\mathcal{Y}$ is directed downward, we may assume that

$$Y_1^n \supset Y_2^n(y_1) \supset \cdots \supset Y_n^n(y_1, \ldots, y_n).$$

Moreover, we may also assume that for all $m \geq n \geq k \geq 1$,

$$Y_k^n(y_1, \ldots, y_{k-1}) \supset Y_k^m(y_1, \ldots, y_{k-1}).$$

Thus the subspaces are nested in the way indicated here:

$$
\begin{array}{ccccc}
Y_1^1 & & & & \\
\cup & & & & \\
Y_1^2 & \supset & Y_2^2(y_1) & & \\
\cup & & \cup & & \\
Y_1^3 & \supset & Y_2^3(y_1) & \supset & Y_3^3(y_1, y_2) \\
\vdots & & \vdots & & \vdots
\end{array}
$$

A sequence $(y_1, \ldots, y_j)$ will be called n-admissible if $j \leq n$ and

$$y_i \in Y_i^n(y_1, \ldots, y_{i-1}) \text{ for every } i = 1, \ldots, j.$$

In these formulas we admit an empty sequence; for example, if $i = 1$, we will require that $y_1 \in Y_1^n$. We observe that if the sequence $(x_1, \ldots, x_n)$ is n-admissible and such that $\Theta_n(x_0, x_1, \ldots, x_n) > \|x_0\| - 1$, then

$$\max\{\|x_1\|, \ldots, \|x_n\|\} \leq M \text{ and } \left\|\sum_{i=0}^{n} x_i\right\| \leq \|x_0\| + CM. \tag{8.8}$$

Indeed, using that $\mathcal{Y}$ is asymptotically c_0 we calculate

$$\|x_0\| - 1 < \Theta_n(x_0, x_1, \ldots, x_n)$$

$$\leq \|x_0\| + C \max\{\|x_1\|, \ldots, \|x_n\|\} - \frac{1}{4} \sum_{i=1}^n \omega(\|x_i\|)$$

$$\leq \|x_0\| + C \max\{\|x_1\|, \ldots, \|x_n\|\} - \frac{1}{4} \omega(\max\{\|x_1\|, \ldots, \|x_n\|\}),$$

which gives $\omega(\max\{\|x_1\|, \ldots, \|x_n\|\}) \leq 4(C \max\{\|x_1\|, \ldots, \|x_n\|\} + 1)$, and hence $\max\{\|x_1\|, \ldots, \|x_n\|\} \leq M$. Using again that $\mathcal{Y}$ is asymptotically c_0 and the already proved first inequality we obtain the second one.

To find an upper bound on $\Theta(x)$, suppose that

$$\Theta(x) > c > \|x\|$$

and find n such that $\Theta_n(x) > c$. By definition of $\Theta_n(x)$, there is $y_1 \in Y_1^n$ such that $\Theta_n(x, y_1) > c$. Continuing recursively, we find points $y_i \in Y_i^n(y_1, \ldots, y_{i-1})$, $1 \leq i \leq n$, such that $\Theta_n(x, y_1, y_2, \ldots, y_i) > c$ for each i. Since the sequence $(y_1, \ldots, y_n)$ is n-admissible, (8.8) implies that $c < \|x\| + CM$. Recalling that the number c was subject only to the condition $\Theta(x) > c > \|x\|$, we infer that

$$\Theta(x) \leq \|x\| + CM.$$

It remains to estimate the modulus of smoothness of Θ. Let $x \in X$ and $t > 0$. Denote

$$\varepsilon = \min\left\{1, \frac{\omega(t)}{4(C+2)}\right\}$$

and choose $n \in \mathbb{N}$ such that $CM + \varepsilon < (n-1)\varepsilon/4$. There is a subspace $Y \in \mathcal{Y}$, $Y \subset Y_1^n$ such that

$$\sup_{y \in Y} \Theta_n(x, y) < \Theta_n(x) + \varepsilon. \tag{8.9}$$

We are going to show that $\Theta(x + y) \leq \Theta(x) + \frac{1}{2}\omega(t)$ whenever $y \in Y$ and $\|y\| \leq t$.

Fix $y \in Y$ with $\|y\| \leq t$. Because $\Theta(x + y)$ is a limit of increasing sequence $\Theta_m(x + y)$, there is $m \geq n$ such that

$$\Theta_m(x + y) > \Theta(x + y) - \varepsilon.$$

We will recursively define points $z_k \in X$, $k = 1, \ldots, m$ such that

(i) $\Theta_m(x + y, z_1, \ldots, z_k) > \Theta(x + y) - \varepsilon$;

(ii) for every choice $1 \leq i_1 < \cdots < i_j \leq k$, the sequence $(z_{i_1}, \ldots, z_{i_j})$ is m-admissible;

(iii) for every $1 \leq j \leq n - 2$ and $1 \leq i_1 < \cdots < i_j \leq k$, the sequence $(y, z_{i_1}, \ldots, z_{i_j})$ is n-admissible and $\Theta_n(x, y, z_{i_1}, \ldots, z_{i_j}) < \Theta(x) + \varepsilon$.

To start the recursion, recall $\Theta_n(x, y) < \Theta_n(x) + \varepsilon \le \Theta(x) + \varepsilon$ by (8.9). Hence there is $Z \in \mathcal{Y}$ such that

$$\sup_{z \in Z} \Theta_n(x, y, z) < \Theta(x) + \varepsilon.$$

Choose $z_1 \in Z \cap Y_2^n(y) \cap Y_1^m$ such that

$$\Theta_m(x + y, z_1) > \Theta(x + y) - \varepsilon.$$

So (i) holds. This choice also guarantees that (z_1) is m-admissible, and hence the validity of (ii). Since we know that $y \in Y_1^n$ and that (y) is n-admissible we see from $z_1 \in Y_2^n(y)$ that (y, z_1) is n-admissible. The inequality

$$\Theta_n(x, y, z_1) < \Theta(x) + \varepsilon$$

follows from the choice of Z.

Suppose now that the points $z_1, \ldots, z_{k-1}$ have been already defined. Since

$$\Theta_n(x, y, z_{i_1}, \ldots, z_{i_j}) < \Theta(x) + \varepsilon$$

for any $1 \le j \le n - 2$ and any sequence $1 \le i_1 < \cdots < i_j \le k - 1$, there are subspaces $Z(z_{i_1}, \ldots, z_{i_j}) \in \mathcal{Y}$ such that

$$\sup_{z \in Z(z_{i_1}, \ldots, z_{i_j})} \Theta_n(x, y, z_1, \ldots, z_{k-1}, z) < \Theta(x) + \varepsilon. \qquad (8.10)$$

Let $Z \in \mathcal{Y}$ be contained in all such $Z(z_{i_1}, \ldots, z_{i_j})$ as well as in all $Y_j^m(z_{i_1}, \ldots, z_{i_j})$, $1 \le i_1 < \cdots < i_j \le k - 1$. Since $Z \in \mathcal{Y}$, the inequality

$$\Theta_m(x + y, z_1, \ldots, z_{k-1}) > \Theta(x + y) - \varepsilon$$

implies that we can find $z_k \in Z$ such that (i) holds. Since Z is contained in all subspaces $Y_j^m(z_{i_1}, \ldots, z_{i_j})$ we get the requirement from (ii). The condition

$$Y_j^m(z_{i_1}, \ldots, z_{i_j}) \subset Y_j^n(z_{i_1}, \ldots, z_{i_j})$$

together with $y \in Y_1^n$ yields that $(y, z_{i_1}, \ldots, z_{i_j})$ is n-admissible. Finally, (8.10) completes the verification of (iii).

Having defined the z_k, we infer from condition (ii) that the sequence $(z_1, \ldots, z_m)$ is m-admissible and from (i) that

$$\Theta_m(x + y, z_1, \ldots, z_m) > \Theta(x + y) - \varepsilon \ge \|x + y\| - 1.$$

Hence (8.8) implies that $\|z_i\| \le M$ for every $i = 1, \ldots, m$. Let

$$A = \{i \mid 1 \le i \le m, \, \omega(\|z_i\|) > \varepsilon\} \text{ and}$$
$$B = \{i \mid 1 \le i \le m, \, \omega(\|z_i\|) \le \varepsilon\}.$$

The set A has no more than $n - 2$ elements, since the opposite case would imply with the help of condition (i) that

$$\|x + y\| \leq \Theta(x + y)$$
$$\leq \Theta_m(x + y, z_1, \ldots, z_m) + \varepsilon$$
$$= \|x + y\| + \sum_{i=1}^{m} \|z_i\| - \frac{1}{4} \sum_{i=1}^{m} \omega(\|z_i\|) + \varepsilon$$
$$\leq \|x + y\| + CM - \frac{1}{4}(n - 1)\varepsilon + \varepsilon < \|x + y\|.$$

Since the $\{z_i \mid i \in B\}$, ordered according to increasing i, is an m-admissible sequence,

$$\left\| \sum_{i \in B} z_i \right\| \leq C \max_{i \in B} \|z_i\| \leq C\varepsilon.$$

Let now $1 \leq i_1 < \cdots < i_j \leq m$, $j \leq n - 2$ be a sequence containing all elements of A. Then by (iii),

$$\Theta(x) + \varepsilon > \Theta_n(x, y, z_{i_1}, \ldots, z_{i_j})$$
$$= \left\| x + y + \sum_{i \in A} z_i \right\| - \frac{1}{4}\omega(\|y\|) - \frac{1}{4}\sum_{i \in A}\omega(\|z_i\|)$$
$$\geq \left\| x + y + \sum_{i=1}^{m} z_i \right\| - \left\| \sum_{i \in B} z_i \right\| - \frac{1}{4}\omega(\|y\|) - \frac{1}{4}\sum_{i=1}^{m}\omega(\|z_i\|)$$
$$\geq \Theta_m(x + y, z_1, \ldots, z_m) - C\varepsilon - \frac{1}{4}\omega(\|y\|)$$
$$\geq \Theta(x + y) - \varepsilon - C\varepsilon - \frac{1}{4}\omega(\|y\|),$$

implying that

$$\Theta(x + y) \leq \Theta(x) + (C + 2)\varepsilon + \frac{1}{4}\omega(t) \leq \Theta(x) + \frac{1}{2}\omega(t).$$

Hence $\rho_{\Theta,Y}(x; t) \leq \omega(t)$, as needed. $\qquad\square$

By Corollary 8.2.8 we have an immediate consequence for renorming of spaces containing an asymptotically c_0 family of subspaces.

Corollary 8.2.12. *Any space containing an asymptotically c_0 family of subspaces may be renormed, for any increasing $\omega \colon (0, \infty) \longrightarrow (0, \infty)$, so that with the new norm $\bar{\rho}_X(t) = o(\omega(t))$.*

Chapter Nine

Preliminaries to main results

Most of this chapter revises some notions and results that will be used in subsequent chapters. In particular, we deepen the concept of regular differentiability and prove several inequalities controlling the increment of functions by the integral of their derivatives. The most important point of this chapter is the crucial lemma on deformation of n-dimensional surfaces that will be basic in all results that we prove in the subsequent chapters. A number of results that should otherwise be here have already been used and therefore proved in the previous chapters, most notably the simple but important Corollary 4.2.9.

9.1 NOTATION, LINEAR OPERATORS, TENSOR PRODUCTS

Much of our work will be done in the spaces $\mathbb{R}^n$, which we will always consider equipped with the Euclidean norm $|\cdot|$ and scalar product $\langle \cdot, \cdot \rangle$. (The notation $\langle \cdot, \cdot \rangle$ will be used in all Hilbert spaces, although the norm will be denoted by $\|\cdot\|$ unless the space is $\mathbb{R}^n$.) In particular, we will use this scalar product to identify the dual of $\mathbb{R}^n$ with itself. Typically, for a function ψ on $\mathbb{R}^n$ we will consider the derivative $\psi'(u)$ as a linear functional, that is, an element of $(\mathbb{R}^n)^*$, and at the same time as an element of $\mathbb{R}^n$, that is, represented by the grad ψ. The standard basis of $\mathbb{R}^n$ will be denoted $e_1, \ldots, e_n$. When $n < p$, we will identify $\mathbb{R}^n$ with $\mathbb{R}^n \times \{0\} \subset \mathbb{R}^p$. For $j \leq n$ we denote

$$B_j(x, r) = \{x + u \mid u \in \mathbb{R}^j, |u| < r\} = B_n(x, r) \cap (x + \mathbb{R}^j).$$

When H is a Hilbert space, the space $L(H, \mathbb{R}^n)$ of bounded linear operators from H to $\mathbb{R}^n$ will be equipped, in addition to the operator norm, with the Hilbert-Schmidt norm $|\cdot|_{\mathrm{H}}$ and the corresponding scalar product $\langle \cdot, \cdot \rangle_{\mathrm{H}}$. Recall that

$$|T|_{\mathrm{H}}^2 = \sum_{i=1}^{\infty} |Tu_i|^2 \quad \text{and} \quad \langle T, S \rangle_{\mathrm{H}} = \sum_{i1}^{\infty} \langle Tu_i, Su_i \rangle$$

for any orthonormal basis (u_i) of H. Recall also the relation between the operator and Hilbert-Schmidt norms for $L \in L(\mathbb{R}^m, \mathbb{R}^n)$:

$$\|L\| \leq |L|_{\mathrm{H}} \leq \sqrt{m}\,\|L\|.$$

The orthogonal projection of a Hilbert space H onto its subspace V will be denoted by π_V. For spaces that have a standard basis (such as $\mathbb{R}^m$ or ℓ_p) we will denote by π_n

the natural projection onto the subspace spanned by the first n basis vectors. So, for example, in $\mathbb{R}^m$ we have $\pi_n = \pi_{\mathbb{R}^n}$ for $0 \leq n \leq m$. Sometimes we will also need notation for the projection onto the subspace spanned by all but the first n basis vectors; for that we will use π^n.

Sometimes it will be convenient to use the identification of the space $L(\mathbb{R}^n, X)$ with the tensor product $X \otimes \mathbb{R}^n$. This is obtained by defining the value of the tensor $\sum_i x_i \otimes u_i \in X \otimes \mathbb{R}^n$ applied to $u \in \mathbb{R}^n$ as

$$\left(\sum_{i=1}^n (x_i \otimes u_i) \right) u = \sum_{i=1}^n x_i \langle u, u_i \rangle.$$

The dual of $L(\mathbb{R}^n, X)$ may be then identified with $X^* \otimes \mathbb{R}^n$, the duality being

$$\left\langle T, \sum_i x_i^* \otimes u_i \right\rangle = \sum_i x_i^*(Tu_i).$$

Notice that, if $T \in L(\mathbb{R}^n, \mathbb{R}^n)$ and $u, v \in \mathbb{R}^n$, then $\langle T, u \otimes v \rangle_{\mathrm{H}} = \langle u, Tv \rangle$. This may be seen, for example, by writing both sides in an orthonormal basis. Similarly easily one sees that $|u \otimes v|_{\mathrm{H}} = \|u \otimes v\| = |u||v|$.

9.2   DERIVATIVES AND REGULARITY

Here we collect various notions and notation related to differentiability. The notation is more or less self-explanatory. Since the the more technical notions will start being seriously used only in Chapter 13, the reader may postpone reading this section till then.

Definition 9.2.1. Let X, Y, U be Banach spaces, $f \colon X \longrightarrow Y$, $T \in L(U, X)$ and $x \in X$.

(i) We say that f is *Fréchet differentiable at x in the direction of T* if there is a $f'(x; T) \in L(U, Y)$ such that for every $\varepsilon > 0$ there is $\delta > 0$ such that

$$\|f(x + Tu) - f(x) - f'(x; T)(u)\| \leq \varepsilon \|u\|$$

whenever $u \in U$ and $\|u\| < \delta$.

(ii) We shall say that f is *regularly Gâteaux differentiable at x in the direction of T* if it is Fréchet differentiable at X in the direction of T and for every $\varepsilon > 0$ and $z \in X$ there is $\delta > 0$ such that

$$\|f(x + tz + Tu) - f(x + tz) - f'(x; T)(u)\| \leq \varepsilon(\|u\| + |t|)$$

whenever $t \in \mathbb{R}$, $u \in U$, and $|t| + \|u\| < \delta$.

(iii) We shall say that f is *regularly Fréchet differentiable at x in the direction of T* if it is Fréchet differentiable at x in the direction of T and for every $\varepsilon > 0$ there is $\delta > 0$ such that

$$\|f(x + z + Tu) - f(x + z) - f'(x; T)(u)\| \leq \varepsilon(\|u\| + \|z\|)$$

whenever $u \in U$, $z \in X$, and $\|u\| + \|z\| < \delta$.

In the special case when T is the identity from a subspace V of X to X, we will write $f'_V(x)$ instead of $f'(x; T)$.

Let us remark that the notion of Fréchet differentiability in the direction of T is not really new: $f'(x; T)$ is the Fréchet derivative of the mapping $u \mapsto f(x + Tu)$ (which maps U to Y) at the point $u = 0$. In particular, the value of $f'(x; T)$ at u is equal to the directional derivative of f at x in the direction of Tu,

$$f'(x; T)(u) = f'(x; Tu).$$

It follows that

$$\|f'(x; T)\| \leq \mathrm{Lip}(f)\,\|T\|.$$

Notice also that if the Fréchet derivative $f'(x)$ exists then

$$f'(x; T) = f'(x) \circ T.$$

Of course, one may also consider the notion of Gâteaux differentiability in the direction of T. We do not do it, since we are interested in the situation when U is finite dimensional and f is Lipschitz, and then the two notions of differentiability coincide.

Observation 9.2.2. *Let X, Y, U be Banach spaces, $f \colon X \longrightarrow Y$, and $T \in L(U, X)$.*

(i) *If f is Fréchet differentiable at x, then it is regularly Fréchet differentiable at x in the direction of T.*

(ii) *If f is Lipschitz and Gâteaux differentiable at x and if U is finite dimensional, then f is regularly Gâteaux differentiable at x in the direction of T.*

(iii) *If X is finite dimensional and f is Lipschitz, then regular Gâteaux and regular Fréchet differentiability are equivalent.*

(iv) *If f is Lipschitz and U is finite dimensional, then f is regularly Gâteaux differentiable at x in the direction of T if and only if its restriction to any finite dimensional subspace of X containing $\{x\} \cup T(U)$ is regularly Fréchet differentiable at x in the direction of T.*

Proof. (i) Given $\varepsilon > 0$, denote $\eta = \varepsilon/(2 + \|T\|)$ and find $\delta > 0$ such that

$$\|f(x + y) - f(x) - f'(x; y)\| \leq \eta\|y\|$$

for $\|y\| < \delta$. Since $f'(x;T) = f'(x) \circ T$, we have

$$\|f(x + z + Tu) - f(x + z) - f'(x;T)(u)\|$$
$$\leq \|f(x + z + Tu) - f(x) - f'(x; z + Tu)\|$$
$$+ \|f(x + z) - f(x) - f'(x; z)\|$$
$$\leq \eta\|z + Tu\| + \eta\|z\| \leq \varepsilon(\|u\| + \|z\|)$$

whenever $u \in U$, $z \in X$, and $\|u\| + \|z\| < \delta/(1 + \|T\|)$.

(ii) The argument is similar to the one used in (i). Let $\varepsilon > 0$ and $z \in X$. We put $\eta = \varepsilon/(2\|z\| + \|T\|)$. Since f is Lipschitz and Gâteaux differentiable at x, one can find $\delta > 0$ such that

$$\|f(x + y) - f(x) - f'(x; y)\| \leq \eta\|y\|$$

for all $y \in \mathrm{span}(TU \cup \{z\})$ with $\|y\| < \delta$. Then for every $t \in \mathbb{R}$ and $u \in U$ with $\|u\| + |t| < \delta/(1 + \|z\| + \|T\|)$ we obtain

$$\|f(x + tz + Tu) - f(x + tz) - f'(x; Tu)\|$$
$$\leq \|f(x + tz + Tu) - f(x) - f'(x; tz + Tu)\|$$
$$+ \|f(x + tz) - f(x) - f'(x; tz)\|$$
$$\leq \eta(2|t|\|z\| + \|T\|\|u\|) \leq \varepsilon(|t| + \|u\|).$$

(iii) It is obvious that regular Fréchet differentiability implies regular Gâteaux differentiability (without any assumptions on f or U). For the opposite direction, suppose that $\varepsilon > 0$, let $\eta = \varepsilon/(1 + 4\,\mathrm{Lip}(f))$, and find a finite η-net S in the unit sphere of X. By regular Gâteaux differentiability there is a $\delta > 0$ such that

$$\|f(x + ts + Tu) - f(x + ts) - f'(x;T)(u)\| \leq \frac{\varepsilon}{2}(\|u\| + |t|)$$

whenever $s \in S$, $u \in U$, $t \in \mathbb{R}$, and $\|u\| + |t| < \delta$. Then for every $u \in U$ and $z \in X$ such that $\|u\| + \|z\| < \delta$ we denote $t = \|z\|$, find $s \in S$ with $\|z - ts\| \leq \eta t$, and estimate

$$\|f(x + z + Tu) - f(x + z) - f'(x;T)(u)\|$$
$$\leq 2\,\mathrm{Lip}(f)\eta t + \|f(x + ts + Tu) - f(x + ts) - f'(x;T)(u)\|$$
$$\leq 2\,\mathrm{Lip}(f)\eta t + \tfrac{1}{2}\varepsilon(\|u\| + |t|) \leq \varepsilon(\|u\| + \|z\|).$$

Finally, (iv) follows immediately from (iii). $\qquad\square$

The following important technical lemma treats the situation when a Lipschitz function $f\colon X \longrightarrow Y$ is, at a given point, regularly Gâteaux differentiable in the direction of $T \in L(U, X)$ but not regularly Fréchet differentiable in the same direction. Rather naturally, it shows that there is a parameter ε with which (an appropriate version of) the formula from the definition of regular Fréchet differentiability just stops being valid. More important, it gives a quantitative version of the fact that the failure of regular Fréchet differentiability cannot occur "close to a finite dimensional subspace." This result will be needed in the most delicate arguments of Chapter 13.

Lemma 9.2.3. *Let X, E, U be Banach spaces, where U is finite dimensional. Let $\overline{x} \in X$, $T \in L(U, X)$, and let $f \colon X \longrightarrow E$ be a Lipschitz function which is at the point $\overline{x}$ regularly Gâteaux differentiable in the direction of T but not regularly Fréchet differentiable in the direction of T. Then there is $0 < \overline{\varepsilon} \leq 2\operatorname{Lip}(f)\|T\|$ such that the following two statements hold.*

(i) *For every $\alpha > 0$ there is a finite codimensional subspace Y of X such that for every finite dimensional subspace Z of X there is $\delta > 0$ such that*

$$\|f(\overline{x} + z + y + Tu) - f(\overline{x} + z + y) - f'(\overline{x}; Tu)\| < 2\overline{\varepsilon}(\|y\| + \|u\|)$$

whenever $y \in Y$, $z \in Z$, $u \in U$, $\alpha\|z\| < \|y\| < \delta$ and $0 < \|u\| < \delta$.

(ii) *There is $\overline{\alpha} > 0$ such that for every finite codimensional subspace Y of X there is a finite dimensional subspace Z of X such that for every $\delta > 0$ there are $y \in Y$, $z \in Z$, and $u \in U$ such that $\overline{\alpha}\|z\| < \|y\| < \delta$, $0 < \|u\| < \delta$, and*

$$\|f(\overline{x} + z + y + Tu) - f(\overline{x} + z + y) - f'(\overline{x}; Tu)\| > \overline{\varepsilon}(\|y\| + \|u\|).$$

Proof. For $\alpha, \delta > 0$ and subspaces Y, Z of X denote

$$\varepsilon(Y, Z; \alpha, \delta) = \sup_{\substack{y \in Y,\, z \in Z,\, u \in U \\ \alpha\|z\| < \|y\| < \delta \\ 0 < \|u\| < \delta}} \frac{\|f(\overline{x} + z + y + Tu) - f(\overline{x} + z + y) - f'(\overline{x}; Tu)\|}{\|y\| + \|u\|}.$$

Observing that $\varepsilon(Y, Z; \alpha, \delta)$ is decreasing in α and increasing in δ, we see that

$$\overline{\varepsilon} = \frac{3}{4} \lim_{\alpha \searrow 0} \inf_{\dim(X/Y) < \infty} \sup_{\dim Z < \infty} \lim_{\delta \searrow 0} \varepsilon(Y, Z; \alpha, \delta)$$

is well defined.

The upper bound on $\overline{\varepsilon}$ follows easily from

$$\varepsilon(Y, Z; \alpha, \delta) \leq \sup_{y \in Y,\, u \in U} \frac{2\operatorname{Lip}(f)\|Tu\|}{\|y\| + \|u\|} \leq 2\operatorname{Lip}(f)\|T\|.$$

Since f is not regularly Fréchet differentiable at $\overline{x}$ in the direction of T, there is $\eta > 0$ such that for every $\delta > 0$ one can find $x \in X$ and $u \in U$ with $\|x\| + \|u\| < \delta$ and

$$\|f(\overline{x} + x + Tu) - f(\overline{x} + x) - f'(\overline{x}; Tu)\| > \eta(\|x\| + \|u\|). \tag{9.1}$$

We let $\tilde{\alpha} = \eta/12\operatorname{Lip}(f)$ and prove that for every finite codimensional $Y \subset X$,

$$\sup_{\dim Z < \infty} \lim_{\delta \searrow 0} \varepsilon(Y, Z; \tilde{\alpha}, \delta) > \frac{\eta}{3}.$$

For this, let $Y \subset X$ be finite codimensional and use Corollary 4.2.9 to find a finite dimensional subspace Z of X such that every $x \in X$ can be written as $x = y + z$ where $y \in Y$, $z \in Z$, $\|y\|, \|z\| \leq 3\|x\|$. We may also assume that

$$\{\overline{x}\} \cup T(U) \subset Z.$$

Hence the restriction of f to Z is regularly Fréchet differentiable at $\overline{x}$ in the direction of T by Observation 9.2.2 (iv). Let $\delta_0 > 0$ be such that

$$\left\| f(\overline{x} + z + Tu) - f(\overline{x} + z) - f'(\overline{x}; Tu) \right\| < \tfrac{1}{6}\eta(\|z\| + \|u\|)$$

whenever $z \in Z$, $u \in U$, and $\|z\| + \|u\| < \delta_0$. Given $\delta > 0$, we choose $x \in X$ and $u \in U$ satisfying $\|x\| + \|u\| < \frac{1}{3}\min\{\delta, \delta_0\}$ such that (9.1) holds. Write $x = y + z$, where $y \in Y$, $z \in Z$, and $\|y\|, \|z\| \leq 3\|x\|$. If $\|y\| \leq \tilde{\alpha}\|z\|$ we would use that $\|z\| + \|u\| \leq 3\|x\| + \|u\| < \delta_0$ to infer the following contradiction.

$$\begin{aligned}
\eta(\|x\| + \|u\|) &< \|f(\overline{x} + x + Tu) - f(\overline{x} + x) - f'(\overline{x}; Tu)\| \\
&\leq \|f(\overline{x} + z + Tu) - f(\overline{x} + z) - f'(\overline{x}; Tu)\| + 2\operatorname{Lip}(f)\|y\| \\
&\leq \tfrac{1}{6}\eta(\|z\| + \|u\|) + 2\operatorname{Lip}(f)\tilde{\alpha}\|z\| \\
&\leq \left(\tfrac{1}{2}\eta + 6\operatorname{Lip}(f)\tilde{\alpha}\right)\|x\| + \tfrac{1}{6}\eta\|u\| \\
&\leq \eta(\|x\| + \|u\|).
\end{aligned}$$

Hence $\tilde{\alpha}\|z\| < \|y\|$ and we have

$$\begin{aligned}
\varepsilon(Y, Z; \tilde{\alpha}, \delta) &\geq \frac{\|f(\overline{x} + z + y + Tu) - f(\overline{x} + z + y) - f'(\overline{x}, Tu)\|}{\|y\| + \|u\|} \\
&\geq \frac{1}{3}\frac{\|f(\overline{x} + z + y + Tu) - f(\overline{x} + z + y) - f'(\overline{x}, Tu)\|}{\|x\| + \|u\|} > \frac{\eta}{3}.
\end{aligned}$$

Since $\varepsilon(Y, Z; \alpha, \delta)$ is decreasing in α, we see that, in particular, $\overline{\varepsilon} > 0$ and

$$\inf_{\dim(X/Y)<\infty} \sup_{\dim Z<\infty} \lim_{\delta \searrow 0} \varepsilon(Y, Z; \alpha, \delta) < 2\overline{\varepsilon}$$

for every $\alpha > 0$. The first statement of the lemma just rewrites this inequality using the definition of limits, suprema, and infima. For the second statement, we choose $\overline{\alpha} > 0$ such that

$$\overline{\varepsilon} < \inf_{\dim(X/Y)<\infty} \sup_{\dim Z<\infty} \lim_{\delta \searrow 0} \varepsilon(Y, Z; \overline{\alpha}, \delta),$$

and again rewrite this inequality using the definition of limits, suprema, and infima. $\quad\square$

9.3 DEFORMATION OF SURFACES CONTROLLED BY ω_n

The following functions will play a crucial role in measuring deformation of surfaces that we will use to find points of differentiability.

Definition 9.3.1. We define $\omega_n \colon [0, \infty) \longrightarrow [0, 1]$ by

$$\omega_n(t) = \begin{cases} 0 & \text{if } t = 0, \\ t^n \log^{n-1}(\tfrac{1}{t}) & \text{if } 0 < t \leq \tfrac{1}{2}, \\ \omega_n(\tfrac{1}{2}) & \text{if } t > \tfrac{1}{2}. \end{cases}$$

Our real interest is in the behavior of these functions close to zero; their extension to the whole real line is just for convenience and was chosen so that ω_n is increasing on $[0, \infty)$.

The point of the following seemingly only technical, but very important lemma is that it allows us to deform the starting flat surface (represented by $\mathbb{R}^n \subset \mathbb{R}^{n+1}$) to a nearby new surface (the graph of ψ) for which the gain in energy (the integral in (iii), where the integrand really should be $\theta(\psi'(u)) - \theta(0)$ to represent the gain) is small provided θ is smooth with modulus ω_n in the sense that

$$\theta(z) + \theta(-z) - 2\theta(0) = o(\omega_n(\|z\|)).$$

In Chapter 14 we will see that ω_n cannot be replaced by a substantially better function (such as t^n), and this will turn out to be the main point behind our result that multi-dimensional mean value estimates for Fréchet derivatives are invalid, for example, for maps of a Hilbert space into $\mathbb{R}^3$.

Deformation lemma 9.3.2. *For each $n \in \mathbb{N}$ and $0 < \kappa \le 1$ there are $\delta > 0$ and an even function $\psi_\kappa \in C^1(\mathbb{R}^n)$ such that*

(i) $\psi_\kappa(0) = 1$, $\psi_\kappa(u) = 0$ *for* $|u| \ge e^{1/\kappa}$, *and* $0 \le \psi_\kappa(u) \le 1$ *for* $u \in \mathbb{R}^n$,

and for every even function $\psi \in C^1(\mathbb{R}^n)$ with $\mathrm{Lip}(\psi - \psi_\kappa) \le \delta$;

(ii) $|\psi'(u)| \le \kappa$ *for all* $u \in \mathbb{R}^n$; *and*

(iii) *if $\theta \colon \mathbb{R}^n \longrightarrow \mathbb{R}$ is bounded, Borel measurable, $\theta(0) = 0$, and $M \subset (0, \kappa]$ is such that $\kappa \in M$ and each point of $(0, \kappa]$ is within $e^{-n/\kappa}$ of M, then*

$$\int_{B(0, e^{1/\kappa})} \theta(\psi'(u)) \, d\mathcal{L}^n(u) \le K_n \sup_{t \in M} \sup_{|z| < t} \frac{\theta(z) + \theta(-z)}{\omega_n(t)},$$

where K_n is a constant depending only on n.

Proof. Let $n \in \mathbb{N}$ and $0 < \kappa \le 1$ be given and denote $s = e^{1/\kappa}$. Observing that $\int_0^s \kappa \min\{1, \frac{1}{t}\} \, dt > \kappa \log s = 1$, we find $0 < \delta < e^{-1/\kappa}$ and a continuous function $\eta \colon \mathbb{R} \longrightarrow \mathbb{R}$ such that

- $0 \le \eta(t) \le \kappa \min\{1, \frac{1}{t}\} - \delta$ for $t \in (0, s)$;

- $\eta(t) = 0$ for $t \notin (0, s)$; and

- $\int_0^s \eta(t) \, dt = 1$.

Let

$$\psi_\kappa(u) = \varphi(|u|), \quad \text{where } \varphi(t) = \int_t^\infty \eta(\tau) \, d\tau.$$

Then both (i) and (ii) are obvious.

Before embarking on the proof of (iii) we estimate, for any unit vector $u \in \mathbb{R}^n$ and $t \in (0, s)$,

$$|\psi'(tu)| \leq |\varphi'(t)| + \operatorname{Lip}(\psi_\kappa - \psi) \leq \kappa \min\left\{1, \frac{1}{t}\right\}.$$

From the definition of ω_n we infer that

$$\int_0^s \omega_n(|\psi'(tu)|)t^{n-1}dt \leq \omega_n(\tfrac{1}{2}) \int_0^{2\kappa} t^{n-1}dt + \int_{2\kappa}^s \omega_n(\kappa/t)t^{n-1}\,dt$$

$$\leq \omega_n(\tfrac{1}{2})\frac{2^n}{n} + \kappa^n \int_{2\kappa}^s \frac{\log^{n-1}(t/\kappa)}{t}dt$$

$$= \frac{\log^{n-1}2}{n} + \frac{\kappa^n}{n}\left(\log^n(s/\kappa) - \log^n 2\right)$$

$$\leq 1 + \log^n\big((s/\kappa)^r\big) \leq 1 + \log^n(e^2) = 1 + 2^n.$$

Integrating this inequality over the sphere $|u| = 1$, we get

$$\int_{|u| \leq s} \omega_n\big(|\psi'(u)|\big)\,d\mathscr{L}^n(u) \leq (1 + 2^n)n\alpha_n. \tag{9.2}$$

We are now ready to show (iii). Suppose that θ and $M \subset (0, \kappa]$ are given. Since $\kappa \in M$, we infer that for every $t \in [0, \kappa]$ there is $\tilde{t} \in M$ such that

$$t \leq \tilde{t} \leq t + 2s^{-n}. \tag{9.3}$$

Denote

$$\xi = \sup_{t \in M} \sup_{|y| \leq t} \frac{\theta(y) + \theta(-y)}{\omega_n(t)}.$$

Let $t \in [0, \kappa]$ and find $\tilde{t} \in M$ satisfying (9.3). Then for $|y| \leq t$

$$\theta(y) + \theta(-y) \leq \xi\,\omega_n(\tilde{t})$$

$$= \xi\big(\omega_n(t) + (\omega_n(\tilde{t}) - \omega_n(t))\big)$$

$$\leq \xi\big(\omega_n(t) + 2\operatorname{Lip}(\omega_n)s^{-n}\big).$$

Since we know that $|\psi'(u)| \leq \kappa$ for all $u \in B(0, s)$, this estimate gives that

$$\theta(\psi'(u)) + \theta(-\psi'(u)) \leq \xi\big(\omega_n(|\psi'(u)|) + 2\operatorname{Lip}(\omega_n)s^{-n}\big)$$

for all such u. Since ψ is even, $\psi'(u) = -\psi'(-u)$. Hence, using also (9.2), we conclude that

$$\int_{B(0,s)} \theta(\psi'(u))\,d\mathscr{L}^n(u) = \frac{1}{2}\int_{|u| \leq s}\big(\theta(\psi'(u)) + \theta(-\psi'(u))\big)\,d\mathscr{L}^n(u)$$

$$\leq \frac{1}{2}\xi\int_{|u| \leq s}\big(\omega_n(|\psi'(u)|) + 2\operatorname{Lip}(\omega_n)s^{-n}\big)\,d\mathscr{L}^n(u)$$

$$\leq \frac{1}{2}\xi\alpha_n\big((1 + 2^n)n + s^n 2\,s^{-n}\operatorname{Lip}(\omega_n)\big) = K_n\xi,$$

where $K_n = \frac{1}{2}\alpha_n\big((1 + 2^n)n + 2\operatorname{Lip}(\omega_n)\big)$. $\qquad\square$

The freedom in the choice of ψ in Lemma 9.3.2 enables us to find ψ with various additional properties. The following corollary serves as an example. It will be used in Chapter 13 in the proof of our main differentiability result.

Corollary 9.3.3. *Suppose that $N \subset \mathbb{R}^{n+1}$ is a Lebesgue null set. Then a $\psi \in C^1(\mathbb{R}^n)$ having the properties from Lemma 9.3.2 may be chosen such that also*

(i) $\psi(0) = 1$, $\psi(u) = 0$ *if* $|u| \geq e^{1/\kappa}$ *and* $|\psi(u)| \leq 1$ *for all* $u \in \mathbb{R}^n$;

(ii) $(u, \psi(u)) \notin N$ *for almost every* $u \in B_n(0, e^{1/\kappa})$.

Proof. Let κ and δ be as in Lemma 9.3.2. Let $\eta \in C^1(\mathbb{R})$ be such that $\eta(t) = 0$ outside $(0, e^{1/\kappa})$ and on $(0, e^{1/\kappa})$ both $0 < \eta(t) < 1$ and $|\eta'(t)| < \delta$. Then for each $0 < s < 1$,

$$\zeta_s(u) := \psi_\kappa(u) - s\eta(|u|)$$

is a function satisfying (i) of the corollary as well as the assumptions made on ψ in Lemma 9.3.2. To satisfy also (ii), it suffices to choose $s \in [0, 1]$ randomly, since by Fubini's theorem,

$$0 = \int_{B_n(0, e^{1/\kappa})} \mathscr{L}^1\{s \in [0, 1] \mid (u, \zeta_s(u)) \in N\} \, d\mathscr{L}^n(u)$$

$$= \int_0^1 \mathscr{L}^n\{u \in B_n(0, e^{1/\kappa}) \mid (u, \zeta_s(u)) \in N\} \, ds. \qquad \square$$

9.4 DIVERGENCE THEOREM

One advantage of the tensor product notation is the neat formulation of a suitable version of the divergence theorem for integral of derivatives of maps from $\mathbb{R}^p$ to vector spaces. We state it for sets with Lipschitz boundaries, since we will need it for simple sets with nonsmooth boundaries such as cylinders and for finite dimensional target space E.

Theorem 9.4.1. *Let $\Omega \subset \mathbb{R}^p$ be a bounded open set with Lipschitz boundary and let ν_Ω be the outer unit normal vector to $\partial\Omega$. Suppose that E is a finite dimensional vector space and $g : \overline{\Omega} \longrightarrow E$ is Lipschitz. Then*

$$\int_\Omega g'(u) \, d\mathscr{L}^p(u) = \int_{\partial\Omega} g(u) \otimes \nu_\Omega \, d\mathscr{H}^{n-1}(u).$$

Proof. To see this, observe that the above equality means that for every $v \in \mathbb{R}^n$,

$$\int_\Omega g'(u; v) \, d\mathscr{L}^n(u) = \int_{\partial\Omega} g(u)\langle v, \nu_\Omega \rangle \, d\mathscr{H}^{n-1}(u). \qquad (9.4)$$

On both sides, the integrated functions are bounded and measurable, so integrable. Given $e^* \in E^*$ and $v \in \mathbb{R}^n$, we introduce an auxiliary vector field $\Phi : \mathbb{R}^n \longrightarrow \mathbb{R}^n$ by

$$\Phi(u) = e^*(g(u))v$$

and notice that

$$\operatorname{div} \Phi(u) = (e^*g)'(u; v) = e^*g'(u; v).$$

So a standard formulation of the divergence theorem yields

$$\int_\Omega e^*(g'(u; v)) \, d\mathcal{L}^n(u) = \int_{\partial\Omega} e^*(g(u))\langle v, \nu_\Omega \rangle \, d\mathcal{H}^{n-1}(u),$$

which gives (9.4). $\qquad\qquad\qquad\qquad\qquad\qquad\qquad\qquad\qquad\qquad$ □

For future reference we state here an immediate corollary of Theorem 9.4.1 which follows by the standard estimate of the norm of the integral and Fubini's theorem.

Corollary 9.4.2. *Let V be a subspace of $\mathbb{R}^n$, and let $Q \subset \mathbb{R}^n$ be the Cartesian product of a bounded open set $\Omega \subset V$ with Lipschitz boundary with a bounded measurable subset I of $V^\perp$. Then for every Lipschitz function $g \colon \overline{Q} \longrightarrow \mathbb{R}^m$,*

$$\left| \int_{\Omega \times I} g'_V(u) \, d\mathcal{L}^n(u) \right|_{\mathrm{H}} \leq \mathcal{H}^{n-1}(\partial\Omega \times I) \max_{u \in \partial\Omega \times I} |g(u)|.$$

The most usual application of this corollary will be with $V = \mathbb{R}^n$, in which case $I = \{0\}$ and $\Omega \times I$ becomes just Ω.

9.5   SOME INTEGRAL ESTIMATES

In this section, we prove some integral estimates of derivatives of Lipschitz maps between Euclidean spaces (not necessarily of the same dimension). However, the finite dimensionality will play a significant role only for the domain, while the target space may well be any Banach space E with the Radon-Nikodým property (or even any Banach space provided we restrict our attention to almost everywhere differentiable mappings). The symbol E will be used to denote such a target; nothing is lost by considering E as a Euclidean space of unspecified dimension. The estimates that we show are in fact valid in much greater generality (e.g., after replacing the norm of the derivative by the metric derivative, even for maps into general metric spaces) or in much finer form (such as Morrey's inequalities). We will, however, state and prove these estimates only in the form needed in the sequel, with no attempt to make these subsidiary results optimal. As in the section on the divergence theorem, we will need to work in simple but possibly nonsmooth domains. To handle the estimates, we introduce the following parameter, called eccentricity. We will not spend much time on it, since we will use it only in special situations when it is easy to compute.

Definition 9.5.1. *The eccentricity $\rho(\Omega)$ of a nonempty bounded open set $\Omega \subset \mathbb{R}^p$ is defined as the ratio of the diameter of Ω to the diameter of the largest ball contained in Ω,*

$$\rho(\Omega) = \frac{\operatorname{diam}\Omega}{2\sup\{r > 0 \mid B_p(x, r) \subset \Omega, \ x \in \mathbb{R}^p\}}.$$

Notice that this definition is normalized such that $\rho(\Omega) \geq 1$ and $\rho(\Omega) = 1$ if and only if Ω is a ball. Notice also that, by compactness, the supremum in this definition is attained.

Lemma 9.5.2. *Suppose that $\Omega \subset \mathbb{R}^p$ is a nonempty bounded open convex set with* $\operatorname{diam} \Omega = d$ *and*

$$0 < \sigma \leq \frac{1}{2\rho(\Omega)}.$$

Then for every $u, v \in \overline{\Omega}$, and $0 < r \leq d$ there are $x, y \in \Omega$ such that

$$B(x, \sigma r) \subset \Omega \cap B(u, r), \quad B(y, \sigma r) \subset \Omega \cap B(v, r),$$

and $y - x = (1 - r/d)(v - u)$.

Proof. Let $z \in \Omega$ be such that $B(z, \sigma d) \subset \Omega$. Denote $x = u + (r/d)(z - u)$ and $y = v + (r/d)(z - v)$. Then

$$B(x, \sigma r) = x + \frac{r}{d} B(0, \sigma d) = \left(1 - \frac{r}{d}\right) u + \frac{r}{d} z + \frac{r}{d} B(0, \sigma d)$$

$$= \left(1 - \frac{r}{d}\right) u + \frac{r}{d} B(z, \sigma d).$$

Since $0 < r/d \leq 1$, the last set is contained in $\operatorname{conv}\big(B(z, \sigma d) \cup \{u\}\big) \setminus \{u\} \subset \Omega$. Moreover, since clearly $|z - u| + \sigma d \leq d$ we conclude that

$$|x - u| = \frac{r}{d}|z - u| \leq (1 - \sigma)r.$$

Hence $B(x, \sigma r) \subset B(u, r) \cap \Omega$. Similarly we see that $B(y, \sigma r) \subset B(v, r) \cap \Omega$. Finally, we observe $y - x = (1 - r/d)(v - u)$, as required. $\qquad \square$

Corollary 9.5.3. *Let $\Omega \subset \mathbb{R}^p$ be a nonempty bounded open convex set. Then for every* $u \in \overline{\Omega}$ *and $r > 0$, $\rho(\Omega \cap B(u, r)) \leq 2\rho(\Omega)$.*

Proof. When $r \geq \operatorname{diam} \Omega$, the statement is trivial since $\Omega \cap B(u, r) = \Omega$. In the opposite case we use Lemma 9.5.2 with $v = u$ and $\sigma = 1/(2\rho(\Omega))$ to get a ball $B(x, \sigma r) \subset \Omega \cap B(u, r)$. Hence

$$\rho(\Omega \cap B(u, r)) \leq \frac{\operatorname{diam}(\Omega \cap B(u, r))}{2\sigma r} \leq \frac{2r}{2\sigma r} = 2\rho(\Omega). \qquad \square$$

One of the key results we need is the following estimate of the increment of a Lipschitz function. As already stated, we do not need more refined versions, and so we made no attempt to make this inequality optimal.

Lemma 9.5.4. *Let Ω be a nonempty bounded open convex set in $\mathbb{R}^n$ and $g \colon \mathbb{R}^n \longrightarrow E$ be Lipschitz. Then for every $1 \leq s < \infty$ and $u, v \in \overline{\Omega}$,*

$$|g(v) - g(u)|^{n-1+s} \leq 3^s K_n \big(\rho(\Omega) \operatorname{Lip}(g)\big)^{n-1} |v - u|^{s-1} \int_\Omega \|g'\|^s \, d\mathscr{L}^n,$$

where $K_n = 12^{n-1}/\alpha_{n-1}$.

Proof. The homogeneity of the inequality in the statement allows us to assume that $\text{Lip}(g) = 1$. Also, we may assume that $g(u) \neq g(v)$.

Let

$$\Omega_0 = \Omega \cap B(\tfrac{1}{2}(u + v), \tfrac{1}{2}|u - v|).$$

Corollary 9.5.3 gives that $\rho(\Omega_0) \leq 2\rho(\Omega)$. Let

$$\sigma = \frac{1}{4\rho(\Omega)} \quad \text{and} \quad r = \frac{1}{3}|g(v) - g(u)|.$$

Since $r \leq \tfrac{1}{3}|v - u| \leq \text{diam}\,\Omega_0$, we may apply Lemma 9.5.2 to get points $x, y \in \Omega_0$ such that

$$B(x, \sigma r) \subset \Omega_0 \cap B(u, r), \quad B(y, \sigma r) \subset \Omega_0 \cap B(v, r),$$

and

$$y - x = \left(1 - \frac{r}{\text{diam}\,\Omega_0}\right)(v - u). \tag{9.5}$$

For every $w \in B(x, \sigma r)$ we use the fact that $B(x, \sigma r) \subset B(u, r)$ to infer that $|g(u) - g(w)| \leq |u - w| \leq r$. Similarly, we see that $|g(z) - g(v)| \leq r$ for every $z \in B(y, \sigma r)$. Hence

$$|g(z) - g(w)| \geq |g(v) - g(u)| - 2r = r$$

for any such w, z. Denote $\tau = |y - x|$ and $e = (y - x)/\tau$. Then $e \in \mathbb{R}^n$ by (9.5) and for every $z \in B(x, \sigma r)$ the point $z + \tau e \in B(y, \sigma r)$. Thus we obtain that

$$r^s \leq |g(z + \tau e) - g(z)|^s = \left|\int_0^\tau \frac{d}{dt} g(z + te)\, dt\right|^s \leq \tau^s \left(\int_0^\tau \|g_n'(z + te)\| \frac{dt}{\tau}\right)^s.$$

Applying Hölder's inequality to the last integral we get the estimate

$$r^s \leq \tau^{s-1} \int_0^\tau \|g_n'(z + te)\|^s\, dt. \tag{9.6}$$

Since the points $z + te$ belong to Ω for all $0 \leq t \leq \tau$, integrating the inequality (9.6) over the set $\{z \in B(x, \sigma r) \mid \langle z, e \rangle = 0\}$ gives

$$\alpha_{n-1}\sigma^{n-1}r^{n-1+s} \leq \tau^{s-1} \int_{\{z \in B(x, \sigma r) \mid \langle z, e \rangle = 0\}} \int_0^\tau \|g_n'(z + te)\|^s\, dt\, d\mathcal{L}^{n-1}(z)$$

$$\leq \tau^{s-1} \int_\Omega \|g_n'\|^s\, d\mathcal{L}^n.$$

It remains to substitute back for $r = \tfrac{1}{3}|g(v) - g(u)|$ and $\sigma = 1/(4\rho(\Omega))$ and use that $\tau \leq |u - v|$ by (9.5) to conclude that

$$|g(v) - g(u)|^{n-1+s} \leq 3^{n-1+s} \frac{(4\rho(\Omega))^{n-1}}{\alpha_{n-1}} |v - u|^{s-1} \int_\Omega \|g_n'\|^s\, d\mathcal{L}^n,$$

which is the desired inequality with $K_n = 12^{n-1}/\alpha_{n-1}$. $\qquad\square$

Finally, we recall the use of the Hardy-Littlewood maximal operator to obtain from Lemma 9.5.4 an estimate of the pointwise Lipschitz constants of Lipschitz functions $g \colon \Omega \subset \mathbb{R}^n \longrightarrow \mathbb{R}^m$. For $u \in \Omega$ the Lipschitz constant of g at $u \in \Omega$ is the least number $\mathrm{Lip}_u(g) = \mathrm{Lip}_{u,\Omega}(g) \in [0, \infty)$ such that

$$|g(v) - g(u)| \leq \mathrm{Lip}_{u,\Omega}(g)|v - u|$$

for every $v \in \Omega$. We will integrate $\mathrm{Lip}_u(g)$ with respect to u, so notice that there are no problems with measurability, since the function $u \mapsto \mathrm{Lip}_u(g)$ is lower semicontinuous.

Recall that the Hardy-Littlewood maximal operator assigns to every measurable function $h \colon \mathbb{R}^n \longrightarrow E$ the function

$$\mathbf{M}h(u) = \sup_{r > 0} \fint_{B_n(u,r)} \|h\| \, d\mathscr{L}^n.$$

In the following chapters we will need to use this concept in a slightly more refined form, but at the moment it is sufficient to know that $\mathbf{M}$ is of the strong type $(2, 2)$. This means that there are constants Λ_n such that for every measurable $h \colon \mathbb{R}^n \longrightarrow E$,

$$\int_{\mathbb{R}^n} (\mathbf{M}h)^2 d\mathscr{L}^n \leq \Lambda_n \int_{\mathbb{R}^n} \|h\|^2 \, d\mathscr{L}^n. \tag{9.7}$$

For a proof see, for example, [33]. By [43] the inequality (9.7) holds even with an absolute constant independent of n. Using this, we could get better formulas for some constants, but unfortunately no simplification of our arguments. We will therefore use (9.7) with the constant Λ_n possibly depending on n, in which form it is more easily accessible in the literature.

Corollary 9.5.5. *There is a constant K depending only on n such that under the assumptions of Lemma 9.5.4,*

$$\int_\Omega \left(\mathrm{Lip}_{u,\Omega}(g)\right)^{2n} d\mathscr{L}^n(u) \leq K \left(\rho(\Omega) \mathrm{Lip}(g)\right)^{2n-2} \int_\Omega \|g'\|^2 \, d\mathscr{L}^n.$$

Proof. Let $u \in \Omega$. For every $v \in \Omega$ we use Lemma 9.5.3 and Lemma 9.5.4 with $s = 1$ and Ω replaced by $\Omega \cap B(u, r)$, where $r = |u - v|$, to infer that

$$|g(v) - g(u)|^n \leq 3K_n \left(2\rho(\Omega) \mathrm{Lip}(g)\right)^{n-1} \int_{\Omega \cap B(u,r)} \|g'\| \, d\mathscr{L}^n$$

$$\leq 3K_n \left(2\rho(\Omega) \mathrm{Lip}(g)\right)^{n-1} \alpha_n |v - u|^n \mathbf{M}(g' \, 1_\Omega)(u).$$

Hence

$$\left(\mathrm{Lip}_{u,\Omega}(g)\right)^{2n} \leq 2^{2n-2} 9 K_n^2 \alpha_n^2 \left(\rho(\Omega) \mathrm{Lip}(g)\right)^{2n-2} \mathbf{M}^2(g' \, 1_\Omega)(u),$$

and the statement follows with $K = 2^{2n-2} 9 K_n^2 \alpha_n^2 \Lambda_n$ by integrating over Ω and using (9.7). $\qquad\square$

Chapter Ten

Porosity, Γ_n- and Γ-null sets

In addition to the porosity notions that have been already defined, we introduce the notion of porosity "at infinity" (which we formally define as porosity with respect to a family of subspaces). Our main result shows that sets porous with respect to a family of subspaces are Γ_n-null provided X admits a continuous bump function whose modulus of smoothness (in the direction of this family) is controlled by $t^n \log^{n-1}(1/t)$. Corollaries include that in spaces with separable dual σ-porous sets are Γ_1-null and, thanks to the logarithmic term, in Hilbert spaces they are Γ_2-null. The first of these results is shown to characterize Asplund spaces: we show that a separable space has separable dual if and only if all its porous sets are Γ_1-null. We finish with a new approach to the study of spaces in which every σ-porous set is Γ-null; its results complete what we have done in Chapter 6. In addition to Chapter 6 the results obtained here are basic for the approach to ε-Fréchet differentiability developed in Chapter 11. In Chapter 14 we will show that the results obtained here cannot be much improved: for example, in Hilbert spaces there are porous sets that are not Γ_3-null and in ℓ_p, $1 < p < 2$, there are porous sets that are not Γ_2-null.

10.1 POROUS AND σ-POROUS SETS

The notion of porous sets was introduced in the Introduction, Section 1.1 and a number of results relating it to differentiability have been proved in the previous chapter. Here we extend the notion of porosity in the direction of a subspace to porosity in the direction of a family of subspaces and study in detail the problem when porous sets are Γ_n- or Γ-null.

We quickly recall the main notions of porosity introduced so far. A set E in a Banach space X is called porous in the direction of a subspace Y if there is $0 < c < 1$ such that, for every $x \in E$ and $\varepsilon > 0$, there is $y \in Y$ such that $0 < \|y\| < \varepsilon$ and

$$B(x + y, c\|y\|) \cap E = \emptyset.$$

In this situation, we also call E porous in the direction of Y with constant c. Because this situation occurs rather often, we will often say instead that E is c-porous. Sets porous in the direction of X are termed porous, sets porous in the direction of the linear span of a vector u are said to be porous in the direction of u, and we also say that E is directionally porous if it is porous in some direction. Finally, the prefix σ- is used to indicate "the countable union of."

Notice that a simple compactness argument shows that we get the same notion of σ-directional porosity if we define directionally porous sets as those that are porous in the direction of a finite dimensional subspace of X. In particular, when X is finite dimensional, σ-porous and σ-directionally porous sets coincide.

By definition, $E \subset X$ is σ-porous if it is a countable union of porous sets, each possibly with a different constant of porosity. However, as observed by Zajíček [48], every σ-porous set can be written for every $0 < c < 1$ as a countable union of sets porous with constant c. Although we will not use this result directly, we include its proof because it contributes to better understanding of σ-porous sets.

Proposition 10.1.1. *Let E be a σ-porous set in a metric space X and let $0 < c < 1$. Then E can be written as a countable union of sets porous with constant c.*

Proof. It is sufficient to prove the statement for E porous. For $i, j \geq 1$ we define

$$M_{i,j} = \bigcup \left\{ B(y, r) \ \Big|\ 0 < r < \frac{1}{j}, \ B(y, c^i r) \cap E = \emptyset \right\}$$

and $M_{0,j} = \emptyset$. We estimate the porosity constant of the set

$$E_{i,j} := \bigcap_{k=1}^{\infty} M_{i,k} \setminus M_{i-1,j}.$$

Let $x \in E_{i,j}$ and $k \geq j$ be any integer. Then $x \in M_{i,k}$ and we can find $0 < r < 1/k$ and $y \in X$ such that

$$x \in B(y, r) \quad \text{and} \quad B(y, c^i r) \cap E = \emptyset.$$

Since $B(y, c^i r) = B(y, c^{i-1} c r)$ we obtain that $B(y, cr) \subset M_{i-1,j} \subset X \setminus E_{i,j}$. Since also $d(x, y) < r$, this shows that $E_{i,j}$ is porous with constant c.

To finish the proof, it is enough to verify that

$$E \subset \bigcup_{i=1}^{\infty} \bigcup_{j=1}^{\infty} E_{i,j}.$$

Let $x \in E$. Since E is porous we find $i \geq 1$ such that for all sufficiently small $r > 0$ there is a point $y \in X$ satisfying

$$d(x, y) < r \quad \text{and} \quad B(y, c^i d(x, y)) \cap E = \emptyset.$$

It follows that $x \in M_{i,j}$ for all $j \geq 1$ and, consequently, x belongs to $\bigcap_{j=1}^{\infty} M_{i,j}$ for some $i \geq 1$. Taking the the smallest such i we conclude that $x \notin M_{i-1,j}$ for some j, hence $x \in E_{i,j}$ as required. $\qquad\square$

The following decomposition property of porous sets is of considerable importance in what follows. It is a variant of [42, Lemma 4.6] (see also [28, Lemma 4.3]).

Lemma 10.1.2. *Let U, V be subspaces of a Banach space X such that for some $\eta > 0$ every $x \in U + V$ may be written as $x = u + v$, where $u \in U$, $v \in V$, and*

$$\max\{\|u\|, \|v\|\} \leq \eta \|x\|.$$

Then, if a set $E \subset X$ is porous in the direction of $U + V$ with constant c, then we can write $E = A \cup B$, where A is σ-porous in the direction of U and B is porous in the direction of V with constant $c/(2\eta)$.

Proof. Denote for every $m \geq 1$ by

$$B_m = \left\{ x \in E \;\middle|\; B\left(x + v, \frac{1}{2\eta}c\|v\|\right) \cap E = \emptyset \text{ for some } v \in V, \; \|v\| < \frac{1}{m} \right\}.$$

Clearly, the set $B = \bigcap_{m=1}^{\infty} B_m$ is porous in the direction of V with constant $c/(2\eta)$. Thus it is sufficient to prove that each set $E \setminus B_m$ is porous in the direction of U and to put $A = \bigcup_m (E \setminus B_m)$.

Let $x \in E \setminus B_m$. By the porosity assumption on E we can find $z \in U + V$ such that $0 < \|z\| < 1/(m\eta)$ and $B(x + z, c\|z\|) \cap E = \emptyset$. Write $z = u + v$ with $u \in U$, $v \in V$, and $\max\{\|u\|, \|v\|\} \leq \eta\|z\|$. We show that

$$B(x + u, \tfrac{1}{2\eta}c\|u\|) \cap (E \setminus B_m) = \emptyset.$$

For this assume that y belongs to the intersection above. Then

$$B\left(y + v, \frac{1}{2\eta}c\|v\|\right) \subset B(x + u + v, c\|u + v\|) = B(x + z, c\|z\|) \subset X \setminus E.$$

Since $\|v\| \leq \eta\|z\| < \frac{1}{m}$ this shows that $y \in B_m$, which contradicts our assumption. $\qquad\square$

We now extend the notions of porosity and σ-porosity in the direction of a subspace to families of subspaces.

Definition 10.1.3. Let $\mathcal{Y}$ be a nonempty family of subspaces of X. We say that a set $E \subset X$ is *c-porous in the direction of* $\mathcal{Y}$ if it is porous with this constant in the direction of every element of $\mathcal{Y}$. A set will be termed *porous in the direction of* $\mathcal{Y}$ if it is porous in the direction of $\mathcal{Y}$ with some constant c, and a set is *σ-porous in the direction of* $\mathcal{Y}$ if it is a countable union of sets porous in the direction of $\mathcal{Y}$.

It should be noted that, when changing the family $\mathcal{Y}$, porosity behaves in the dual way to smoothness: while adding to $\mathcal{Y}$ all subspaces contained in an element of $\mathcal{Y}$ preserves the notion of smoothness, the notion of porosity is preserved by adding to $\mathcal{Y}$ all subspaces that contain an element of $\mathcal{Y}$.

Some of our results are based on rather detailed information on descriptive set theoretic character of certain sets. For that the following simple fact will be of crucial importance.

Lemma 10.1.4. *Every set c-porous in the direction of a subspace Y is, for every $0 < d < c$, contained in a G_δ set which is d-porous in the direction of Y.*

Proof. Let $E \subset X$ be c-porous in the direction of Y. For $k \in \mathbb{N}$, let D_k be the set of those elements $x \in X$ for which there are $0 < \varepsilon < c - d$ and $y \in Y$ such that $0 < \|y\| < 1/k$ and

$$B(x + y, (c - \varepsilon)\|y\|) \cap E = \emptyset.$$

The sets D_k are clearly open and $E \subset D_k$. Hence $\bigcap_{k=1}^{\infty} D_k$ is a G_δ set containing E and, since $d < c - \varepsilon$, it is d-porous in the direction of Y. □

It is obvious that Lemma 10.1.4 holds also for sets porous in the direction of a countable family of subspaces: just intersect the countably many G_δ sets obtained for each of the subspaces. Since we assume that our spaces are separable, this simple fact is enough for most of our purposes (although this is not obvious, as families of subspaces may be very far from countable). We may also extend this lemma to families of subspaces having the property that countably many subspaces from the family are enough to determine whether a set is porous or not. The first step toward this aim is the next lemma.

Lemma 10.1.5. *Let $\mathcal{Y}_0 \subset \mathcal{Y}$ be two families of subspaces of X having the property that for every $Y \in \mathcal{Y}$ and every $\eta > 0$ there is $Y_0 \in \mathcal{Y}_0$ such that every norm one vector from Y_0 is in distance less than η from Y. Then for every $0 < d < c$, every set c-porous in the direction of $\mathcal{Y}_0$ is d-porous in the direction of $\mathcal{Y}$.*

Proof. Let E be c-porous in the direction of $\mathcal{Y}_0$ and let $Y \in \mathcal{Y}$. We put

$$\eta = \frac{c - d}{1 + d}$$

and find $Y_0 \in \mathcal{Y}_0$ such that every norm one vector from Y_0 is in distance less than η from Y. By the porosity assumption on E in the direction of Y_0, for every $x \in E$ and $\varepsilon > 0$ there is $y_0 \in Y_0$ such that

$$0 < \|y_0\| < \frac{\varepsilon}{1 + \eta} \text{ and } B(x + y_0, c\|y_0\|) \cap E = \emptyset.$$

By the assumption connecting $\mathcal{Y}$ and $\mathcal{Y}_0$, for this y_0 there is $y \in Y$ such that

$$\left\| \frac{y_0}{\|y_0\|} - \frac{y}{\|y_0\|} \right\| < \eta, \text{ i.e., } \|y_0 - y\| < \eta\|y_0\|.$$

Then $0 < \|y\| < (1 + \eta)\|y_0\| < \varepsilon$. Since

$$\|y - y_0\| + d\|y\| < (\eta + d(1 + \eta))\|y_0\| = c\|y_0\|,$$

we see that

$$B(x + y, d\|y\|) \cap E \subset B(x + y_0, c\|y_0\|) \cap E = \emptyset. \qquad \square$$

Lemma 10.1.5 leads to the following notion. A family $\mathcal{Y}$ of subspaces of X will be called *separable* if there is a countable subfamily $\mathcal{Y}_0$ of $\mathcal{Y}$ such that for every $Y \in \mathcal{Y}$ and every $\eta > 0$ there is $Y_0 \in \mathcal{Y}_0$ such that every norm one vector from Y_0 is in distance less than η from Y. Examples of separable families $\mathcal{Y}$ include the trivial (but important) case $\mathcal{Y} = \{X\}$ and the family of all finite codimensional subspaces of a space X with separable dual. In the latter case, the required countable family $\mathcal{Y}_0$ is

$$\mathcal{Y}_0 = \left\{ \bigcap_{n \in F} \text{Ker}(x_n^*) \; \middle| \; F \subset \mathbb{N} \text{ finite} \right\},$$

where (x_n^*) is a countable dense subset of the dual unit ball B_{X^*}.

Lemma 10.1.5 and the discussion following Lemma 10.1.4 immediately give

Lemma 10.1.6. *Every set c-porous in the direction of a separable family $\mathcal{Y}$ of sub-spaces of X is, for every $0 < d < c$, contained in a G_δ set which is d-porous in the direction of $\mathcal{Y}$.*

10.2   A CRITERION OF Γ_n-NULLNESS OF POROUS SETS

Here we show that sets porous in the direction of a family of subspaces are Γ_n-null provided the space admits a continuous bump function that is smooth in the direction of the given family with modulus controlled by $t^n \log^{n-1}(1/t)$.

In the most delicate arguments of this section it will be extremely important that some sets or functions on $\Gamma_n(X)$ are of low class with respect to the maximum norm. We recall that $\| \cdot \|_{C^1}$ denotes the C^1 norm on $\Gamma_n(X)$,

$$\| \cdot \|_{C^1} = \max\{\|\gamma\|_\infty, \|\gamma'\|_\infty\}.$$

Except when specifically indicated, all topological notions are with respect to the C^1 norm. For example, $B(\gamma, r)$ denotes a C^1 ball while the $\| \cdot \|_\infty$ ball is denoted by $B_\infty(\gamma, r)$.

We will need to approximate surfaces from $\Gamma_n(X)$ by surfaces γ that are locally affine a.e., or, in other words, have the property that almost every point is contained in an open set on which the surface is affine. Because of the rather special nature of the following lemma, we include these results here, even though at least Corollary 10.2.2 would perhaps more naturally belong to Chapter 5.

Lemma 10.2.1. *Suppose that $\gamma \in \Gamma_n(X)$, $A \subset U \subset [0,1]^n$, where U is open in $[0,1]^n$, $\varepsilon > 0$, and $r_k \colon A \longrightarrow (0, \infty)$ are such that $r_k(u) \to 0$ for every $u \in [0,1]^n$. Then there are $\xi \in \Gamma_n(X)$, finite sequences $u_1, \ldots, u_m \in [0,1]^n$ of points and $k_1, \ldots, k_m \in \mathbb{N}$ of numbers such that*

  (i) $\|\xi - \gamma\|_{C^1} \leq \varepsilon$,

  (ii) $\xi(u) = \gamma(u)$ and $\xi'(u) = \gamma'(u)$ for $u \notin U$,

  (iii) *the balls $B(u_i, r_{k_i}(u_i))$ are disjoint,*

  (iv) $\mathcal{L}^n\big(A \setminus \bigcup_i B(u_i, r_{k_i}(u_i))\big) < \varepsilon$,

  (v) *on $B(u_i, r_{k_i}(u_i))$, $\xi(u) = \gamma(u_i) + \gamma'(u_i)(u - u_i)$.*

Proof. Let $\tau > 0$ satisfy $(1+\tau)^n < 1 + \frac{1}{2}\varepsilon$ and let $\zeta \in C^1(\mathbb{R}^n)$ be such that $0 \leq \zeta \leq 1$, $\zeta(u) = 1$ on $B(0,1)$, and $\zeta(u) = 0$ outside $B(0, 1+\tau)$. Let

$$\eta = \frac{\varepsilon}{1 + \|\zeta'\|_\infty},$$

and for each $u \in \mathbb{R}^n$ find $0 < \delta(u) < 1$ such that $\|\gamma'(u) - \gamma'(v)\| < \eta$ whenever $|u - v| < \delta(u)$.

The family of balls $B(u, (1 + \tau)r_k(u))$, $u \in A$, that are contained in U and such that $(1 + \tau)r_k(u) < \delta(u)$ forms a Vitali cover of A. Hence there are $u_1, \ldots, u_m \in A$ and corresponding $k_1, \ldots, k_m \in \mathbb{N}$ such that

$$(1 + \tau)r_{k_i}(u_i) < \delta(u_i),$$

the balls $B(u_i, (1 + \tau)r_{k_i}(u_i))$ are disjoint and contained in U, and

$$\mathscr{L}^n \left(A \setminus \bigcup_{i=1}^{m} B(u_i, (1 + \tau)r_{k_i}(u_i)) \right) < \frac{\varepsilon}{2}.$$

Hence (iii) holds. At the same time due to the choice of τ we also have

$$\mathscr{L}^n \left(\bigcup_{i=1}^{m} B(u_i, (1 + \tau)r_{k_i}(u_i)) \setminus \bigcup_{i=1}^{m} B(u_i, r_{k_i}(u_i)) \right) < \frac{\varepsilon}{2},$$

which gives (iv).

Define

$$\xi(u) = \gamma(u) + \sum_{i=1}^{m} \zeta \left(\frac{u - u_i}{r_{k_i}(u_i)} \right) \big(\gamma(u_i) + \gamma'(u_i)(u - u_i) - \gamma(u) \big).$$

Since the terms of the sum and their derivatives are nonzero only on mutually disjoint balls $B(u_i, (1 + \tau)r_{k_i}(u_i))$, we see that (ii) and (v) hold. In fact we showed more than (ii): we proved that γ and ξ and their derivatives coincide outside the union of balls $B(u_i, (1 + \tau)r_{k_i}(u_i))$. To complete the proof we estimate the differences $\xi - \gamma$ and $\xi' - \gamma'$ on balls $B(u_i, (1 + \tau)r_{k_i}(u_i))$. For each $u \in B(u_i, (1 + \tau)r_{k_i}(u_i))$ the mean value estimate shows that for some v lying on the segment $[u_i, u]$ we have

$$\|\xi(u) - \gamma(u)\| = \|\gamma(u) - \gamma(u_i) - \gamma'(u_i)(u - u_i)\|$$
$$\leq \|\gamma'(v) - \gamma'(u_i)\| \|u - u_i\| < \eta \, r_{k_i}(u_i) \leq \varepsilon,$$

and

$$\|\xi'(u) - \gamma'(u)\| \leq \frac{\|\zeta'\|_\infty}{r_{k_i}(u_i)} \|\gamma(u_i) + \gamma'(u_i)(u - u_i) - \gamma(u)\| + \|\gamma'(u) - \gamma'(u_i)\|$$
$$< \|\zeta'\|_\infty \, \eta + \eta = \varepsilon,$$

which is (i). $\qquad \square$

Corollary 10.2.2. *The set of a.e. affine surfaces is dense in* $\Gamma_n(X)$.

Proof. Suppose $\gamma \in \Gamma_n(X)$ and $\varepsilon > 0$. Using Lemma 10.2.1 with $U_0 = A_0 = [0, 1]^n$, we find a surface $\xi_0 \in \Gamma_n(X)$ and a set $V_0 \subset U_0$ such that $\|\xi_0 - \gamma\|_{C^1} < \frac{1}{2}\varepsilon$, $\mathscr{L}^n(U_0 \setminus V_0) < \frac{1}{2}$ and V_0 is a finite union of open balls on each of which ξ_0 is affine. Continuing recursively, we let

$$U_k = A_k = [0, 1]^n \setminus \bigcup_{0 \leq j < k} \overline{V_j}$$

and use Lemma 10.2.1 to find $\xi_k \in \Gamma_n(X)$ and a set $V_k \subset U_k$, which is a finite union of open balls on each of which ξ_k is affine, such that $\|\xi_k - \xi_{k-1}\|_{C^1} < 2^{-k-1}\varepsilon$, $\xi_k(u) = \xi_{k-1}(u), \xi_k'(u) = \xi_{k-1}'(u)$ for $u \notin U_k$, $\mathscr{L}^n(U_k \setminus V_k) < 2^{-k-1}$.

The surfaces ξ_k form a sequence converging to a surface $\xi \in \Gamma_n(X)$ such that $\|\xi - \gamma\|_{C^1} < \varepsilon$. If u belongs to one of the sets V_k, that $\xi = \xi_k$ is affine on an open ball containing u. Hence points around which ξ is not affine belong either to the boundary of some V_k or to all sets U_k. Since the boundaries of V_k have measure zero and $\mathscr{L}^n(U_k) \le 2^{-k}$, this shows that ξ is locally affine a.e. $\qquad\square$

The following lemma is our main technical tool. It allows us to modify an arbitrary surface from $\Gamma_n(X)$ so that many points in which the original surface passes through the given porous set are moved outside the closure of the porous set. The "energy" Φ introduced below whose increment is controlled is exactly the one we need in the application to the proof of Proposition 10.2.4. The key point of the lemma is that the increase of the energy is smaller than the increase of the measure of the points whose image is inside the holes. Also, it will be important that the difference depends only on r, although the particular form of this dependence will play no role.

Lemma 10.2.3. *Suppose that $\Theta\colon L(\mathbb{R}^n, X) \longrightarrow \mathbb{R}$ is a bounded continuous function smooth in the direction of a family $\mathcal{Y}$ of subspaces of X with modulus controlled by ω_n. Suppose further that $(\gamma_k)_{k=0}^\infty$ are a.e. locally affine surfaces from $\Gamma_n(X)$, $\sum_{k=0}^\infty \lambda_k$ is a convergent series of positive numbers, and define $\Phi\colon \Gamma_n(X) \longrightarrow \mathbb{R}$ by*

$$\Phi(\gamma) = \int_{[0,1]^n} \sum_{k=0}^\infty \lambda_k \Theta\big(\gamma'(u) - \gamma_k'(u)\big)\, d\mathscr{L}^n(u).$$

Then for every $\alpha > 0$ and $\widetilde{\gamma} \in \Gamma_n(X)$ one can find $\delta > 0$ with the following property:

For every $0 < r < \delta$, $\zeta > 0$, every set $E \subset X$ which is c-porous in the direction of $\mathcal{Y}$, and every open set $U \subset [0,1]^n$ with

$$\mathscr{L}^n(U \cap \widetilde{\gamma}^{-1}(E)) > \alpha, \tag{10.1}$$

there is $\gamma \in \Gamma_n(X)$ satisfying

(i) $\gamma(u) = \widetilde{\gamma}(u)$ and $\gamma'(u) = \widetilde{\gamma}'(u)$ for $u \notin U$;

(ii) $\|\gamma(u) - \widetilde{\gamma}(u)\| \le \zeta$ for all $u \in [0,1]^n$;

(iii) $\|\gamma'(u) - \widetilde{\gamma}'(u)\| \le r$ for all $u \in [0,1]^n$;

and, letting

$$\chi(r) = \frac{c^n \alpha}{4(1 + \|\widetilde{\gamma}\|)^n}\, e^{-2n/r},$$

(iv) $\mathscr{L}^n \gamma^{-1}(\overline{E}) + \Phi(\gamma) \le \mathscr{L}^n \widetilde{\gamma}^{-1}(\overline{E}) + \Phi(\widetilde{\gamma}) - \chi(r)$.

Proof. Let A_0 be the set of points of $\tilde{\gamma}^{-1}(E)$ around which each γ_i is affine (of course, each possibly on a different neighborhood). If $\mathscr{L}^n A_0 \leq \alpha$, there is nothing to prove since then it is impossible to satisfy (10.1). Hence we assume that $\mathscr{L}^n A_0 > \alpha$. Denote

$$\Lambda_k = 2 \sum_{i=k}^{\infty} \lambda_i \|\Theta\|_\infty, \quad C = 1 + \|\tilde{\gamma}\|, \quad \text{and} \quad \varepsilon := \frac{\alpha_n}{8K_n} \frac{c^n \alpha}{4C^n},$$

where K_n is the constant from Lemma 9.3.2. Also, it will be convenient to express $\Phi(\gamma)$ as

$$\Phi(\gamma) = \int_{[0,1]^n} \Theta_u(\gamma'(u)) \, d\mathscr{L}^n(u), \quad \text{where} \quad \Theta_u(T) := \sum_{i=0}^{\infty} \lambda_i \Theta(T - \gamma_i'(u)).$$

By Lemma 8.2.5 and Observation 8.2.4, for each $u \in [0,1]^n$ the function Θ_u is smooth in the direction of the family $\mathcal{Y}$ with modulus controlled by ω_n. Hence there is $0 < \delta_u < 1$ such that

$$\bar{\rho}_{\Theta_u, \mathcal{Y}}(\tilde{\gamma}'(u); t) < \varepsilon \omega_n(t) \tag{10.2}$$

for $0 < t \leq \delta_u$. We fix $0 < \delta < 1$ for which the set $A := \{u \in A_0 \mid \delta_u > \delta\}$ has (outer) measure $\mathscr{L}^n A > \mathscr{L}^n A_0 - \frac{1}{2}\alpha$. This will be the δ for which we show the statement.

Suppose that $0 < r < \delta$, $\zeta > 0$, and the open set U such that (10.1) holds are given. Observing that

$$\mathscr{L}^n(U \cap A) \geq \mathscr{L}^n(U \cap \tilde{\gamma}^{-1}(E)) - (\mathscr{L}^n A_0 - \mathscr{L}^n A) > \frac{\alpha}{2},$$

we find an open set V, $A \subset V \subset U$, such that $\mathscr{L}^n V < \mathscr{L}^n A + \eta$, where

$$\eta = \frac{\chi(r)}{16(1 + \Lambda_0)}.$$

We also find a finite set $M \subset (0, \frac{1}{2}r]$ such that $\frac{1}{2}r \in M$ and each point of $(0, \frac{1}{2}r)$ is within $e^{-2n/r}$ of M. By Lemma 9.3.2 (with r replaced by $\frac{1}{2}r$) there is a C^1 function $\psi \colon \mathbb{R}^n \longrightarrow \mathbb{R}$ having the following properties:

(a) $\psi(0) = 1$, $\psi(u) = 0$ for $|u| \geq e^{2/r}$ and $0 \leq \psi(u) \leq 1$ for $u \in \mathbb{R}^n$;

(b) $\|\psi'(u)\| \leq \frac{1}{2}r$ for all $u \in \mathbb{R}^n$;

(c) if $\theta \colon \mathbb{R}^n \longrightarrow \mathbb{R}$ is bounded, Borel measurable, and $\theta(0) = 0$, then

$$\int_{\mathbb{R}^n} \theta(\psi'(u)) \, d\mathscr{L}^n(u) \leq K_n \sup_{t \in M} \sup_{|z| < t} \frac{\theta(z) + \theta(-z)}{\omega_n(t)}.$$

Fix an index $q \in \mathbb{N}$ such that $\Lambda_q \leq \frac{1}{8}\chi(r)$. For each $u \in A$ the estimate of the modulus of smoothness established in (10.2) allows us to choose a subspace $Y_u \in \mathcal{Y}$ such that for every $t \in M$,

$$\rho_{\Theta_u, Y_u}(\tilde{\gamma}'(u); t) < \varepsilon \omega_n(t). \tag{10.3}$$

Using that E is c-porous at $\tilde{\gamma}(u)$ in the direction of Y_u and that the function $v \to \Theta_v(\tilde{\gamma}'(v))$ is continuous, we choose a sequence $y_j(u) \in Y_u$, $y_j(u) \to 0$ for $j \to \infty$, such that

(α) $0 < \|y_j(u)\| \le \frac{1}{2}\varsigma$;

(β) $B(\tilde{\gamma}(u) + y_j(u), c\|y_j(u)\|) \cap E = \emptyset$;

(γ) $\Theta_v(\tilde{\gamma}'(v)) > \Theta_u(\tilde{\gamma}'(u)) - \frac{1}{8}\chi(r)$ for $v \in B(u, e^{2/r}\|y_j(u)\|)$;

(δ) each of the functions γ_i, $0 \le i \le q$, is affine on $B(u, e^{2/r}\|y_j(u)\|)$.

By Lemma 10.2.1 we find a surface $\xi \in \Gamma_n(X)$ and finite sequences of points $u_1, \ldots, u_m \in A$ and numbers $k_1, \ldots, k_m \in \mathbb{N}$ such that, simplifying the notation by letting $r_j = e^{2/r}\|y_{k_j}(u_j)\|$, we have

(A) $\|\xi - \tilde{\gamma}\|_{C^1} \le \frac{1}{2}\min\{\varsigma, r\}$;

(B) $\xi(u) = \tilde{\gamma}(u)$ and $\xi'(u) = \tilde{\gamma}'(u)$ for $u \notin V$;

(C) the balls $B(u_j, r_j)$ are disjoint subsets of V;

(D) $\mathscr{L}^n\left(\bigcup_{j=1}^{m} B(u_j, r_j)\right) > \max\{\frac{1}{2}a, \mathscr{L}^n A - \eta\}$;

(E) on $B(u_j, r_j)$, $\xi(u) = \tilde{\gamma}(u_j) + \tilde{\gamma}'(u_j)(u - u_j)$.

We further simplify the notation by letting $y_j = y_{k_j}(u_j)$ and $s_j = \|y_{k_j}(u_j)\|$ and show that

$$\gamma(u) := \xi(u) + \sum_{j=1}^{m} \psi\left(\frac{u - u_j}{s_j}\right) y_j. \tag{10.4}$$

is the surface we are looking for.

Notice that by (a) the jth term of the sum in (10.4) is nonzero only on the ball $B(u_j, r_j)$ and that on this ball $\|\gamma(u) - \xi(u)\| \le \|y_j(u)\| \le \frac{1}{2}\varsigma$. Also, the difference

$$\gamma'(u) - \tilde{\gamma}'(u_j) = \frac{y_j}{s_j} \otimes \psi'\left(\frac{u - u_j}{s_j}\right) = \frac{y_j}{\|y_j\|} \otimes \psi'\left(\frac{u - u_j}{s_j}\right)$$

has by (b) norm at most $\frac{1}{2}r$ on $B(u_j, r_j)$. We will often use these observations together with the disjointness of the balls $B(u_j, r_j)$ without any further mention.

The statement (i) follows immediately from (C) and (B). The statements (ii) and (iii) follow directly from the corresponding estimates in the construction of γ: (A) and the just observed facts give

$$\|\gamma(u) - \tilde{\gamma}(u)\| \le \|\gamma(u) - \xi(u)\| + \|\xi(u) - \tilde{\gamma}(u)\| \le \frac{\varsigma}{2} + \frac{\varsigma}{2} = \varsigma$$

and

$$\|\gamma'(u) - \tilde{\gamma}'(u)\| \le \|\gamma'(u) - \xi'(u)\| + \|\xi'(u) - \tilde{\gamma}'(u)\| \le \frac{r}{2} + \frac{r}{2} = r.$$

The proof of (iv) is more involved. Since $\xi(u_j) = \tilde{\gamma}(u_j)$ by (E), it follows from (10.4) that $\gamma(u_j) = \tilde{\gamma}(u_j) + y_j$. Using this fact together with

$$\|\gamma'(u)\| \leq \|\gamma'(u) - \tilde{\gamma}'(u)\| + \|\tilde{\gamma}'(u)\| \leq r + \|\tilde{\gamma}\| \leq C,$$

which gives that $\mathrm{Lip}(\gamma) \leq C$, we get

$$\gamma\big(B(u_j, c\|y_j\|/C)\big) \subset B(\tilde{\gamma}(u_j) + y_j, c\|y_j\|).$$

The choice of y_j thus implies that $\gamma^{-1}(\overline{E})$ does not meet $B(u_j, c\|y_j\|/C)$. Denote $B = \bigcup_{j=1}^m B(u_j, r_j)$ and $D = \bigcup_{j=1}^m B(u_j, c\|y_j\|/C)$. Since

$$\frac{c\|y_j\|}{C} = \frac{ce^{-2/r}r_j}{C} \leq r_j,$$

both unions are disjoint and

$$\mathscr{L}^n(D) = e^{-2n/r}\left(\frac{c}{C}\right)^n \mathscr{L}^n(B) > e^{-2n/r}\left(\frac{c}{C}\right)^n \frac{\alpha}{2} = 2\chi(r).$$

Hence, using that $D \subset V \setminus \gamma^{-1}(\overline{E})$, $A \subset \tilde{\gamma}^{-1}(\overline{E}) \cap V$, $\mathscr{L}^n V < \mathscr{L}^n A + \eta$, and $\eta \leq \frac{1}{2}\chi(r)$, we get

$$\mathscr{L}^n(\gamma^{-1}(\overline{E}) \cap V) < \mathscr{L}^n V - 2\chi(r) \leq \mathscr{L}^n(\tilde{\gamma}^{-1}(\overline{E}) \cap V) - \frac{3}{2}\chi(r).$$

Since the surfaces γ and $\tilde{\gamma}$ coincide outside V, this gives

$$\begin{aligned}
\mathscr{L}^n\gamma^{-1}(\overline{E}) &= \mathscr{L}^n(\gamma^{-1}(\overline{E}) \setminus V) + \mathscr{L}^n(\gamma^{-1}(\overline{E}) \cap V) \\
&\leq \mathscr{L}^n(\tilde{\gamma}^{-1}(\overline{E}) \setminus V) + \mathscr{L}^n(\tilde{\gamma}^{-1}(\overline{E}) \cap V) - \frac{3}{2}\chi(r) \\
&= \mathscr{L}^n\tilde{\gamma}^{-1}(\overline{E}) - \frac{3}{2}\chi(r).
\end{aligned} \tag{10.5}$$

In order to finish the proof of the last statement, we now fix $1 \leq j \leq m$. By (δ), γ'_i is constant on $B(u_j, r_j)$ for each $0 \leq i \leq q$. Hence we have for all $u \in B(u_j, r_j)$,

$$\begin{aligned}
\Theta_u(T) &= \Theta_{u_j}(T) + \sum_{i=q+1}^{\infty} \lambda_i\big(\Theta(T - \gamma'_i(u)) - \Theta(T - \gamma'_i(u_i))\big) \\
&\leq \Theta_{u_j}(T) + \Lambda_{q+1} \leq \Theta_{u_j}(T) + \frac{1}{8}\chi(r).
\end{aligned} \tag{10.6}$$

We estimate the integral

$$\begin{aligned}
&\int_{B(u_j, r_j)} \Theta_{u_j}(\gamma'(u)) - \Theta_{u_j}(\tilde{\gamma}'(u_j)) \, d\mathscr{L}^n(u) \\
&= s_j^n \int_{B(0, e^{2/r})} \Theta_{u_j}\left(\tilde{\gamma}'(u_j) + \frac{y_j}{s_j} \otimes \psi'(u)\right) - \Theta_{u_j}(\tilde{\gamma}'(u_j)) \, d\mathscr{L}^n(u).
\end{aligned}$$

This is an integral of the form treated in (c) with

$$\theta(z) = \Theta_{u_j}\left(\tilde{\gamma}'(u_j) + \frac{y_j \otimes z}{s_j}\right) - \Theta_{u_j}(\tilde{\gamma}'(u_j)).$$

If $t \in M$ and $|z| < t$, then $\|y_j \otimes z/s_j\| = |z| < t$ and so (10.3) implies

$$\theta(z) + \theta(-z)$$
$$= \Theta_{u_j}\left(\tilde{\gamma}'(u_j) + \frac{y_j \otimes z}{s_j}\right) + \Theta_{u_j}\left(\tilde{\gamma}'(u_j) - \frac{y_j \otimes z}{s_j}\right) - 2\Theta_{u_j}(\tilde{\gamma}'(u_j))$$
$$\leq \rho_{\Theta_{u_j}, Y_{u_j}}(\tilde{\gamma}'(u_j); t) < \varepsilon\omega_n(t).$$

Hence by (c),

$$\int_{B(u_j, r_j)} \Theta_{u_j}(\gamma'(u)) - \Theta_{u_j}(\tilde{\gamma}'(u_j))\, d\mathscr{L}^n(u) \leq K_n \varepsilon s_j^n,$$

which together with (10.6) shows that

$$\int_{B(u_j, r_j)} \Theta_u(\gamma'(u)) - \Theta_{u_j}(\tilde{\gamma}'(u_j))\, d\mathscr{L}^n(u) \leq K_n \varepsilon s_j^n + \frac{1}{8}\chi(r)\,\alpha_n r_j^n. \qquad (10.7)$$

Further, by (γ), $\Theta_u(\tilde{\gamma}'(u)) \geq \Theta_{u_j}(\tilde{\gamma}'(u_j)) - \frac{1}{8}\chi(r)$ for every $u \in B_n(u_j, r_j)$. Hence

$$\alpha_n r_j^n \Theta_{u_j}(\tilde{\gamma}'(u_j)) \leq \int_{B(u_j, r_j)} \Theta_u(\tilde{\gamma}'(u))\, d\mathscr{L}^n(u) + \frac{1}{8}\chi(r)\,\alpha_n r_j^n. \qquad (10.8)$$

Adding (10.7) and (10.8) and then summing them for $j = 1, \ldots, m$ we get

$$\int_B \Theta_u(\gamma'(u))\, d\mathscr{L}^n(u)$$
$$\leq \int_B \Theta_u(\tilde{\gamma}'(u))\, d\mathscr{L}^n(u) + K_n \varepsilon \sum_{j=1}^m s_j^n + \frac{1}{4}\chi(r) \sum_{j=1}^m \alpha_n r_j^n.$$

Since the disjointness of the balls $B(u_j, r_j)$ implies $\sum_{j=0}^m \alpha_n r_j^n \leq 1$, we obtain

$$K_n \varepsilon \sum_{j=0}^m s_j^n = \frac{K_n \varepsilon}{\alpha_n} e^{-2n/r} \sum_{j=0}^m \alpha_n r_j^n \leq \frac{K_n \varepsilon}{\alpha_n} e^{-2n/r} = \frac{1}{8}\chi(r).$$

Thus

$$\int_B \Theta_u(\gamma'(u))\, d\mathscr{L}^n(u) \leq \int_B \Theta_u(\tilde{\gamma}'(u))\, d\mathscr{L}^n(u) + \frac{3}{8}\chi(r). \qquad (10.9)$$

For the integral over $V \setminus B$ we just use the trivial estimate $\|\Theta_u\|_\infty \leq \frac{1}{2}\Lambda_0$ together with (D),

$$\mathscr{L}^n(V \setminus B) < 2\eta = \frac{1}{8(1 + \Lambda_0)}\chi(r),$$

to infer that

$$\int_{V \setminus B} \Theta_u(\gamma'(u)) \, d\mathscr{L}^n(u) \leq \frac{1}{16} \chi(r)$$

$$\leq \int_{V \setminus B} \Theta_u(\tilde{\gamma}'(u)) \, d\mathscr{L}^n(u) + \frac{1}{8} \chi(r). \qquad (10.10)$$

On $[0,1]^n \setminus V$, the estimate is even simpler since there $\gamma'(u) = \tilde{\gamma}'(u)$:

$$\int_{[0,1]^n \setminus V} \Theta_u(\gamma'(u)) \, d\mathscr{L}^n(u) = \int_{[0,1]^n \setminus V} \Theta_u(\tilde{\gamma}'(u)) \, d\mathscr{L}^n(u). \qquad (10.11)$$

Adding the estimates (10.9), (10.10), and (10.11), we finally get

$$\Phi(\gamma) \leq \Phi(\tilde{\gamma}) + \frac{1}{2} \chi(r),$$

which together with (10.5) proves the last statement. $\qquad \square$

We are now ready for the crucial proposition of this section, which combines Lemma 10.1.6 with the variational principle of Theorem 7.2.1.

Proposition 10.2.4. *Suppose that a Banach space X admits a continuous bump function smooth in the direction of a separable family $\mathcal{Y}$ of subspaces of X with modulus controlled by ω_n. Then every subset E of X which is porous in the direction of $\mathcal{Y}$ is contained in a Γ_n-null G_δ set.*

Proof. We first adjust the assumptions to our needs. By Lemma 10.1.6 it suffices to fix $c > 0$, assume that $E \subset X$ is a G_δ set which is c-porous in the direction of $\mathcal{Y}$, and then show that E is Γ_n-null. To this aim we will assume that E is not Γ_n-null, and infer that the set

$$\left\{ \gamma \in \Gamma_n(X) \mid \mathscr{L}^n \gamma^{-1}(E) > 0 \right\} = \bigcup_{j=1}^{\infty} \left\{ \gamma \in \Gamma_n(X) \mid \mathscr{L}^n \gamma^{-1}(E) \geq \frac{1}{j} \right\}$$

is not of the first category. It follows that we can find $\alpha > 0$ such that the set

$$S := \left\{ \gamma \in \Gamma_n(X) \mid \mathscr{L}^n \gamma^{-1}(E) \geq \alpha \right\}$$

is not nowhere dense in $\Gamma_n(X)$. Hence there are $\tilde{\gamma}_0 \in \Gamma_n(X)$ and $0 < \tilde{r}_0 \leq 1$ such that S is dense in the ball $B(\tilde{\gamma}_0, \tilde{r}_0)$. Using that by Lemma 5.4.1, S is G_δ in the topology induced by $\| \cdot \|_\infty$, we write $S = \bigcap_{k=0}^{\infty} S_k$, where S_k are $\| \cdot \|_\infty$-open sets. Since in the space $\Gamma_n(X)$ the $\| \cdot \|_{C^1}$-closed balls are also $\| \cdot \|_\infty$-closed, and since S is dense in $B(\tilde{\gamma}_0, \tilde{r}_0)$, the sets

$$G_k = S_k \cup \left(\Gamma_n(X) \setminus \overline{B(\tilde{\gamma}_0, \tilde{r}_0)} \right)$$

are $\| \cdot \|_\infty$-open and $\| \cdot \|_{C^1}$-dense in $\Gamma_n(X)$. Consequently, Corollary 10.2.2 implies that for each k the set $\mathcal{A}_k$ of a.e. locally affine surfaces belonging to G_k is dense in $\Gamma_n(X)$.

We also recall that by Proposition 8.2.1 and Observation 8.2.4 the space $L(\mathbb{R}^n, X)$ admits a bounded, continuous function Θ smooth in the direction of the family $\mathcal{Y}$ with modulus controlled by ω_n, such that for every $r > 0$,

$$\theta(r) := \inf_{\|T\| > r} \Theta(T) > \Theta(0) = 0.$$

Notice also that the function

$$\Psi(\gamma) = \int_{[0,1]^n} \Theta(\gamma'(u)) \, d\mathscr{L}^n(u)$$

is well defined on $\Gamma_n(X)$, bounded, continuous, and $\Psi(0) = 0$.

Before coming to the main part of the proof, the use of the perturbation game of Section 7.2 in Chapter 7, we introduce some auxiliary functions and parameters. Let

$$\widetilde{\chi}(r) = \frac{\alpha c^n}{8(1 + \widetilde{r}_0 + \|\widetilde{\gamma}_0\|)^n} \, e^{-2n/r} \quad \text{and} \quad \xi(t) = \max\{0, \min\{1, 2t - 1\}\}.$$

For $k \geq 0$ denote $r_k = 2^{-k-1}\widetilde{r}_0$ and let $\sigma_k > 0$ be numbers satisfying

$$\sigma_0 = 3, \quad \sum_{j=k+1}^{\infty} \sigma_j \leq \frac{1}{4} \chi(r_k/4), \quad \sum_{j=k+1}^{\infty} \sigma_j \leq \frac{1}{2} \widetilde{\chi}(r_k/4).$$

Finally, we let $\varepsilon_k > 0$ and $\lambda_k > 0$ be such that $\varepsilon_0 = 2$, $\varepsilon_k < \sigma_k$ for all $k \geq 0$, $\sum_{j=0}^{\infty} \lambda_j < \infty$ and

$$\sum_{j=0}^{\infty} \frac{\varepsilon_j}{\lambda_j \, \theta(r_j/4)} < \frac{\alpha}{2}.$$

It is our intention to play the perturbation game in the space $\Gamma_n(X)$ with the function

$$f(\gamma) = \mathscr{L}^n \gamma^{-1}(\overline{E}).$$

We describe a strategy for player (β). In the kth move, player (β) is faced with the function

$$h_k(\gamma) = f(\gamma) + \sum_{0 \leq j < k} F_j(\gamma) \tag{10.12}$$

and the value $\eta_k > 0$ chosen by player (α). Then player (β) notices that h_k is upper semicontinuous on $\Gamma_n(X)$ and so, since $\mathcal{A}_k$ is dense in $\Gamma_n(X)$, infers that

$$\inf_{\gamma \in \mathcal{A}_k} h_k(\gamma) = \inf_{\gamma \in \Gamma_n(X)} h_k(\gamma).$$

This allows (β) to move according to the following rules:

- let r_k and ε_{k+1} be as defined above;
- choose:

⋆ any $\gamma_0 \in \mathcal{A}_0 \cap B(\tilde{\gamma}_0, r_0)$ if $k = 0$,

⋆ a $\gamma_k \in \mathcal{A}_k$ such that $h_k(\gamma_k) < \eta_k + \inf_{\gamma \in \Gamma_n(X)} h_k(\gamma)$ if $k \geq 1$;

- find $0 < s_k < r_k$ such that

⋆ $B_\infty(\gamma_k, 2s_k) \subset G_k$,

⋆ $f(\gamma) < f(\gamma_k) + 2^{-k}$ on $B_\infty(\gamma_k, 2s_k)$;

- let $F_k(\gamma) = \lambda_k \Psi(\gamma - \gamma_k) + \sigma_k \xi\left(\dfrac{\|\gamma - \gamma_k\|_\infty}{s_k}\right) + \sigma_k \xi\left(\dfrac{\|\gamma' - \gamma_k'\|_\infty}{r_k}\right).$

Notice that for all $k \geq 0$, $F_k(\gamma_k) = 0$, $F_k \geq 0$, and $F_k(\gamma) \geq \sigma_k > \varepsilon_k$ for $\|\gamma\|_{C^1} > r_k$. Hence the above choices form an allowed move of (β) in the perturbation game. Since $\varepsilon_0 = 2$ and $0 \leq f(\gamma) \leq 1$ we have

$$f(\gamma_0) < \varepsilon_0 + \inf_{\gamma \in \Gamma_n(X)} f(\gamma),$$

which allows an application of Theorem 7.2.4. The game in which (β) uses the above strategy and (α) the strategy from Theorem 7.2.4 leads to a sequence (γ_k) such that $\gamma_k \in \mathcal{A}_k$ and for every $0 \leq j \leq k < \infty$,

(a) $h_k(\gamma_k) < \varepsilon_j + \inf_{\gamma \in \Gamma_n(X)} h_j(\gamma)$,

(b) $F_j(\gamma_k) < \varepsilon_j$.

Let $0 \leq j \leq k$. Since $F_k(\gamma) \geq \sigma_j > \varepsilon_j$ if either $\|\gamma' - \gamma_j'\|_\infty > r_j$ or $\|\gamma - \gamma_j\|_\infty > s_j$, we infer from (b) that

$$\|\gamma_k' - \gamma_j'\|_\infty \leq r_j \quad \text{and} \quad \|\gamma_k - \gamma_j\|_\infty \leq s_j.$$

It follows that γ_k converge to some $\gamma_\infty \in \Gamma_n(X)$. Moreover, $\|\gamma_\infty' - \gamma_j'\|_\infty \leq r_j$ and $\|\gamma_\infty - \gamma_j\|_\infty \leq s_j$ for each j. In particular, for each k,

$$\gamma_k \in \overline{B(\gamma_0, r_0)} \subset B(\tilde{\gamma}_0, \tilde{r}_0),$$

which yields that $\gamma_\infty \in \overline{B(\tilde{\gamma}_0, \tilde{r}_0)}$. Recalling that $B_\infty(\gamma_k, 2s_k) \subset G_k$ and the just established fact $\|\gamma_\infty - \gamma_k\|_\infty \leq s_k$ we see that

$$\gamma_\infty \in G_k \cap \overline{B(\tilde{\gamma}_0, \tilde{r}_0)} \subset S_k$$

for every k. Hence γ_∞ belongs to S. In other words,

$$\mathcal{L}^n(\gamma_\infty^{-1}(E)) \geq \alpha. \tag{10.13}$$

The facts $\gamma_\infty \in B(\gamma_k, 2s_k)$ and $f(\gamma_\infty) \leq f(\gamma_k) + 2^{-k}$ together imply that $f(x_\infty) \leq \liminf_{k \to \infty} f(x_k)$. Since F_k are continuous, the assumptions of Observation 7.2.2 hold, and we infer that the function h_∞ attains its minimum on $\Gamma_n(X)$ at γ_∞.

The rest of the proof consists in finding a contradiction with the just obtained conclusion. We begin by using Lemma 10.2.3 to find $\delta > 0$ having the property that for every open set $U \subset [0,1]^n$ with

$$\mathcal{L}^n(U \cap \gamma_\infty^{-1}(E)) > \frac{\alpha}{2},$$

every $0 < r < \delta$, and $\zeta > 0$, there is $\gamma \in \Gamma_n(X)$ satisfying

(i) $\gamma(u) = \gamma_\infty(u)$ and $\gamma'(u) = \gamma'_\infty(u)$ for $u \notin U$,

(ii) $\|\gamma - \gamma_\infty\|_\infty \leq \zeta$,

(iii) $\|\gamma' - \gamma'_\infty\|_\infty \leq r$, and

(iv) $\mathcal{L}^n \gamma^{-1}(\overline{E}) + \Phi(\gamma) \leq \mathcal{L}^n \gamma_\infty^{-1}(\overline{E}) + \Phi(\gamma_\infty) - \chi(r)$,

where

$$\Phi(\gamma) = \sum_{k=0}^\infty \lambda_k \Psi(\gamma - \gamma_k) \quad \text{and} \quad \chi(r) = \frac{c^n \alpha}{8(1 + \|\gamma_\infty\|)^n} e^{-2n/r}.$$

We need to pick U, r, and ζ suitable to our situation. For this we fix $k \in \mathbb{N}$ such that $r_k < \delta$ and show that the open set

$$U = \left\{ u \in (0,1)^n \mid \|\gamma'_\infty(u) - \gamma'_j(u)\| < \frac{r_j}{4} \text{ for } j = 0, 1, \ldots, k \right\}$$

contains most of $[0,1]^n$. Since

$$\int_{[0,1]^n} \theta(\|\gamma'_\infty(u) - \gamma'_j(u)\|) \, d\mathcal{L}^n(u) \leq \int_{[0,1]^n} \Theta(\gamma'_\infty(u) - \gamma'_j(u)) \, d\mathcal{L}^n(u)$$

$$= \Psi(\gamma_\infty - \gamma_j) \leq \frac{1}{\lambda_j} F_j(\gamma_\infty) \leq \frac{\varepsilon_j}{\lambda_j},$$

we have

$$\mathcal{L}^n([0,1]^n \setminus U) \leq \sum_{j=0}^k \mathcal{L}^n \left\{ u \in [0,1]^n \mid \|\gamma'_\infty(u) - \gamma'_j(u)\| \geq \frac{r_j}{4} \right\}$$

$$\leq \sum_{j=0}^k \mathcal{L}^n \left\{ u \in [0,1]^n \mid \theta(\|\gamma'_\infty(u) - \gamma'_j(u)\|) \geq \theta(r_j/4) \right\}$$

$$\leq \sum_{j=0}^k \frac{1}{\theta(r_j/4)} \int_{[0,1]^n} \theta(\|\gamma'_\infty(u) - \gamma'_j(u)\|) \, d\mathcal{L}^n(u)$$

$$\leq \sum_{j=0}^k \frac{\varepsilon_j}{\lambda_j \theta(r_j/4)} < \frac{\alpha}{2},$$

where the last inequality follows from the choice of ε_j and λ_j. Since $\gamma_\infty \in S$, we infer from (10.13) that

$$\mathcal{L}^n(U \cap \gamma_\infty^{-1}(E)) > \frac{\alpha}{2},$$

as required. Finally, we let $r = \frac{1}{4}r_k$ and find $\zeta > 0$ such that

$$2\zeta \sum_{j=0}^{k} \frac{\sigma_j}{s_j} < \frac{1}{4} \chi(r).$$

Having defined r, ζ and U, we find $\gamma \in \Gamma_n(X)$ satisfying (i)–(iv). The statement (iv) is close to what we need, but we still need to estimate the sum of the remaining terms of the F_j. For the penultimate term we observe that $\xi \leq 1$, $\mathrm{Lip}(\xi) \leq 2$, and $\|\gamma - \gamma_\infty\|_\infty \leq \zeta$ by (ii), and so we have

$$\sum_{j=0}^{\infty} \sigma_j \xi\left(\frac{\|\gamma - \gamma_j\|_\infty}{s_j}\right) \leq \sum_{j=0}^{k} \sigma_j \xi\left(\frac{\|\gamma - \gamma_j\|_\infty}{s_j}\right) + \sum_{j=k+1}^{\infty} \sigma_j$$

$$\leq \sum_{j=0}^{k} \sigma_j \xi\left(\frac{\|\gamma_\infty - \gamma_j\|_\infty}{s_j}\right) + \sum_{j=0}^{k} \sigma_j \frac{\mathrm{Lip}(\xi)\zeta}{s_j} + \sum_{j=k+1}^{\infty} \sigma_j.$$

Recalling the choices of ζ and σ_j we obtain

$$< \sum_{j=0}^{\infty} \sigma_j \xi\left(\frac{\|\gamma_\infty - \gamma_j\|_\infty}{s_j}\right) + \frac{1}{4} \chi(r) + \frac{1}{4} \chi(r)$$

$$= \sum_{j=0}^{\infty} \sigma_j \xi\left(\frac{\|\gamma_\infty - \gamma_j\|_\infty}{s_j}\right) + \frac{1}{2} \chi(r). \tag{10.14}$$

To estimate the sum of the last terms of the F_j we have to argue differently for the terms coming from $j \leq k$ and for the rest. For $0 \leq j \leq k$ we can control the difference of the derivatives $\gamma'(u)$ and $\gamma_j'(u)$: if $u \notin U$ then (i) implies that

$$\|\gamma'(u) - \gamma_j'(u)\| = \|\gamma_\infty'(u) - \gamma_j'(u)\|.$$

If, on the other hand, $u \in U$, then by (iii)

$$\|\gamma'(u) - \gamma_j'(u)\| \leq \|\gamma'(u) - \gamma_\infty'(u)\| + \|\gamma_\infty'(u) - \gamma_j'(u)\|$$

$$< r + \frac{r_j}{4}$$

$$= \frac{r_k}{4} + \frac{r_j}{4} \leq \frac{r_j}{2}.$$

Now it follows that

$$\|\gamma' - \gamma_j'\|_\infty \leq \max\{\tfrac{1}{2}r_j, \|\gamma_\infty' - \gamma_j'\|_\infty\},$$

which, since ξ is increasing and vanishes on $[0, \frac{1}{2}]$, gives that for $j \leq k$,

$$\xi\Big(\frac{\|\gamma' - \gamma_j'\|_\infty}{r_j}\Big) \leq \xi\Big(\frac{\|\gamma_\infty' - \gamma_j'\|_\infty}{r_j}\Big).$$

For $j > k$ we use again that $\xi \leq 1$ to estimate

$$\sum_{j=k+1}^{\infty} \sigma_j \xi\Big(\frac{\|\gamma' - \gamma_j'\|_\infty}{r_j}\Big) \leq \sum_{j=k+1}^{\infty} \sigma_j < \frac{\widetilde{\chi}(r)}{2}.$$

Together,

$$\sum_{j=0}^{\infty} \sigma_j \xi\Big(\frac{\|\gamma' - \gamma_j'\|_\infty}{r_j}\Big) < \sum_{j=0}^{\infty} \sigma_j \xi\Big(\frac{\|\gamma_\infty' - \gamma_j'\|_\infty}{r_j}\Big) + \frac{\widetilde{\chi}(r)}{2}. \qquad (10.15)$$

Finally, the inequality

$$\|\gamma_\infty\| \leq \|\gamma_\infty - \widetilde{\gamma}_0\| + \|\widetilde{\gamma}_0\| \leq \widehat{r}_0 + \|\widehat{\gamma}_0\|$$

implies that $\widetilde{\chi}(r) \leq \chi(r)$. Using this, we get by adding (iv), (10.14), and (10.15) that

$$\begin{aligned}
h(\gamma) &= f(\gamma) + \Phi(\gamma) + \sum_{j=0}^{\infty} \sigma_j \xi\Big(\frac{\|\gamma - \gamma_j\|_\infty}{s_j}\Big) + \sum_{j=0}^{\infty} \sigma_j \xi\Big(\frac{\|\gamma' - \gamma_j'\|_\infty}{r_j}\Big) \\
&< f(\gamma_\infty) + \Phi(\gamma_\infty) - \chi(r) \\
&\quad + \sum_{j=0}^{\infty} \sigma_j \xi\Big(\frac{\|\gamma - \gamma_j\|_\infty}{s_j}\Big) + \frac{1}{2}\chi(r) + \sum_{j=0}^{\infty} \sigma_j \xi\Big(\frac{\|\gamma' - \gamma_j'\|_\infty}{r_j}\Big) + \frac{1}{2}\widetilde{\chi}(r) \\
&\leq h(\gamma_\infty),
\end{aligned}$$

contradicting the fact that h attains its minimum at $\gamma = \gamma_\infty$.   □

The main result of this section is an easy corollary.

Theorem 10.2.5. *Suppose that $\mathcal{Y}$ is a family of subspaces of a separable Banach space X for which there is a bump function uniformly continuous on bounded sets and smooth in the direction of $\mathcal{Y}$ with modulus controlled by $t^n \log^{n-1}(1/t)$. Then every set in X that is porous in the direction of $\mathcal{Y}$ is contained in a Γ_n-null G_δ set.*

Proof. Let Θ be the bump function whose existence is assumed. By Proposition 8.1.11, $\mathcal{Y}$ contains a countable subfamily $\mathcal{Y}_0$ such that Θ is smooth in the direction of $\mathcal{Y}_0$ with modulus controlled by $t^n \log^{n-1}(1/t)$. Since a set porous in the direction of $\mathcal{Y}$ is also porous in the direction of $\mathcal{Y}_0$, the statement follows from Proposition 10.2.4.   □

10.3 DIRECTIONAL POROSITY AND Γ_n-NULLNESS

Here we prove several technical statements, mostly of topological nature, that we need to understand when every subset of X porous in the direction of a subspace Y of X is Γ_n-null in X. We show that once this holds with $X = \mathbb{R}^n \oplus Y$, it holds with every X. It is not clear whether the assumption that every porous subset of Y is Γ_n-null in Y suffices; in particular, it is not known whether every porous subset of $\mathbb{R}^4$ is Γ_3-null or not.[*]

We will need a simple corollary of the fact that the image of a set residual in a (topologically) complete metric space under a continuous, open surjection is residual. For the proof of this fact see, for example, [37, Lemma 4.25].

Lemma 10.3.1. *Let $\pi\colon U \longrightarrow V$ be continuous and open, where U, V are topologically complete separable metric spaces. If $Q \subset U$ is a second category G_δ subset of U, then the set*

$$\left\{ v \in V \mid \pi^{-1}(v) \cap Q \text{ is of the second category in } \pi^{-1}(v) \right\}$$

is second category in V.

Proof. Let $G \subset U$ be a nonempty open set such that $Q \cap G$ is dense in G. Since Q is G_δ in G, it is also residual in G. Let $(G_i)_{i \geq 1}$ form a countable basis of the topology of G. Then the restriction of π to each G_i is a continuous open map of G_i to $\pi(G_i)$. Hence $\pi(Q \cap G_i)$ is residual in $\pi(G_i)$ by [37]. Since $\pi(G_i)$ are open subsets of V, the sets

$$\big(\pi(G_i) \setminus \pi(G_i \cap Q)\big) \cup \partial\pi(G_i)$$

are first category in V. Hence their complements

$$\pi(Q \cap G_i) \cup (V \setminus \overline{\pi(G_i)})$$

are residual in V. Let

$$R := \bigcap_{i \geq 1} \Big(\pi(Q \cap G_i) \cup \big(V \setminus \overline{\pi(G_i)}\big) \Big).$$

Then R is residual in V.

Let $v \in \pi(G) \cap R$. Then $\pi^{-1}(v) \cap Q \cap G_i \neq \emptyset$ whenever $\pi^{-1}(v) \cap G_i \neq \emptyset$. Hence $\pi^{-1}(v) \cap Q$ is dense in $\pi^{-1}(v) \cap G$. Since Q is G_δ in G, it implies that $\pi^{-1}(v) \cap Q$ is residual in $\pi^{-1}(v) \cap G$ and, consequently, $\pi^{-1}(v) \cap Q$ is second category in $\pi^{-1}(v)$. This conclusion holds for all $v \in \pi(G) \cap R$, which is a second category set in V. $\square$

Lemma 10.3.2. *Let $Q \subset \Gamma_n(X)$ be a G_δ set which is not nowhere dense in $\Gamma_n(X)$, and let Y be a subspace of X. Then there is $\gamma \in \Gamma_n(X)$ such that the set*

$$\{\beta \in \Gamma_n(Y) \mid \gamma + \beta \in Q\}$$

is a second category subset of $\Gamma_n(Y)$.

---

[*] Added in proof: Gareth Speight recently decomposed $\mathbb{R}^{n+1}$, $n \geq 3$, into a union of a σ-porous and a Γ_n-null set. In particular, for $n \geq 3$ there are non-Γ_n-null porous subsets of $\mathbb{R}^{n+1}$.

Proof. The quotient mapping $\pi\colon \Gamma_n(X) \longrightarrow \Gamma_n(X)/\Gamma_n(Y)$ is continuous and open. The set Q is dense in some open set $G \subset \Gamma_n(X)$, and, since it is a G_δ set, Q is residual in G. In particular, Q is second category. By Lemma 10.3.1 the set

$$\left\{\widehat{\gamma} \in \Gamma_n(X)/\Gamma_n(Y) \ \middle| \ \pi^{-1}(\widehat{\gamma}) \cap Q \text{ is of the second category in } \pi^{-1}(\widehat{\gamma})\right\}$$

is nonempty. Take any $\widehat{\gamma}$ from this set. Writing $\pi^{-1}(\widehat{\gamma}) = \gamma + \Gamma_n(Y)$ for some $\gamma \in \Gamma_n(X)$, we see that

$$\pi^{-1}(\widehat{\gamma}) \cap Q = \left(\gamma + \Gamma_n(Y)\right) \cap Q$$

is a second category set in $\gamma + \Gamma_n(Y)$. Since any shift does not affect the category of a set, we obtain that

$$\left(\pi^{-1}(\widehat{\gamma}) \cap Q\right) - \gamma = \Gamma_n(Y) \cap (Q - \gamma) = \{\beta \in \Gamma_n(Y) \mid \gamma + \beta \in Q\}$$

is a second category set in $\Gamma_n(Y)$. $\qquad\square$

Lemma 10.3.3. *Let Y be a separable Banach space and let E be Γ_n-null in $\mathbb{R}^n \oplus Y$. Then the set of $\beta \in \Gamma_n(Y)$ such that*

$$\mathscr{L}^n\{u \in [0,1]^n \mid u \oplus \beta(u) \in E\} = 0$$

is residual in $\Gamma_n(Y)$.

Proof. Let H be the set of those $\gamma = \gamma_1 \oplus \gamma_2 \in \Gamma_n(\mathbb{R}^n \oplus Y) \approx \Gamma_n(\mathbb{R}^n) \oplus \Gamma_n(Y)$ such that γ_1 is a diffeomorphism and $\gamma_1((0,1)^n) \supset [0,1]^n$. Recalling that the set of all diffeomorphisms of $[0,1]^n$ into $\mathbb{R}^n$ is an open set in $\Gamma_n(\mathbb{R}^n)$, we infer that H is an open set in $\Gamma_n(\mathbb{R}^n \oplus Y)$. We show that the map $P\colon H \longrightarrow \Gamma_n(Y)$ given as

$$P(\gamma_1 \oplus \gamma_2) = \gamma_2 \circ \gamma_1^{-1}$$

is an open map of H onto $\Gamma_n(Y)$. To see the surjectivity of P, suppose that $\beta \in \Gamma_n(Y)$. By Lemma 5.3.1 we extend β to a map $\widetilde{\beta} \in C^1(\mathbb{R}^n, Y)$. Choose any diffeomorphism $\alpha\colon [0,1]^n \longrightarrow \mathbb{R}^n$ with $\alpha((0,1)^n) \supset [0,1]^n$ and put

$$\gamma = \alpha \oplus (\widetilde{\beta} \circ \alpha). \qquad (10.16)$$

Then $\gamma \in H$ and $P(\gamma) = \beta$. It remains to verify that P is open. Notice first that for a fixed diffeomorphism $\alpha\colon [0,1]^n \longrightarrow \mathbb{R}^n$ with $\alpha((0,1)^n) \supset [0,1]^n$ the map $P_\alpha\colon H \longrightarrow \Gamma_n(Y)$ defined as

$$P_\alpha(\gamma_1 \oplus \gamma_2) = \gamma_2 \circ \alpha^{-1}$$

is open. Indeed, P_α is a composition

$$\gamma_1 \oplus \gamma_2 \to \gamma_1 \circ \alpha^{-1} \oplus \gamma_2 \circ \alpha^{-1} \to \gamma_2 \circ \alpha^{-1},$$

where the first map is a homeomorphism and the second a projection. Using it we prove that for every $\gamma \in H$ and $r > 0$ with $B(\gamma, r) \subset H$, the image $P(B(\gamma, r))$ contains an open neighborhood of $P(\gamma)$ in $\Gamma_n(Y)$. This is clearly sufficient for P being open. Let $B(\gamma, r) \subset H$, where $\gamma = \gamma_1 \oplus \gamma_2$ is given. Since $P_{\gamma_1}(B(\gamma, r))$ is an open neighborhood of $P(\gamma) = P_{\gamma_1}(\gamma)$, we need to show that

$$P_{\gamma_1}(B(\gamma, r)) \subset P(B(\gamma, r)).$$

Take any element in $P_{\gamma_1}(B(\gamma, r))$. It is of the form

$$\beta_2 \circ \gamma_1^{-1} = P_{\gamma_1}(\beta_1 \oplus \beta_2)$$

for some $\beta_1 \oplus \beta_2 \in B(\gamma, r)$. So, in particular, $\|\beta_1 - \gamma_1\|_{C^1} + \|\beta_2 - \gamma_2\|_{C^1} < r$. Consider the surface $\gamma_1 \oplus \beta_2$. Clearly, $P_{\gamma_1}(\beta_1 \oplus \beta_2) = P(\gamma_1 \oplus \beta_2)$, and

$$\|\gamma_1 \oplus \beta_2 - \gamma_1 \oplus \gamma_2\|_{C^1} = \|\beta_2 - \gamma_2\|_{C^1} < r,$$

implying that $\gamma_1 \oplus \beta_2 \in B(\gamma, r)$. So $P_{\gamma_1}(B(\gamma, r)) \subset P(B(\gamma, r))$ and the map P is open.

Since E is Γ_n-null in $\mathbb{R}^n \oplus Y$, the set $\left\{ \gamma \in H \mid \mathscr{L}^n \gamma^{-1}(E) = 0 \right\}$ is residual in H and the P-image of this set is residual in $\Gamma_n(Y)$ by [37]. If β belongs to this image then $\beta = \gamma_2 \circ \gamma_1^{-1}$ for some $\gamma = \gamma_1 \oplus \gamma_2 \in H$ with $\mathscr{L}^n \gamma^{-1}(E) = 0$. Now

$$
\begin{aligned}
\left\{ u \in [0,1]^n \mid u \oplus \beta(u) \in E \right\} &= \left\{ u \in [0,1]^n \mid u \oplus \gamma_2 \circ \gamma_1^{-1}(u) \in E \right\} \\
&= \left\{ t \in \gamma_1^{-1}([0,1]^n) \mid \gamma_1(t) \oplus \gamma_2(t) \in E \right\} \\
&\subset \left\{ t \in [0,1]^n \mid \gamma(t) \in E \right\}.
\end{aligned}
$$

The last set is $\mathscr{L}^n$-null, which implies that

$$\mathscr{L}^n \left\{ u \in [0,1]^n \mid u \oplus \beta(u) \in E \right\} = 0. \qquad \square$$

Proposition 10.3.4. *Suppose that every subset of $\mathbb{R}^n \oplus Y$ porous in the direction of Y is Γ_n-null. Then, whenever Y is a subspace of X, every subset of X porous in the direction Y is Γ_n-null in X.*

Proof. Let $E \subset X$ be c-porous in the direction of Y. By Lemma 10.1.4 we may assume that E is G_δ. Supposing that E is not Γ_n-null, we find $\alpha > 0$ such that

$$Q := \left\{ \gamma \in \Gamma_n(X) \mid \mathscr{L}^n \gamma^{-1}(E) \geq \alpha \right\}$$

is not nowhere dense in $\Gamma_n(X)$. Since Lemma 5.4.1 gives that Q is G_δ, there is, by Lemma 10.3.2, $\gamma \in \Gamma_n(X)$ such that the set

$$S = \left\{ \beta \in \Gamma_n(Y) \mid \gamma + \beta \in Q \right\}$$

is a second category subset of $\Gamma_n(Y)$.

Define $\Phi\colon [0,1]^n \oplus Y \longrightarrow X$ by $\Phi(u \oplus y) = \gamma(u) + y$. We show that $\Phi^{-1}(E)$ is porous in $\mathbb{R}^n \oplus Y$ in the direction of Y. Let $u_0 \oplus y_0 \in \Phi^{-1}(E)$. One can find points $y \in Y \setminus \{0\}$ with arbitrarily small norm such that

$$B\big(\gamma(u_0) + y_0 + y, c\|y\|\big) \cap E = \emptyset.$$

Take any such y with $0 < \|y\| \le 1$. Continuity of γ' allows us to find $\delta > 0$ such that $\|\gamma'(u)\| \le \|\gamma'(u_0)\| + 1$ for $\|u - u_0\| < \delta$ and, in addition,

$$\delta\big(\|\gamma'(u_0)\| + 2\big) < c.$$

We show that
$$B\big(u_0 \oplus y_0 + u_0 \oplus y, \delta\|y\|\big) \cap \Phi^{-1}(E) = \emptyset, \tag{10.17}$$

which will yield the required porosity. Let $u \oplus z \in B\big(u_0 \oplus y_0 + u_0 \oplus y, \delta\|y\|\big)$. Then

$$\|u \oplus z - (u_0 \oplus y_0 + u_0 \oplus y)\| = \|u - u_0\| + \|z - (y_0 + y)\| < \delta\|y\|.$$

Using it we estimate the distance

$$
\begin{aligned}
\|\Phi(u \oplus z) - (\gamma(u_0) + y_0 + y)\| &= \|\gamma(u) - \gamma(u_0) + z - (y_0 + y)\| \\
&\le \|\gamma(u) - \gamma(u_0)\| + \|z - (y_0 + y)\| \\
&\le \big(\|\gamma'(u_0)\| + 1\big)\|u - u_0\| + \|z - (y_0 + y)\| \\
&\le \big(\|\gamma'(u_0)\| + 1\big)\delta\|y\| + \delta\|y\| \\
&= \big(\|\gamma'(u_0)\| + 2\big)\delta\|y\| < c\|y\|.
\end{aligned}
$$

It follows that $u \oplus z \notin \Phi^{-1}(E)$, and so (10.17) holds. The set $\Phi^{-1}(E)$ is Γ_n-null by assumption of the proposition. Hence for any $\beta \in S$,

$$
\begin{aligned}
\alpha &\le \mathscr{L}^n\big\{u \in [0,1]^n \mid \gamma(u) + \beta(u) \in E\big\} \\
&= \mathscr{L}^n\big\{u \in [0,1]^n \mid u \oplus \beta(u) \in \Phi^{-1}(E)\big\},
\end{aligned}
$$

which contradicts Lemma 10.3.3. $\square$

10.4 σ-POROSITY AND Γ_n-NULLNESS

We now turn our attention to the question in which spaces, and for what values of n, porous sets are Γ_n-null. For higher values of n ($n \ge 3$) the question has two different subquestions: the question whether directionally porous sets are Γ_n-null, which is left open because of the problem mentioned at the beginning of the previous section,[*] and the question when sets porous "at infinity" are Γ_n-null, which we relate to smoothness.

Theorem 10.4.1. *Every σ-porous subset of a separable Banach space with separable dual is Γ_1-null.*

---

[*] Added in proof: Gareth Speight answered this question negatively; see footnote on page 186.

Proof. If X has a separable dual, it can be renormed in an equivalent way by a Fréchet smooth norm $\|\cdot\|$; see 3.2.1. We can easily verify that the function $\Theta(x) = \|x\|^2$ is smooth in the direction of $\mathcal{Y} = \{X\}$ with modulus controlled by $\omega_1(t) = t$: Since Θ is Lipschitz on bounded sets, we may apply Theorem 10.2.5 with $n = 1$. $\qquad\square$

For directionally porous sets, we have the following result which is valid in all separable Banach spaces.

Theorem 10.4.2. *Every σ-directionally porous subset of a separable Banach space X is Γ_1-null as well as Γ_2-null.*

Proof. Let us start with the case of Γ_1-null sets. Assume that E is a subset of X porous in the direction of a one-dimensional subspace Y. Since by Theorem 10.4.1 every subset of $\mathbb{R} \oplus Y \approx \mathbb{R}^2$ porous in the direction of Y is Γ_1-null, we may apply Proposition 10.3.4 with the above Y and $n = 1$. It follows that E is Γ_1-null.

We proceed similarly in the case of Γ_2-nullness. By Proposition 10.3.4, to show that every subset of X porous in the direction of a one-dimensional subspace Y is Γ_2-null, it suffices to show that every subset of $\mathbb{R}^2 \oplus Y \approx \mathbb{R}^3$ porous in the direction of Y is Γ_2-null. But it follows from Theorem 10.2.5 that every porous subset of $\mathbb{R}^3$ is Γ_2-null, since the square of the Euclidean norm on $\mathbb{R}^3$, $\Theta(x) = |x|^2$, is a bump function, Lipschitz on bounded sets, and smooth in the direction of $\mathcal{Y} = \{\mathbb{R}^3\}$ with modulus controlled by $t^2 \log(1/t)$. $\qquad\square$

The following simple extension of decomposition results for porous sets proved in the first section of this chapter divides a porous set into a set porous "at infinity" and a σ-directionally porous set.

Lemma 10.4.3. *Suppose that $\mathcal{Y}$, $\mathcal{Z}$ are families of subspaces of X, $\mathcal{Z}$ is countable, and for every $Z \in \mathcal{Z}$ there is $Y \in \mathcal{Y}$ such that $Z \subset Y$ and $\dim(Y/Z) < \infty$. Then every set in X that is porous in the direction of $\mathcal{Y}$ is contained in a union of a σ-directionally porous set and a G_δ set porous in the direction of $\mathcal{Z}$.*

Proof. Let $\mathcal{Z} = \{Z_1, Z_2, \dots\}$ and for each k find $Y_k \in \mathcal{Y}$ such that $Z_k \subset Y_k$ and $\dim(Y_k/Z_k) < \infty$. By Lemma 4.2.8 there are finite dimensional subspaces U_k of Y_k such that every $y \in Y_k$ can we written as $y = u + z$, where $u \in U_k$, $z \in Z_k$, and $\max\{\|u\|, \|z\|\} \le 3\|y\|$.

Let E be c-porous in the direction of $\mathcal{Y}$. Since E is c-porous in the direction of Y_k, we may use Lemmas 10.1.2 and 10.1.4 to write

$$E \subset A_k \cup B_k,$$

where A_k is σ-porous in the direction of U_k and B_k is a G_δ set which is $c/7$-porous in the direction of Z_k. Hence

$$E \subset \bigcup_{k=1}^{\infty} A_k \cup \bigcap_{k=1}^{\infty} B_k$$

is the required cover of E by a σ-directionally porous set and a G_δ set porous in the direction of $\mathcal{Z}$. $\qquad\square$

It is now a simple task to combine previous results to get smoothness criteria for Γ_n smallness of porous sets.

Corollary 10.4.4. *Suppose that* $\mathcal{Y}$, $\mathcal{Z}$ *are families of* X *such that for every* $Z \in \mathcal{Z}$ *there is* $Y \in \mathcal{Y}$ *such that* $Z \subset Y$ *and* $\dim Y/Z < \infty$. *Assume further that* X *admits a bump function* Θ *smooth in the direction of* $\mathcal{Z}$ *with modulus controlled by* $t^n \log^{n-1}(1/t)$. *If either* Θ *is continuous and* $\mathcal{Z}$ *is separable or* Θ *is locally uniformly continuous and* X *is separable, then every set in* X *that is porous in the direction of* $\mathcal{Y}$ *is contained in a union of a* σ-directionally porous set and a Γ_n-null G_δ set.*

Proof. By Lemma 10.4.3, every set porous in the direction of $\mathcal{Y}$ is contained in a union of a σ-directionally porous set and a G_δ set porous in the direction of $\mathcal{Z}$. The latter set is Γ_n-null by Proposition 10.2.4 in the first case and by Theorem 10.2.5 in the second. $\square$

Theorem 10.4.5. *Suppose that a separable Banach space* X *admits, for some* $n \in \mathbb{N}$, *a locally uniformly continuous bump function asymptotically smooth with modulus controlled by* $t^n \log^{n-1}(1/t)$. *Then every porous set in* X *is contained in the union of a* σ-directionally porous set and a Γ_n-null G_δ set.*

Proof. This is a special case of Corollary 10.4.4 with $\mathcal{Y} = \{X\}$ and $\mathcal{Z}$ the family of finite codimensional subspaces of X. $\square$

Corollary 10.4.6. *Let* X *be a separable Banach space such that, for some* $n \in \mathbb{N}$, *its modulus of asymptotic uniform smoothness satisfies*

$$\bar{\rho}_X(t) = o\big(t^n \log^{n-1}(1/t)\big), \; t \searrow 0.$$

Then every porous set in X *is contained in a union of a* σ-directionally porous set and a Γ_n-null G_δ set.*

Proof. By Corollary 8.2.8 there is a continuous convex bump function asymptotically smooth with modulus controlled by $t^n \log^{n-1}(1/t)$. Since a continuous convex function is locally uniformly continuous, Theorem 10.4.5 applies. $\square$

Combining the last result with Theorem 10.4.2, we obtain a smoothness criterion for all σ-porous sets to be Γ_2-null.

Corollary 10.4.7. *Let* X *be a separable Banach space such that its modulus of asymptotic uniform smoothness satisfies*

$$\bar{\rho}_X(t) = o(t^2 \log(1/t)), \; t \searrow 0.$$

Then every σ-porous subset of X is Γ_2-null. In particular, all σ-porous subsets of a separable Hilbert space are Γ_2-null.*

Proof. By Corollary 10.4.6 with $n = 2$, every porous set is contained in the union of a σ-directionally porous set and a Γ_2-null set. It suffices to notice that the former set is also Γ_2-null by Theorem 10.4.2. $\square$

Finally, we show how one may handle the σ-directionally porous part of the porous sets that appeared in a number of results proved above. Since σ-directionally porous sets are Haar null by Example 2.2.5, we prove a stronger result by showing that even unions of a Γ_n-null G_δ set with a Haar null set are small on many curves.

Lemma 10.4.8. *Suppose that $E \subset X$ can be covered by a union of a Haar null set and a Γ_n-null G_δ set. Then for every $\tau > 0$ there is a dense set of surfaces $\gamma \in \Gamma_n(X)$ such that $\mathscr{L}^n \gamma^{-1}(E) < \tau$.*

Proof. Let $\gamma_0 \in \Gamma_n(X)$ and $\varepsilon > 0$ be given. By assumption, E can be covered by a union of a (Borel) Haar null set F and a Γ_n-null G_δ set H, $E \subset F \cup H$.

By Lemma 5.4.1, the set

$$\left\{ \gamma \in \Gamma_n(X) \mid \mathscr{L}^n \gamma^{-1}(H) \geq \tau \right\}$$

is G_δ. Moreover, H is also Γ_n-null, which implies that the above set is of the first category. Any G_δ set of the first category in a Baire space is nowhere dense. It follows that there are $\gamma_1 \in \Gamma_n(X)$ and $0 < \eta < \frac{1}{2}\varepsilon$ such that

$$\|\gamma_1 - \gamma_0\|_{C^1} < \frac{\varepsilon}{2} \quad \text{and} \quad \mathscr{L}^n \gamma^{-1}(H) < \tau \ \text{ for all } \ \|\gamma - \gamma_1\|_{C^1} < \eta.$$

By Proposition 2.2.3 there is $x \in X$ with $\|x\| < \eta$ such that

$$\mathscr{L}^n \{ u \in [0,1]^n \mid x + \gamma_1(u) \in F \} = 0.$$

Let $\gamma(t) = x + \gamma_1(t)$. Then $\|\gamma - \gamma_0\|_{C^1} \leq \|\gamma - \gamma_1\|_{C^1} + \|\gamma_1 - \gamma_0\|_{C^1} < \varepsilon$. Moreover, $\mathscr{L}^n \gamma^{-1}(H) < \tau$ since $\|\gamma - \gamma_1\|_{C^1} < \eta$ and also

$$\mathscr{L}^n \gamma^{-1}(F) = \mathscr{L}^n \{ t \in [0,1]^n \mid x + \gamma_1(t) \in F \} = 0.$$

Hence $\mathscr{L}^n \gamma^{-1}(E) \leq \mathscr{L}^n \gamma^{-1}(F) + \mathscr{L}^n \gamma^{-1}(H) < \tau.$ $\qquad\square$

10.5 Γ_1-NULLNESS OF POROUS SETS AND ASPLUNDNESS

In this section we complete the information gained in Theorem 10.4.1 by proving the following result.

Theorem 10.5.1. *Every separable Banach space with nonseparable dual contains a σ-porous set with Γ_1-null complement.*

As an immediate corollary we get a new characterization of separability of the dual in terms of smallness of porous sets.

Theorem 10.5.2. *A separable Banach space X has separable dual if and only if every porous subset of X is Γ_1-null.*

Proof. We have seen in Theorem 10.4.1 that every σ-porous set in a Banach space with separable dual is Γ_1-null. The converse is Theorem 10.5.1. $\qquad\square$

Before coming to the proof of Theorem 10.5.1, we fix the setup and prove some auxiliary lemmas. For the rest of this section we assume that X is a separable Banach space with nonseparable dual. By Theorem 3.2.3 we may also assume that there is $0 < c_0 < 1$ (in fact, we can choose any $0 < c_0 < 1$) such that for every $x \in X, r > 0$, and every finite codimensional subspace Y of X,

$$\sup_{y \in Y, \|y\| \leq r} (\|x + y\| - \|x\|) > c_0 r.$$

Lemma 10.5.3. *Suppose $e_0 \in X$, $e_0^* \in X^*$, and $\|e_0\| = \|e_0^*\| = e_0^*(e_0) = 1$. Then for every $r > 0$ there are unit vectors $y_i \in X$, $i \in \mathbb{N}$, such that $e_0^*(y_i) = 0$ and $\|e_0 + ty_i + sy_j\| \geq 1 + c_0 t$ for every $i < j$, $s \in \mathbb{R}$, and $t \geq r$.*

Proof. Pick u_0 such that $\|u_0\| \leq r$, $e_0^*(u_0) = 0$, and $\|e_0 + u_0\| - \|e_0\| > c_0 r$, and choose u_0^* in the subdifferential of the norm at the point $e_0 + u_0$. Continuing recursively, we assume that u_i and u_i^* have been defined for $i < k$, and choose $u_k \in \operatorname{Ker} e_0^* \cap \bigcap_{i=0}^{k-1} \operatorname{Ker} u_i^*$ such that $\|u_k\| \leq r$ and

$$\|e_0 + u_k\| - \|e_0\| > c_0 r.$$

Pick u_k^* in the subdifferential of the norm at the point $e_0 + u_k$. Notice that $\|u_k^*\| = 1$ and, since

$$-c_0 r > \|e_0 + u_k - u_k\| - \|e_0 + u_k\| \geq u_k^*(-u_k),$$

we have $u_k^*(u_k) > c_0 r$. Let $\alpha_i = 1/\|u_i\|$ and $y_i = \alpha_i u_i$. Then $\alpha_i r \geq 1$, and for every $i < j$ and $t, s \in \mathbb{R}$ such that $t \geq r$ we use that $t\alpha_i - 1 \geq 0$ to estimate

$$
\begin{aligned}
\|e_0 + ty_i + sy_j\| &= \|e_0 + u_i + (t\alpha_i - 1)u_i + sy_j\| \\
&\geq \|e_0 + u_i\| + u_i^*((t\alpha_i - 1)u_i + sy_j) \\
&= \|e_0 + u_i\| + u_i^*((t\alpha_i - 1)u_i) \\
&\geq 1 + c_0 r + c_0(t\alpha_i - 1)r = 1 + c_0 t\alpha_i r \geq 1 + c_0 t. \qquad \square
\end{aligned}
$$

Lemma 10.5.4. *Let $e_0 \in X$ be a unit vector, $x_0 \in X$, and $0 < r \leq c_0/12$. Then there is a finite family of balls $B(z_i, r)$ such that, if we denote $\kappa = 6/c_0$ and $G = \bigcup_i B(z_i, r)$, the following hold:*

(i) $[x_0, x_0 + e_0] \subset \bigcup_i B(z_i, \kappa r)$;

(ii) *whenever $\gamma \colon [a, b] \subset \mathbb{R} \longrightarrow X$ is such that for some unit vector e*

$$\|\gamma(t) - \gamma(s) - (t - s)e\| \leq \frac{|t - s|}{2} \text{ for all } t, s \in [a, b],$$

the set $[a, b] \setminus \gamma^{-1}(G)$ contains a disjoint union of no more than $1/r$ intervals $[a_i, b_i]$ such that

$$\sum_i \|\gamma(b_i) - \gamma(a_i)\| \geq \|\gamma(b) - \gamma(a)\| - \kappa r. \tag{10.18}$$

Proof. We first notice that, however we define the points z_i, the second statement holds if the curve γ meets no more than one of the balls $B(z_i, r)$. Indeed, if γ meets no such ball, we can take $[a_0, b_0] = [a, b]$, and if it meets exactly one ball $B(z_i, r)$, we let $a_0 = a$, $b_1 = b$, $b_0 = \inf \gamma^{-1}(B(z_i, r))$, and $a_1 = \sup \gamma^{-1}(B(z_i, r))$. Since $\|\gamma(a_1) - \gamma(b_0)\| \leq 2r$, we have that

$$\|\gamma(b_0) - \gamma(a_0)\| + \|\gamma(b_1) - \gamma(a_1)\| \geq \|\gamma(b) - \gamma(a)\| - 2r.$$

Hence we have two intervals (so less than $1/r$), namely, $[a_0, b_0]$ and $[a_1, b_1]$, having all the desired properties provided they are nondegenerate; otherwise we obtain the desired set of intervals by deleting the degenerate ones. (The resulting family of intervals may be empty, unlike the family of balls, which has to be nonempty because of (i).)

We notice that $\kappa r \leq \frac{1}{2}$, denote $\Delta = \kappa r$, and find the greatest integer n such that $(n-1)\Delta \leq 1$. Let $e_0^* \in X^*$ be such that $\|e_0\| = \|e_0^*\| = e_0^*(e_0) = 1$, and let y_i, $i \geq 1$, be the unit vectors from Lemma 10.5.3. We show that the statement holds with

$$z_i = x_0 + \frac{2i-1}{2}\Delta e_0 + \frac{\kappa r}{3} y_i, \ 1 \leq i \leq n.$$

Since

$$\frac{\kappa r}{(j-i)\Delta} \geq \frac{\kappa r}{(n-1)\Delta} \geq \kappa r \geq r$$

for $1 \leq i < j \leq n$, we infer from Lemma 10.5.3 that for every $u \in \overline{B(z_i, r)}$ and $v \in \overline{B(z_j, r)}$,

$$\begin{aligned}
\|v - u\| &\geq \|z_j - z_i\| - 2r \\
&= \|(j-i)\Delta e_0 + \kappa r y_j - \kappa r y_i\| - 2r \\
&\geq (j-i)\Delta + c_0 \kappa r - 2r \geq (j-i)\Delta.
\end{aligned} \tag{10.19}$$

To prove (i), we notice that for every $x \in [x_0, x_0 + e_0]$ there is $1 \leq i \leq n$ such that $\|x - (x_0 + i\Delta e_0)\| \leq \frac{1}{2}\Delta$. Hence $\|x - z_i\| \leq \frac{1}{2}\Delta + \frac{1}{3}\kappa r < \kappa r$, implying that $x \in B(z_i, \kappa r)$.

Suppose now that $\gamma \colon [a, b] \subset \mathbb{R} \longrightarrow X$ and e are as in (ii). Since neither the statement nor its assumptions depend on the orientation of γ, we will also assume that $e_0^*(e) \geq 0$. Also, by what we have already proved above, we will assume that γ meets at least two of the balls $B(z_i, r)$. We let N be the set of indexes $1 \leq i \leq n$ for which γ meets $B(z_i, r)$ and index its elements as $N = \{i_1 < i_2 < \cdots < i_m\}$ where $m \geq 2$. Denote $a_0 = a$, $b_m = b$,

$$a_k = \sup \gamma^{-1}(B(z_{i_k}, r)) \ \text{for } 0 < k \leq m, \text{ and}$$
$$b_k = \inf \gamma^{-1}(B(z_{i_{k+1}}, r)) \ \text{for } 0 \leq k < m.$$

Since $\gamma(a_k) \in \overline{B(z_{i_k}, r)}$ and $\gamma(b_k) \in \overline{B(z_{i_{k+1}}, r)}$, we have

$$\left\|\gamma(a_k) - (x_0 + (i_k - \tfrac{1}{2})\Delta e_0)\right\| \leq \left(1 + \frac{\kappa}{3}\right) r \ \text{for } 0 < k \leq m \tag{10.20}$$

and

$$\|\gamma(b_k) - (x_0 + (i_{k+1} - \tfrac{1}{2})\Delta e_0)\| \leq \left(1 + \frac{\kappa}{3}\right)r \quad \text{for } 0 \leq k < m. \tag{10.21}$$

These inequalities imply that for $0 < k < m$,

$$e_0^*(\gamma(b_k) - \gamma(a_k)) \geq (i_{k+1} - i_k)\Delta - 2\left(1 + \frac{\kappa}{3}\right)r > 0.$$

It follows that $a_k < b_k$ for every $0 < k < m$. Indeed, if $a_k \geq b_k$ for some such k, the inequality

$$|e_0^*(\gamma(b_k) - \gamma(a_k)) - (b_k - a_k)e_0^*(e)| \leq \frac{|b_k - a_k|}{2}$$

would imply that $e_0^*(\gamma(b_k) - \gamma(a_k)) \leq \tfrac{1}{2}(b_k - a_k) \leq 0$.
Hence we have

$$a = a_0 \leq b_0 < a_1 < b_1 < a_2 < \cdots < b_{m-1} < a_m \leq b_m = b,$$

and we show that the statement holds with the intervals $[a_i, b_i]$, after removing the degenerate ones if there are any. By the choice of N we have $[a_i, b_i] \subset \gamma^{-1}([a, b] \setminus G)$. (For the degenerate intervals this may be false, which is why we removed them.) Moreover, the number of these intervals is at most $m + 1 \leq n + 1 \leq 1/\Delta + 2 \leq 1/r$.

It remains to show (10.18). Recalling that $m \geq 2$, we infer from (10.19) that

$$\sum_{k=1}^{m-1} \|\gamma(b_k) - \gamma(a_k)\| \geq \sum_{k=1}^{m-1} (i_{k+1} - i_k)\Delta = (i_m - i_1)\Delta.$$

Using (10.20) for $k = m$ and (10.21) for $k = 0$, we get

$$\|\gamma(a_m) - \gamma(b_0)\| \leq 2\left(1 + \frac{\kappa}{3}\right)r + (i_m - i_1)\Delta.$$

Hence we conclude that

$$\sum_{k=1}^{m-1} \|\gamma(b_k) - \gamma(a_k)\| \geq \|\gamma(a_m) - \gamma(b_0)\| - 2\left(1 + \frac{\kappa}{3}\right)r.$$

Adding $\|\gamma(b_0) - \gamma(a_0)\|$ and/or $\|\gamma(b_m) - \gamma(a_m)\|$ (depending or whether the intervals $[a_0, b_0]$ and/or $[a_m, b_m]$ are degenerate), and noticing that $2(1 + \tfrac{1}{3}\kappa) \leq \kappa$, we get (10.18). $\qquad\square$

Proof of Theorem 10.5.1. Let (x_k, e_k) be a sequence of pairs dense in the product of X with the unit sphere in X with the following properties:

- each pair repeats infinitely often in the sequence;

- for every (x_k, e_k), the pair $(x_k + e_k, e_k)$ is also among the pairs (x_j, e_j).

Let $r_0 = c_0/12$ and for $k \geq 1$ choose recursively $0 < r_k \leq 2^{-k}r_0$ such that

$$\kappa r_k \prod_{i=1}^{k-1} \frac{1}{r_i} \leq 2^{-k},$$

where κ is the constant from Lemma 10.5.4. For every $k \geq 1$ we use this lemma with vectors e_k, x_k, and number r_k to find the corresponding finite family of balls $\{B(z_{k,i}, r_k) \mid i = 1, \ldots, m_k\}$.

Let $G_k = \bigcup_{i=1}^{m_k} B(z_{k,i}, r_k)$, $S_j = \bigcup_{k=j}^{\infty} G_k$, and $S = \bigcap_{j=1}^{\infty} S_j$. Consider any $\gamma \in C^1([a, b], X)$ such that for some unit vector e,

$$\|\gamma(t) - \gamma(s) - (t - s)e\| \leq \frac{|t - s|}{2} \quad \text{for all } t, s \in [a, b].$$

Let $\varepsilon > 0$ and, denoting $\|\gamma(I)\| = \|\gamma(\beta) - \gamma(\alpha)\|$ for $I = [\alpha, \beta]$, find a (finite) partition $\mathcal{I}$ of $[a, b]$ into nonoverlapping closed intervals such that

$$\sum_{I \in \mathcal{I}} \|\gamma(I)\| > \int_a^b \|\gamma'(t)\| \, dt - \frac{\varepsilon}{4}.$$

Let $j \geq 1$ be such that the cardinality of $\mathcal{I}$ is less than $1/r_j$ and

$$\sum_{k=j}^{\infty} \kappa r_k \prod_{i=j}^{k-1} \frac{1}{r_i} < \frac{\varepsilon}{4}. \tag{10.22}$$

For $k \geq j$ we will recursively construct families $\mathcal{I}_k$ of subintervals of $[a, b]$. We start by letting $\mathcal{I}_j := \mathcal{I}$. Applying Lemma 10.5.4 to the vectors x_{j+1}, e_{j+1}, number r_{j+1}, and γ restricted to each interval $I \in \mathcal{I}_j$, we find in every such I at most $1/r_{j+1}$ disjoint subintervals $J_1, \ldots, J_{m(I)}$ such that

$$\bigcup_{i=1}^{m(I)} J_i \subset \gamma^{-1}(X \setminus G_{j+1}) \quad \text{and} \quad \sum_{i=1}^{m(I)} \|\gamma(J_i)\| \geq \|\gamma(I)\| - \kappa r_{j+1}.$$

Put $\mathcal{I}_{j+1} = \bigcup_{I \in \mathcal{I}_j} \{J_1, \ldots, J_{m(I)}\}$. Thus cardinality of the family $\mathcal{I}_{j+1}$ is at most $1/(r_j r_{j+1})$.

We repeat the above application of Lemma 10.5.4 recursively for $k \geq j + 2$ to the vectors x_k, e_k, number r_k, and γ restricted to each of the intervals from the previous family $\mathcal{I}_{k-1}$. In this way we obtain families of mutually disjoint intervals $\mathcal{I}_k$ of cardinality at most $\prod_{i=j}^{k} 1/r_i$ such that their unions $M_k := \bigcup \mathcal{I}_k$ satisfy $M_k \subset M_{k-1}$ and

$$M_k \subset \gamma^{-1}(X \setminus G_k) \quad \text{and} \quad \sum_{I \in \mathcal{I}_k} \|\gamma(I)\| \geq \sum_{I \in \mathcal{I}_{k-1}} \|\gamma(I)\| - \kappa r_k \prod_{i=j}^{k-1} \frac{1}{r_i}.$$

Let $M = \bigcap_{k=j}^{\infty} M_k$. The condition (10.22) yields

$$\int_M \|\gamma'(t)\| \, dt \geq \lim_{k \to \infty} \sum_{I \in \mathcal{I}_k} \|\gamma(I)\| > \sum_{I \in \mathcal{I}} \|\gamma(I)\| - \frac{\varepsilon}{4} > \int_a^b \|\gamma'(t)\| \, dt - \frac{\varepsilon}{2}.$$

Since $\|\gamma'(t)\| \geq 1/2$ for every t, we conclude that

$$\mathcal{L}\gamma^{-1}(S) \leq \mathcal{L}\gamma^{-1}(S_j) \leq \mathcal{L}([a,b] \setminus M) \leq 2 \int_{[a,b] \setminus M} \|\gamma'(t)\| \, dt < \varepsilon.$$

Hence $\mathcal{L}\gamma^{-1}(S) = 0$ for every γ that we have considered. In particular, this holds for every $\gamma \in C^1([a,b], X)$ with the properties $\gamma'(a) \neq 0$ and $\|\gamma'(t) - \gamma'(a)\| < \frac{1}{2}\|\gamma'(a)\|$ for $t \in [a,b]$. Indeed, by a reparametrization we may achieve that $\|\gamma'(a)\| = 1$, and so

$$\|\gamma(t) - \gamma(s) - (t-s)\gamma'(a)\| \leq \int_s^t \|\gamma'(\tau) - \gamma'(a)\| \, d\tau < \frac{|t-s|}{2}.$$

Since for any $\gamma \in \Gamma_1(X)$ the set $\{t \in [0,1] \mid \gamma'(t) \neq 0\}$ can be covered by countably many intervals $[a,b]$ on which γ has the above properties, we have that

$$\mathcal{L}\{t \in [0,1] \mid \gamma(t) \in S, \ \gamma'(t) \neq 0\} = 0$$

for every $\gamma \in \Gamma_1(X)$. By Lemma 5.3.5 this implies that S is Γ_1-null.

Define now $H_k = \bigcup_{i=1}^{m_k} B(z_{k,i}, \kappa r_k)$ and $Q_j = \bigcup_{k=j}^{\infty} H_k$. We show that each $X \setminus Q_j$ is Γ_1-null. For this we employ Corollary 5.3.4. Suppose that $\varepsilon > 0$ and $\gamma_0 \colon \mathbb{R} \longrightarrow X$ is affine with $\gamma_0' = u$. We find $p \in \mathbb{N}$ such that

$$\|x_p - \gamma_0(0)\| < \frac{\varepsilon}{2} \quad \text{and} \quad \|e_p\|u\| - u\| < \frac{\varepsilon}{2}.$$

Define $\tilde{\gamma} \in \Gamma_1(X)$ by $\tilde{\gamma}(t) = x_p + te_p\|u\|$ for $t \in [0,1]$. Then $\|\tilde{\gamma} - \gamma_0\|_{\leq 1} < \varepsilon$. Let m be the least integer such that $m > \|u\|$. The choice of pairs (x_k, e_k) guarantees that given any $j \in \mathbb{N}$, there are integers $k_0, \ldots, k_{m-1} > j$ such that

$$(x_{k_i}, e_{k_i}) = (x_p + ie_p, e_p), \quad i = 0, \ldots, m-1.$$

By the first statement of Lemma 10.5.4,

$$[x_p, x_p + me_p] = [x_{k_0}, x_{k_0} + e_{k_0}] \cup \cdots \cup [x_{k_{m-1}}, x_{k_{m-1}} + e_{k_{m-1}}] \subset Q_j.$$

Hence the image of $\tilde{\gamma}$ lies in Q_j, which, since Q_j is open and the image of $\tilde{\gamma}$ is compact, implies that the image of any γ is contained in Q_j once $\|\gamma - \tilde{\gamma}\|_{\leq 1}$ is small enough. Corollary 5.3.4 now implies that $X \setminus Q_j$ is Γ_1-null.

Letting $Q = \bigcap_{j=1}^{\infty} Q_j$, we see that $S \cup (X \setminus Q)$ is Γ_1-null. It remains to show that its complement $(X \setminus S) \cap Q = \bigcup_{j=1}^{\infty}(Q \setminus S_j)$ is σ-porous. We prove that each set $Q \setminus S_j$ is porous. Let $x \in Q \setminus S_j$. For every $m > j$ the fact $x \in Q_m$ implies that there is $k \geq m$ with $x \in H_k$. This means that $x \in B(z_{k,i}, \kappa r_k)$ for some i, and so

$$B(z_{k,i}, \|z_{k,i} - x\|/\kappa) \subset B(z_{k,i}, r_k) \subset S_j.$$

Since the distance $\|z_{k,i} - x\| \leq \kappa r_k \leq \kappa 2^{-m}$ can be made arbitrarily small, we conclude that $Q \setminus S_j$ is porous at x. $\qquad\square$

10.6   SPACES IN WHICH σ-POROUS SETS ARE Γ-NULL

In this section we present a new condition on a Banach space guaranteeing that its σ-porous subsets are Γ-null. Instead of assuming existence of a suitable asymptotically c_0 sequence of subspaces (see Definition 8.2.10) as in [28], we assume the existence of bump functions arbitrarily smooth in the direction of such subspaces (see Definition 10.6.4). In Proposition 8.2.11 we proved that the new condition is weaker than the old one, and so our results contain those of [28].

The key to our new approach to the problem of Γ-nullness of porous sets is the following simple statement reducing it to showing their Γ_n-nullness.

Proposition 10.6.1. *Let E be a subset of a separable Banach space X. Assume that for infinitely many values $n \in \mathbb{N}$ the set E is contained in a union of a σ-directionally porous set and a Γ_n-null G_δ set. Then E is Γ-null.*

Proof. Let $N \subset \mathbb{N}$ be the set of those n for which we have $E \subset A_n \cup B_n$, where A_n is σ-directionally porous and B_n is a Γ_n-null G_δ set. Then

$$E \subset \bigcup_{n \in N} A_n \cup \bigcap_{n \in N} B_n.$$

Since every directionally porous set is Γ-null by Remark 5.2.4, the union of sets A_n is Γ-null. Hence it suffices to observe that Theorem 5.4.2 implies that the intersection of B_n, being a G_δ set which is Γ_n null for every n, is Γ-null as well. $\square$

Various simple combinations of this proposition with the results of Section 10.4 lead to smoothness conditions guaranteeing that every σ-porous subset of X is Γ-null. For example, Corollary 10.4.6 and Proposition 10.6.1 immediately imply the following statement, which gives concrete examples of spaces having the property that every σ-porous subset in them is Γ-null.

Theorem 10.6.2. *Suppose that the asymptotic modulus of smoothness of a Banach space X satisfies, for every $n \in \mathbb{N}$,*

$$\lim_{t \searrow 0} \frac{\bar{\rho}_X(t)}{t^n} = 0.$$

Then every σ-porous subset of X is Γ-null.

The property of a space X required here is clearly inherited by subspaces. In particular, c_0 and so any subspace of c_0 have this property. Hence any porous set in a subspace of c_0 is Γ-null.

A somewhat more general criterion than the one given in Theorem 10.6.2 follows immediately from Corollary 10.4.5 and Proposition 10.6.1.

Proposition 10.6.3. *Suppose a separable Banach space X admits, for every $n \in \mathbb{N}$, a locally uniformly continuous bump function asymptotically smooth with modulus controlled by t^n. Then any σ-porous subset of X is Γ-null.*

Finally, we discuss iteration arguments that may allow us to extend the class of spaces in which porous sets are Γ-null. We are really interested here in the property "every σ-porous subset of X is Γ-null." However, it is not clear how this property behaves even under c_0 sums. We therefore strengthen it slightly, which will allow iteration arguments to prove, for example, that any space $C(K)$ with K countable compact has this property. This was done in [28] using the concept of asymptotically c_0 sequences. Here we use a somewhat more general concept introduced it the following definition.

Definition 10.6.4. Let $\mathcal{Y}$ be a nonempty family of subspaces of X, which is closed under finite intersections. We say that X is *arbitrarily smooth in the direction of* $\mathcal{Y}$ if it admits, for each $n \in \mathbb{N}$, a locally uniformly continuous bump function smooth in the direction of $\mathcal{Y}$ with modulus controlled by t^n.

We first show that asymptotically c_0 families (see Definition 8.2.10) have this property.

Proposition 10.6.5. *Let $\mathcal{Y}$ be an asymptotically c_0 family of subspaces of X. Then X is arbitrarily smooth in the direction of $\mathcal{Y}$.*

Proof. Let $\omega \colon (0, \infty) \longrightarrow (0, \infty)$ be an increasing function such that

$$\lim_{t \to \infty} \frac{\omega(t)}{t} = \infty \text{ and for every } p > 0, \ \lim_{t \searrow 0} \frac{\omega(t)}{t^p} = 0.$$

By Proposition 8.2.11 there is a function $\Theta \colon X \longrightarrow \mathbb{R}$ such that $\mathrm{Lip}(\Theta) \leq 1$, $\|x\| \leq \Theta(x) \leq \|x\| + K$, where K is a constant, and for every $t > 0$ there is $Y \in \mathcal{Y}$ such that $\rho_{\Theta, Y}(x; t) \leq \omega(t)$ for every $x \in X$. It follows that Θ is a locally uniformly continuous bump function smooth in the direction of $\mathcal{Y}$ with modulus controlled by t^n, for every n. $\qquad \square$

If follows immediately from Proposition 10.6.3 that in a space arbitrarily smooth in the direction of a family of finite codimensional subspaces every porous subset is Γ-null. Similarly easily, Theorem 10.4.5 and Proposition 10.6.1 imply the following fact.

Lemma 10.6.6. *Suppose that a separable space X is arbitrarily smooth in the direction of a family $\mathcal{Y}$ of its subspaces. Then every subset E of X which is porous in the direction of $\mathcal{Y}$ is Γ-null.*

We also observe that, similar to other notions of smoothness, arbitrary smoothness (in separable spaces) is reducible to countable families of subspaces: if X is arbitrarily smooth in the direction of $\mathcal{Y}$, Proposition 8.1.11 shows that for each n there is a locally uniformly continuous bump function smooth in the direction of a countable subfamily $\mathcal{Y}_n$ of $\mathcal{Y}$ with modulus controlled by t^n. It now suffices to put the $\mathcal{Y}_n$ together and add countably many subspaces from Y to make the resulting family directed downward.

After these remarks, we are ready to state the iteration argument.

Lemma 10.6.7. *Consider the following property of a Banach space Y.*

(**E**) *Whenever Y is a complemented subspace of X and $E \subset X$ is σ-porous in the direction of Y, then E is Γ-null.*

*Any finite or c_0 (infinite) direct sum of spaces having (**E**) also has (**E**).*

*Moreover, let Y be arbitrarily smooth in the direction of a family $\mathcal{V}$ of its subspaces, and let $\eta > 0$ be such that for each $V \in \mathcal{V}$ there is a complemented subspace U of Y having (**E**) and every $y \in Y$ can be written as $y = u + v$, where $u \in U$, $v \in V$ with $\max\{\|u\|, \|v\|\} \leq \eta\|y\|$. Then Y has property (**E**).*

Proof. Assume that Y_1 and Y_2 have (**E**), $Y = Y_1 \oplus Y_2$ is complemented in X, and $E \subset X$ is σ-porous in the direction of Y. By Lemma 10.1.2, $E = E_1 \cup E_2$, where E_i is σ-porous in the direction of Y_i. It follows from (**E**) that each E_i is Γ-null and thus E is Γ-null. Hence (**E**) is stable under finite direct sums.

Notice that the "moreover" statement and the stability of (**E**) under finite direct sums imply that (**E**) is also closed under c_0 direct sums. Indeed, let $Y = \left(\sum_{i=1}^{\infty} \oplus Y_i\right)_{c_0}$ be complemented in X. We put

$$U_k = \left(\sum_{i=1}^{k} \oplus Y_i\right)_{c_0}, \quad V_k = \left(\sum_{i=k+1}^{\infty} \oplus Y_i\right)_{c_0}.$$

Then every U_k is complemented in Y and by the already proved stability under finite sums, all U_k have the property (**E**). Since (V_k) is clearly asymptotically c_0, Y is arbitrarily smooth in the direction of $\mathcal{V} = \{V_k \mid k \geq 1\}$ by Proposition 10.6.5. Hence stability under c_0 sums follows.

To prove the "moreover" statement let E be a porous subset of X in the direction of Y with constant c. As pointed out before the statement of the lemma, we may assume that $\mathcal{V}$ is countable, $\mathcal{V} = \{V_1, V_2, \dots\}$. For each k we find U_k as in the assumptions, and by Lemma 10.1.2 we write $E = A_k \cup B_k$, where A_k is σ-porous in the direction of U_k, and B_k is porous in the direction of V_k with constant $c/2\eta$. By assumption all sets A_k are Γ-null. Since the set $B = \bigcap_{k=1}^{\infty} B_k$ is porous in the direction of each V_k with the same constant $c/2\eta$, it is porous in the direction of $\mathcal{V}$. Hence by Lemma 10.6.6, B is Γ-null. Consequently, so is $E = B \cup \bigcup_{k=1}^{\infty} A_k$. $\square$

Recall that any countable compact Hausdorff topological space K is homeomorphic to the space K_α of ordinals not exceeding α in the order topology for some countable ordinal α. Recall also that $C(K_\omega)$ is isomorphic to c_0 and that every $C(K_\alpha)$, $\alpha < \omega_1$, is isomorphic to $\left(\sum_{i=1}^{\infty} \oplus C(K_{\beta_i})\right)_{c_0}$ for a suitable sequence of ordinals $(\beta_i)_{i=1}^{\infty}$ smaller than α. Hence we get from Lemma 10.6.7 by (countable) transfinite induction that each such $C(K)$ has property (**E**).

Summing up the preceding observations we get the result that has been already used in the discussion of differentiability Γ-almost everywhere in Section 6.4.

Theorem 10.6.8. *The following spaces have the property that all their σ-porous subsets are Γ-null: c_0, $C(K)$ with K compact countable, the Tsirelson space, subspaces of c_0.*

The Tsirelson space (as first defined in [45]) is a Banach space with an unconditional basis which admits an asymptotically c_0 sequence of finite codimensional subspaces. An important feature of this space (in general, and for us here) is that it is reflexive.

The class of spaces in Theorem 10.6.8 may seem to be rather small. We will see in Chapter 14 that this is in the heart of the matter: in many spaces, including the ℓ_p spaces, there are porous sets that are *not* Γ-null.

Chapter Eleven

Porosity and ε-Fréchet differentiability

We show that every slice of the set of Gâteaux derivatives of a Lipschitz function $f\colon X \longrightarrow Y$, where $\dim Y = n$, contains an ε-Fréchet derivative provided certain porous sets associated with f are small in the sense that each of them can be covered by a union of Haar null sets and a Γ_n-null G_δ set. By results of Chapter 10, this condition holds when X admits a bump function which is uniformly continuous on bounded sets and asymptotically smooth with modulus controlled by ω_n. In Chapter 13 we replace, under more restrictive assumptions, ε-Fréchet derivatives by Fréchet derivatives. The advantage of the present approach is not only its greater generality, but also its (relative) simplicity and the fact that it relates the problem to smallness of porous sets.

11.1 INTRODUCTION

Here we investigate a natural question stemming from Theorem 6.3.9: can the results of Chapter 10 about smallness of porous sets, and so also of sets of irregularity points of a given Lipschitz function, be used to show existence of points of (at least) ε-Fréchet differentiability of vector-valued functions? By combining this new idea with the basic idea that points of ε-Fréchet differentiability should appear in small slices of the set of Gâteaux derivatives, we will see that this is indeed the case and that this approach leads to very precise results on existence of points of ε-Fréchet differentiability for Lipschitz maps with finite dimensional range. Our main result, Theorem 11.3.6, applies, in particular, when every porous set is contained in the unions of a σ-directionally porous (and hence Haar null) set and a Γ_n-null G_δ set. Hence Theorem 10.4.5 immediately implies the following corollary.

Theorem 11.1.1. *Suppose that an Asplund space X admits a bump function that is uniformly continuous on bounded sets and asymptotically smooth with modulus controlled by ω_n. Then for every Lipschitz map f of a nonempty open set $G \subset X$ to a Banach space Y of dimension not exceeding n, every $x_0 \in G$ at which f is Gâteaux differentiable, $\varepsilon, \eta > 0$, $u_1, \ldots, u_m \in X$, and $y_1^*, \ldots, y_m^* \in Y^*$, there is a point $x \in G$ at which f is Gâteaux differentiable, ε-Fréchet differentiable, and*

$$\sum_{i=1}^{m} y_i^*(f'(x;u_i)) > \sum_{i=1}^{m} y_i^*(f'(x_0;u_i)) - \eta.$$

In other words, under the given conditions the multidimensional mean value theorem holds, for any $\varepsilon > 0$, for Gâteaux derivatives that are also ε-Fréchet derivatives.

Theorem 11.1.1 is an easy consequence of the more general Theorem 11.3.5, whose main points are that only certain porous sets need to have the required decomposition and that separability of the dual is replaced by norm separability of the set of Gâteaux derivatives.

We have seen in Chapter 4 that points at which f is Gâteaux differentiable as well as ε-Fréchet differentiable exist when X is just asymptotically uniformly smooth without any restriction on the dimension of the target space Y (as long as it is finite dimensional). As far as is known, it may even hold in all Asplund spaces. However, as we will see in Chapter 14, validity of Theorem 11.1.1 needs a considerably stronger assumption than just asymptotic uniform smoothness.

In Chapter 6 we have already seen an important connection between smallness of porous sets and differentiability: once we know that porous sets in X are Γ-null, Fréchet differentiability of Lipschitz maps into finite dimensional spaces follows. It is therefore natural to conjecture that in spaces in which porous sets are Γ_n-null, Fréchet differentiability of Lipschitz maps into at least spaces of small finite dimension can be shown. We may even hope that in such situation the multidimensional mean value estimate holds, that is, every slice of the set of Gâteaux derivatives of any Lipschitz map into a space with dimension not exceeding n contains a Fréchet derivative. (The restriction on the dimension of the target is necessary: see Theorem 14.1.1 in Chapter 14.) Unfortunately, for $n \geq 3$ we do not know whether there is a space of dimension greater than n in which every porous set is Γ_n-null (see the beginning of Section 10.3), so such results may bring little new.[*] However, from Chapter 10 we know that there are spaces in which every porous set is contained in the union of a σ-directionally porous set and a Γ_n-null G_δ set. Whether the multidimensional mean value estimate for Fréchet derivatives holds in all spaces in which porous sets are small in this way remains open. Here we prove that with ε-Fréchet derivatives instead of Fréchet derivatives it holds even under the weaker condition that every porous set is contained in the union of a Haar null set and a Γ_n-null G_δ set. (Notice that the G_δ requirement is essential.) In Chapter 13 we show its validity with Fréchet derivatives, but under a smoothness assumption that is stronger than the one needed to obtain results on decomposition of porous sets.

Throughout this section, $G \subset X$ will be an open subset of a separable Banach space X and $f \colon G \longrightarrow Y$ a Lipschitz map to a Banach space Y. We will be mostly interested in the case when Y is finite dimensional, and in this case we will usually consider $Y = \mathbb{R}^n$ only. Since our results do not depend on the particular choice of norm in the target space, we equip $\mathbb{R}^n$, in addition to the Euclidean norm $|\cdot|$, also with the maximum norm $\|y\| = \max(|y_1|, \ldots, |y_n|)$, and will usually work with this norm. The balls in this norm will be called cubes; so all cubes are, by this definition, axis parallel.

11.2 FINITE DIMENSIONAL APPROXIMATION

As in all differentiability results, we first need to understand the finite dimensional situation. So suppose that $g = (g_1, \ldots, g_n) \colon [0,1]^n \longrightarrow \mathbb{R}^n$ is a Lipschitz map with

---

[*] Added in proof: By the result of Gareth Speight (see footnote on page 186), there are no such spaces.

non-negative divergence on $[0, 1]^n$ and with $\int_{[0,1]^n} \operatorname{div} g \, d\mathscr{L}^n$ close to zero. For $n = 1$ this obviously implies that g is almost constant, and this fact was somewhere in the deep background of Fréchet differentiability results for real-valued Lipschitz functions (see [39], [27]). In the vector-valued case, however, nonconstant functions with divergence zero show that an analogously simple statement is hopelessly wrong. Fortunately, Lemma 11.2.2 will tell us that, quantifying the words "small" and "close" and assuming that a suitable slice of Gâteaux derivatives $\{g'(t) \mid \operatorname{div} g(t) < \delta\}$ is contained in a small neighborhood of the origin, we can still deduce near constancy of g. Of course, we have to pay for the additional assumptions, and this will appear in subsequent chapters as the need for special smoothness assumptions on the domain space. This causes a substantial difference between our differentiability results for vector-valued functions and those for scalar-valued functions. However, we will see in Chapter 14 that this is not a fault of our method, but that without such smoothness assumptions the differentiability results cannot be as strong as the ones we prove in this book as they cannot include validity of the multidimensional mean value estimates.

Before stating the lemma alluded to above, we recall some properties of *shears*, which are simple Lebesgue measure preserving maps of $\mathbb{R}^n$ into itself that we will use to control certain integrals.

Lemma 11.2.1. *Assume that Q is a cube in $\mathbb{R}^n$, $a, b \in Q$, $d = \|b - a\| > 0$, and $0 < \kappa \le 1/2$. There is a Lebesgue measure preserving affine map $\varphi \colon \mathbb{R}^n \longrightarrow \mathbb{R}^n$ which maps the box $[0, \kappa d]^{n-1} \times [0, d]$ into Q such that*

(a) *the restriction of φ to each hyperplane $\mathbb{R}^{n-1} \times \{\sigma\}$ is an isometry;*

(b) $a \in \varphi\big([0, \kappa d]^{n-1} \times \{0\}\big)$ *and* $b \in \varphi\big([0, \kappa d]^{n-1} \times \{d\}\big)$;

(c) $\|\varphi'\| \le 2$.

Proof. We may assume that $d = 1$ and use shifts, reflexions, and permutations of coordinates to reduce the proof to the special case when $a = 0$, $0 \le b_i \le 1$ for $i = 1, \ldots, n$ and $b_n = 1$. Then we let for $(\omega, \sigma) \in [0, \kappa]^{n-1} \times [0, 1]$

$$\varphi(w, \sigma) = (w + \psi(\sigma), \sigma),$$

where $\psi = (\psi_1, \ldots, \psi_{n-1}) \colon \mathbb{R} \longrightarrow \mathbb{R}^{n-1}$ is a linear map. Its components ψ_i are defined according to the following rule. Since the cases below are not in general mutually exclusive, we define $\psi_i(\sigma)$ by the first rule whose condition is satisfied:

$$\psi_i(\sigma) = \begin{cases} \sigma b_i & \text{if } b + \kappa e_i \in Q, \\ \sigma(b_i - \kappa) & \text{if } b - \kappa e_i \in Q. \end{cases} \tag{11.1}$$

Since for any fixed $\sigma \in [0, 1]$ the map φ is a shift, we see that (a) holds. In particular, φ is measure preserving. It is also easy to check that $\varphi(0, 0) = 0$ and $\varphi(w, 1) = b$ for a suitable choice of $\omega \in [0, \kappa]^{n-1}$. Hence (b) follows. As for (c), notice first that

$$\psi' = \big(b_1 - q_1, \ldots, b_{n-1} - q_{n-1}\big),$$

where $q_i = 0$ or $q_i = \kappa$ depending on the choice in (11.1). Then

$$\|\varphi'\| \leq \max\{1 + |b_i - q_i| \mid i = 1, \ldots, n - 1\} \leq 2. \qquad \square$$

We are now ready to state and prove the announced finite dimensional result on affine approximation of Lipschitz functions.

Lemma 11.2.2. *For every $n \in \mathbb{N}$ and $\theta > 0$ there is $\tau > 0$ such that for every $\alpha > 0$ the following statement holds.*

Let $g: [0, 1]^n \longrightarrow \mathbb{R}^n$ be a Lipschitz map with $\mathrm{Lip}(g) \leq 1$ such that $\mathrm{div}\, g(t) \geq 0$ for almost all $t \in [0, 1]^n$. Assume that there is a set $N \subset [0, 1]^n$, $\mathscr{L}^n N < \tau$, satisfying

(i) $\displaystyle\int_{[0,1]^n \setminus N} \mathrm{div}\, g(t)\, d\mathscr{L}^n(t) < \alpha\tau$, *and*

(ii) $\|g'(t)\| < \tau$ *for each $t \in [0, 1]^n \setminus N$ at which $\mathrm{div}\, g(t) < \alpha$.*

Then

$$\mathscr{L}^n\left\{t \in [0, 1]^n \mid \|g(s) - g(t)\| > \theta\|s - t\| \text{ for some } s \in [0, 1]^n\right\} < \theta.$$

In particular, $\|g(s) - g(t)\| \leq 2\theta$ for all $s, t \in [0, 1]^n$.

Before proving this lemma, we discuss its strengths and weaknesses, which are very much related to all our Fréchet differentiability results for vector-valued functions. We neglect technical points (multiplicative constants and the set N), and first notice that one clearly has to have $\tau \leq \theta$. Validity of (ii) is controlled by the relation of the height of a slice $\{g'(t) \mid \mathrm{div}\, g(t) < \alpha\}$ and its diameter, for which the necessary condition is, in general, $\alpha \leq \tau$ (in the Hilbertian case the condition is $\alpha \leq \tau^2$). Hence the largest value of $\alpha\tau$ for which we can still get a meaningful result is θ^p, where $p \geq 2$. To see where this leads, we rescale the statement. Denoting for a moment

$$\widetilde{g}(t) = \frac{1}{r}\, g(rt),$$

the first assumption becomes

$$\fint_{B(0,r)} \mathrm{div}\, g(t)\, d\mathscr{L}^n(t) = \int_{B(0,1)} \mathrm{div}\, \widetilde{g}(t)\, \mathscr{L}^n(t) < \theta^p,$$

and the conclusion becomes

$$\|g(t) - g(0)\| = r\|\widetilde{g}(t/r) - \widetilde{g}(0)\| \leq \theta r$$

for $t \in B(0, r)$. Now imagine that the domain of g is $\mathbb{R}^{n+1}$ and that $g = 0$ on $\mathbb{R}^n$. Then, to prove that zero is almost a derivative of g, we need in particular an estimate of the form

$$\|g(u + \xi e_{n+1}) - g(\xi e_{n+1})\| \leq \varepsilon\xi$$

if $u \in \mathbb{R}^n$, $\|u\| = \xi$ and ξ small. To use the lemma, we choose a suitable $r \geq \xi$ and $\theta = \varepsilon\xi/r$. So we need to get an estimate of the type

$$\fint_{B(0,r)} \operatorname{div} g(t + \xi e_{n+1}) \, d\mathscr{L}^n(t) < \left(\frac{\varepsilon\xi}{r}\right)^p.$$

Using the divergence theorem and Lipschitz estimates on g, the left side is bounded by ξ/r; but this leads to the requirement $\xi/r \leq (\varepsilon\xi/r)^p$, which is impossible even if $p = 1$. We can improve upon this estimate: given $\eta > 0$, one can show that

$$\fint_{B(0,r)} \operatorname{div} g(t + \xi e_{n+1}) \, d\mathscr{L}^n(t) < \frac{\eta\xi}{r},$$

provided ξ and ξ/r are small enough. This leads to the requirement $\eta\xi/r \leq (\varepsilon\xi/r)^p$, which can be satisfied if $p = 1$; for $p > 1$ its validity depends on the exact estimate of the dependence of the permitted size of ξ/r on η. Most of the main results obtained in this book are, in some sense, based on having a better estimate of this dependence. The negative results obtained in Chapter 14 depend, on the other hand, on showing that the estimate cannot be too good. In this chapter, however, we obtain a better estimate of $\fint_{B(0,r)} \operatorname{div} g(t + \xi e_{n+1}) \, d\mathscr{L}^n(t)$, thanks to porous sets being small, which allows us to assume that 0 is a point of regularity of g, and so the divergence theorem gives as good estimates as one may wish.

Proof. Since the conclusion is trivially true for $\theta \geq 1$, let $\theta < 1$. We show that the statement holds with

$$\tau = \frac{1}{2} \, 3^{-n-1} 5^{-n} \theta^{2n+1}.$$

For a cube $Q \subset [0,1]^n$, denote by $r(Q)$ the side-length of Q and let

$$\mathcal{Q} = \left\{ \text{cube } Q \subset [0,1]^n \mid \|g(s) - g(t)\| > \theta r(Q) \text{ for some } s, t \in Q \right\}.$$

We show that the measure of each $Q \in \mathcal{Q}$ may be estimated from above by a constant multiple of $\int_Q \|g'\| \, d\mathscr{L}^n$, namely,

$$\mathscr{L}^n Q < 2 \cdot 3^n \theta^{-2n} \int_Q \|g'\| \, d\mathscr{L}^n. \tag{11.2}$$

To prove this we fix, for a moment, a cube $Q \in \mathcal{Q}$, and find $a, b \in Q$ such that $\|g(b) - g(a)\| > \theta r(Q)$. Denoting $d = \|b - a\|$, we notice that $\mathscr{L}^n Q < d^n \theta^{-n}$. Let $\varphi \colon \mathbb{R}^n \longrightarrow \mathbb{R}^n$ be the shear from Lemma 11.2.1 with $\kappa = \frac{1}{3}\theta$. The first two properties of φ imply that

$$\|g(\varphi(w,0)) - g(a)\| \leq \|\varphi(w,0) - a\| \leq \frac{\theta d}{3}$$

for each $w \in [0, \frac{1}{3}\theta d]^{n-1}$. The same argument shows that for these w also

$$\|g(\varphi(w,d)) - g(b)\| \leq \frac{\theta d}{3}.$$

Hence

$$
\begin{aligned}
\|g(\varphi(w,d)) &- g(\varphi(w,0))\| \\
&\geq \|g(b) - g(a)\| - \|g(\varphi(w,d)) - g(b)\| - \|g(a) - g(\varphi(w,0))\| \\
&> \theta r(Q) - \frac{\theta d}{3} - \frac{\theta d}{3} \geq \frac{\theta d}{3}
\end{aligned}
$$

for every $w \in [0, \frac{1}{3}\theta d]^{n-1}$. Using also that φ is measure preserving and the last property of the shear φ, we infer that

$$
\begin{aligned}
\int_Q \|g'\| \, d\mathscr{L}^n &\geq \int_{\varphi([0,\theta d/3]^{n-1} \times [0,d])} \|g'(t)\| \, d\mathscr{L}^n(t) \\
&= \int_{[0,\theta d/3]^{n-1} \times [0,d]} \|g'(\varphi(s))\| \, d\mathscr{L}^n(s) \\
&\geq \int_{[0,\theta d/3]^{n-1} \times [0,d]} \|(g \circ \varphi)'(s)\| \frac{d\mathscr{L}^n(s)}{\|\varphi'(s)\|} \\
&\geq \frac{1}{2} \int_{[0,\theta d/3]^{n-1}} \int_0^d \|(g \circ \varphi)'(w,\sigma)\| \, d\sigma \, d\mathscr{L}^{n-1}(w) \\
&\geq \frac{1}{2} \int_{[0,\theta d/3]^{n-1}} \|g(\varphi(w,d)) - g(\varphi(w,0))\| \, d\mathscr{L}^{n-1}(w) \\
&> \frac{1}{2} \frac{\theta d}{3} \left(\frac{\theta d}{3}\right)^{n-1} = \frac{1}{2}\left(\frac{\theta d}{3}\right)^n > \frac{1}{2} 3^{-n} \theta^{2n} \mathscr{L}^n Q,
\end{aligned}
$$

which is (11.2).

Denote by M the union of all cubes from $\mathcal{Q}$, $M = \bigcup \mathcal{Q}$. Noting that

$$
\left\{ t \in [0,1]^n \mid \|g(s) - g(t)\| > \theta \|s - t\| \text{ for some } s \in [0,1]^n \right\} \subset M,
$$

we see that the proof will be finished once we show that $\mathscr{L}^n M < \theta$. By the $5r$-covering theorem (see, e.g., [33]), there is a disjoint at most countable subfamily $(Q_j) \subset \mathcal{Q}$ such that for every $Q \in \mathcal{Q}$ one can find Q_j with $Q \cap Q_j \neq \emptyset$ and $r(Q) < 2r(Q_j)$. Consequently, any $Q \in \mathcal{Q}$ is covered by the fivefold enlargement of some Q_j. This, (11.2), and the disjointness of the cubes Q_j allow us to estimate the measure of M by

$$
\begin{aligned}
\mathscr{L}^n M \leq 5^n \sum_j \mathscr{L}^n Q_j &< C_n \sum_j \int_{Q_j} \|g'\| \, d\mathscr{L}^n \\
&\leq C_n \int_{[0,1]^n} \|g'\| \, d\mathscr{L}^n,
\end{aligned}
\tag{11.3}
$$

where $C_n = 2 \cdot 3^n 5^n \theta^{-2n}$.

To estimate the integral of $\|g'(t)\|$ over $[0,1]^n$ fix $\alpha > 0$ and denote

$$
N_0 = \{t \notin N \mid \operatorname{div} g(t) < \alpha\} \text{ and } N_1 = \{t \notin N \mid \operatorname{div} g(t) \geq \alpha\}.
$$

Then

$$\int_{[0,1]^n} \|g'\| \, d\mathscr{L}^n = \int_N \|g'\| \, d\mathscr{L}^n + \int_{N_0} \|g'\| \, d\mathscr{L}^n + \int_{N_1} \|g'\| \, d\mathscr{L}^n.$$

The first two integrals are less than τ in view of the assumptions $\|g'\| \le 1$, $\mathscr{L}^n N < \tau$, and (ii). As for the last integral, we recall that $\operatorname{div} g(t) \ge 0$ a.e. and employ assumption (i):

$$\int_{N_1} \|g'\| \, d\mathscr{L}^n \le \frac{1}{\alpha} \int_{[0,1]^n \setminus N} \operatorname{div} g(t) \, d\mathscr{L}^n(t) < \tau.$$

Together with (11.3) we obtain

$$\mathscr{L}^n M < 3\tau C_n = 2 \cdot 3^{n+1} 5^n \theta^{-2n} \tau = \theta,$$

as required.

For the additional "in particular" statement notice that the just proved part of the lemma and $\theta < 1$ guarantee the existence of $t_0 \in [0,1]^n$ such that $\|g(s) - g(t_0)\| \le \theta \|s - t_0\| \le \theta$ for all $s \in [0,1]^n$. Hence we get

$$\|g(s) - g(t)\| \le \|g(s) - g(t_0)\| + \|g(t) - g(t_0)\| \le 2\theta. \qquad \square$$

11.3 SLICES AND ε-DIFFERENTIABILITY

Since in all statements and proofs in this section (with the single exception of Theorem 11.3.5) the domain will be a fixed nonempty open set $G \subset X$ and the target space will be $\mathbb{R}^n$, we slightly simplify the notation for slices of the set of Gâteaux derivatives. For a Lipschitz map $f = (f_1, \dots, f_n)$ of G to $\mathbb{R}^n$ we denote by $D(f)$ the set of points of G at which f is Gâteaux differentiable. When $u = (u_1, \dots, u_n) \in X^n$, we also let

$$M(f; u) = \sup_{x \in D(f)} \sum_{i=1}^n f_i'(x; u_i),$$

and, for $\eta > 0$, we denote the slice determined by u and η by

$$S(f; u; \eta) = \left\{ f'(x) \ \middle|\ x \in D(f), \ \sum_{i=1}^n f_i'(x; u_i) > M(f; u) - \eta \right\}.$$

The set of points at which the derivative belongs to the slice is denoted by

$$D(f; u; \eta) = \left\{ x \in D(f) \mid f'(x) \in S(f; u; \eta) \right\}.$$

For future reference, we make a simple observation on the dependence of these sets on a change of vectors.

Observation 11.3.1. *Suppose that $G \subset X$ is an open set, $f \colon G \longrightarrow \mathbb{R}^n$ a Lipschitz function, $u = (u_1, \ldots, u_n)$, $v = (v_1, \ldots, v_n) \in X^n$, and $\kappa = \mathrm{Lip}(f) \sum_{i=1}^n \|v_i - u_i\|$. Then for every $\eta > 0$,*

$$D(f; v; \eta) \subset D(f; u; \eta + 2\kappa)$$

or, equivalently,

$$S(f; v; \eta) \subset S(f; u; \eta + 2\kappa).$$

Proof. If $x \in D(f)$, then $\|f'(x)\| \leq \mathrm{Lip}(f)$ and so

$$\left| \sum_{i=1}^n f_i'(x; v_i) - \sum_{i=1}^n f_i'(x; u_i) \right| \leq \mathrm{Lip}(f) \sum_{i=1}^n \|v_i - u_i\| = \kappa.$$

Hence $M(f; v) \geq M(f; u) - \kappa$. If now $x \in D(f; v; \eta)$, then

$$\sum_{i=1}^n f_i'(x; u_i) \geq \sum_{i=1}^n f_i'(x; v_i) - \kappa > M(f; v) - \eta - \kappa$$
$$\geq M(f; u) - (\eta + 2\kappa),$$

which implies the statement. □

After these preliminaries, we come to our first main lemma, which shows that ε-Fréchet differentiability of the function f at a point x follows from its rather weak approximability by an element of a small slice of its set of Gâteaux derivatives.

Lemma 11.3.2. *Let $G \subset X$ be an open subset of a separable Banach space X, and let $f \colon G \longrightarrow \mathbb{R}^n$ be Lipschitz with $\mathrm{Lip}(f) \leq 1$. Assume further that*

- *$\varepsilon > 0$, $\chi > 0$, and the vectors $w_1, \ldots, w_n \in X$ of norm at most one are chosen such that $\mathrm{diam}\, S(f; w_1, \ldots w_n; \chi) \leq \frac{1}{4}\varepsilon$;*

- *$x \in G$, $\delta > 0$, and $T \in S(f; w_1, \ldots, w_n; \frac{1}{16}\varepsilon\chi)$ are such that $B(x, \delta) \subset G$ and, putting $\zeta = \dfrac{\varepsilon\chi}{16n^2(n+2)}$,*

$$\|f(x + v + rw_i) - f(x + v) - T(rw_i)\| \leq \zeta(\|v\| + |r|) \tag{11.4}$$

  for every $i = 1, \ldots, n$, $v \in X$, and $r \in \mathbb{R}$ with $\|v\| + |r| < \delta$.

Then

$$\|f(x + v) - f(x) - Tv\| \leq \varepsilon\|v\|$$

for every $v \in X$ with $\|v\| < \delta/(n+2)$.

Proof. We assume that $\varepsilon < 2$ since otherwise the conclusion follows trivially from $\mathrm{Lip}(f) \leq 1$ and $\|T\| \leq 1$. We may also assume, without loss of generality, that $\chi \leq 2n$, since otherwise $S(f; w_1, \ldots, w_n; \chi)$ contains all Gâteaux derivatives of f, hence $\mathrm{Lip}(f - T) \leq \frac{1}{4}\varepsilon$ and the conclusion follows as well.

Let $v \in X$, $\|v\| < \delta/(n+2)$. Then

$$x + s_{n+1}v + \|v\| \sum_{i=1}^{n} s_i w_i \in G$$

for all $s = (s_1, \ldots, s_{n+1}) \in [-1,1]^{n+1}$. Since it suffices to prove the statement for v belonging to a dense subset of $\{v \in X \mid \|v\| < \delta/(n+2)\}$, in view of Lemma 2.3.2 we will consider only those $v \neq 0$ satisfying

$$x + s_{n+1}v + \|v\| \sum_{i=1}^{n} s_i w_i \in D(f)$$

for almost every $s \in [-1,1]^{n+1}$.

For $\sigma \in [0,1]$ define $\gamma_\sigma \in \Gamma_n(X)$ and $g_\sigma : [0,1]^n \longrightarrow \mathbb{R}^n$ by

$$\gamma_\sigma(t) = x + \sigma v + \|v\| \sum_{i=1}^{n} t_i w_i \quad \text{and} \quad g_\sigma(t) = (f - T)(\gamma_\sigma(t)).$$

The main part of our argument is based on estimates of the divergence of g_σ. Writing $f = (f_1, \ldots, f_n)$ and $T = (T_1, \ldots, T_n)$, we see that for every $\sigma \in [0,1]$ and $s \in [0,1]^n$ such that $\gamma_\sigma(s) \in D(f)$,

$$\operatorname{div} g_\sigma(s) = \sum_{i=1}^{n} f_i'\big(\gamma_\sigma(s); \gamma_\sigma'(s; e_i)\big) - \sum_{i=1}^{n} T_i\big(\gamma_\sigma'(s; e_i)\big)$$

$$= \|v\| \sum_{i=1}^{n} f_i'(\gamma_\sigma(s); w_i) - \|v\| \sum_{i=1}^{n} T_i w_i. \tag{11.5}$$

Recalling that $T \in S(f; w_1, \ldots, w_n; \frac{1}{16}\varepsilon\chi)$ we obtain the estimate

$$\operatorname{div} g_\sigma(s) \leq \|v\| \, M(f; w_1, \ldots, w_n) - \|v\| \left(M(f; w_1, \ldots, w_n) - \frac{\varepsilon\chi}{16} \right)$$

$$= \frac{\varepsilon\chi}{16} \|v\|. \tag{11.6}$$

Since the divergence theorem reduces the problem of estimating the integral of $\operatorname{div} g_\sigma$ to that of estimating the norm of differences $g_\sigma(s) - g_\sigma(t)$, we now turn our attention to the latter. Let $\sigma \in [0,1]$ and $s, t \in [0,1]^n$ and write

$$g_\sigma(s) - g_\sigma(t) = \sum_{j=1}^{n} \Big(f(x + v_{j+1}) - f(x + v_j) - \|v\| T((s_j - t_j)w_j) \Big), \tag{11.7}$$

where vectors $v_1, \ldots, v_{n+1}$ are defined recursively by

$$v_1 = \sigma v + \|v\| \sum_{i=1}^{n} t_i w_i \quad \text{and} \quad v_{j+1} = v_j + \|v\|(s_j - t_j)w_j;$$

so $x + v_1 = \gamma_\sigma(t)$ and $x + v_{n+1} = \gamma_\sigma(s)$, which justifies (11.7).

We estimate the norm of each of the terms on the right-hand side of (11.7). Since the explicit form of v_j is

$$v_j = \sigma v + \|v\| \sum_{i=1}^{j-1} s_i w_i + \|v\| \sum_{i=j}^{n} t_i w_i,$$

we obtain that $\|v_j\| \leq (n+1)\|v\|$. Hence

$$\|v_j\| + \|v\|\,|s_j - t_j| \leq (n+2)\|v\|. \tag{11.8}$$

Recalling that $(n+2)\|v\| < \delta$, we may use the assumption (11.4) to infer that

$$\left\| f(x + v_{j+1}) - f(x + v_j) - \|v\| T((s_j - t_j) w_j) \right\| \leq \zeta \left(\|v_j\| + \|v\|\,|s_j - t_j| \right)$$
$$\leq \zeta(n+2)\|v\|.$$

Hence (11.7) and the triangle inequality give

$$\|g_\sigma(s) - g_\sigma(t)\| \leq n(n+2)\|v\|\zeta. \tag{11.9}$$

As already indicated, we now use this inequality together with the divergence theorem to obtain a lower estimate of the integral of the divergence of g_σ, for each $\sigma \in [0,1]$.

$$\int_{[0,1]^n} \operatorname{div} g_\sigma(t)\, d\mathscr{L}^n(t)$$
$$\geq -\sum_{i=1}^{n} \int_{\{t \in [0,1]^n \,|\, t_i = 0\}} \|g_\sigma(t) - g_\sigma(t + e_i)\|\, d\mathscr{L}^{n-1}(t)$$
$$\geq -\sum_{i=1}^{n} n(n+2)\|v\|\zeta = -n^2(n+2)\|v\|\zeta.$$

Integrating this inequality over $\sigma \in [0,1]$ and using Fubini's theorem, we obtain

$$\int_{[0,1]^n} \int_0^1 \operatorname{div} g_\sigma(t)\, d\sigma\, d\mathscr{L}^n(t) = \int_0^1 \int_{[0,1]^n} \operatorname{div} g_\sigma(t)\, d\mathscr{L}^n(t)\, d\sigma$$
$$\geq -n^2(n+2)\|v\|\zeta.$$

Hence

$$\int_0^1 \operatorname{div} g_\sigma(t)\, d\sigma \geq -n^2(n+2)\|v\|\zeta \tag{11.10}$$

for t belonging to a subset of $[0,1]^n$ of positive measure. It follows that there is an $s \in [0,1]^n$ such that $\gamma_\sigma(s) \in D(f)$ for a.e. $\sigma \in [0,1]$ and, at the same time, (11.10) holds true. For the rest of the proof we fix one such s.

Let $Q = \{\sigma \in [0,1] \mid \operatorname{div} g_\sigma(s) \leq -\frac{1}{2}\chi\|v\|\}$. The reason for proving the previous divergence estimates was that they show that Q has small measure. Using (11.10) and (11.6), we have

$$-n^2(n+2)\|v\|\zeta \leq \int_0^1 \operatorname{div} g_\sigma(s)\,d\sigma \leq -\frac{\chi\|v\|}{2}\mathscr{L}^1 Q + \frac{\varepsilon\chi}{16}\|v\|,$$

hence

$$\mathscr{L}^1 Q \leq \frac{2}{\chi}\left(n^2(n+2)\zeta + \frac{\varepsilon\chi}{16}\right) = \frac{\varepsilon}{4}, \tag{11.11}$$

by the choice of ζ.

For those σ for which $\gamma_\sigma(s) \in D(f)$ we now find an upper estimate of $\left\|\frac{\partial}{\partial\sigma}g_\sigma(s)\right\|$. A crude estimate, using that $\operatorname{Lip}(f) \leq 1$, and so also $\|T\| \leq 1$, gives

$$\left\|\frac{\partial}{\partial\sigma}g_\sigma(s)\right\| = \left\|\frac{\partial}{\partial\sigma}(f-T)\gamma_\sigma(s)\right\| = \|f'(\gamma_\sigma(s);v) - Tv\| \leq 2\|v\|. \tag{11.12}$$

If $\sigma \notin Q$, we need a better estimate. In that case $\operatorname{div} g_\sigma(s)/\|v\| > -\frac{1}{2}\chi$ and we obtain from the identity (11.5) that

$$\sum_{i=1}^n f_i'(\gamma_\sigma(s); w_i) = \sum_{i=1}^n T_i w_i + \frac{1}{\|v\|}\operatorname{div} g_\sigma(s)$$

$$> M(f; w_1, \ldots, w_n) - \frac{\varepsilon\chi}{16} - \frac{\chi}{2}$$

$$> M(f; w_1, \ldots, w_n) - \chi.$$

Hence $f'(\gamma_\sigma(s))$ and T belong to the same slice $S(f; w_1, \ldots, w_n; \chi)$, and we see that $\|f'(\gamma_\sigma(s)) - T\| \leq \frac{1}{4}\varepsilon$. Consequently, the same calculation as in (11.12) above yields

$$\left\|\frac{\partial}{\partial\sigma}g_\sigma(s)\right\| = \|f'(\gamma_\sigma(s);v) - Tv\| \leq \frac{\varepsilon\|v\|}{4}. \tag{11.13}$$

Combining (11.12) with the estimate (11.11) of the measure of Q and (11.13), we get

$$\|g_1(s) - g_0(s)\| \leq \int_0^1 \left\|\frac{\partial}{\partial\sigma}g_\sigma(s)\right\|d\sigma$$

$$\leq \int_Q \left\|\frac{\partial}{\partial\sigma}g_\sigma(s)\right\|d\sigma + \int_{[0,1]^n\setminus Q}\left\|\frac{\partial}{\partial\sigma}g_\sigma(s)\right\|d\sigma$$

$$\leq 2\|v\|\mathscr{L}^1 Q + \frac{\varepsilon}{4}\|v\| \leq \frac{3}{4}\varepsilon\|v\|.$$

We are now ready for the final estimate. Recalling (11.9) we obtain

$$\|f(x+v) - f(x) - Tv\|$$

$$= \|f(x+v) - T(x+v) - (f(x) - Tx)\|$$

$$= \|g_1(0) - g_0(0)\|$$

$$\leq \|g_1(0) - g_1(s)\| + \|g_1(s) - g_0(s)\| + \|g_0(s) - g_0(0)\|$$

$$< n(n+2)\|v\|\zeta + \frac{3}{4}\varepsilon\|v\| + n(n+2)\|v\|\zeta \leq \varepsilon\|v\|. \qquad \square$$

Lemma 11.3.2 allows us to deduce ε-Fréchet differentiability under the assumption of a certain type of regularity; see (11.4). We now complement it by a statement showing that points of such regularity exist provided the porous sets associated with the function f by Lemma 6.2.5 are small on many surfaces from $\Gamma_n(X)$ in a sense that is loosely related to the notion of Γ_n-null sets. (We do not know the precise relation of these two notions of smallness.)

Recall that in Definition 6.2.4 we denoted, for $f \colon G \longrightarrow \mathbb{R}^n$, by $R(f; u, y; \zeta, \rho)$ the set of all $x \in G$ such that at least one of the following requirements holds:

(i) $B(x, \rho) \not\subset G$;

(ii) there is $r \in \mathbb{R}$ such that $\|ru\| < \rho$ and $\|f(x + ru) - f(x) - ry\| > \zeta\|ru\|$;

(iii) there is $\delta > 0$ such that $\|f(x + v + ru) - f(x + v) - ry\| \leq \zeta(\|v\| + \|ru\|)$ whenever $\|v\| + \|ru\| \leq \delta$.

Lemma 11.3.3. *For every $\zeta > 0$ there is $\tau > 0$ such that the following statement holds. Let G be a nonempty open subset of a separable Banach space X, $f \colon G \longrightarrow \mathbb{R}^n$ a Lipschitz map with $\mathrm{Lip}(f) \leq 1$, and $H \subset G$ a Haar null set. Assume further that*

- *$\xi > 0$ and the vectors $w_1, \dots, w_n \in X$ of norm at most one are such that $\mathrm{diam}\, S(f; w_1, \dots, w_n; \xi) \leq \tau$;*

- *for any choice of $y_1, \dots, y_n \in \mathbb{R}^n$ and $\rho > 0$ there is a dense set of surfaces $\gamma \in \Gamma_n(X)$ such that*

$$\mathcal{L}^n\left\{t \in [0,1]^n \;\middle|\; \gamma(t) \in H \cup \left(G \setminus \bigcap_{i=1}^{n} R(f; w_i, y_i; \zeta, \rho)\right)\right\} < \tau. \qquad (11.14)$$

Then there are $x \in D(f; w_1, \dots, w_n; \xi) \setminus H$, $T \in S(f; w_1, \dots, w_n; \xi)$, and $\delta > 0$ such that $B(x, \delta) \subset G$ and

$$\|f(x + v + rw_i) - f(x + v) - T(rw_i)\| \leq \zeta(\|v\| + |r|) \qquad (11.15)$$

for every $i = 1, \dots, n$, $v \in X$ and $r \in \mathbb{R}$ with $\|v\| + |r| < \delta$.

Proof. We start with a sketch of our plan of the proof, in which we assume for simplicity that $H = \emptyset$. First, we choose $T = f'(x_0)$ for some point $x_0 \in D(f; w_1, \dots, w_n; \beta)$, where β is much smaller than ξ. Then, for a suitably small $\rho > 0$, we pick another point $x \in \bigcap_{i=1}^{n} R(f; w_i, Tw_i; \zeta, \rho)$ in such a way that the first two alternatives from the definition of the sets $R(f; w_i, Tw_i; \zeta, \rho)$ are false. Hence the third alternative must hold, implying (11.15). The main difficulty in our plan is to negate the second alternative in the definition of $R(f; w_i, Tw_i; \zeta, \rho)$. It requires showing that at x the map f is close to being affine in the directions $w_1, \dots, w_n$ on a segment of length ρ. To handle this, we intend to choose a surface $\gamma \in \Gamma_n(X)$ in such a way that many points $\gamma(t)$ belong to

$$D(f; w_1, \dots, w_n; \xi) \cap \bigcap_{i=1}^{n} R(f; w_i, Tw_i; \zeta, \rho).$$

Then we employ Lemma 11.2.2 to a suitably transformed $f \circ \gamma$ to show that, for some t, our argument will work for $x = \gamma(t)$.

After this informal description, we now begin the real proof. Without loss of generality we may assume that H contains all points at which f is not Gâteaux differentiable, all $w_i \neq 0$, $i = 1, \ldots, n$, and $\zeta \leq 1$. Denote

$$\theta = \min\left\{ \frac{1}{3} 2^{-n}, \; \frac{\zeta \|w_1\|}{2(n+1)}, \ldots, \frac{\zeta \|w_n\|}{2(n+1)} \right\}.$$

We show that the lemma holds with any $0 < \tau \leq \theta$ for which the conclusion of Lemma 11.2.2 holds. The additional requirement $\tau \leq \theta$ is achieved by observing that, once the conclusion of Lemma 11.2.2 holds for one τ, it holds for all smaller τ.

Assume that $0 < \xi \leq 1$ and w_i with the properties from the lemma are given. Denote $U = \text{span}\{w_1, \ldots, w_n\}$. We put

$$\alpha = \frac{\xi}{2n+1} \quad \text{and} \quad \beta = \frac{\alpha \tau}{4}.$$

We will also use various simple upper estimates of β that follow immediately from the definition: $\beta \leq \xi$, $\beta \leq \theta/4(2n+1)$, and, clearly, $\beta \leq \frac{1}{12}$.

Pick an $x_0 \in D(f; w_1, \ldots, w_n; \beta)$ and let $T = f'(x_0)$. Since $\beta \leq \xi$, we see that $T \in S(f; w_1, \ldots, w_n; \xi)$. Further, we can find $0 < \sigma \leq 1$ small enough so that both $B(x_0, (n+2)\sigma) \subset G$ and

$$\|f(x_0 + u) - f(x_0) - Tu\| < \beta \sigma \text{ for each } u \in U \text{ with } \|u\| \leq n\sigma.$$

Hence, for any $u \in U$ and $x \in x_0 + U$ satisfying $\|x - x_0\| + \|u\| \leq n\sigma$ we have the estimate

$$\begin{aligned}
\|f(x + u) - f(x) - Tu\| &\leq \|f(x_0 + (x + u - x_0)) - f(x_0) - T(x + u - x_0)\| \\
&\quad + \|f(x_0 + (x - x_0)) - f(x_0) - T(x - x_0)\| \\
&< 2\beta\sigma.
\end{aligned} \tag{11.16}$$

Let $\gamma_0 \in \Gamma_n(X)$ be a surface defined by

$$\gamma_0(t) = x_0 + \sigma \sum_{i=1}^{n} t_i w_i,$$

and let

$$\rho = \frac{\sigma}{4} \min\{\|w_1\|, \ldots, \|w_n\|\}.$$

Using assumption (11.14) of the lemma we find $\gamma \in \Gamma_n(X)$ such that $\|\gamma - \gamma_0\|_{C^1} < \beta\sigma$ and

$$\mathscr{L}^n\left\{ t \in [0,1]^n \;\middle|\; \gamma(t) \in H \cup \left(X \setminus \bigcap_{i=1}^{n} R(f; w_i, Tw_i; \zeta, \rho) \right) \right\} < \tau. \tag{11.17}$$

Notice that in (11.17) we replaced G by X. This is justified by observing that the image of γ is contained in G: Indeed, $\|\gamma - \gamma_0\|_{C^1} < \beta\sigma \leq \sigma$ implies that

$$\|\gamma(t) - x_0\| \leq \|\gamma_0(t) - x_0\| + \|\gamma(t) - \gamma_0(t)\| < (n+1)\sigma. \tag{11.18}$$

We are going to show that the required conclusion holds with $x = \gamma(t)$ for a suitable $t \in [0,1]^n$. This t should be outside certain sets of small measure on which the behavior of f is irregular (for example, the sets from (11.17)), but, at the same time, be such that $f(\gamma(s)) - f(\gamma(t))$ is well approximated by $T(\gamma(s) - \gamma(t))$. In order to guarantee that the points having the latter property form a set of large measure we will use Lemma 11.2.2. To this aim, we transform the function $f(\gamma(t))$ into the function $g \colon [0,1]^n \longrightarrow \mathbb{R}^n$ defined by

$$g(t) = \frac{1}{(2n+1)\sigma}(T - f)(\gamma(t)) + \frac{\beta}{n}\,\mathrm{Id}(t).$$

Let us examine the property of g. We begin by showing that

$$\operatorname{div} g(t) \geq 0 \text{ for almost every } t \in [0,1]^n. \tag{11.19}$$

In fact we show that $\operatorname{div} g(t) \geq 0$ for every t at which $f \circ \gamma$ is differentiable, and so, in particular, f is differentiable in the direction of $\gamma'(t)(\mathbb{R}^n)$. For this, notice first that $\gamma_0'(t; e_i/\sigma) = w_i$ and so

$$\|\gamma'(t; e_i/\sigma) - w_i\| \leq \frac{1}{\sigma}\|\gamma - \gamma_0\|_{C^1} < \beta.$$

Since $\operatorname{Lip}(f) \leq 1$, so $\|T\| \leq 1$, and we have

$$
\begin{aligned}
\sum_{i=1}^{n} T_i(\gamma'(t; e_i/\sigma)) &= \sum_{i=1}^{n} T_i(w_i) + \sum_{i=1}^{n} T_i\big(\gamma'(t; e_i/\sigma) - w_i\big) \\
&> M(f; w_1, \ldots, w_n) - \beta - \sum_{i=1}^{n} \|\gamma'(t; e_i/\sigma) - w_i\| \\
&> M(f; w_1, \ldots, w_n) - (n+1)\beta.
\end{aligned}
\tag{11.20}
$$

Also, by Proposition 2.4.3,

$$\sum_{i=1}^{n} f_i'(\gamma(t); \gamma'(t; e_i/\sigma)) \leq M(f; \gamma'(t; e_1/\sigma), \ldots, \gamma'(t; e_n/\sigma)),$$

and we obtain

$$
\begin{aligned}
\sum_{i=1}^{n} f_i'(\gamma(t); \gamma'(t; e_i/\sigma)) &\leq M(f; w_1, \ldots, w_n) + \sum_{i=1}^{n} \|\gamma'(t; e_i/\sigma) - w_i\| \\
&< M(f; w_1, \ldots, w_n) + n\beta.
\end{aligned}
$$

Hence

$$\operatorname{div} g(t) = \frac{1}{2n+1} \left(\sum_{i=1}^{n} T_i(\gamma'(t; e_i/\sigma)) - \sum_{i=1}^{n} f_i'(\gamma(t); \gamma'(t; e_i/\sigma)) \right) + \beta$$

$$\geq \frac{M(f; w_1, \ldots, w_n) - (n+1)\beta}{2n+1} - \frac{M(f; w_1, \ldots, w_n) + n\beta}{2n+1} + \beta$$

$$= 0.$$

Similar arguments will be used to show that

$$\gamma(t) \in D(f; w_1, \ldots, w_n; \xi) \text{ if both } \gamma(t) \in D(f) \text{ and } \operatorname{div} g(t) < \alpha. \quad (11.21)$$

Indeed, $\|f'(\gamma(t))\| \leq 1$ and $\|\gamma'(t; e_i/\sigma) - w_i\| < \beta$ imply

$$\sum_{i=1}^{n} f_i'(\gamma(t); w_i) \geq \sum_{i=1}^{n} f_i'(\gamma(t); \gamma'(t; e_i)/\sigma) - n\beta$$

$$= \sum_{i=1}^{n} T_i(\gamma'(t; e_i/\sigma)) - (2n+1) \operatorname{div} g(t) + (2n+1)\beta - n\beta.$$

Using (11.20) and $\operatorname{div} g(t) < \alpha$ we conclude

$$\sum_{i=1}^{n} f_i'(\gamma(t); w_i) > M(f; w_1, \ldots, w_n) - (2n+1)\alpha = M(f; w_1, \ldots, w_n) - \xi.$$

Let us now verify that the function g, the constant α defined above, and the set

$$N = \left\{ t \in [0,1]^n \;\middle|\; \gamma(t) \notin H \cup \left(X \setminus \bigcap_{i=1}^{n} R(f; w_i, Tw_i; \varsigma, \rho) \right) \right\} \quad (11.22)$$

satisfy the assumptions of Lemma 11.2.2.

The requirement that $\operatorname{Lip}(g) \leq 1$ follows from $\operatorname{Lip}(f) \leq 1$, $\|T\| \leq 1$, and $\beta \leq \frac{1}{12}$:

$$\operatorname{Lip}(g) \leq \frac{1}{(2n+1)\sigma} \left(\|T\| + \operatorname{Lip}(f) \right) \operatorname{Lip}(\gamma) + \frac{\beta}{n}$$

$$\leq \frac{1}{(2n+1)\sigma} 2 \operatorname{Lip}(\gamma) + \frac{\beta}{n}$$

$$\leq \frac{2}{(2n+1)\sigma} \left(\operatorname{Lip}(\gamma_0) + \|\gamma - \gamma_0\| \right) + \frac{\beta}{n}$$

$$\leq \frac{2}{(2n+1)\sigma} (n\sigma + \beta\sigma) + \frac{\beta}{n} \leq 1.$$

Further, $\mathscr{L}^n N < \tau$ because of (11.17). To estimate the integral of $\operatorname{div} g(t)$, we first invoke the divergence theorem to get

$$\int_{[0,1]^n} \operatorname{div} g(t) \, d\mathscr{L}^n(t) = \sum_{i=1}^{n} \int_{\{t \in [0,1]^n \,|\, t_i = 0\}} \left(g_i(t+e_i) - g_i(t) \right) d\mathscr{L}^{n-1}(t). \quad (11.23)$$

In order to calculate the norm $\|g(t + e_i) - g(t)\|$ we proceed in the following way.

$$\|g(t + e_i) - g(t)\|$$

$$= \left\| \frac{1}{(2n+1)\sigma} \Big((T-f)(\gamma(t+e_i)) - (T-f)(\gamma(t)) \Big) + \frac{\beta}{n} e_i \right\|$$

$$\leq \frac{1}{(2n+1)\sigma} \left\| (T-f)(\gamma_0(t+e_i)) - (T-f)(\gamma_0(t)) \right\|$$

$$+ \frac{2}{(2n+1)\sigma} \left(\|T\| + \mathrm{Lip}(f) \right) \|\gamma_0 - \gamma\|_{C^1} + \frac{\beta}{n}$$

$$\leq \frac{1}{(2n+1)\sigma} \left\| T(\sigma w_i) - f(\gamma_0(t) + \sigma w_i) + f(\gamma_0(t)) \right\| + \frac{4\|\gamma_0 - \gamma\|}{(2n+1)\sigma} + \frac{\beta}{n}$$

$$\leq \frac{1}{(2n+1)\sigma} \left\| T(\sigma w_i) - f(\gamma_0(t) + \sigma w_i) + f(\gamma_0(t)) \right\| + \beta \left(\frac{4}{2n+1} + \frac{1}{n} \right).$$

Since for $t \in \{[0,1]^n \mid t_i = 0\}$ we have $\|\gamma_0(t) - x_0\| + \|\sigma w_i\| \leq n\sigma$, we can use (11.16) to finish the estimate:

$$< \frac{1}{(2n+1)\sigma} 2\beta\sigma + \beta \left(\frac{4}{2n+1} + \frac{1}{n} \right) \leq \frac{4\beta}{n} = \frac{\alpha\tau}{n}.$$

Using this in (11.23) and recalling that $\mathrm{div}\, g(t) \geq 0$ almost everywhere, we obtain that

$$\int_{[0,1]^n \setminus N} \mathrm{div}\, g(t) \, d\mathscr{L}^n(t) \leq \int_{[0,1]^n} \mathrm{div}\, g(t) \, d\mathscr{L}^n(t) < \alpha\tau.$$

Finally, if $t \in [0,1]^n \setminus N$ and $\mathrm{div}\, g(t) < \alpha$, we infer from (11.21) that

$$\gamma(t) \in D(f; w_1, \ldots, w_n; \xi).$$

This means that $f'(\gamma(t))$ belongs to the same slice $S(f; w_1, \ldots, w_n; \xi)$ as T, and so $\|f'(\gamma(t)) - T\| \leq \tau$. Consequently,

$$\|g'(t)\| \leq \frac{1}{(2n+1)\sigma} \left\| T - f'(\gamma(t)) \right\| \|\gamma'(t)\| + \frac{\beta}{n}$$

$$\leq \frac{\tau}{(2n+1)\sigma} \left(\|\gamma_0'(t)\| + \|\gamma'(t) - \gamma_0'(t)\| \right) + \frac{\beta}{n}$$

$$\leq \frac{\tau}{(2n+1)\sigma} (n\sigma + \beta\sigma) + \frac{\beta}{n} < \frac{\tau}{2} + \frac{\tau}{4} < \tau,$$

and the last assumption of Lemma 11.2.2 is verified. Denoting

$$E = \left\{ t \in [0,1]^n \mid \|g(s) - g(t)\| > \theta \|s - t\| \text{ for some } s \in [0,1]^n \right\},$$

we thus conclude from Lemma 11.2.2 that $\mathscr{L}^n E < \theta$. We also denote

$$F = \left\{ t \in [0,1]^n \setminus N \mid \mathrm{div}\, g(t) \geq \alpha \right\}$$

and use that $\int_{[0,1]^n \setminus N} \operatorname{div} g(t) \, d\mathscr{L}^n(t) t < \alpha\tau$ to infer that

$$\mathscr{L}^n F \le \frac{1}{\alpha} \int_F \operatorname{div} g(t) \, dt < \frac{1}{\alpha} \alpha\tau = \tau \le \theta.$$

The estimates of measures of the sets E, F, and N (N is defined in (11.22) and its measure is estimated in (11.17)), together with the choice of θ, allow us to conclude that the set

$$\left[\frac{1}{4}, \frac{3}{4}\right]^n \setminus (E \cup F \cup N)$$

has positive measure. We choose t in this set and show that $x := \gamma(t)$ is the point we have been looking for.

Since $t \notin N$, we have $x \in D(f)$. From $t \notin F$ we see that $\operatorname{div} g(t) < \alpha$, and so (11.21) implies $x \in D(f; w_1, \dots, w_n; \xi)$, which is the first part of the conclusion of the lemma. Also, T was chosen in $S(f; w_1, \dots, w_n; \xi)$. The remaining statement (11.15) will follow easily from the fact that x belongs to $\bigcap_{i=1}^n R(f; w_i, Tw_i; \zeta, \rho)$ once we show that for each i, the first two of the three alternatives defining the set $R(f; w_i, Tw_i; \zeta, \rho)$ cannot occur.

The failure of condition (i) of the definition of $R(f; w_i, Tw_i; \zeta, \rho)$ follows from (11.18), which implies that

$$B(x, \rho) \subset B(x_0, \rho + (n+1)\sigma) \subset B(x_0, (n+2)\sigma) \subset G. \tag{11.24}$$

To see that condition (ii) of this definition fails as well needs a little more work. Let $1 \le i \le n$ and $|r| \, \|w_i\| < \rho$. Then $|r|/\sigma \le \frac{1}{4}$, and so $t + re_i/\sigma \in [0,1]^n$. Hence

$$\begin{aligned}
f(x + rw_i) - f(x) - T(rw_i) =&\, f(\gamma(t) + rw_i) - f(\gamma(t + re_i/\sigma)) \\
&+ T\big(\gamma(t + re_i/\sigma) - \gamma(t) - rw_i\big) \\
&- (2n+1)\sigma\big(g(t + re_i/\sigma) - g(t)\big) \\
&+ \frac{2n+1}{n}\beta r e_i.
\end{aligned} \tag{11.25}$$

We estimate each of the four terms on the right-hand side. First notice that by the mean value theorem

$$\begin{aligned}
\|\gamma(t + re_i/\sigma) &- \gamma(t) - rw_i\| \\
&= \big\|\big(\gamma(t + re_i/\sigma) - \gamma_0(t + re_i/\sigma)\big) - \big(\gamma(t) - \gamma_0(t)\big)\big\| \\
&\le \max_{s \in [0,1]} \left\|\frac{d}{ds}\big(\gamma(t + sre_i/\sigma) - \gamma_0(t + sre_i/\sigma)\big)\right\| < \beta\sigma \frac{|r|}{\sigma} \le \frac{\theta|r|}{4},
\end{aligned}$$

since $\|\gamma' - \gamma_0'\|_\infty < \beta\sigma$ and $\beta \le \theta/4$. Hence the sum of the first two terms on the right-hand side of (11.25) may be estimated by

$$\begin{aligned}
\big\|f(\gamma(t) + rw_i) &- f(\gamma(t + re_i/\sigma)) + T\big(\gamma(t + re_i/\sigma) - (\gamma(t) + rw_i)\big)\big\| \\
&\le \big(\operatorname{Lip}(f) + \|T\|\big)\|\gamma(t + re_i/\sigma) - \gamma(t) - rw_i\| \le 2\frac{\theta|r|}{4} = \frac{\theta|r|}{2}.
\end{aligned}$$

For the third term we use $t \notin E$ to get

$$(2n+1)\sigma\|(g(t+re_i/\sigma) - g(t))\| \leq (2n+1)\theta|r|,$$

and the fourth term is clearly bounded by

$$\frac{2n+1}{n}\beta|r| \leq \frac{\theta|r|}{4n} \leq \frac{\theta|r|}{4}.$$

All together

$$\|f(x+rw_i) - f(x) - T(rw_i)\|$$
$$\leq \frac{\theta|r|}{2} + (2n+1)\theta|r| + \frac{\theta|r|}{4}$$
$$\leq 2(n+1)\theta|r| \leq \zeta|r|\min\{\|w_1\|, \ldots, \|w_n\|\} \leq \zeta|r|\|w_i\|$$

for all $i = 1, \ldots, n$ and all r with $|r|\,\|w_i\| < \rho$. This shows that condition (ii) of the definition of $R(f; w_i, Tw_i; \zeta, \rho)$ fails as well.

We are now ready to finish the proof. For each $i = 1, \ldots, n$ we have that x belongs to $R(f; w_i, Tw_i; \zeta, \rho)$, but the first two alternatives that define this fail. Hence the third alternative occurs, giving that there are $\delta_i > 0$, $i = 1, \ldots, n$, such that

$$\|f(x+v+rw_i) - f(x+v) - T(rw_i)\| \leq \zeta(\|v\| + |r|\|w_i\|) \qquad (11.26)$$

for $v \in X$ and $r \in \mathbb{R}$ such that $\|v\| + |r|\|w_i\| < \delta_i$. Let

$$\delta = \min\{\rho, \delta_1, \ldots, \delta_n\}.$$

Then in view of (11.24), $B(x, \delta) \subset G$ and (11.26) implies the last required conclusion (11.15) of the lemma. $\qquad \square$

To find slices of small diameter, we will use the next proposition, which can be found in [27, Proposition 1].

Proposition 11.3.4. *Let $D \subset Z^*$ be a nonempty bounded subset of a dual space Z^* such that the w^*-closure of D is norm separable. Then for every $u \in Z$ and every $\varepsilon > 0$ there is a w^*-slice of D which is determined by a vector v with $\|v - u\| < \varepsilon$ and has diameter at most ε.*

This fact allows us to take a small slice contained in the previous slice, and so combine Lemmas 11.3.2 and 11.3.3 into results guaranteeing the existence of ε-Fréchet differentiability points in some slices.

Theorem 11.3.5. *Let $G \subset X$ be a nonempty open subset of a separable Banach space X and let $f : G \longrightarrow Y$ be a Lipschitz map to a Banach space Y of dimension not exceeding n, such that the w^*-closure of the set of Gâteaux derivatives of f is norm separable in $L(X, Y)$. Suppose further that*

- *for every $u \in X$, $y \in Y$, and $\zeta, \rho > 0$ the set $G \setminus R(f; u, y; \zeta, \rho)$ can be covered by a union of a Haar null set and a Γ_n-null G_δ set.*

Then for every $x_0 \in D(f)$, $\varepsilon, \eta > 0$, $u_1, \dots, u_m \in X$, $y_1^*, \dots, y_m^* \in Y^*$ and every Haar null set H there is a point $x \in D(f) \setminus H$ at which f is ε-Fréchet differentiable and

$$\sum_{i=1}^{m} y_i^*(f'(x; u_i)) > \sum_{i=1}^{m} y_i^*(f'(x_0; u_i)) - \eta.$$

Proof. We may assume that $Y = \mathbb{R}^n$ and $\mathrm{Lip}(f) \leq 1$. Writing the tensor $\sum_{i=1}^{m} y_i^* \otimes u_i$ as $\sum_{i=1}^{n} e_i^* \otimes \tilde{u}_i$ and denoting $\tilde{u} = (\tilde{u}_1, \dots, \tilde{u}_n)$, we see that our task is to find a point $x \in D(f, \tilde{u}, \eta) \setminus H$ at which f is ε-Fréchet differentiable. To avoid writing unnecessary constants, we will also assume that $\|\tilde{u}_i\| \leq \frac{1}{2n}$, and so also $\varepsilon, \eta \leq 1$.

It will be convenient to equip X^n, the nth Cartesian power of X, with the norm $\|z\| = \|(z_1, \dots, z_n)\| = \sum_{i=1}^{n} \|z_i\|$. Notice that

$$\{f'(x) \mid x \in D(f)\} \subset L(X, \mathbb{R}^n) \approx (X^*)^n \approx (X^n)^*,$$

where $(x_1^*, \dots, x_n^*) \in (X^*)^n$ is identified with the linear functional on X^n,

$$(x_1, \dots, x_n) \to \sum_{i=1}^{n} x_i^*(x_i).$$

Hence Proposition 11.3.4 used with $Z = X^n$ for the set $\{f'(x) \mid x \in D(f)\}$ provides us with a $v \in X^n$ such that $\|v - \tilde{u}\| < \frac{1}{4}\eta$ and, for some $0 < \chi < \frac{1}{8}\eta$,

$$\mathrm{diam}\, S(f; v; 3\chi) \leq \frac{\varepsilon}{4}.$$

With the future use of Lemma 11.3.2 in mind, we denote

$$\zeta = \frac{\varepsilon \chi}{16n^2(n+2)}$$

and use Lemma 11.3.3 to find the τ corresponding to this ζ. In order to apply this lemma, we use Proposition 11.3.4 once more to find $w \in X^n$ and $0 < \xi < \frac{1}{16}\varepsilon\chi$ such that $\|w - v\| < \chi$ and

$$\mathrm{diam}\, S(f; w; \xi) \leq \tau.$$

These choices, together with the inequality $\|w_i\| \leq \|\tilde{u}_i\| + \chi + \frac{1}{4}\eta \leq 1$, guarantee that the first assumption of Lemma 11.3.3 holds. An application of Lemma 10.4.8 to the set

$$H \cup \left(G \setminus \bigcap_{i=1}^{n} R(f; w_i, y_i; \zeta, \rho) \right)$$

implies that the second assumption of Lemma 11.3.3 holds as well. Hence there are $x \in D(f; w; \xi) \setminus H$, $T \in S(f; w; \xi)$, and $\delta > 0$ such that $B(x, \delta) \subset G$ and

$$\|f(x + v + rw_i) - f(x + v) - T(rw_i)\| \leq \zeta(\|v\| + r) \tag{11.27}$$

for each $i = 1, \ldots, n$, $v \in X$ and $r \in \mathbb{R}$ with $\|v\| + |r| < \delta$. Since $x \notin H$, g is Gâteaux differentiable at x.

Notice that Observation 11.3.1 and $\|w - v\| < \chi$ imply that

$$S(f; w; \chi) \subset S(f; v; 3\chi)$$

and hence diam $S(f; w; \chi) \leq \frac{1}{4}\varepsilon$. This is the first of the assumptions we have to verify for our intended application of Lemma 11.3.2. The remaining assumption of this lemma follows from

$$T \in S(f; w; \xi) \subset S(f; w; \tfrac{1}{16}\varepsilon\chi),$$

from the choice of ζ and from (11.27). Hence Lemma 11.3.2 is applicable, implying that f is ε-Fréchet differentiable at x.

Finally, since $\xi < \frac{1}{16}\varepsilon\chi$ and

$$\frac{\varepsilon\chi}{16} + 2(\|w - v\| + \|v - \tilde{u}\|) < \frac{\chi}{16} + 2\left(\chi + \frac{\eta}{4}\right) < \eta,$$

we see from Observation 11.3.1 that

$$x \in D(f; w; \xi) \subset D(f; w; \tfrac{1}{16}\varepsilon\chi) \subset D(f; \tilde{u}; \eta).$$

So $x \in D(f; \tilde{u}; \eta)$ which, since $x \notin H$, finishes the proof. $\qquad\square$

The main result of this section is an immediate corollary of Theorem 11.3.5.

Theorem 11.3.6. *Suppose that the Banach space X with separable dual has the property that every porous set in X can be covered by a union of a Haar null set and a Γ_n-null G_δ set. Let f be a Lipschitz map of a nonempty open set $G \subset X$ to a Banach space of dimension not exceeding n. Then for every w^*-slice S of the set of Gâteaux derivatives of f and every $\varepsilon > 0$ there is $x \in X$ at which f is Gâteaux differentiable, ε-Fréchet differentiable, and $f'(x) \in S$.*

Proof. In order to apply Theorem 11.3.5 there are only three points to notice. First, if $f \colon X \longrightarrow Y$ and $\dim Y < \infty$, the separability of the dual X^* implies the separability of $L(X, Y)$. Second, since $\dim Y \leq n$, every w^*-slice of any subset of $L(X, Y)$ has rank at most n. Third, invoke Lemma 6.2.5. $\qquad\square$

Remark 11.3.7. Once again, we point out that the strength of the results of this chapter depends on the interplay between the smoothness condition and the dimension of the target space. Nevertheless, thanks to the Haar null set in Theorem 11.3.5, one may strengthen the Gâteaux differentiability part of these results. For example, if, under the assumptions of Theorem 11.3.6, we are given a mapping g of G into an RNP space Z and an $L \in L(Z, Y)$, where $\dim Y \leq n$, we can find a point x at which g is Gâteaux differentiable, $L \circ g$ is ε-Fréchet differentiable, and $L \circ g'(x)$ belongs to any given slice of the Gâteaux derivatives of $L \circ g$.

Chapter Twelve

Fréchet differentiability of real-valued functions

We prove in this chapter that cone-monotone functions on Asplund spaces have points of Fréchet differentiability, the appropriate version of the mean value estimates holds, and, moreover, the corresponding point of Fréchet differentiability may be found outside any given σ-porous set. This is a new result which considerably strengthens known Fréchet differentiability results for real-valued Lipschitz functions on such spaces. The avoidance of σ-porous sets is new even in the Lipschitz case. To explain the new ideas, in particular the use of the variational principle, and to introduce the reader to the proofs of more special but much harder differentiability results in the next chapter, we first discuss simpler proofs of two special (already known) cases.

12.1 INTRODUCTION AND MAIN RESULTS

Our aim here is to provide a substantial strengthening of Fréchet differentiability results for real-valued Lipschitz functions of [39]. If one considers our arguments as a new proof of this result as was done in [29], one could say that the basic ideas still come from [39], but many difficulties become much easier to overcome because the differentiability point is found by direct use of a variational principle and not by a special iterative procedure, which was needed both in [39] and in the simpler proof [27].

Theorem 12.1.1. *Let $f\colon G \longrightarrow \mathbb{R}$ be a Lipschitz function defined on a nonempty open subset G of an Asplund space X. Then f has a point of Fréchet differentiability outside any given σ-porous set.*

Moreover, for any $a, b \in G$ for which the segment $[a, b]$ lies entirely in G, any σ-porous set $P \subset G$, and any $\varepsilon > 0$ there is $x \in G \setminus P$ at which f is Fréchet differentiable and $f'(x; b - a) < f(b) - f(a) + \varepsilon$.

As we have already indicated, the existence and mean value part of this theorem has a number of proofs in the literature. However, the result that the point of differentiability may be found outside any given σ-porous set is new. The only situation in which such a result was known was the case of c_0-like spaces from [28], which was improved to spaces admitting very smooth norms in Section 10.6. A weaker result, with almost Fréchet derivatives instead of Fréchet derivatives, but also avoiding a σ-porous set, may be obtained by methods developed in Chapter 11.

To state the stronger result alluded to in the beginning of the Introduction, we recall the following notion.

Definition 12.1.2. Let $\mathcal{C}$ be an open cone in X. A function f from an open subset G of X to $\mathbb{R}$ is said to be $\mathcal{C}$-monotone on G if $f(y) \geq f(x)$ whenever $x, y \in G$ and $y - x \in \mathcal{C}$.

The function f is said to be cone-monotone if there is a nonempty open cone $\mathcal{C}$ with respect to which it is $\mathcal{C}$-monotone.

The main result of this chapter is the following new differentiability result for cone-monotone functions. Its earliest predecessor is probably the result of [10] that coordinatewise monotone functions in the plane are differentiable almost everywhere. More recently, a Gâteaux differentiability result for cone-monotone functions was proved in [8].

Theorem 12.1.3. *Let $f : G \longrightarrow \mathbb{R}$ be a cone-monotone function defined on a nonempty open subset G of an Asplund space X. Then f has a point of Fréchet differentiability outside any given σ-porous set.*

More precisely, if f is $\mathcal{C}$-monotone with respect to a nonempty open cone $\mathcal{C}$, then for any $a, b \in G$ for which the segment $[a, b]$ lies entirely in G and whose direction $b - a$ belongs to the cone $\mathcal{C}$, for any σ-porous set $P \subset G$, and for any $\varepsilon > 0$ there is $x \in G \setminus P$ at which f is Fréchet differentiable and $f'(x; b - a) < f(b) - f(a) + \varepsilon$.

Theorem 12.1.3 is easily seen to imply Theorem 12.1.1: since a Lipschitz function becomes cone-monotone after adding a suitable continuous linear form x^*, the existence of a point of Fréchet differentiability follows immediately. For the additional statement, we pick x^* such that

$$x^*(b - a) > \mathrm{Lip}(f)\|b - a\|.$$

Then the function $h(x) = x^*(x) + f(x)$ is $\mathcal{C}$-monotone with respect to an open cone $\mathcal{C}$ containing $b - a$. Theorem 12.1.3 provides a point $x \in G$ at which h, and so f, is Fréchet differentiable and $h'(x; b - a) < h(b) - h(a) + \varepsilon$. It follows that

$$
\begin{aligned}
f'(x; b - a) &= h'(x; b - a) - x^*(b - a) \\
&< h(b) - h(a) + \varepsilon - x^*(b - a) \\
&= f(b) - f(a) + \varepsilon.
\end{aligned}
$$

This chapter has three main sections. The first, purely motivational, provides a simple proof of Theorem 12.1.1 in a special case only slightly more general than assuming that f is additionally everywhere Gâteaux differentiable. This very special assumption avoids many of the technical difficulties of more general cases, yet it allows us to illustrate how we intend to use the variational principle to prove Fréchet differentiability. (As explained later, it also illustrates other important points.) The key point is that if X is equipped with a Fréchet smooth norm and the function $(x, u) \mapsto f'(x; u)$, or its slight perturbation, attains its minimum at (x_0, u_0), then f is Fréchet differentiable at x_0. The function $(x, u) \mapsto f'(x; u)$ does not satisfy the assumptions of the usual perturbational variational principles. However, it satisfies the assumptions of our bimetric variational principle.

The second section treats a one-dimensional mean value problem similar to the one treated for Lipschitz functions in all existing proofs of the main statement of Theorem 12.1.1 in [39], [27], and [29]. It says, roughly: if a Lipschitz function $h\colon [a, b] \longrightarrow \mathbb{R}$ has equal values at the endpoints but is not constant, then there is a point ξ at which $h'(\xi)$ is, up to a constant multiple, as large as what would be predicted by the mean value statement, and all slopes $\frac{h(t) - h(\xi)}{t - \xi}$ are, in absolute value, majorized by a certain function of $h'(\xi)$. Although we give and use a particular formula for this function of $h'(\xi)$, its only important property is that it tends to zero as its argument approaches zero.

To treat the cone-monotone case, we need the above result also for functions h whose upper derivative is bounded from above. Here the proofs from the papers referred to above cannot be immediately repeated, since the corresponding "one-sided estimate of the maximal operator," Lemma 12.3.2, is not readily available. We therefore also provide a (very simple) proof of Lemma 12.3.2. Then we prove Lemma 12.3.3, which contains the mean value result mentioned above, and finally modify it to the somewhat technical Corollary 12.3.4 in order to avoid handling technicalities in the main part of the argument.

In the third section we prove Theorem 12.1.3. We will again minimize a suitable perturbation of the map $(x, u) \mapsto f'(x; u)$. But, unlike in the case of an everywhere Gâteaux differentiable function f, this time the domain M of the map does not have the completeness properties it had in the special case. The main point is that M is not in general topologically complete (admits no complete metric inducing its topology), since it is only an $F_{\sigma\delta}$ and not a G_δ subset of a complete metric space. To solve this difficulty, we employ an idea stemming from the descriptive set theory: since M is a Borel set, it is a one-to-one continuous image of a complete metric space Z, and so there are many ways of changing the metric of M in a natural (although *not topologically equivalent*) way so that it becomes complete. Alternately, we could have transferred the whole minimization problem to Z, in which case we would not even need the map to be one-to-one. We do not do it, since this more abstract approach does not seem to bring any additional advantages.

Of course, the remetrization of M that we want to use cannot be obtained just by the above abstract principle, but has to be chosen in a way that allows us to deduce Fréchet differentiability of f at x_0 from the assumption that the perturbed function $(x, u) \mapsto f'(x; u)$ attains its minimum at (x_0, u_0). In fact, our remetrization will possess one additional advantage, namely that the space M equipped with the new metric is complete and the function $(x, u) \mapsto f'(x; u)$ is continuous. This means that if we wish to prove only existence of points of Fréchet differentiability and the mean value statement, we can, unlike in the special case of Section 12.2, use a more classical version of the perturbational variational principle. However, we still have to use our general version of the variational principle because we need a very special choice of perturbation functions, even nonsymmetric ones, that allow some delicate estimates not only at the point where the minimum is attained but also at the approximating points. To prove the full statement of our results, that is, to find the required differentiability point outside the given σ-porous set, we again use the bimetric variant of the variational principle.

It turns out that the remetrization argument needed to prove Theorem 12.1.3 is considerably more involved than the one used to prove Theorem 12.1.1 (which can be found in [29]). We therefore give two remetrization arguments, one suitable for the case when f is Lipschitz and the other when f is cone-monotone. The rest of the argument is written so that it is possible to skip the latter case completely and read the whole proof as showing Theorem 12.1.1 only.

12.2 AN ILLUSTRATIVE SPECIAL CASE

To demonstrate the variational approach we prove the following theorem. Its special case, existence of points of Fréchet differentiability for everywhere Gâteaux differentiable Lipschitz functions on Asplund spaces, was originally proved in [38] and was the first general infinite dimensional Fréchet differentiability result for Lipschitz functions. This special case was also used in [29] to illustrate the variational technique. Readers wishing to have only a similar illustration may assume that in the assumptions of Theorem 12.2.1 the set $E = X$. In view of Proposition 2.4.1, this would give a statement only slightly stronger than the already mentioned special case. We have, however, decided to present a more general result, since it does not need any significant change in the argument and it illustrates two other points:

- The role of descriptive complexity of the set of points of Gâteaux or directional differentiability. For a Lipschitz function on a separable space this set is $F_{\sigma\delta}$; but if it happens to be a G_δ set or at least to contain a sufficiently large G_δ set, Theorem 12.2.1 already provides points of Fréchet differentiability.

- The role of the mean value estimate, since it is the validity of this estimate and not the size of the set of points of differentiability that makes the result true.

Theorem 12.2.1. *Let X be an Asplund space and $f\colon X \longrightarrow \mathbb{R}$ be Lipschitz. Suppose further that there is a G_δ set $E \subset X$ such that the following condition holds.*

(D) *The directional derivative of f exists at every point of E and every direction, and satisfies the mean value estimate: for every $a, b \in X$, every open set $G \supset [a, b]$, and every $\varepsilon > 0$ there is $x \in G \cap E$ such that $f'(x; b - a) > f(b) - f(a) - \varepsilon$.*

Then f has points of Fréchet differentiability.

By following the proof, the reader may notice that we could have assumed that f is defined only on a nonempty open subset G of X, and that the statement is actually stronger: the mean value estimate holds also with the set E_0 of points of E at which f is Fréchet differentiable.

Besides the variational principle, which is its main ingredient, the proof requires two additional observations.

Observation 12.2.2. *Suppose that $\Theta\colon X \longrightarrow \mathbb{R}$ is upper Fréchet differentiable, ψ is a continuous real-valued function on X, and $f\colon X \longrightarrow \mathbb{R}$ is Lipschitz and satisfies, for* *some $E \subset X$, the condition* (**D**). *Suppose further that the function* $h: E \times X \longrightarrow \mathbb{R}$ *given as*

$$h(x, u) = f'(x; u) + \Theta(u) + \psi(x)$$

attains its minimum at (x_0, u_0). *Then* f *is Fréchet differentiable at* x_0.

Proof. Although it is not necessary, we first notice that after fixing the variable $x = x_0$ in $h(x, u)$, the resulting function of u is upper Fréchet differentiable and attains its minimum at $u = u_0$. Hence its derivative is zero, giving that $f'(x_0) + T = 0$, where $T \in X^*$ is the upper Fréchet derivative of Θ at u_0. So it should be no surprise that the formulas below actually show that the Fréchet derivative of f at x_0 is equal to $-T$.

Let $\varepsilon > 0$ and find $\Delta > 0$ such that

$$\Theta(u) - \Theta(u_0) \le T(u - u_0) + \frac{\varepsilon}{4} \|u - u_0\|$$

for $\|u - u_0\| \le \Delta$. By continuity of ψ, there is $\delta_0 > 0$ such that

$$|\psi(x) - \psi(x_0)| < \frac{\varepsilon \Delta}{4}$$

for $x \in E$, $\|x - x_0\| \le \delta_0$. Finally, let $0 < \delta < \delta_0 / (1 + \|u_0\|/\Delta)$ be such that

$$\left| f(x_0 + tu_0) - f(x_0) - f'(x_0; tu_0) \right| \le \frac{\varepsilon \Delta}{4} |t| \tag{12.1}$$

for $|t| \le \delta / \Delta$.

Assume that $v \in X$ with $\|v\| < \delta$. We denote for a moment $t = \|v\|/\Delta$ and $u = \frac{1}{t} v + u_0$. Since the segment $[x_0 + v, x_0 - tu_0]$ lies in the ball

$$B(x_0, \max\{\|v\|, t\|u_0\|\}) \subset B(x_0, \delta_0),$$

the mean value assumption implies that there is $x \in E$ such that $\|x - x_0\| < \delta_0$ and

$$f'(x; tu) = f'(x; v + tu_0) < f(x_0 + v) - f(x_0 - tu_0) + \frac{\varepsilon}{4} \|v\|. \tag{12.2}$$

Since $h(x, u)$ attains its minimum at (x_0, u_0), we have

$$f'(x; u) + \Theta(u) + \psi(x) \ge f'(x_0; u_0) + \Theta(u_0) + \psi(x_0).$$

Hence noticing that $\|u - u_0\| = \Delta$ we get

$$\begin{aligned}
f'(x; u) &\ge f'(x_0; u_0) - (\Theta(u) - \Theta(u_0)) - (\psi(x) - \psi(x_0)) \\
&\ge f'(x_0; u_0) - T(u - u_0) - \frac{\varepsilon}{4} \|u - u_0\| - \frac{\varepsilon \Delta}{4} \\
&= f'(x_0; u_0) - \frac{1}{t} Tv - \frac{\varepsilon}{4t} \|v\| - \frac{\varepsilon}{4t} \|v\| \\
&= f'(x_0; u_0) - \frac{1}{t} Tv - \frac{\varepsilon}{2t} \|v\|.
\end{aligned}$$

Together with (12.2) we obtain

$$f(x_0 + v) - f(x_0 - tu_0) > tf'(x; u) - \frac{\varepsilon}{4} \|v\| \geq f'(x_0; tu_0) - Tv - \frac{3\varepsilon}{4} \|v\|. \quad (12.3)$$

Employing the approximation of the derivative (12.1) and $|t| \leq \delta/\Delta$, we conclude that

$$f(x_0 - tu_0) - f(x_0) \geq -f'(x_0; tu_0) - \frac{\varepsilon\Delta}{4} t = -f'(x_0; tu_0) - \frac{\varepsilon}{4} \|v\|.$$

Adding this to (12.3), we get

$$f(x_0 + v) - f(x_0) \geq -Tv - \varepsilon\|v\|.$$

To obtain the upper estimate of this increment, we proceed in a completely symmetric way. Let t be as above but this time we let $u = -\frac{1}{t}v + u_0$. The mean value assumption provides us with a point $x \in E$ such that $\|x - x_0\| < \delta_0$ and

$$f'(x; tu) = f'(x; tu_0 - v) < f(x_0 + tu_0) - f(x_0 + v) + \frac{\varepsilon}{4} \|v\|.$$

Using again that

$$f'(x; u) + \Theta(u) + \varphi(x) \geq f'(x_0; u_0) + \Theta(u_0) + \varphi(x_0),$$

we get

$$f'(x; u) \geq f'(x_0; u_0) - (\Theta(u) - \Theta(u_0)) - (\psi(x) - \psi(x_0))$$
$$\geq f'(x_0; u_0) - T(u - u_0) - \frac{\varepsilon}{4} \|u - u_0\| - \frac{\varepsilon\Delta}{4}$$
$$= f'(x_0; u_0) + \frac{1}{t} Tv - \frac{\varepsilon}{2t} \|v\|.$$

So

$$f(x_0 + tu_0) - f(x_0 + v) \geq tf'(x; u) - \frac{\varepsilon}{4} \|v\| \geq f'(x_0, tu_0) + Tv - \frac{3\varepsilon}{4} \|v\|.$$

Subtracting this from

$$f(x_0 + tu_0) - f(x_0) \leq f'(x_0; tu_0) + \frac{\varepsilon\Delta}{4} t$$

we get

$$f(x_0 + v) - f(x_0) \leq -Tv + \frac{3\varepsilon}{4} \|v\| + \frac{\varepsilon\Delta}{4} t = -Tv + \varepsilon\|v\|. \qquad \square$$

The observation below represents the specific feature of our approach. The directional derivative $f'(x; u)$ is not continuous as a function in two variables x, u, but it is (d, d_0)-continuous for a suitable choice of metrics d and d_0.

Observation 12.2.3. *Let X be a Banach space, $f\colon X \longrightarrow \mathbb{R}$ a Lipschitz function, and $M \subset X \times X$ such that $f'(x; u)$ exists for every $(x, u) \in M$. Then the function*

$$(x, u) \to f'(x; u)$$

is (d, d_0)-continuous on M, where d is the metric

$$d((x, u), (y, v)) = \sqrt{\|x - y\|^2 + \|u - v\|^2},$$

and d_0 is the pseudometric

$$d_0((x, u), (y, v)) = \|x - y\|.$$

Proof. We define the following strategies $\delta_j \colon M^{j+1} \longrightarrow (0, \infty)$. Let $\delta_0(x_0, u_0) = 1$. Given $j \geq 1$ and the pairs $(x_0, u_0), \ldots, (x_j, u_j)$, we find

$$0 < \delta \leq \tfrac{1}{2}\delta_{j-1}\big((x_0, u_0), \ldots, (x_{j-1}, u_{j-1})\big)$$

such that

$$\big|f(x_j + tu_j) - f(x_j) - f'(x_j; tu_j)\big| \leq \frac{|t|}{j}$$

whenever $|t| \leq j\delta$. Then we put $\delta_j\big((x_0, u_0), \ldots, (x_j, u_j)\big) = \delta$.

We have to show that $f'(x_j; u_j)$ converge to $f'(x; u)$ whenever $(x_j, u_j) \in M$ d-converge to $(x, u) \in M$ and

$$\|x_{j+1} - x_j\| \leq \delta_j\big((x_0, u_0), \ldots, (x_j, u_j)\big).$$

To simplify the notation we let $\delta_j = \delta_j\big((x_0, u_0), \ldots, (x_j, u_j)\big)$.

Since $\delta_{j+1} \leq \tfrac{1}{2}\delta_j$, we have that $\|x - x_j\| \leq 2\delta_j$ and also $j\delta_j \to 0$. Let $\varepsilon > 0$. One can find $j \in \mathbb{N}$ such that

$$\mathrm{Lip}(f)\|u - u_j\| < \varepsilon, \qquad \frac{1 + 4\,\mathrm{Lip}(f)}{j} < \varepsilon,$$

and

$$\big|f(x + tu) - f(x) - f'(x; tu)\big| \leq \varepsilon|t| \ \text{ for } |t| \leq j\delta_j.$$

Let $t \in \mathbb{R}$ with $|t| = j\delta_j$. Then we obtain the following estimate

$$|f'(x; tu) - f'(x_j; tu_j)|$$

$$\leq \big|f(x + tu) - f(x) - f(x_j + tu_j) + f(x_j)\big| + \varepsilon|t| + \frac{|t|}{j}$$

$$\leq |f(x + tu) - f(x_j + tu_j)| + |f(x_j) - f(x)| + \varepsilon|t| + \frac{|t|}{j}$$

$$\leq 2\,\mathrm{Lip}(f)\,\|x - x_j\| + \mathrm{Lip}(f)\,\|u - u_j\|\,|t| + \varepsilon|t| + \frac{|t|}{j}$$

$$\leq \left(\frac{4\,\mathrm{Lip}(f)\delta_j}{|t|} + \mathrm{Lip}(f)\,\|u - u_j\| + \varepsilon + \frac{1}{j}\right)|t|$$

$$= \left(\frac{1 + 4\,\mathrm{Lip}(f)}{j} + \mathrm{Lip}(f)\,\|u - u_j\| + \varepsilon\right)|t|$$

$$\leq 3\varepsilon|t|. \qquad\qquad \square$$

Proof of Theorem 12.2.1. Since the dual space X^* is separable, the space X admits a Fréchet smooth norm $\| \cdot \|$, see Theorem 3.2.1. We plan to use the variational principle of Theorem 7.3.5 in the metric space $M = E \times X$ equipped with the metric d,

$$d((x, u), (y, v)) = \sqrt{\|x - y\|^2 + \|u - v\|^2},$$

and with the pseudometric d_0,

$$d_0((x, u), (y, v)) = \|x - y\|.$$

Since $(X \times X, d)$ is complete and M is a G_δ subset of $(X \times X, d_0)$, (M, d) is (d, d_0)-complete by Lemma 7.3.4.

We choose the functions

$$F_j((x, u), (y, v)) := 2^{-j} d^2((x, u), (y, v)) = 2^{-j}(\|x - y\|^2 + \|u - v\|^2)$$

and constants $r_j = 2^{-j}$, $j \geq 0$. Then clearly

$$\inf\left\{ F_j((x, u), (y, v)) \mid d((x, u), (y, v)) > r_j \right\} > 0.$$

To apply Theorem 7.3.5 to find a suitable minimum attaining perturbation of the function

$$g(x, u) := f'(x; u) + \|u\|^2,$$

we still need to check the remaining assumptions. First, Observation 12.2.3 guarantees that the function g is (d, d_0)-continuous. It is also bounded from below, since $g(x, u) \geq -\operatorname{Lip}(f)\|u\| + \|u\|^2$; this was the reason for adding $\|u\|^2$. The choice of the starting point and of the parameters ε_j controlling the speed of convergence is irrelevant in our situation, with the exception of the case $j = 0$ when we have an assumption to verify. Thus we set, for example, $\varepsilon_j = 2^{-j}$ and we find the starting point (x_0, u_0) such that

$$g(x_0, u_0) < \varepsilon_0 + \inf_{(x, u) \in M} g(x, u).$$

Then Theorem 7.3.5 provides us with a sequence of pairs (x_j, u_j) converging to some (x_∞, u_∞) and a d_0-continuous function $\varphi \colon G \times X \longrightarrow \mathbb{R}$ such that the function

$$h(x, u) = f'(x, u) + \|u\|^2 + \varphi(x, u) + \sum_{j=0}^{\infty} 2^{-j}(\|x - x_j\|^2 + \|u - u_j\|^2)$$

attains its minimum at (x_∞, u_∞). Notice that d_0-continuity of φ means that the function φ depends only on the variable x. Finally, an appeal to Observation 12.2.2 with $\psi(x) = \varphi(x) + \sum_{j=0}^{\infty} 2^{-j}\|x - x_j\|^2$ and

$$\Theta(u) = \|u\|^2 + \sum_{j=0}^{\infty} 2^{-j}\|u - u_j\|^2$$

gives that f is Fréchet differentiable at x_∞. $\qquad\square$

12.3   A MEAN VALUE ESTIMATE

We start with a slightly more general but easy version of the classical one-dimensional mean value theorem, which we include here only for the sake of completeness.

Let $h\colon [a,b] \longrightarrow \mathbb{R}$ be a function. We denote the upper (resp. lower) derivative of h at $t \in [a,b]$ by

$$\overline{D}h(t) = \limsup_{\substack{s \to t \\ s \in [a,b]}} \frac{h(s) - h(t)}{s - t} \quad \left(\text{resp. } \underline{D}h(t) = \liminf_{\substack{s \to t \\ s \in [a,b]}} \frac{h(s) - h(t)}{s - t} \right).$$

Lemma 12.3.1. *Let* $h\colon [a,b] \longrightarrow \mathbb{R}$ *be a function such that* $\underline{D}h(t) \geq 0$ *on* $[a,b]$. *Then* h *is increasing on* $[a,b]$.

Proof. It is clearly sufficient to show that, for every $\varepsilon > 0$, the function $g(t) = h(t) + \varepsilon t$ is increasing. Suppose it is not the case. Then there are $a_1 < b_1$ in $[a,b]$ such that $g(a_1) > g(b_1)$. Let ξ be the midpoint of $[a_1, b_1]$ and denote by $[a_2, b_2]$ one of the intervals $[a_1, \xi]$ and $[\xi, b_1]$ at the endpoints of which the function g satisfies

$$g(a_1) > g(\xi), \text{ or } g(\xi) > g(b_1).$$

Proceeding in the above manner we obtain a decreasing sequence of intervals $[a_n, b_n]$ and a point $s \in \bigcap_n [a_n, b_n]$. Then either $g(s) - g(a_n) < 0$ for infinitely many n or $g(b_n) - g(s) < 0$ for infinitely many n. In both cases we obtain that $\underline{D}g(s) \leq 0$ which contradicts the fact that $\underline{D}g(s) \geq \varepsilon$. $\square$

We will also need a certain variant of standard estimates of maximal operators. For Lipschitz h this result can be easily deduced from the weak type $(1,1)$ inequality for the Hardy-Littlewood maximal operator (applied to the derivative of h). In this inequality as well as in a number of other estimates we tend to give numerical constant. Although their particular value plays no role in what follows, it is hoped that this will make it easier to track the source of particular estimates. However, no effort was made to make these constants optimal. For example, the reader may easily improve the constant 10 in the following lemma to 4 just by using a more efficient covering theorem.

Lemma 12.3.2. *Let* $h\colon [a,b] \longrightarrow \mathbb{R}$ *be a function such that* $h(a) = h(b) = 0$ *and* $\sup_{t \in [a,b]} \overline{D}h(t) < \infty$. *Denote for* $t \in [a,b]$

$$H(t) = \sup \left\{ \frac{|h(\beta) - h(\alpha)|}{\beta - \alpha} \;\middle|\; a \leq \alpha \leq t \leq \beta \leq b, \; \alpha < \beta \right\}.$$

Then for every $\lambda > 0$,

$$\mathscr{L}^1 \{ t \in [a,b] \mid H(t) > \lambda \} \leq \frac{10}{\lambda} \int_a^b \max\{0, h'(t)\} \, dt.$$

Proof. Denote $\kappa = \sup_{t \in [a,b]} \overline{D}h(t)$ and observe that by Lemma 12.3.1 the function $\kappa t - h(t)$ is increasing. (In fact this is an equivalent way of expressing the condition that $\sup_{t\in[a,b]}\overline{D}h(t)=\kappa<\infty$.) Consequently h is differentiable a.e., its derivative is Lebesgue integrable, and

$$h(\beta)-h(\alpha)\le\int_\alpha^\beta h'(t)\,dt$$

for any $\alpha,\beta\in[a,b]$, $\alpha<\beta$.

Let $\lambda>0$ and $A=\{t\in[a,b]\mid H(t)>\lambda\}$. For each $t\in A$ choose an interval $[\alpha_t,\beta_t]\subset[a,b]$ containing t such that

either $h(\beta_t)-h(\alpha_t)>\lambda(\beta_t-\alpha_t)$ or $h(\beta_t)-h(\alpha_t)<-\lambda(\beta_t-\alpha_t)$.

Denote by A_0 the set of $t\in A$ for which the former case occurred and let $A_1=A\setminus A_0$. We distinguish two cases.

Assume that $\mathscr{L}^1A_0\ge\frac12\mathscr{L}^1A$. We apply the $5r$-covering theorem to the system $\mathcal{V}_0=\{[\alpha_t,\beta_t]\mid t\in A_0\}$ to find intervals $[\alpha_i,\beta_i]\in\mathcal{V}_0$, $i=1,\dots,m$, such that the intervals (α_i,β_i) are pairwise disjoint and $\mathscr{L}^1A_0\le5\sum_{i=1}^m(\beta_i-\alpha_i)$. Hence

$$\lambda\mathscr{L}^1A\le10\lambda\sum_{i=1}^m(\beta_i-\alpha_i)\le10\sum_{i=1}^m(h(\beta_i)-h(\alpha_i))$$

$$\le10\sum_{i=1}^m\int_{\alpha_i}^{\beta_i}\max\{0,h'(t)\}\,dt\le10\int_a^b\max\{0,h'(t)\}\,dt.$$

In the case $\mathscr{L}^1A_1\ge\frac12\mathscr{L}^1A$, we use the $5r$-covering theorem, this time for the system $\mathcal{V}_1=\{[\alpha_t,\beta_t]\mid t\in A_1\}$, to find intervals $[\alpha_i,\beta_i]\in\mathcal{V}_1$, $i=1,\dots,n$, such that the intervals (α_i,β_i) are pairwise disjoint and $\mathscr{L}^1A_1\le5\sum_{i=1}^n(\beta_i-\alpha_i)$. Let (c_j,d_j), $j=1,\dots,k$, be the components of $(a,b)\setminus\bigcup_{i=1}^n[\alpha_i,\beta_i]$. Since $h(a)=h(b)=0$,

$$\sum_{i=1}^n(h(\beta_i)-h(\alpha_i))=\sum_{j=1}^k(h(d_j)-h(c_j)),$$

and we obtain

$$\lambda\mathscr{L}^1A\le10\lambda\sum_{i=1}^n(\beta_i-\alpha_i)\le-10\sum_{i=1}^n(h(\beta_i)-h(\alpha_i))$$

$$=10\sum_{j=1}^k(h(d_j)-h(c_j))\le10\sum_{j=1}^k\int_{c_j}^{d_j}\max\{0,h'(t)\}\,dt$$

$$\le10\int_a^b\max\{0,h'(t)\}\,dt.\qquad\square$$

Lemma 12.3.3. *Let $h\colon[a,b]\longrightarrow\mathbb{R}$ be such that $h(b)=h(a)=0$. Assume that $\kappa:=\sup_{t\in[a,b]}\overline{D}h(t)<\infty$. Then there is a set $S\subset[a,b]$ such that $3\kappa\,\mathscr{L}^1S\ge\|h\|_\infty$ and for every $\xi\in S$,*

(i) h is differentiable at ξ;

(ii) $h'(\xi) \geq \dfrac{\|h\|_\infty}{3(b-a)}$;

(iii) $|h(t) - h(\xi)| \leq 60\sqrt{\kappa h'(\xi)}|t - \xi|$ for every $t \in [a, b]$.

Proof. We may assume that $\kappa > 0$; otherwise Lemma 12.3.1 gives that $h \equiv 0$. More-over, the same lemma implies the monotonicity of $\kappa t - h(t)$; hence h is differentiable a.e. in $[a, b]$. Recall also that by Lemma 12.3.2 the function

$$H(t) = \sup\left\{\frac{|h(\beta) - h(\alpha)|}{\beta - \alpha} \,\Big|\, a \leq \alpha \leq t \leq \beta \leq b, \, \alpha < \beta\right\}$$

satisfies the maximal operator estimate

$$\mathscr{L}^1\{t \in [a, b] : H(t) > \lambda\} \leq \frac{10}{\lambda} \int_a^b \max\{0, h'(t)\} \, dt.$$

To show the statement of the lemma, denote by $S \subset [a, b]$ the set of all points $\xi \in (a, b)$ for which (i)–(iii) hold and suppose to the contrary that the measure $\mathscr{L}^1 S < \frac{1}{3\kappa}\|h\|_\infty$. Then for a.e. $t \in [a, b] \setminus S$, either

$$h'(t) < \frac{\|h\|_\infty}{3(b-a)} \quad \text{or} \quad h'(t) \geq \frac{\|h\|_\infty}{3(b-a)} \text{ and } H(t) > 60\sqrt{\kappa h'(t)}.$$

Recalling also that $h' \leq \kappa$, we see that for a.e. $t \in [a, b]$,

$$\max\{0, h'(t)\} < \max\{0, h'(t)\}\mathbf{1}_S(t) + \frac{\|h\|_\infty}{3(b-a)} + \min\left\{\kappa, \frac{1}{3600\kappa}H^2(t)\right\},$$

where $\mathbf{1}_S$ denotes the indicator function of S. Hence we get a contradiction by esti-mating

$$\int_a^b \max\{0, h'(t)\} \, dt < \int_S \max\{0, h'(t)\} \, dt + \int_a^b \frac{\|h\|_\infty}{3(b-a)} \, dt$$

$$+ \int_a^b \min\left\{\kappa, \frac{1}{3600\kappa}H^2(t)\right\} dt$$

$$\leq \kappa\mathscr{L}^1 S + \frac{1}{3}\|h\|_\infty + \int_0^\kappa \mathscr{L}^1\left\{t \in [a, b] \,\Big|\, \frac{1}{3600\kappa}H^2(t) > \lambda\right\} d\lambda$$

$$\leq \frac{2}{3}\|h\|_\infty + \int_0^\kappa \frac{10}{60\sqrt{\kappa\lambda}} \, d\lambda \int_a^b \max\{0, h'(t)\} \, dt$$

$$\leq \frac{2}{3}\int_a^b \max\{0, h'(t)\} \, dt + \frac{20}{60}\int_a^b \max\{0, h'(t)\} \, dt$$

$$= \int_a^b \max\{0, h'(t)\} \, dt. \qquad \square$$

Corollary 12.3.4. *Let* $0 \leq \tau < \kappa < \infty$, $0 < a \leq \delta < \infty$. *Suppose further that* $g: [-2\delta, 2\delta] \longrightarrow \mathbb{R}$ *is such that* $\sup_{t \in [-a,a]} \overline{D}g(t) \leq \kappa$, $g(0) > 13\tau a$, *and* $|g(t)| \leq \frac{1}{\kappa}\tau^2|t|$ *for* $a \leq |t| \leq 2\delta$. *Then there is* $\xi \in (-a, a) \setminus \{0\}$ *such that*

(i) *g is differentiable at ξ;*

(ii) $\kappa \geq g'(\xi) \geq \tau$;

(iii) $|g(t) - g(\xi)| \leq 123\sqrt{\kappa g'(\xi)}\,|t - \xi|$ *for every* $t \in [-2\delta, 2\delta]$.

Proof. Define $h: [-a, a] \longrightarrow \mathbb{R}$ by

$$h(t) = g(t) - \frac{a - t}{2a}g(-a) - \frac{a + t}{2a}g(a).$$

We apply Lemma 12.3.3 to the function h with $\kappa_0 := \sup_{t \in [-a,a]} \overline{D}h(t)$ to get a subset $S \subset [-a, a]$ such that $3\kappa_0 \mathscr{L}^1 S \geq \|h\|_\infty$ and for every $\xi \in S$,

(a) *h is differentiable at ξ;*

(b) $h'(\xi) \geq \dfrac{\|h\|_\infty}{6a}$;

(c) $|h(t) - h(\xi)| \leq 60\sqrt{\kappa_0 h'(\xi)}|t - \xi|$ *for every* $t \in [-a, a]$.

In order to see what (a), (b), and (c) imply for the original function g, observe that

$$\|h - g\|_\infty = \max\{|g(a)|, |g(-a)|\} \leq \frac{\tau^2}{\kappa}a \leq \tau a.$$

Also, for $t, s \in [-a, a]$,

$$|(g(t) - g(s)) - (h(t) - h(s))| = \frac{|t - s|}{2a}|g(a) - g(-a)| \leq \tau|t - s|. \qquad (12.4)$$

In particular, we have $\|h\|_\infty \geq \|g\|_\infty - \tau a \geq |g(0)| - \tau a > 13\tau a - \tau a = 12\tau a$, and $0 < \kappa_0 \leq \kappa + \tau \leq 2\kappa$. It follows that

$$\mathscr{L}^1\Big([-a, a] \setminus [-a + \tau a/\kappa, a - \tau a/\kappa]\Big) = \frac{2\tau a}{\kappa} < \frac{\|h\|_\infty}{6\kappa} \leq \frac{\|h\|_\infty}{3\kappa_0}.$$

Thus we can find a point $\xi \in S \cap [-a + \tau a/\kappa, a - \tau a/\kappa] \setminus \{0\}$. Since $h'(\xi)$ exists, $g'(\xi)$ exists as well. By (12.4) and (b) we also have

$$g'(\xi) \geq h'(\xi) - \tau \geq \frac{\|h\|_\infty}{6a} - \tau \geq \tau,$$

which, together with $g'(\xi) \leq \overline{D}g(\xi) \leq \kappa$, gives (ii). Moreover, it also gives that

$$h'(\xi) \leq g'(\xi) + \tau \leq 2g'(\xi).$$

Let $t \in [-a, a]$. Then by (12.4) and (c) we obtain

$$
\begin{aligned}
|g(t) - g(\xi)| &\leq |h(t) - h(\xi)| + \tau \, |t - \xi| \\
&\leq \left(60 \sqrt{\kappa_0 h'(\xi)} + \sqrt{\kappa \tau} \right) |t - \xi| \\
&\leq \left(60 \sqrt{2\kappa \, 2g'(\xi)} + \sqrt{\kappa g'(\xi)} \right) |t - \xi| \\
&= 121 \sqrt{\kappa g'(\xi)} |t - \xi|.
\end{aligned}
\tag{12.5}
$$

Hence (iii) holds for $t \in [-a, a]$, and it remains to show only that it holds also for $t \in [-2\delta, 2\delta] \setminus [-a, a]$. Let s be the (unique) point of $(\xi, t) \cap \{-a, a\}$. Observing that $\xi \in [-a + \tau a / \kappa, a - \tau a / \kappa]$, we have $|s - \xi| \geq \tau |s| / \kappa$ and so

$$
|t - \xi| = |t - s| + |s - \xi| \geq \frac{\tau}{\kappa} |t - s| + \frac{\tau}{\kappa} |s| = \frac{\tau}{\kappa} |t|.
$$

Hence

$$
\begin{aligned}
|g(t) - g(s)| = |g(t) - g(s)| &\leq \frac{\tau^2}{\kappa} (|t| + |s|) \leq \frac{2\tau^2}{\kappa} |t| \\
&\leq 2\tau |t - \xi| \leq 2 \sqrt{\kappa g'(\xi)} |t - \xi|.
\end{aligned}
$$

Since (12.5) implies $|g(s) - g(\xi)| \leq 121 \sqrt{\kappa g'(\xi)} |s - \xi| \leq 121 \sqrt{\kappa g'(\xi)} |t - \xi|$, we get the required

$$
|g(t) - g(\xi)| \leq |g(t) - g(s)| + |g(s) - g(\xi)| \leq 123 \sqrt{\kappa g'(\xi)} |t - \xi|. \qquad \square
$$

12.4   PROOF OF THEOREMS 12.1.1 AND 12.1.3

We will prove the following variant of Theorems 12.1.1 and 12.1.3. The main differences are that we have replaced the mean value statement by a simpler one and the assumption that X be Asplund by existence of a suitably smooth bump function. In fact, our arguments could use even slightly weaker assumptions, but in view of the separable reduction arguments, this would not add anything new. However, as we will explain at the end of this chapter, it may be useful in studying other derivatives.

Although, as we have seen, the Lipschitz case follows immediately from the cone-monotone one, it can be proved by an easier argument, and so we will continue treating both cases so that readers interested only in differentiability of Lipschitz functions may read our arguments as a (relatively simple) proof of this case only.

Proposition 12.4.1. *Assume that a Banach space X admits a Lipschitz and everywhere upper Fréchet differentiable bump function. Let $f : G \longrightarrow \mathbb{R}$ be a function defined on a nonempty open subset G of X which is either Lipschitz or cone-monotone. Then f has a point of Fréchet differentiability outside any given σ-porous set.*

More precisely, if P is σ-porous in G, $x_0 \in G$ and $u_0 \in X$ are such that $f'(x_0; u_0)$ exists, and, in the cone-monotone case, u_0 belongs to a cone with respect to which f is monotone, then for every $\varepsilon > 0$ and $r_0 > 0$ one may find a point $x \in G \setminus P$ and a vector $u \in B(u_0, r_0)$ such that f is Fréchet differentiable at x and $f'(x; u) < f'(x_0; u_0) + \varepsilon$.

The existence of a bump function with the properties required in this proposition follows from the assumptions of our main results. (Recall that for us Asplund spaces are separable; of course, the differentiability results may be easily extended to nonseparable Asplund spaces by separable reduction.)

The first statements of Theorems 12.1.1 and 12.1.3 follow immediately from this. For the second statement, we let $u_0 = b - a$ and notice that the function $t \mapsto f(a + tu_0)$ is almost everywhere differentiable on $[0, 1]$, its derivative is Lebesgue integrable, and

$$\int_0^1 \tfrac{d}{dt} f(a + tu_0)\, dt \le f(b) - f(a).$$

Hence there is a point $x_0 \in [a, b]$ such that $f'(x_0; u_0)$ exists and

$$f'(x_0; u_0) \le f(b) - f(a).$$

To finish the proof of Theorem 12.1.1 we use Proposition 12.4.1 with ε replaced by $\tfrac{1}{2}\varepsilon$ and any $r_0 > 0$ such that $r_0 \operatorname{Lip}(f) < \tfrac{1}{2}\varepsilon$. Then the point x and vector u from Proposition 12.4.1 satisfy

$$f'(x; u_0) \le f'(x; u) + \operatorname{Lip}(f)\|u - u_0\| < f'(x_0; u_0) + \varepsilon \le f(b) - f(a) + \varepsilon,$$

which is the required mean value statement.

For Theorem 12.1.3 we choose $0 < \eta < 1$ with $\eta f'(x_0; u_0) < \tfrac{1}{3}\varepsilon$ and use Proposition 12.4.1, with ε replaced by $\tfrac{1}{3}\varepsilon$ and any $r_0 > 0$ such that the ball $B(u_0, (1+\eta)r_0/\eta)$ is contained in a cone $\mathcal{C}$ with respect to which f is monotone. Then for the points $x \in G$ and $u \in B(u_0, r_0)$ obtained from the proposition the vector $v := u + \tfrac{1}{\eta}(u - u_0)$ belongs to $B(u_0, (1+\eta)r_0/\eta) \subset \mathcal{C}$. Hence for small enough $t > 0$,

$$f(x + tu_0) - f(x) \le f(x + tu_0 + \eta tv) - f(x) = f(x + (1+\eta)tu) - f(x),$$

which gives that

$$f'(x; u_0) \le (1 + \eta)f'(x; u) < (1 + \eta)\left(f'(x_0; u_0) + \frac{\varepsilon}{3}\right)$$
$$< f'(x_0; u_0) + \varepsilon \le f(b) - f(a) + \varepsilon.$$

One could join the two cases treated in Proposition 12.4.1 into one by introducing the notion of functions Lipschitz with respect to a nonempty open cone $\mathcal{C}$ by requiring that there be $C < \infty$ such that

$$f(y) - f(x) \ge -C\|y - x\|$$

whenever $x, y \in G$ and $y - x \in \mathcal{C}$. However, since such functions become cone-monotone (with respect to a possibly different cone) after we add a suitable linear functional, the Fréchet differentiability results for them follow from Theorem 12.1.3 by the same easy argument as for Lipschitz functions. We will therefore not use this formally more general concept.

The rest of this section is devoted to the proof of Proposition 12.4.1. Without loss of generality, we shall assume that $\|u_0\| = 1$ and, recalling Lemma 8.2.5, we fix a Lipschitz, everywhere upper Fréchet differentiable function $\Theta\colon X \longrightarrow [0, 1]$ such that

$$\inf_{\|z\|>s} \Theta(z) > 0 \text{ for every } s > 0.$$

We will also assume that $r_0 < \frac{1}{6}$ (in the Lipschitz case we will use only that $r_0 < 1$), $B(x_0, 3r_0) \subset G$, and, in the cone-monotone case, $B(u_0, 3r_0)$ is contained in a cone $\mathcal{C}$ with respect to which f is monotone.

The space M

The basic idea of the proof of Proposition 12.4.1 is to minimize a suitable perturbation of the function $(x, u) \mapsto f'(x; u)$ on a suitable set

$$M \subset \left\{ (x, u) \in G \times X \mid f'(x; u) \text{ exists} \right\}.$$

Most of the differences between the Lipschitz and cone-monotone settings are related to the choice of the metric on M. While in the former case this choice is rather simple, in the latter it is more involved and we in fact embed M into a space on which the distance may attain infinite values. In this section we therefore describe the construction and properties of the space M in general terms and state the properties we need in the proof of differentiability, leaving the details of the choice of parameters as well as proofs of these properties to separate sections for each of the two settings.

We fix a small enough $0 < r < 1$ such that, in particular, $B(x_0, 3r) \subset G$ and in the cone-monotone case $B(u_0, 3r)$ is contained in a cone with respect to which f is monotone; denote

$$B = \{x \in X \mid \|x - x_0\| \le r\} \text{ and } U = \{u \in X \mid \|u - u_0\| \le r\},$$

and observe that the above inclusion conditions imply that for every pair $(x, u) \in B \times U$ the function

$$f_{x,u}(t) = f(x + tu) - f(x)$$

is well defined on the interval $(-r, r)$. We choose a space $\mathcal{P}$ of real-valued functions on $(-r, r)$ that contains all these $f_{x,u}$ and define a pseudometric ϱ on $\mathcal{P}$. Then we define a new distance on $B \times U$ by

$$d((x, u), (y, v)) = \max\left\{ \|x - y\|, \|u - v\|, \varrho(f_{x,u}, f_{y,v}) \right\}, \tag{12.6}$$

let

$$M = \left\{ (x, u) \in B \times U \mid f'(x; u) \text{ exists} \right\}, \tag{12.7}$$

and consider M as a metric space equipped with the distance d.

Notice that $\mathcal{P}$ could been have defined as the set of the functions $f_{x,u}$, where $(x, u) \in B \times U$. However, choosing a bigger and more natural space reveals better what is going on and simplifies some arguments.

The idea behind the choice of the pseudometric ϱ is that $\varrho(f_{x,u}, f_{y,v})$ should control the difference of the slopes

$$\frac{f(x + tu) - f(x)}{t} - \frac{f(y + tv) - f(y)}{t}.$$

This will allow us to prove that the function mapping $(x, u) \in M$ to $f'(x; u)$ is continuous. Also, the space $(\mathcal{P}, \varrho)$ will be complete, and this will be reflected in completeness of (M, d). We state the completeness and continuity properties of M in the following lemma.

Lemma 12.4.2. *The space (M, d) is complete and*

(i) *the function $(x, u) \mapsto f'(x; u)$ is d-continuous and lower bounded on M;*

(ii) *for any $(x, u) \in M$ the function $(y, v) \mapsto \varrho(f_{x,u}, f_{y,u})$ is d-lower semicontinuous on M.*

We will also need some relations between ordinary topological or metric notions in $B \times U$ and those coming from the metric ϱ. They are stated in the following lemmas. The first is quite natural, but the second is rather technical and geared toward its use in the main part of the proof of the differentiability result. Some (but not all) of its technicalities are needed only in the cone-monotone case, and so readers interested in the Lipschitz case may consider the simplified form of the second statement explained after the lemma. Also, the simplified form is what we actually prove in the section devoted to the construction of M in the Lipschitz case.

Lemma 12.4.3. *With a suitable constant $C \in (0, \infty)$ the pseudometric ϱ has the following properties.*

(i) *For any $(x, u), (x, v) \in B \times U$, $\varrho(f_{x,u}, f_{x,v}) \leq C\|u - v\|$.*

(ii) *For every $x, y, z \in B$, $u, v \in U$ and $0 < \delta \leq r$,*

$$\varrho(f_{x,u}, f_{y,u}) \leq \max \left\{ \begin{array}{l} \varrho(f_{y,u}, f_{z,u}) + \dfrac{C}{\delta}\, e^{\varrho(f_{y,u}, f_{z,u})}\|x - z\|, \\[2mm] \varrho(f_{y,u}, f_{z,u}) + \dfrac{C}{\delta} \displaystyle\sup_{|t| \leq C\|x-z\|} |f_{z,v}(t)|, \\[2mm] \displaystyle\sup_{0 < |t| < \delta} \inf_{w \in U} \left(C\|u - w\| + \left| \frac{f_{x,w}(t) - f_{y,u}(t)}{t} \right| \right) \end{array} \right\}.$$

In the Lipschitz case the metric $\varrho(f_{y,u}, f_{z,u})$ is bounded on $B \times U$, and so the expression $e^{\varrho(f_{y,u}, f_{z,u})}$ in the first term of the estimate of $\varrho(f_{x,u}, f_{y,u})$ in (ii) may be deleted (at the cost of increasing C). Also, in this case the supremum in the second term is at most $C\operatorname{Lip}(f)\|x - z\|(1 + r)$; hence the second term is bounded by the first (with a possibly bigger C) and so the whole second term may be deleted.

Construction of the space M in the Lipschitz case

We fix any $0 < r < r_0$ and define $\mathcal{P}$ as the set of real-valued functions g on $(-r, r)$ for which one may find $C < \infty$ such that $|g(t)| \leq C|t|$ for all $t \in (-r, r)$. We equip $\mathcal{P}$ with the distance $\varrho(g, h)$ defined by

$$\varrho(g, h) = \sup_{t \in (-r,r) \setminus \{0\}} \frac{|g(t) - h(t)|}{|t|}.$$

It is obvious that ϱ is a complete metric on $\mathcal{P}$ and that the sets $B \times U$ and M defined in the previous section are metrized by the metric d from (12.6). Also, since f is Lipschitz, for every pair $(x, u) \in B \times U$ the function $f_{x,u}$ is also Lipschitz and so it belongs to $\mathcal{P}$.

Proof of Lemma 12.4.2. Suppose that a sequence $(x_k, u_k) \in M$ is d-Cauchy. Then the points x_k norm converge to some $x \in B$, the directions u_k norm converge to some $u \in U$, and the functions f_{x_k, u_k} converge in the space $\mathcal{P}$ to some function g. The last assertion means that

$$\sup_{t \in (-r,r) \setminus \{0\}} \frac{|f_{x_k, u_k}(t) - g(t)|}{|t|} \to 0. \tag{12.8}$$

Since on the set $(-r, r)$ the functions f_{x_k, u_k} converge pointwise to $f_{x,u}$, we see that $g(t) = f_{x,u}(t)$. Hence f_{x_k, u_k} converge to $f_{x,u}$ in $(\mathcal{P}, \varrho)$, which implies that (x_k, u_k) converge to (x, u) in $(B \times U, d)$.

To finish the proof of completeness of M and also to prove (i), we have to verify that $f'(x; u)$ exists and that $f'(x_k; u_k) \to f'(x; u)$. The fact that

$$\lim_{t \to 0} \frac{f_{x_k, u_k}(t)}{t} = f'(x_k; u_k)$$

exists for each k and the condition (12.8), which says that $\frac{f_{x_k, u_k}(t)}{t}$ converge to $\frac{f_{x,u}(t)}{t}$ uniformly on $(-r, r) \setminus \{0\}$, imply that the limit

$$\lim_{t \to 0} \frac{f_{x,u}(t)}{t} = f'(x; u)$$

exists and is equal to $\lim_{k \to \infty} f'(x_k; u_k)$. It follows that $(x, u) \in M$ and the function $(x, u) \to f'(x; u)$ is d-continuous on M. Boundedness is an immediate consequence of the Lipschitz condition.

(ii) We show a stronger statement, that for any fixed $(x, u) \in B \times U$ the function $y \to \varrho(f_{x,u}, f_{y,u})$ is norm lower semicontinuous. To see this, it suffices to observe that it is the supremum, over $t \in (-r, r) \setminus \{0\}$, of the norm continuous functions

$$y \to \frac{|f_{x,u}(t) - f_{y,u}(t)|}{|t|}. \qquad \square$$

Proof of Lemma 12.4.3. (i) The required estimate of $\varrho(f_{x,u}, f_{x,v})$ is guaranteed by choosing $C \geq \mathrm{Lip}(f)$,

$$\varrho(f_{x,u}, f_{x,v}) = \sup_{t \in (-r,r) \setminus \{0\}} \frac{|f(x+tu) - f(x+tv)|}{|t|} \leq C \|u - v\|.$$

(ii) We show a better estimate, namely, that for every $x, y, z \in B$, $u \in U$, and $\delta > 0$,

$$\varrho(f_{x,u}, f_{y,u}) \leq \max \left\{ \begin{array}{l} \varrho(f_{y,u}, f_{z,u}) + \dfrac{C}{\delta} \|x - z\|, \\[2mm] \sup_{0 < |t| < \delta} \inf_{w \in U} \left(C\|u - w\| + \left| \dfrac{f_{x,w}(t) - f_{y,u}(t)}{t} \right| \right) \end{array} \right\}.$$

To prove it, we combine two estimates of the ratio defining $\varrho(f_{x,u}, f_{y,u})$.

- If $|t| \geq \delta$, then

$$\frac{|f_{x,u}(t) - f_{y,u}(t)|}{|t|} \leq \frac{|f_{y,u}(t) - f_{z,u}(t)|}{|t|} + \frac{2\,\mathrm{Lip}(f)}{|t|} \|x - z\|$$

$$\leq \varrho(f_{y,u}, f_{z,u}) + \frac{2\,\mathrm{Lip}(f)}{\delta} \|x - z\|.$$

- If $0 < |t| < \delta$, then for every $w \in U$,

$$\frac{|f_{x,u}(t) - f_{y,u}(t)|}{|t|} \leq \mathrm{Lip}(f)\|u - w\| + \frac{|f_{x,w}(t) - f_{y,u}(t)|}{|t|}.$$

Hence the statement holds for any constant $C \geq 2\,\mathrm{Lip}(f)$. $\square$

Construction of the space M in the cone-monotone case

We recall that $\mathcal{C}$ denotes a cone with respect to which the function f is monotone, and $0 < r_0 < \frac{1}{6}$ is such that $B(x_0, 3r_0) \subset G$ and $B(u_0, 3r_0) \subset \mathcal{C}$ where $\|u_0\| = 1$. Based on these data we choose r, the radius of the ball in which we will work, as $r = r_0^2$. The motivation for this choice of r is explained by the estimate of η needed in the proof of the following lemma, which proves an inequality that will be used as a replacement for Lipschitz estimates.

Lemma 12.4.4. *Suppose that $x, y \in B$, $u, v \in U$, $t \in (-r, r)$, and*

$$\eta = \frac{1}{r_0} \big(\|x - y\| + |t| \|u - v\| \big).$$

Then the points $y + tv$ and $x + (t \pm \eta)u$ belong to $B(x_0, 3r_0) \subset G$ and

$$f(x + (t - \eta)u) \leq f(y + tv) \leq f(x + (t + \eta)u). \tag{12.9}$$

Proof. The first statement follows from the inequalities

$$\|y + tv - x_0\| \le r + r(1 + r) \le 3r < 3r_0,$$

$$0 \le \eta < \frac{2r + 2r^2}{r_0} = 2r_0(1 + r_0^2), \text{ and}$$

$$\|x + (t \pm \eta)u - x_0\| \le r + \big(r + 2r_0(1 + r_0^2)\big)(1 + r) < 3r_0.$$

Since the point

$$(y + tv) - (x + (t - \eta)u) = \eta u + ((y - x) + t(v - u))$$

has distance from ηu_0 at most $\eta r + \eta r_0 \le 2r_0\eta$, it belongs to $\mathcal{C}$. Consequently, $f(x + (t - \eta)u) \le f(y + tv)$. The proof of the second inequality in (12.9) is similar. □

It would immensely simplify the following arguments if we could conclude, instead of (12.9), that $f_{x,u}(t - \eta) \le f_{y,v}(t) \le f_{x,u}(t + \eta)$. In the case when $y = x$, this follows by subtracting $f(x)$. However, in the case when $y \ne x$ it is, in general, false, since it requires subtracting different values $f(x)$ and $f(y)$ from the middle and outer terms, respectively. Unlike in the Lipschitz case, here we do not have any estimate of $|f(y) - f(x)|$ in terms of $\|y - x\|$. The following corollary estimates this difference using just monotonicity, and it is this difference that will lead to the middle term in the estimate from Lemma 12.4.3 (ii).

Corollary 12.4.5. *For any $x, y \in B$,*

$$|f(y) - f(x)| \le \sup_{|s| \le \|x - y\|/r_0} \inf_{w \in U} \min\big\{|f_{x,w}(s)|, |f_{y,w}(-s)|\big\}.$$

Proof. Use (12.9) with $t = 0$ and $u = v = w$, and so with $\eta = \|x - y\|/r_0$, to infer that $f(y) \le f(x + \eta w)$ and $f(y - \eta w) \le f(x)$. Subtracting $f(x)$ from the first of these inequalities and $f(y)$ from the second gives

$$f(y) - f(x) \le f_{x,w}(\eta) \quad \text{and} \quad f_{y,w}(-\eta) \le f(x) - f(y).$$

Hence

$$f(y) - f(x) \le \min\{f_{x,w}(\eta), -f_{y,w}(-\eta)\} = \min\big\{|f_{x,w}(\eta)|, |f_{y,w}(-\eta)|\big\}.$$

Taking the infimum over $w \in U$ and using this inequality also with x, y exchanged gives the result. □

Finally, a combination of the two estimates we have just proved gives the most useful replacement of Lipschitz inequalities in our case.

Corollary 12.4.6. *Let x, y, u, v, t, and η be as in Lemma 12.4.4. Denote*

$$\beta = \sup_{|s| \le \|x - y\|/r_0} \inf_{w \in U} \min\big\{|f_{x,w}(s)|, |f_{y,w}(-s)|\big\}.$$

Then

$$f_{x,u}(t - \eta) - \beta \le f_{y,v}(t) \le f_{x,u}(t + \eta) + \beta.$$

Proof. Subtracting $f(y)$ from (12.9) gives

$$f_{x,u}(t-\eta) + (f(x) - f(y)) \leq f_{y,v}(t) \leq f_{x,u}(t+\eta) + (f(x) - f(y)).$$

Now it suffices to use Corollary 12.4.5. □

Although we will not use it, it may be helpful to recall that the space of increasing functions on an interval (a,b), equipped with the topology $\mathcal{T}_p$ of pointwise convergence at all points of (a,b) except countably many, is pseudometrizable. Recall also that in this topology convergence to g is equivalently described as pointwise convergence to g at every point of continuity of g, or as convergence to g at all points of a dense subset of (a,b). We need, however, a pseudometric adjusted to our purposes; in particular, values at points near zero will play a special role and the topology it generates will be finer than $\mathcal{T}_p$.

We let

$$\mathcal{P} = \big\{g\colon (-r,r) \longrightarrow \mathbb{R} \,\big|\, g \text{ is increasing, } g(0) = 0\big\}.$$

It will be technically convenient to define our pseudometric first on the subset of $\mathcal{P}$,

$$\mathcal{P}^+ = \big\{g \in \mathcal{P} \,\big|\, g(t) = 0 \text{ for } t \leq 0\big\},$$

and then extend it to $\mathcal{P}$ in a straightforward way: every function from $\mathcal{P}$ can be written in a unique way as

$$g(t) - h(-t), \text{ where } g, h \in \mathcal{P}^+. \tag{12.10}$$

This identifies $\mathcal{P}$ with $\mathcal{P}^+ \times \mathcal{P}^+$, and we use this identification to equip $\mathcal{P}$ with the maximum distance.

We define the distance $\varrho(g,h)$ of $g, h \in \mathcal{P}^+$ as the infimum of the set

$$E = \big\{\alpha \geq 0 \,\big|\, g(t) \geq h(e^{-\alpha}t) - \alpha t, \; h(t) \geq g(e^{-\alpha}t) - \alpha t \text{ for all } t \in [0,r)\big\}.$$

Observe that if the conditions defining the set E hold for some α, then they hold for any bigger α. However, the set E may well be empty, in which case we have $\varrho(g,h) = \infty$.

We need to establish that $(\mathcal{P}, \varrho)$ is a pseudometric space and relate ϱ to questions we are interested in.

Lemma 12.4.7. *The distance ϱ on the space $\mathcal{P}$ has the following properties.*

(i) *$\varrho(g,h) = 0$ if and only if $g = h$ except on a countable set.*

(ii) *$(\mathcal{P}, \varrho)$ is a complete pseudometric space.*

(iii) *If a sequence (g_k) converges to g in $(\mathcal{P}, \varrho)$ and if, for each k, the limit*

$$l_k := \lim_{t \to 0} \frac{g_k(t)}{t}$$

exists, then

- *there are $\delta_0 > 0$ and $K_0 < \infty$ such that $|g_k(t)| \leq K_0|t|$ for each $k \in \mathbb{N}$ and $|t| < \delta_0$,*
- *the limit $l := \lim_{k \to \infty} l_k$ exists, and*
- $\lim_{t \to 0} \dfrac{g(t)}{t} = l.$

Proof. We prove analogous statements for the space $(\mathcal{P}^+, \varrho)$; in particular, we restrict ourselves to $t > 0$ in (iii). The general case follows from the identification of $\mathcal{P}$ with $\mathcal{P}^+ \times \mathcal{P}^+$.

(i) If $g = h$ except on a countable set, then for every $\alpha > 0$ and every $0 < t < r$,

$$g(t) \geq h(e^{-\alpha}t) \geq h(e^{-\alpha}t) - \alpha t.$$

Similarly with g, h exchanged; hence $\varrho(g, h) = 0$.

If $\varrho(g, h) = 0$ and $0 < t < r$ is a point of continuity of h, then the definition of $\varrho(g, h)$ implies that

$$g(t) \geq \lim_{\alpha \searrow 0} (h(e^{-\alpha}t) - \alpha t) = h(t).$$

Similarly with g, h exchanged; so $g(t) = h(t)$ at every point at which they are both continuous. Recalling that a monotone function is continuous except for a countable set, we are done.

(ii) For the pseudometric part of the assertion only the triangle inequality

$$\varrho(g_0, g_2) \leq \varrho(g_0, g_1) + \varrho(g_1, g_2)$$

may need an argument. If $\varrho(g_0, g_1) + \varrho(g_1, g_2) < \alpha$, there are α_1 and α_2 such that

$$\varrho(g_0, g_1) < \alpha_1, \ \varrho(g_1, g_2) < \alpha_2 \ \text{ and } \ \alpha_1 + \alpha_2 < \alpha.$$

Then for every $0 < t < r$,

$$g_0(t) \geq g_1(e^{-\alpha_1}t) - \alpha_1 t \geq g_2(e^{-\alpha_2}e^{-\alpha_1}t) - \alpha_2 e^{-\alpha_1}t - \alpha_1 t \geq g_2(e^{-\alpha}t) - \alpha t.$$

Similarly,

$$g_2(t) \geq g_1(e^{-\alpha_2}t) - \alpha_2 t \geq g_0(e^{-\alpha_1}e^{-\alpha_2}t) - \alpha_1 e^{-\alpha_2}t - \alpha_2 t \geq g_0(e^{-\alpha}t) - \alpha t.$$

Hence $\varrho(g_0, g_2) \leq \alpha$.

To prove the completeness of $(\mathcal{P}^+, \varrho)$, assume that (g_i) is a ϱ-Cauchy sequence. It means that there are $\alpha_k \searrow 0$ such that

$$\varrho(g_i, g_j) < \alpha_k \quad \text{for } i, j \geq k.$$

Hence for $i, j \geq k$ and $0 < t < r$,

$$g_j(t) \geq g_i(e^{-\alpha_k}t) - \alpha_k t. \tag{12.11}$$

Notice that for any $t \in (-r, r)$ the sequence $(g_i(t))$ is bounded. Indeed, for $t \leq 0$ we have $g_i(t) = 0$, and for any $0 < t < r$ we may find k and $t < s < r$ such that $e^{\alpha_k} t < s$ and infer from (12.11) that

$$g_i(t) \leq g_k(e^{\alpha_k} t) + \alpha_k e^{\alpha_k} t \leq g_k(s) + \alpha_k s.$$

Denote
$$g_-(t) = \liminf_{i \to \infty} g_i(t) \quad \text{and} \quad g_+(t) = \limsup_{i \to \infty} g_i(t).$$

Clearly, $g_-, g_+ \in \mathcal{P}^+$ and (12.11) implies that for each k,

$$g_+(t) \geq g_-(t) \geq g_+(e^{-\alpha_k} t) - \alpha_k t.$$

Consequently, $\varrho(g_-, g_+) = 0$. By (i) we have $g_- = g_+$ except for a countable set. Taking in (12.11) on one hand $\liminf$ as $j \to \infty$ and on the other hand $\liminf$ as $i \to \infty$, we conclude that $\varrho(g_-, g_j) \leq \alpha_k$ for $j \geq k$; hence the completeness part of (ii).

(iii) Let $q \in \mathbb{N}$ be such that $\varrho(g_i, g_j) < 1$ for all $i, j \geq q$, and choose $0 < \delta_0 \leq r$ such that $|g_i(t) - l_i t| \leq t$ for $0 \leq i \leq q$ and $0 \leq t \leq e\delta_0$. We put

$$K_0 = 1 + \max\{l_0, \ldots, l_q, (l_q + 1)e + e\}$$

and show that $g_i(t) \leq K_0 t$ for all i and $0 \leq t \leq \delta_0$. This is obvious for $i \leq q$, and for $i > q$ it follows from

$$g_i(t) \geq g_q(e^{-1} t) - t \geq ((l_q - 1)e^{-1} - 1)t \quad \text{and}$$
$$g_i(t) \leq g_q(et) + et \leq ((l_q + 1)e + e)t.$$

To prove the second part of the statement, we divide (12.11) by t and take the limit as $t \searrow 0$ to obtain

$$l_j \geq e^{-\alpha_k} l_i - \alpha_k, \quad i, j \geq k.$$

Since the previous part of the proof shows that the sequence (l_i) of positive real numbers is bounded, we conclude that it is Cauchy. Hence the limit $l := \lim_{k \to \infty} l_k$ exists. Given $\varepsilon > 0$ and k, we find $\delta > 0$ such that

$$\left| \frac{g_k(t)}{t} - l_k \right| < \varepsilon$$

whenever $0 < t < \delta$. Then for $0 < t < e^{-\alpha_k} \delta$

$$e^{\alpha_k}(l_k + \varepsilon) + \alpha_k e^{\alpha_k} \geq \frac{g_k(e^{\alpha_k} t)}{t} + \alpha_k e^{\alpha_k}$$
$$\geq \frac{g(t)}{t} \geq \frac{g_k(e^{-\alpha_k} t)}{t} - \alpha_k \geq e^{-\alpha_k}(l_k - \varepsilon) - \alpha_k.$$

Since $\alpha_k \to 0$ and $\varepsilon > 0$ may be arbitrarily small, this implies that

$$\lim_{t \searrow 0} \frac{g(t)}{t} = l. \qquad \square$$

Proof of Lemma 12.4.2. We first prove that for any sequence $(x_k, u_k) \in M$ which is d-Cauchy and norm converging to some $(x, u) \in B \times U$, the functions f_{x_k, u_k} converge to $f_{x,u}$ in the pseudometric ϱ. This will show both completeness of (M, d) and the statement (i) of Lemma 12.4.2. Indeed, using Lemma 12.4.7 (iii) we infer that $f'(x; u)$ exists, so $(x, u) \in M$ and so (M, d) is complete. Moreover, Lemma 12.4.7 (iii) implies that $f'(x_k; u_k) \to f'(x; u)$. Hence the function $(x, u) \in M \mapsto f'(x; u)$ is d-continuous. Its lower boundedness is obvious since $f'(x; u) \geq 0$ for all $(x, u) \in M$.

So suppose that a sequence $(x_k, u_k) \in M$ is d-Cauchy and norm converging to $(x, u) \in B \times U$. Since $(\mathcal{P}, \varrho)$ is complete, the sequence (f_{x_k, u_k}) converges in the pseudometric ϱ to some $g \in \mathcal{P}$. Moreover, by Lemma 12.4.7 (iii) the limit

$$l := \lim_{k \to \infty} f'(x_k; u_k)$$

exists and there are $\delta_0 > 0$ and $K_0 < \infty$ such that $|f_{x_k, u_k}(t)| \leq K_0 |t|$ for each $k \in \mathbb{N}$ and $|t| < \delta_0$.

Let $\alpha_k \to 0$ be such that $\varrho(f_{x_k, u_k}, g) < \alpha_k$ for every k. Define also

$$\eta_k = \frac{1}{r_0}(\|x - x_k\| + r\|u - u_k\|) \quad \text{and} \quad \beta_k = \sup_{|s| \leq \|x - x_k\|/r} |f_{x_k, u_k}(s)|.$$

Clearly, $\eta_k \to 0$, and since $\beta_k \leq K_0 \|x - x_k\|/r$ once $\|x - x_k\| < r\delta_0$, we get that $\beta_k \to 0$ as well.

By Lemma 12.4.7 (i) it suffices to show that $g(s) = f_{x,u}(s)$ for every $0 < |s| < r$ at which $f_{x,u}$ is continuous. We will treat only the case $s > 0$, as the case $s < 0$ follows by symmetry. Given any $\varepsilon > 0$ we first find $0 < \delta < \min\{s, r - s\}$ such that

$$|f_{x,u}(t) - f_{x,u}(s)| < \frac{\varepsilon}{2}$$

for $|t - s| < \delta$, and then $k \in \mathbb{N}$ such that

$$s - \delta < e^{-\alpha_j}s - \eta_j \leq e^{\alpha_j}s + \eta_j < s + \delta \quad \text{and} \quad \beta_j + \alpha_j e^{\alpha_j} r < \frac{\varepsilon}{2}$$

for every $j \geq k$. Then Corollary 12.4.6 shows that for every $j \geq k$,

$$g(s) \geq f_{x_j, u_j}(e^{-\alpha_j}s) - \alpha_j s \geq f_{x,u}(e^{-\alpha_j}s - \eta_j) - \beta_j - \alpha_j s > f_{x,u}(s) - \varepsilon$$

and

$$g(s) \leq f_{x_j, u_j}(e^{\alpha_j}s) + \alpha_j e^{\alpha_j}s \leq f_{x,u}(e^{\alpha_j}s + \eta_j) + \beta_j + \alpha_j e^{\alpha_j}s < f_{x,u}(s) + \varepsilon.$$

Consequently, $g(s) = f_{x,u}(s)$ and we are done.

(ii) We show a stronger statement, namely, that if $(x, u), (y, v) \in M$ and $y_k \in B$ norm converge to y, then $\varrho(f_{x,u}, f_{y,u}) \leq \liminf_{k \to \infty} \varrho(f_{x,u}, f_{y_k, u})$.

Let $\alpha < \tau < \varrho(f_{x,u}, f_{y,u})$. By the definition of $\varrho(f_{x,u}, f_{y,u})$, this is witnessed by some $s \in (-r, r) \setminus \{0\}$. Assuming that this s is positive (negative s is handled similarly), we have

$$f_{x,u}(s) < f_{y,u}(e^{-\tau}s) - \tau s \quad \text{or} \quad f_{y,u}(s) < f_{x,u}(e^{-\tau}s) - \tau s. \qquad (12.12)$$

Let $\eta_k = \|y - y_k\|/r_0$ and $\beta_k = \sup_{|s| \le \|y - y_k\|/r_0} |f_{y,v}(s)|$. Clearly, $\eta_k \to 0$ and, since $\beta_k \le (f'(y; v) + 1)\|y - y_k\|/r_0$ once $\|y - y_k\|$ is small enough, $\beta_k \to 0$ as well. Hence there is k such that $\eta := \max_{j \ge k} \eta_j$ and $\beta := \max_{j \ge k} \beta_j$ satisfy

$$e^{-\tau}s + \eta < e^{-\alpha}(s - \eta) \quad \text{and} \quad \beta - \tau s < -\alpha s.$$

If the first inequality in (12.12) holds, Corollary 12.4.6 shows that for every $j \ge k$,

$$f_{x,u}(s) < f_{y,u}(e^{-\tau}s) - \tau s \le f_{y_j,u}(e^{-\tau}s + \eta) + \beta - \tau s \le f_{y_j,u}(e^{-\alpha}s) - \alpha s,$$

while if the second inequality in (12.12) holds, we let $t = s - \eta$ and use Corollary 12.4.6 to show that for every $j \ge k$,

$$f_{y_j,u}(t) \le f_{y,u}(s) + \beta < f_{x,u}(e^{-\tau}s) - \tau s + \beta \le f_{x,u}(e^{-\alpha}t) - \alpha t.$$

It follows that $\varrho(f_{x,u}, f_{y_j,u}) > \alpha$ for $j \ge k$, as required. $\qquad\square$

Proof of Lemma 12.4.3. (i) Let $C > 0$ be such that $e^{-Cr_0 s} \le 1 - s$ for all $s \in [0, \frac{1}{2}]$. For any $x \in B$, $u, v \in U$, and $0 < t < r$ we notice that $s = \|u - v\|/r_0$ and $\eta = ts$ satisfy

$$0 \le s \le \frac{2r}{r_0} \le \frac{1}{2} \quad \text{and} \quad t - \eta = t(1 - s) \ge e^{-Cr_0 s}t = e^{-C\|u-v\|}t.$$

Hence Corollary 12.4.6 in which now $\beta = 0$ implies that

$$f_{x,v}(t) \ge f_{x,u}(t - \eta) \ge f_{x,u}(e^{-C\|u-v\|}t)$$

and

$$f_{x,u}(t) \ge f_{x,v}(t - \eta) \ge f_{x,v}(e^{-C\|u-v\|}t).$$

An analogous conclusion holds also for $-r < t < 0$, and we conclude that

$$\varrho(f_{x,u}, f_{x,v}) \le C\|u - v\|.$$

(ii) By the preceding point there is $C \ge 4/r_0$ such that $\varrho(f_{x,u}, f_{x,v}) \le C\|u - v\|$ whenever $x \in B$ and $u, v \in U$.

We have to show that, whenever $x, y, z \in B$, $u, v \in U$, and $0 < \delta \le r$,

$$\varrho(f_{x,u}, f_{y,u}) \le \max \left\{ \begin{array}{l} \varrho(f_{y,u}, f_{z,u}) + \dfrac{C}{\delta} \, e^{\varrho(f_{y,u}, f_{z,u})} \|x - z\|, \\[2mm] \varrho(f_{y,u}, f_{z,u}) + \dfrac{C}{\delta} \sup\limits_{|t| \le C\|x-z\|} |f_{z,v}(t)|, \\[2mm] \sup\limits_{0 < |t| < \delta} \inf\limits_{w \in U} \left(C\|u - w\| + \left| \dfrac{f_{x,w}(t) - f_{y,u}(t)}{t} \right| \right) \end{array} \right\}.$$

Suppose that the right-hand side of this inequality is strictly smaller than some $\tau < \infty$. We show that for every $0 < t < r$,

$$f_{x,u}(t) \ge f_{y,u}(e^{-\tau}t) - \tau t \quad \text{and} \quad f_{y,u}(t) \ge f_{x,u}(e^{-\tau}t) - \tau t. \qquad (12.13)$$

We will treat first the case $\delta \leq t < r$. Let $\beta = \sup_{0 \leq |s| \leq C \|x-z\|} |f_{z,v}(s)|$. Noticing that $C \geq 2$, find $\alpha > \varrho(f_{y,u}, f_{z,u})$ such that

$$\alpha + \frac{C}{\delta} e^{\alpha} \|x - z\| < \tau \quad \text{and} \quad \alpha + \frac{2}{\delta}\beta < \tau.$$

Also denote $\eta = \frac{1}{r_0}\|x - z\|$ and observe that the inequality $C\|x - z\| \geq 4\eta$ and Corollary 12.4.6 imply that for every $0 < t \leq 3\eta$,

$$f_{z,u}(t) \leq f_{z,v}(t + t\|u - v\|/r_0) \leq f_{z,v}(4\eta) \leq \beta. \tag{12.14}$$

To show the first inequality in (12.13) we distinguish two cases. First, if $t \leq 2\eta$, we have

$$
\begin{aligned}
f_{x,u}(t) &\geq f_{z,u}(t + \eta) - \beta && \text{since the right side is } \leq 0 \text{ by (12.14),} \\
&\geq f_{y,u}(t) - 2\beta && \text{by Corollary 12.4.6,} \\
&\geq f_{y,u}(e^{-\tau}t) - \tau t && \text{since } 2\beta/t \leq C\beta/\delta \leq \tau.
\end{aligned}
$$

Second, if $t > 2\eta$, we use that $\alpha + Cr_0\eta/t \leq \alpha + Ce^{\alpha}\|x - z\|/\delta \leq \tau$ to get

$$e^{-\alpha}(t - \eta) = e^{-\alpha}\left(1 - \frac{\eta}{t}\right)t \geq e^{-\alpha - Cr_0\eta/t}t \geq e^{-\tau}t, \tag{12.15}$$

and we conclude that

$$
\begin{aligned}
f_{x,u}(t) &\geq f_{z,u}(t - \eta) - \beta && \text{by Corollary 12.4.6,} \\
&\geq f_{y,u}(e^{-\alpha}(t - \eta)) - \alpha(t - \eta) - \beta && \text{since } \varrho(f_{y,u}, f_{z,u}) < \alpha, \\
&\geq f_{y,u}(e^{-\tau}t) - \tau t && \text{by (12.15) and } \alpha + \beta/t \leq \tau.
\end{aligned}
$$

Notice that the inequality we are proving is not symmetric in x, y. Hence, although the proof of the second inequality in (12.13) uses similar ideas, they have to be used slightly differently, and so we give all details. Also, we now divide the two cases based on the value of $e^{-\alpha}t$ instead of t. First, if $e^{-\alpha}t \leq 2\eta$, we have

$$
\begin{aligned}
f_{y,u}(t) &\geq f_{z,u}(3\eta) - \beta && \text{since the right side is } \leq 0 \text{ by (12.14),} \\
&\geq f_{x,u}(2\eta) - 2\beta && \text{by Corollary 12.4.6,} \\
&\geq f_{x,u}(e^{-\tau}t) - \tau t && \text{since } e^{-\tau}t \leq 2\eta \text{ and } 2\beta/t \leq 2\beta/\delta \leq \tau.
\end{aligned}
$$

Finally, if $e^{-\alpha}t > 2\eta$, we use $\alpha + Cr_0 e^{\alpha}\eta/t \leq \alpha + Ce^{\alpha}\|x - z\|/\delta \leq \tau$ to get

$$e^{-\alpha}t - \eta = e^{-\alpha}\left(1 - \frac{e^{\alpha}\eta}{t}\right)t \geq e^{-\alpha - Cr_0 e^{\alpha}\eta/t}t \geq e^{-\tau}t, \tag{12.16}$$

and we conclude that

$$
\begin{aligned}
f_{y,u}(t) &\geq f_{z,u}(e^{-\alpha}t) - \alpha t && \text{since } \varrho(f_{y,u}, f_{z,u}) < \alpha, \\
&\geq f_{x,u}(e^{-\alpha}t - \eta) - \alpha t - \beta && \text{by Corollary 12.4.6,} \\
&\geq f_{x,u}(e^{-\tau}t) - \tau t && \text{by (12.16) and } \alpha + \beta/t \leq \tau.
\end{aligned}
$$

It remains to treat the case when $0 < t < \delta$. Here we use the last term in the required estimate of $\varrho(f_{x,u}, f_{y,u})$. As it is slightly simpler, we start by showing the second inequality in (12.13). Let $w \in U$ (depending on t) be such that

$$\tau > C\|u - w\| + \left|\frac{f_{x,w}(t) - f_{y,u}(t)}{t}\right|.$$

Recalling that $\varrho(f_{x,u}, f_{x,w}) \leq C\|u - w\|$ by the choice of C, we find $a > \varrho(f_{x,u}, f_{x,w})$ and $b > |f_{x,w}(t) - f_{y,u}(t)|$ such that $a + b/t < \tau$. Then

$$
\begin{aligned}
f_{y,u}(t) &\geq f_{x,w}(t) - b && \text{by definition of } b, \\
&\geq f_{x,u}(e^{-a}t) - at - b && \text{since } \varrho(f_{x,u}, f_{x,w}) < a, \\
&\geq f_{x,u}(e^{-\tau}t) - \tau t && \text{since } a + b/t < \tau.
\end{aligned}
$$

To prove the first inequality in (12.13), we argue similarly, but this time we choose w for $e^{-\tau}t$, so

$$\tau > C\|u - w\| + \left|\frac{f_{x,w}(e^{-\tau}t) - f_{y,u}(e^{-\tau}t)}{e^{-\tau}t}\right|.$$

We choose a, b such that $a > \varrho(f_{x,u}, f_{x,w})$, $b > |f_{x,w}(e^{-\tau}t) - f_{y,u}(e^{-\tau}t)|$ and $a + b/t < \tau$. Then

$$
\begin{aligned}
f_{x,u}(t) &\geq f_{x,w}(e^{-a}t) - at && \text{since } \varrho(f_{x,u}, f_{x,w}) < a, \\
&\geq f_{x,w}(e^{-\tau}t) - at && \text{since } a \leq \tau, \\
&\geq f_{y,u}(e^{-\tau}t) - b - at && \text{by definition of } b, \\
&\geq f_{y,u}(e^{-\tau}t) - \tau t && \text{since } a + b/t < \tau.
\end{aligned}
$$

These estimates, together with symmetric estimates for $t < 0$, imply the estimate $\varrho(f_{x,u}, f_{y,u}) \leq \tau$, which completes the proof. $\qquad\square$

Use of the variational principle

Recall that in the previous sections we have chosen $0 < r < r_0$ such that $B(x_0, 3r)$ is contained in the domain of f and, in the cone-monotone case, $B(u_0, 3r)$ is contained in a cone with respect to which f is monotone. We have also denoted

$$B = \{x \in X \mid \|x - x_0\| \leq r\}, \quad U = \{u \in X \mid \|u - u_0\| \leq r\}$$

and

$$M = \{(x, u) \in B \times U \mid f'(x; u) \text{ exists}\}.$$

The space $B \times U$ is equipped with pseudometrics ϱ and d, where

$$d((x, u), (y, v)) = \max\{\|x - y\|, \|u - v\|, \varrho(f_{x,u}, f_{y,v})\}$$

in such a way that Lemmas 12.4.2 and 12.4.3 hold.

We choose a number of parameters to define the perturbations. The particular formulas defining our parameters are often not important, but their sometimes delicate interrelation is crucial. As it is not always clear that these relations can be satisfied, we describe specific choices.

In order to avoid working with unnecessary numerical estimates, we start by letting $\tau = \frac{1}{30}$. This particular value of τ comes from the last line in the proof of Lemma 12.4.17, although any sufficiently small constant would do as well. We also choose a constant $C \geq 1$ for which both estimates of Lemma 12.4.3 hold, find $\varepsilon_0 > 0$ such that

$$f'(x_0; u_0) < \varepsilon_0 + \inf_{(x,u) \in M} f'(x; u), \tag{12.17}$$

and choose $\kappa \geq 8(1 + |f'(x_0; u_0)| + \varepsilon_0 + \varepsilon)$ such that

$$f(x + tu) - f(x) \geq -\frac{1}{8}\kappa t$$

whenever $x \in B$, $u \in U$ and $0 \leq t < 1$. (Of course, in the cone-monotone case the last condition holds for any $\kappa \geq 0$.) Then we let

$$\sigma_0 = \frac{\tau}{8\kappa} \quad \text{and} \quad s_0 = \max\left\{274\sqrt{\kappa(\kappa + \varepsilon_0 + \varepsilon)}, \sqrt{\frac{8\varepsilon_0}{\sigma_0}}\right\}.$$

Finally, we notice that $s_0 > \kappa > \varepsilon_0$ and choose

$$\lambda_0 > \max\left\{\varepsilon_0, \frac{2\varepsilon_0}{r}, \frac{\varepsilon_0}{s_0}\right\}$$

large enough that

$$\Theta(z) > \frac{\varepsilon_0}{\lambda_0} \text{ for } \|z\| \geq \min\left\{\frac{r}{2}, \frac{s_0}{16C}\right\}.$$

Remark 12.4.8. For readers interested only in the Lipschitz case or for those familiar with the Lipschitz case argument from [29] we explain here a serious technical difference between the Lipschitz and cone-monotone case. The formula for s_0 is used only at the end of the proof of the main technical result, Proposition 12.4.14. There the constant κ appearing in the definition of s_0 estimates the magnitude of the derivatives $f'(x; u)$ at some pairs $(x, u) \in M$ at which we are interested. In the Lipschitz case we can estimate these derivatives using the Lipschitz property, and so define

$$s_0 = 274\sqrt{2\operatorname{Lip}(f)(2\operatorname{Lip}(f) + \varepsilon_0 + \varepsilon)}.$$

This means that s_0 and, more important, that λ_0 may be defined before κ, and so we may also require that $\kappa \geq 4(\operatorname{Lip}(f) + 4\lambda_0 \operatorname{Lip}(\Theta))$. The main effect of this would be trivialization of the proof of Lemma 12.4.12, which we point out when stating it. In the cone-monotone case this lemma needs a slightly roundabout argument, although its proof is still simple.

Having defined the parameters $\sigma_i, s_i, \varepsilon_i, \lambda_i$ for $i = 0$, we now define them also for $i \geq 1$. We let $\sigma_i = 2^{-i}\sigma_0$, $s_i = 2^{-i}s_0$, and $\lambda_i = 2^{-i}\lambda_0$, and we finish by recursively choosing

$$0 < \varepsilon_i < \min\left\{\lambda_i, s_i, \lambda_i s_i, \frac{\lambda_i r}{2}, \frac{\sigma_i s_i^2}{64}, \frac{s_i^2}{274^2 \kappa}, \frac{\varepsilon_{i-1}}{4}\right\}$$

such that

$$\Theta(z) > \frac{\varepsilon_i}{\lambda_i} \text{ for } \|z\| \geq \min\left\{\frac{r}{2}, \frac{s_i}{16C}\right\}.$$

Notice the difference in the order of definitions for $i = 0$ and for $i > 0$. The point is that ε_i has to be small in comparison with λ_i and s_i. When $i = 0$, we achieve this by making s_0 and λ_0 large since ε_0 has to satisfy (12.17), which is needed to apply the variational principle, and so cannot be made small. However, when $i > 0$ we can choose ε_i as small as we wish.

As in the illustrative example, to find a point of differentiability of f, we could minimize a suitable perturbation of the function $(x, u) \mapsto f'(x; u)$. This would also show the validity of the mean value estimate. However, to place the point of differentiability outside the given σ-porous set, we will minimize a perturbation of the function $(x, u) \mapsto f'(x; u) + \psi(x)$, where ψ is a suitable upper semicontinuous function. Readers interested only in finding points of differentiability and validity of the mean value estimate may assume that ψ is identically zero, in which case they may use in the following proof the more classical variational principle of Corollary 7.2.4 instead of its bimetric variant from Theorem 7.3.5.

From now on we will therefore assume that $\psi \colon X \longrightarrow [0, \varepsilon)$ is upper semicontinuous. We intend to use the variational principle to the function $(x, u) \mapsto f'(x; u) + \psi(x)$ with perturbation functions $F_i : M \times M \longrightarrow [0, \infty)$,

$$F_i((x, u), (y, v)) = \Phi_i(x, y) + \Psi_i(u, v) + Q_i((x, u), (y, v)) + \Delta_i((x, u), (y, v)),$$

where

$$\Phi_i(x, y) = \lambda_i \|y - x\|, \quad \Psi_i(u, v) = \lambda_i \Theta(v - u),$$

$$Q_i((x, u), (y, v)) = \sigma_i\left(f'(y; v) - f'(x; u)\right)^2,$$

and

$$\Delta_i((x, u), (y, v)) = \min\{\lambda_i, \max\{0, \varrho(f_{x,u}, f_{y,u}) - s_i\}\}.$$

Notice that the peculiarity in the definition of Δ_i is not a misprint: Δ_i really does not depend on v.

We have to verify that these functions satisfy the assumptions of Theorem 7.3.5.

Lemma 12.4.9. *The functions F_i are positive, d-lower semicontinuous in the second variable, satisfy $F_i((x, u), (x, u)) = 0$, and*

$$\inf\left\{F_i((x, u), (y, v)) \mid d((x, u), (y, v)) > \tau_i\right\} > \varepsilon_i \tag{12.18}$$

where $\tau_i = (C + 2)s_i$.

Proof. Clearly, $F_i \geq 0$ and $F_i((x, u), (x, u)) = 0$. Lower semicontinuity in metric d of the functions Δ_i in the second variable follows directly from Lemma 12.4.2 (ii). Since the remaining functions of which F_i consists are continuous (the functions Q_i by Lemma 12.4.2 (i)), this shows the lower semicontinuity of F_i in the second variable.

To show (12.18), we consider various possibilities in which

$$d((x, u), (y, v)) = \max\{\|x - y\|, \|u - v\|, \varrho(f_{x,u}, f_{y,v})\}$$

can exceed τ_i. If $\|x - y\| > s_i$ then $F_i((x, u), (y, v)) > \lambda_i s_i > \varepsilon_i$, and if $\|u - v\| \geq s_i$ then $F_i((x, u), (y, v)) > \lambda_i \Theta(v - u) > \varepsilon_i$. Finally, if $\|x - y\| < s_i$ and $\|u - v\| < s_i$, then

$$\varrho(f_{x,u}, f_{y,v}) > (C + 2)s_i$$

and Lemma 12.4.3 (i) implies that

$$\varrho(f_{x,u}, f_{y,u}) \geq \varrho(f_{x,u}, f_{y,v}) - \varrho(f_{y,u}, f_{y,v}) > (C + 2)s_i - C\|u - v\| > 2s_i.$$

Hence $\Delta_i((x, u), (y, v)) > \min\{\lambda_i, s_i\} > \varepsilon_i$. $\qquad\square$

Letting $d_0((x, u), (y, v)) = \|x - y\|$, we see from Lemma 12.4.2 that M, being d-complete, is (d, d_0)-complete, and, using also Lemma 7.3.3, that the function $(x, u) \to f'(x; u) + \psi(x)$ is (d, d_0)-lower semicontinuous. Hence by the variational principle of Theorem 7.3.5, (x_0, u_0) is the starting term of a sequence $(x_j, u_j) \in M$ which d-converges to some $(x_\infty, u_\infty) \in M$ such that

$$F_j((x_j, u_j), (x_k, u_k)) \leq \varepsilon_j \text{ for } 0 \leq j < \infty, j \leq k \leq \infty, \tag{12.19}$$

$$f'(x_j; u_j) + \psi(x_j) \leq f'(x_0; u_0) + \psi(x_0) \text{ for } 0 \leq j \leq \infty, \tag{12.20}$$

and for some norm continuous $\varphi \colon X \longrightarrow [0, \infty)$, the functions

$$h_j(x, u) := f'(x; u) + \psi(x) + \varphi(x) + \sum_{i=0}^{j-1} F_i((x_i, u_i), (x, u)) \tag{12.21}$$

satisfy, for every $0 \leq j < \infty$, the estimates

$$h_\infty(x_\infty, u_\infty) \leq \varepsilon_j + \inf_{(x,u) \in M} h_j(x, u). \tag{12.22}$$

Notice that the last inequality can be considered as a quantitative way of saying that h_∞ attains its minimum on M at (x_∞, u_∞), as this follows from it immediately by taking the limit as $j \to \infty$.

Recalling the definition of the functions F_i, we have

$$h_\infty(x, u) = f'(x; u) + \psi(x) + \Phi(x) + \Psi(u) + Q(x, u) + \Delta(x), \tag{12.23}$$

where

$$\Phi(x) = \varphi(x) + \sum_{i=0}^{\infty} \Phi_i(x_i, x) = \varphi(x) + \sum_{i=0}^{\infty} \lambda_i \|x - x_i\|,$$

$$\Psi(u) = \sum_{i=0}^{\infty} \Psi_i(u_i, u) = \sum_{i=0}^{\infty} \lambda_i \Theta(u - u_i),$$

$$Q(x, u) = \sum_{i=0}^{\infty} Q_i((x_i, u_i), (x, u)) = \sum_{i=0}^{\infty} \sigma_i \big(f'(x; u) - f'(x_i; u_i) \big)^2,$$

$$\Delta(x) = \sum_{i=0}^{\infty} \Delta_i((x_i, u_i), (x, u));$$

to justify the last definition we recall the independence of Δ_i on the last variable. Observe that all these functions are positive and finite, Φ, Ψ, and Q are d-continuous on M, and Ψ is everywhere upper Fréchet differentiable.

Since

$$f'(x_\infty; u_\infty) \leq f'(x_0; u_0) + \psi(x_0) < f'(x_0; u_0) + \varepsilon$$

by (12.20), the proof of the proposition will be completed once we show that f is Fréchet differentiable at x_∞ however the function ψ is chosen, and that $x_\infty \notin P$ for a particular choice of ψ.

We first collect a number of simple estimates, all following from the choice of parameters and from (12.19)–(12.22), which will be needed in what follows.

Lemma 12.4.10. *For each* $0 \leq i \leq \infty$, $\|x_\infty - x_i\| < \frac{1}{2}r$, $\|u_\infty - u_i\| < \frac{1}{2}r$, $\|u_\infty - u_i\| < \frac{1}{16C}s_i$, $|f'(x_\infty; u_\infty) - f'(x_i; u_i)| < \frac{1}{8}s_i$, $\varrho(f_{x_i,u_i}, f_{x_\infty,u_i}) \leq s_i + \varepsilon_i$, and $|f'(x_i; u_i)| \leq \frac{1}{8}\kappa$.

Proof. Since $F_i((x_i, u_i), (x_\infty, u_\infty)) \leq \varepsilon_i$ by (12.19), we have

$$\|x_\infty - x_i\| = \frac{1}{\lambda_i} \Phi_i(x_i, x_\infty) \leq \frac{\varepsilon_i}{\lambda_i} < \frac{r}{2},$$

$$\Theta(u_\infty - u_i) = \frac{1}{\lambda_i} \Psi_i(u_i, u_\infty) \leq \frac{\varepsilon_i}{\lambda_i},$$

which implies that $\|u_\infty - u_i\| < \min\{\frac{1}{2}r, \frac{1}{16C}s_i\}$, and

$$\big(f'(x_\infty; u_\infty) - f'(x_i; u_i) \big)^2 = \frac{1}{\sigma_i} Q_i((x_i, u_i), (x_\infty, u_\infty)) \leq \frac{\varepsilon_i}{\sigma_i} < \left(\frac{s_i}{8}\right)^2.$$

To estimate $\varrho(f_{x_i,u_i}, f_{x_\infty,u_i})$, we notice that $\lambda_i > \varepsilon_i$, and so the inequality

$$\min\{\lambda_i, \varrho(f_{x_i,u_i}, f_{x_\infty,u_i}) - s_i\} \leq \Delta_i((x_i, u_i), (x_\infty, u_\infty)) \leq \varepsilon_i$$

yields that $\varrho(f_{x_i,u_i}, f_{x_\infty,u_i}) - s_i \leq \varepsilon_i$.

For the last statement we use the choice of κ and (12.20) to infer that

$$-\frac{1}{8}\kappa \leq \inf_{(x,u)\in M} f'(x; u) \leq f'(x_i; u_i) \leq f'(x_0; u_0) + \varepsilon \leq \frac{1}{8}\kappa. \qquad \square$$

Assuming that $f'(x_\infty)$ exists, we easily guess its value from the fact that the function $H(u) = h_\infty(x_\infty, u)$ attains its minimum at $u = u_\infty$. Let $L_\Psi \in X^*$ be an upper Fréchet derivative of Ψ at u_∞. By (12.23), H is upper differentiable at $u = u_\infty$, with upper derivative

$$f'(x_\infty) + L_\Psi + R\,f'(x_\infty),$$

where

$$R := 2\sum_{i=0}^{\infty} \sigma_i(f'(x_\infty; u_\infty) - f'(x_i; u_i)).$$

Since this upper derivative must be equal to zero, $(1 + R)f'(x_\infty) = -L_\Psi$. Hence to guess $f'(x_\infty)$ we just need to show that $R \neq -1$. This holds since the final statement of Lemma 12.4.10 implies the estimate (which is better than what we need now, since it will also be used later)

$$|R| \leq \kappa \sum_{i=0}^{\infty} \sigma_i = 2\kappa\sigma_0 \leq \tau. \tag{12.24}$$

Since $\tau < 1$, this allows us to define

$$L := \frac{-L_\Psi}{1 + R}, \tag{12.25}$$

and our task now is to show that indeed $f'(x_\infty) = L$.

We first make the simple but important observation that the above heuristic argument is correct when restricted to the direction u_∞, giving that the derivative of f at the point x_∞ in the direction u_∞ agrees with the value of the linear form L at u_∞.

Lemma 12.4.11. $f'(x_\infty; u_\infty) = L(u_\infty)$.

Proof. Since $\|u_\infty - u_0\| < \frac{1}{2}r$ by Lemma 12.4.10, the vector su_∞ belongs to U for $s \in (1-\delta, 1+\delta)$ and suitable $0 < \delta < 1$. Hence $(x_\infty, su_\infty) \in M$ for these s, implying that the function $\zeta \colon (1 - \delta, 1 + \delta) \longrightarrow \mathbb{R}$,

$$\zeta(s) = f'(x_\infty; su_\infty) + \psi(x_\infty) + \Phi(x_\infty) + \Psi(su_\infty) + \Delta(x_\infty)$$
$$+ \sum_{i=0}^{\infty} \sigma_i\big(f'(x_\infty; su_\infty) - f'(x_i; u_i)\big)^2$$

attains its minimum at $s = 1$. Since ζ is upper differentiable at $s = 1$ with upper derivative $f'(x_\infty; u_\infty) + L_\Psi u_\infty + Rf'(x_\infty; u_\infty)$, we get

$$f'(x_\infty; u_\infty) + L_\Psi u_\infty + Rf'(x_\infty; u_\infty) = 0,$$

yielding

$$f'(x_\infty; u_\infty) = -\frac{L_\Psi u_\infty}{1 + R} = Lu_\infty$$

as required. $\square$

Before embarking on the main part of the proof, we notice that

$$\|L\| \le 2\|L_\Psi\| \le 2\,\mathrm{Lip}(\Psi) \le 4\lambda_0\,\mathrm{Lip}(\Theta), \tag{12.26}$$

and we show the following estimate, whose main point is to enable the use of Corollary 12.3.4 along suitable paths γ. Notice that if, in the Lipschitz case, we have defined κ as explained in Remark 12.4.8, this lemma has a simple proof:

$$\mathrm{Lip}(h) \le (\mathrm{Lip}(f) + \|L\|)\,\mathrm{Lip}(\gamma) \le 2(\mathrm{Lip}(f) + 4\lambda_0\,\mathrm{Lip}(\Theta)) \le \tfrac{1}{2}\kappa.$$

Lemma 12.4.12. *Let* $0 < \beta < \tfrac{1}{4}r$, $0 < \delta < \tfrac{1}{8}r$, *and a Lipschitz curve* $\gamma\colon \mathbb{R} \longrightarrow X$ *be such that* $16\lambda_0\,\mathrm{Lip}(\Theta)\beta \le \kappa$, $\|\gamma(0) - x_\infty\| \le \tfrac{1}{4}r$, *and also* $\|\gamma'(t) - u_\infty\| \le \beta$ *for almost all* $t \in \mathbb{R}$. *Then the function* $h\colon [-2\delta, 2\delta] \longrightarrow \mathbb{R}$,

$$h(t) = -f(\gamma(t)) + L(\gamma(t) - x_\infty)$$

is well defined and satisfies $\overline{D}h(t) \le \tfrac{1}{2}\kappa$ *for all* $t \in [-\delta, \delta]$.

Proof. For $t \in [-2\delta, 2\delta]$ we have

$$\|\gamma(t) - x_0\| \le \|\gamma(t) - \gamma(0)\| + \|\gamma(0) - x_\infty\| + \|x_\infty - x_0\|$$

$$\le 2\delta(\|u_\infty\| + \beta) + \frac{r}{4} + \frac{r}{2} \le \frac{r}{4}\Big(1 + \frac{r}{4} + \frac{r}{2}\Big) + \frac{3r}{4} < 3r.$$

Hence h is well defined. Using (12.26) and the estimate $f'(x_\infty; u_\infty) \le \tfrac{1}{8}\kappa$ obtained in Lemma 12.4.10, we get for $-2\delta \le t < s \le 2\delta$,

$$L(\gamma(s)) - L(\gamma(t)) \le f'(x_\infty; u_\infty)(s - t) + \|L\|\|\gamma(s) - \gamma(t) - (s - t)u_\infty\|$$

$$\le \frac{1}{8}\kappa\,(s - t) + 4\lambda_0\,\mathrm{Lip}(\Theta)\|\gamma(s) - su_\infty - (\gamma(t) - tu_\infty)\|$$

$$\le \frac{1}{8}\kappa\,(s - t) + 4\lambda_0\,\mathrm{Lip}(\Theta)\beta(s - t) \le \frac{3}{8}\kappa(s - t).$$

Letting $v := (\gamma(s) - \gamma(t))/(s - t)$ and observing that the inequalities $\beta \le \tfrac{1}{2}r$ and $\|u_\infty - u_0\| < \tfrac{1}{2}r$ imply that $v \in U$, we get

$$f(\gamma(s)) - f(\gamma(t)) = f(\gamma(t) + (s - t)v) - f(\gamma(t)) \ge -\frac{1}{8}\kappa(s - t).$$

Hence

$$h(t) - h(s) \le -\big(f(\gamma(s)) - f(\gamma(t))\big) + L(\gamma(s)) - L(\gamma(t)) \le \frac{1}{2}\kappa(t - s). \qquad \square$$

Main estimate

Our goal here is to prove the rather technical Proposition 12.4.14, whose main aim is to transform the results of the mean value estimates of Section 12.3 to suitable paths in the space X. Again, we need to introduce a number of parameters. For each $k \ge 1$, we define parameters a_k, δ_k, β_k, which will serve to control the speed of differentiability of f at x_∞.

(α) We choose $0 < a_k < 1$ such that $4k\kappa C^2 e^{s_0+\varepsilon_0} a_k < \varepsilon_k$.

(β) We find $0 < \delta_k < \frac{1}{8}r$ such that

- $\psi(x) < \psi(x_\infty) + \varepsilon_k$ for $\|x - x_\infty\| \le 4\delta_k$;
- $\Phi(x) < \Phi(x_\infty) + \tau\varepsilon_k$ for $\|x - x_\infty\| \le 4\delta_k$;
- $|f_{x_\infty,u_\infty}(t)| \le \kappa|t|$ for $|t| \le 4C\delta_k$;
- $|f_{x_i,u_i}(t) - tf'(x_i; u_i)| \le \frac{1}{8}s_i|t|$ for $0 \le i \le k$ and $|t| \le \delta_k$.

(γ) Finally, we choose the largest number $\beta_k \in [0, \frac{1}{4}r]$ such that

$$16\lambda_0 \operatorname{Lip}(\Theta)\beta_k \le \kappa, \ 16C\beta_k \le s_k, \ 4\sigma_0\|L\|^2\beta_k^2 \le \tau\varepsilon_k, \ 8\|L\|\beta_k \le s_k,$$

and for every $\|u - u_\infty\| \le \beta_k$,

$$\Psi(u) \le \Psi(u_\infty) + L_\Psi(u - u_\infty) + \tau\varepsilon_k.$$

Since the last inequality arises from the estimate of Ψ by its upper Fréchet derivative at u_∞, one may expect the error term to be a small multiple of $\|u - u_\infty\|$. In the above formula $\|u - u_\infty\|$ has been incorporated into ε_k. A corollary of this choice is the fact, which will become crucial at the very last stage of the proof of Fréchet differentiability, that $\varepsilon_k = o(\beta_k)$. Because of its importance, we state it and give a (simple) proof.

Observation 12.4.13. $\lim_{k\to\infty} \varepsilon_k/\beta_k = 0$.

Proof. Given any $\eta > 0$ there is $\beta > 0$ such that

$$\Psi(u) \le \Psi(u_\infty) + L_\Psi(u - u_\infty) + \tau\eta\|u - u_\infty\|$$

for $\|u - u_\infty\| \le \beta$. Since $\varepsilon_k \to 0$, it follows that for all sufficiently large k the number ε_k/η has all properties required from β_k. Hence $\beta_k \ge \varepsilon_k/\eta$, and so $\varepsilon_k/\beta_k \le \eta$. $\qquad\square$

Provided the function f is not Fréchet differentiable at x_∞ or that x_∞ does not belong to the given porous set, the mean value estimates of Section 12.3 used along a suitable path γ can provide a point at which we can control both the derivative and the increment along γ. The following statement gives a method of using such control to finish the proof of our main results.

Proposition 12.4.14. *Let $k \ge 1$, $0 < \delta \le \delta_k$, $0 < a \le a_k\delta$, and $\gamma\colon \mathbb{R} \longrightarrow X$ be a Lipschitz curve such that $\|\gamma(0) - x_\infty\| \le a_k\delta$ and $\|\gamma'(t) - u_\infty\| \le \beta_k$ for almost all $t \in \mathbb{R}$. Define $h\colon [-2\delta, 2\delta] \longrightarrow \mathbb{R}$ by the same formula as in Lemma 12.4.12,*

$$h(t) = -f(\gamma(t)) + L(\gamma(t) - x_\infty), \tag{12.27}$$

and suppose that $\xi \in [-a, a]$ is such that

(i) *both γ and h are differentiable at ξ and $-\varepsilon_k \le h'(\xi) \le \kappa$;*

(ii) $|h(t) - h(\xi)| \le 137\sqrt{\kappa(\varepsilon_k + h'(\xi))}|t - \xi|$ *for every* $t \in [-2\delta, 2\delta]$.

Then

$$h'(\xi) + \psi(x_\infty) - \psi(\gamma(\xi)) < 3\varepsilon_k. \tag{12.28}$$

The rest of this section is devoted to the proof of this lemma. We will continue using its notation and assumptions, and also denote $x = \gamma(\xi)$ and $u = \gamma'(\xi)$. Notice that

$$\|x - x_\infty\| \le \|\gamma(0) - x_\infty\| + \|x - \gamma(0)\| \le a_k\delta + (2 + \beta_k)a_k\delta$$
$$\le 4a_k\delta. \tag{12.29}$$

In particular, since $4a_k\delta \le 4\delta_k$, the first condition in the choice of δ_k in (β) implies that $\psi(x) < \psi(x_\infty) + \varepsilon_k$.

We will argue by contradiction, and so will assume that (12.28) is false, that is, that

$$H := 2\varepsilon_k + h'(\xi) - \psi(x) + \psi(x_\infty) \ge 5\varepsilon_k.$$

The $2\varepsilon_k$ have been added to make $H = (\varepsilon_k + h'(\xi)) + (\varepsilon_k - \psi(x) + \psi(x_\infty))$ positive and also to guarantee that $\varepsilon_k + h'(\xi) \le H$, which allows us to replace the term $\sqrt{\kappa(\varepsilon_k + h'(\xi))}$ in (ii) by $\sqrt{\kappa H}$.

Lemma 12.4.15. *For every* $0 \le i \le k$,

$$\varrho(f_{x_i,u_i}, f_{x,u_i}) \le \max\left\{\varrho(f_{x_i,u_i}, f_{x_\infty,u_i}) + \frac{\varepsilon_k}{k}, \; 137\sqrt{\kappa H} + \frac{s_i}{2}\right\}.$$

Proof. By Lemma 12.4.3 (ii)

$$\varrho(f_{x,u_i}, f_{x_i,u_i}) \le \max\left\{\begin{array}{l} \varrho(f_{x_i,u_i}, f_{x_\infty,u_i}) + \dfrac{C}{\delta}e^{\varrho(f_{x_i,u_i},f_{x_\infty,u_i})}\|x - x_\infty\|, \\[2ex] \varrho(f_{x_i,u_i}, f_{x_\infty,u_i}) + \dfrac{C}{\delta}\displaystyle\sup_{|t|\le C\|x-x_\infty\|}|f_{x_\infty,u_\infty}(t)|, \\[2ex] \displaystyle\sup_{0<|t|<\delta}\inf_{w\in U}\left(C\|u - w\| + \left|\dfrac{f_{x,w}(t) - f_{x_i,u_i}(t)}{t}\right|\right) \end{array}\right\}.$$

Since (12.29) gives

$$\|x - x_\infty\| \le 4a_k\delta \le \delta e^{-(s_0+\varepsilon_0)}\frac{\varepsilon_k}{kC},$$

and the penultimate inequality of Lemma 12.4.10 shows that

$$\varrho(f_{x_i,u_i}, f_{x_\infty,u_i}) \le s_0 + \varepsilon_0,$$

the first term in the maximum is at most $\varrho(f_{x_i,u_i}, f_{x_\infty,u_i}) + \varepsilon_k/k$.

We obtain the same estimate also for the second term (which is redundant in the Lipschitz case). Assume that $|t| \le C\|x - x_\infty\|$. Then (12.29) gives that $|t| \le 4Ca_k\delta$, which due to the choice of δ_k implies

$$|f_{x_\infty,u_\infty}(t)| \le \kappa|t|.$$

Also, the choice of a_k implies that

$$|t| \leq C\|x - x_\infty\| \leq \frac{\delta \varepsilon_k}{\kappa k C}.$$

Hence

$$\frac{C}{\delta} \sup_{|t| \leq C\|x - x_0\|} |f_{x_\infty, u_\infty}(t)| \leq \frac{C}{\delta} \kappa \frac{\delta \varepsilon_k}{\kappa k C} = \frac{\varepsilon_k}{k},$$

and we obtain the bound $\varrho(f_{x_i, u_i}, f_{x_\infty, u_i}) + \varepsilon_k/k$ also for the second term in the maximum.

To estimate the last term, we consider any $0 < |t| < \delta$ and estimate the expression

$$C\|u_i - w\| + \left| \frac{f_{x_i, u_i}(t) - f_{x, w}(t)}{t} \right|$$

for $w = \frac{1}{t}(\gamma(\xi + t) - \gamma(\xi))$.

Observe first that $\|u_\infty - w\| \leq \beta_k$ by the conditions imposed on γ. Together with Lemma 12.4.10 this gives

$$C\|u_i - w\| \leq C\|u_i - u_\infty\| + C\|u_\infty - w\| \leq \frac{s_i}{16} + C\beta_k \leq \frac{s_i}{8}.$$

Since $f_{x, w}(t) = -(h(\xi + t) - h(\xi)) + tL(w)$, we can control the second summand by the following sum:

$$\left| \frac{f_{x, w}(t) - f_{x_i, u_i}(t)}{t} \right| \leq \left| \frac{h(\xi + t) - h(\xi)}{t} \right| + |L(w) - f'(x_\infty; u_\infty)|$$

$$+ |f'(x_\infty; u_\infty) - f'(x_i; u_i)| + \left| f'(x_i; u_i) - \frac{f_{x_i, u_i}(t)}{t} \right|.$$

We estimate each of the four terms in the sum above:

$$\left| \frac{h(\xi + t) - h(\xi)}{t} \right| \leq 137\sqrt{\kappa H} \qquad\qquad \text{by assumption;}$$

$$|L(w) - f'(x_\infty; u_\infty)| = |L(w) - L(u_\infty)| \leq \|L\|\beta_k \leq \frac{s_i}{8} \quad \text{by } (\gamma);$$

$$|f'(x_\infty; u_\infty) - f'(x_i; u_i)| \leq \frac{s_i}{8} \qquad\qquad \text{by Lemma 12.4.10;}$$

$$\left| \frac{f_{x_i, u_i}(t)}{t} - f'(x_i; u_i) \right| \leq \frac{s_i}{8} \qquad\qquad \text{by } (\beta).$$

Adding these inequalities, we get the required

$$C\|u_i - w\| + \left| \frac{f_{x, w}(t) - f_{x_i, u_i}(t)}{t} \right| \leq 137\sqrt{\kappa H} + \frac{s_i}{2}. \qquad \square$$

Lemma 12.4.16. $(x, u) \in M$.

Proof. We have $\|x - x_0\| < 4a_k\delta + \frac{1}{2}r < r$ and, since $\|u - u_\infty\| \leq \beta_k < \frac{1}{2}r$, we see that also $\|u - u_0\| \leq \|u_0 - u_\infty\| + \|u - u_\infty\| < r$. Since $f'(x; u)$ exists, $(x, u) \in M$. $\square$

Our ultimate aim is to arrive to a contradiction by using the inequality (12.22), which is just a quantitative way of saying that h_∞ attains its minimum on M at the point (x_∞, u_∞). For this, we first compare the terms of the function $h_\infty(y, v)$ at (x, u) and (x_∞, u_∞) that do not contain the function Δ. Let us denote them

$$G(y, v) = f'(y; v) + \psi(y) + \Phi(y) + \Psi(v) + Q(y, v).$$

Lemma 12.4.17. $G(x, u) < -\frac{1}{2} H + G(x_\infty, u_\infty).$

Proof. We estimate the last three terms $Q(x, u)$, $\Psi(u)$, and $\Phi(x)$ in $G(x, u)$. With the help of Lemma 12.4.11 we write $Q(x, u)$ as

$$\sum_{i=0}^{\infty} \sigma_i \Big[f'(x_i; u_i) - f'(x_\infty; u_\infty) + f'(x_\infty; u_\infty) - f'(x; u) \Big]^2$$

$$= Q(x_\infty, u_\infty) + R(f'(x; u) - f'(x_\infty; u_\infty)) + 2\sigma_0(f'(x; u) - f'(x_\infty; u_\infty))^2$$

$$= Q(x_\infty, u_\infty) + R(-h'(\xi) + L(u - u_\infty)) + 2\sigma_0\big(-h'(\xi) + L(u - u_\infty)\big)^2$$

$$\leq Q(x_\infty, u_\infty) - Rh'(\xi) + RL(u - u_\infty) + 4\sigma_0\big(h'(\xi)\big)^2 + 4\sigma_0\big(L(u - u_\infty)\big)^2.$$

Recall that by assumption (i) of Proposition 12.4.14 (which we are proving) we have $|h'(\xi)| \leq \max\{\varepsilon_k, \kappa\} = \kappa$ and $|h'(\xi)| \leq 2\varepsilon_k + h'(\xi) \leq \varepsilon_k + H$. Hence the choice of σ_0 gives

$$4\sigma_0\big(h'(\xi)\big)^2 \leq 4\sigma_0\kappa(\varepsilon_k + H) \leq \tau(\varepsilon_k + H).$$

Also, the choice of β_k in (γ) implies $4\sigma_0(L(u - u_\infty))^2 \leq 4\sigma_0\|L\|^2\beta_k^2 \leq \tau\varepsilon_k$, and the estimate $|R| \leq \tau$ from (12.24) shows that $|Rh'(\xi)| \leq \tau H + \tau\varepsilon_k$. This allows us to infer that

$$Q(x, u) \leq Q(x_\infty, u_\infty) + RL(u - u_\infty) + 2\tau H + 3\tau\varepsilon_k.$$

Since $\|u - u_\infty\| \leq \beta_k$, we see from (γ) that

$$\Psi(u) \leq \Psi(u_\infty) + L_\Psi(u - u_\infty) + \tau\varepsilon_k = \Psi(u_\infty) - (1 + R)L(u - u_\infty) + \tau\varepsilon_k.$$

Finally, the last needed inequality

$$\Phi(x) < \Phi(x_\infty) + \tau\varepsilon_k$$

follows from (β), since $\|x - x_\infty\| \leq 4\delta_k$.

Using all three just established estimates and recalling at the end that $5\varepsilon_k \leq H$, we get

$$G(x, u) = f'(x; u) + \psi(x) + \Phi(x) + \Psi(u) + Q(x, u)$$

$$< f'(x; u) + \psi(x_\infty) + \Phi(x_\infty) + \Psi(u_\infty) + Q(x_\infty, u_\infty)$$

$$\quad + \psi(x) - \psi(x_\infty) - L(u - u_\infty) + 2\tau H + 5\tau\varepsilon_k$$

$$= f'(x_\infty; u_\infty) + \psi(x_\infty) + \Phi(x_\infty) + \Psi(u_\infty) + Q(x_\infty, u_\infty)$$

$$\quad - (h'(\xi) - \psi(x) + \psi(x_\infty)) + 2\tau H + 5\tau\varepsilon_k$$

$$= G(x_\infty, u_\infty) - (1 - 2\tau)H + (2 + 5\tau)\varepsilon_k$$

$$\leq G(x_\infty, u_\infty) - \Big(1 - 2\tau - \frac{2 + 5\tau}{5}\Big)H = G(x_\infty, u_\infty) - \frac{1}{2} H. \qquad \square$$

Proof of Proposition 12.4.14. We show by induction that $H \leq 4\varepsilon_i$ for all $1 \leq i \leq k$. For $i = k$ this will contradict our assumption that $H \geq 5\varepsilon_k$.

So assume that $1 \leq i \leq k$ and $H \leq 4\varepsilon_j$ for all $1 \leq j < i$. (In this way we include the starting step $i = 1$ of the induction.) We first infer from Lemma 12.4.17 and (12.22) that

$$
\begin{aligned}
\frac{1}{2} H &\leq G(x_\infty, u_\infty) - G(x, u) \\
&= h_\infty(x_\infty, u_\infty) - \Delta(x_\infty) - G(x, u) \\
&\leq \varepsilon_i + f'(x; u) + \psi(x) + \varphi(x) + \sum_{j=0}^{i-1} F_j((x_j, u_j), (x, u)) \\
&\quad - \Delta(x_\infty) - G(x, u) \\
&\leq \varepsilon_i + \sum_{j=0}^{i-1} \Big(\Delta_j((x_j, u_j), (x, u)) - \Delta_j((x_j, u_j), (x_\infty, u_\infty)) \Big).
\end{aligned}
$$

To estimate the difference of the Δ_j, we show that $137\sqrt{\kappa H} \leq \frac{1}{2} s_j$ for each $0 \leq j < i$. For $j = 0$ we use that $0 \leq \psi \leq \varepsilon$ to get

$$
H \leq \kappa + 2\varepsilon_k + \varepsilon \leq \kappa + \varepsilon_0 + \varepsilon.
$$

Hence $137\sqrt{\kappa H} \leq 137\sqrt{\kappa(\kappa + \varepsilon_0 + \varepsilon)} \leq \frac{1}{2} s_0$. For $1 \leq j < i$ we obtain the required estimate from the induction assumption,

$$
137\sqrt{\kappa H} \leq 274\sqrt{\kappa \varepsilon_j} \leq \frac{1}{2} s_j.
$$

With these estimates, Lemma 12.4.15 implies that

$$
\begin{aligned}
\varrho(f_{x_j, u_j}, f_{x, u_j}) - s_j &\leq \max\Big\{ \varrho(f_{x_j, u_j}, f_{x_\infty, u_j}) - s_j + \frac{\varepsilon_k}{k}, \ 137\sqrt{\kappa H} - \frac{s_j}{2} \Big\} \\
&\leq \max\{ \varrho(f_{x_j, u_j}, f_{x_\infty, u_j}) - s_j, \ 0 \} + \frac{\varepsilon_k}{k}.
\end{aligned}
$$

Hence $\Delta_j((x_j, u_j), (x, u)) - \Delta_j((x_j, u_j), (x_\infty, u_\infty)) \leq \varepsilon_k/k$ for each $0 \leq j < i$, implying that

$$
H \leq 2\varepsilon_i + 2 \sum_{j=0}^{i-1} \frac{\varepsilon_k}{k} \leq 2\varepsilon_i + 2\varepsilon_k \leq 4\varepsilon_i. \qquad \square
$$

Fréchet differentiability of f at x_∞

Given any $k \geq 1$, we let $\tau = 6\varepsilon_k$, $\eta = 13\tau/\beta_k$ and use Lemma 12.4.11 to find $0 < \delta \leq \delta_k$ such that

$$
|f(x_\infty + tu_\infty) - f(x_\infty) - tLu_\infty| \leq \tau^2 |t|/\kappa \tag{12.30}
$$

for $|t| \leq 2\delta$. We show that

$$|f(x_\infty + v) - f(x_\infty) - L(v)| \leq \eta\|v\|$$

whenever $\|v\| < a_k\beta_k\delta$. Since $\eta = 78\varepsilon_k/\beta_k \to 0$ as $k \to \infty$, this will show that f is Fréchet differentiable at x_∞ and $f'(x_\infty) = L$.

We will argue by contradiction, and so assume that

$$|f(x_\infty + v) - f(x_\infty) - L(v)| > \eta\|v\|. \tag{12.31}$$

for some $0 < \|v\| < a_k\beta_k\delta$. Let $a = \frac{1}{\beta_k}\|v\|$ and define $\gamma\colon \mathbb{R} \longrightarrow X$ by

$$\gamma(t) = x_\infty + tu_\infty + \max\{0, 1 - |t|/a\}v.$$

Let $h\colon [-2\delta, 2\delta] \longrightarrow \mathbb{R}$ be as in (12.27). We intend to use Corollary 12.3.4 with the τ, κ, a, δ defined above and

$$g(t) = f(x_\infty) + h(t) = f(x_\infty) - f(\gamma(t)) + L(\gamma(t) - x_\infty).$$

To verify the assumptions of this corollary, we clearly have $\tau < \kappa$, $a < \delta$, and the inequalities $|g(0)| > \eta\|v\| = 13\tau a$ and

$$|g(t)| = |f(x_\infty + tu_\infty) - f(x_\infty) - tLu_\infty| \leq \frac{\tau^2}{\kappa}|t| \text{ for } a \leq |t| \leq 2\delta$$

follow directly from (12.31) and (12.30). The remaining assumption, namely, that $\sup_{t\in[-a,a]} \overline{D}g(t) \leq \kappa$ has been proved in Lemma 12.4.12, which is applicable because $\|\gamma(0) - x_\infty\| = \|v\| < a\delta_k \leq \frac{1}{4}r$ and $\|\gamma'(t) - u_\infty\| \leq \|v\|/a = \beta_k < \frac{1}{4}r$. So Corollary 12.3.4 provides us with a $\xi \in (-a, a) \setminus \{0\}$ such that

(a) g is differentiable at ξ;

(b) $g'(\xi) \geq 6\varepsilon_k$;

(c) $|g(t) - g(\xi)| \leq 123\sqrt{\kappa g'(\xi)}|t - \xi|$ for every $t \in [-2\delta, 2\delta]$.

Since h differs from g only by a constant, statements (a)–(c) hold also with g replaced by h. Hence Proposition 12.4.14 implies that

$$h'(\xi) < 3\varepsilon_k + \psi(\gamma(\xi)) - \psi(x_\infty).$$

Since $\|\gamma(\xi) - x_\infty\| \leq \delta_k$, we have $\psi(\gamma(\xi)) \leq \psi(x_\infty) + \varepsilon_k$, and we conclude that $g'(\xi) = h'(\xi) < 6\varepsilon_k$ in contradiction to (b).

Putting the point x_∞ outside the given σ-porous set

We now describe a special choice of the function ψ that will force the point x_∞ to avoid a given σ-porous set P. Find $q \geq 1$ such that $\sum_{i=q}^{\infty} 8\varepsilon_i < \varepsilon$ and write $P = \bigcup_{i=q}^{\infty} P_i$, where P_i is c_i-porous. Denoting by Q_i the closure of P_i we see that the function

$$\psi(x) := \sum_{i=q}^{\infty} 8\varepsilon_i \, 1_{Q_i}(x)$$

is upper semicontinuous and $0 \le \psi(x) < \varepsilon$ for every $x \in X$.

We show that, if our construction is done with the above ψ, the point x_∞ is outside P. Suppose, for a contradiction, that $x_\infty \in P$, and find the least $k \ge q$ for which $x_\infty \in P_k$. We let $\tau = \varepsilon_k/14$ and define η by the requirement $\eta(4/c_k + \|u_\infty\|) = \tau^2/\kappa$. Since f is Fréchet differentiable at x_∞ we find $0 < \delta < \frac{1}{5}\delta_k$ such that

$$|f(x_\infty + u) - f(x_\infty) - L_\infty u| \le \eta\|u\|$$

for every u with $\|u\| \le 5\delta$. We take δ small enough that also $B(x_\infty, 2\delta) \cap Q_i = \emptyset$ for every $q \le i < k$ for which $x_\infty \notin Q_i$.

Find $0 < \|v\| < a_k\delta$ such that $B(x_\infty + v, c_k\|v\|) \cap P_k = \emptyset$ and define a curve $\gamma\colon \mathbb{R} \longrightarrow X$ by

$$\gamma(t) = x_\infty + v + tu_\infty.$$

Let $h\colon [-2\delta, 2\delta] \longrightarrow \mathbb{R}$ be as in (12.27), and define

$$\begin{aligned} g(t) &= f(x_\infty + v) - Lv + h(t) + \varepsilon_k \max\{0, a - |t|\} \\ &= f(x_\infty + v) - Lv - f(\gamma(t)) + L(\gamma(t) - x_\infty) + \varepsilon_k \max\{0, a - |t|\}. \end{aligned}$$

The choice of δ guarantees that $\gamma(t) \in B$ for $|t| \le \delta$.

We intend to use Corollary 12.3.4 with this function g, numbers κ and δ defined above, $\tau = \varepsilon_k/14$ and $a = \frac{1}{2}c_k\|v\|$. Clearly, $\tau < \kappa$, $a < \delta$, and $|g(0)| = \varepsilon_k a > 13\tau a$. Also, by Lemma 12.4.12 we have $\overline{D}g(t) \le \frac{1}{2}\kappa + \varepsilon_k \le \kappa$ for $t \in [-a, a]$. Finally, for $a \le |t| \le 2\delta$ we get

$$\begin{aligned} |g(t) - g(0)| &\le |f(x_\infty + v) - f(x_\infty) - Lv| \\ &\quad + |f(x_\infty + v + tu_\infty) - f(x_\infty) - L(v + tu_\infty)| \\ &\le \eta(2\|v\| + |t|\|u_\infty\|) = \eta\left(\frac{4a}{c_k} + |t|\|u_\infty\|\right) \le \frac{\tau^2}{\kappa}|t|. \end{aligned}$$

Hence by Corollary 12.3.4 there is $\xi \in (-a, a) \setminus \{0\}$ such that g is differentiable at ξ,

(i) $\tau \le g'(\xi) \le \kappa$;

(ii) $|g(t) - g(\xi)| \le 123\sqrt{\kappa g'(\xi)}|t - \xi|$ for every $t \in [-2\delta, 2\delta]$.

This implies that $h'(\xi) \ge -\varepsilon_k$ and

$$\begin{aligned} |h(t) - h(\xi)| &\le |g(t) - g(\xi)| + 14\tau|t - \xi| \\ &\le 137\sqrt{\kappa g'(\xi)}|t - \xi| \le 137\sqrt{\kappa(\varepsilon_k + h'(\xi))}|t - \xi| \end{aligned}$$

for every $t \in [-2\delta, 2\delta]$. Hence, letting $x = \gamma(\xi)$, we have from Proposition 12.4.14 that

$$\psi(x) > \psi(x_\infty) + h'(\xi) - 3\varepsilon_k \ge \psi(x_\infty) - 4\varepsilon_k.$$

To deduce the opposite inequality we recall that $1_{Q_k}(x_\infty) = 1$, while $1_{Q_k}(x) = 0$ since $x \in B(x_\infty + v, c\|v\|)$. The inequality $\|x - x_\infty\| < 2\delta$ and the choice of δ imply that $1_{Q_i}(x) \le 1_{Q_i}(x_\infty)$ for $q \le i < k$. It follows that

$$\psi(x) \le \sum_{i=q}^{k-1} 8\varepsilon_i \, 1_{Q_i}(x_\infty) + \sum_{i=k+1}^{\infty} 8\varepsilon_i < \sum_{i=q}^{k-1} 8\varepsilon_i \, 1_{Q_i}(x_\infty) + 4\varepsilon_k \le \psi(x_\infty) - 4\varepsilon_k,$$

and we have reached our final contradiction.

12.5   GENERALIZATIONS AND EXTENSIONS

Straightforward extensions of results on existence of points of Fréchet differentiability have in fact little to do with differentiability. One looks for conditions under which for a given function there is a nonempty open set on which it is Lipschitz or cone-monotone; a differentiability result then can be immediately stated by combining this with the corresponding result of the previous sections.

For example, it is easy to see that a continuous, everywhere Gâteaux differentiable function $f \colon X \longrightarrow \mathbb{R}$ is Lipschitz on a nonempty open set; so in the illustrative example of Section 12.2 the assumption that f be Lipschitz was redundant. More generally, if a continuous function $f \colon X \longrightarrow Y$ satisfies

$$\limsup_{t \to 0} \frac{\|f(x + tu) - f(x)\|}{|t|} < \infty$$

for every x in a nonempty open subset of a Banach space and for every unit vector u, then it is Lipschitz on a nonempty open set. This is proved by a standard application of the Baire category theorem. Similar generalizations may be obtained also for cone monotonicity.

In a very different direction, one may generalize the above results to other derivatives than just the Fréchet derivative. For example, it is rather easy to see that, if the function Θ in Proposition 12.4.1 is assumed to be everywhere upper Gâteaux (instead of Fréchet) differentiable, our proof provides a point of Gâteaux differentiability of every Lipschitz or even cone-monotone function. Of course, there are better Gâteaux differentiability results for separable spaces, but this statement has its interest in the nonseparable situation, when it was proved before only for Lipschitz functions and only on spaces with Gâteaux smooth norms.

Chapter Thirteen

Fréchet differentiability of vector-valued functions

We prove that if a space X admits a bump function which is upper Fréchet differentiable, Lipschitz on bounded sets, and asymptotically smooth with modulus controlled by $t^n \log^{n-1}(1/t)$, then every Lipschitz map of X to a space of dimension not exceeding n has points of Fréchet differentiability. We also show that in this situation the multidimensional mean value estimate for Fréchet derivatives of locally Lipschitz maps of open subsets of X to spaces of dimension not exceeding n holds. In Chapter 14 we will see that this mean value statement is close to optimal. Particular situations in which our results apply include maps of Hilbert spaces to $\mathbb{R}^2$ and maps of ℓ^p to $\mathbb{R}^n$ for $1 < p < \infty$ and $n \leq p$.

13.1 MAIN RESULTS

We prove a Fréchet differentiability result for $\mathbb{R}^n$-valued functions on Banach spaces satisfying, for the given number n, one of the smoothness assumptions of order n introduced in Chapter 8.

Theorem 13.1.1. *Suppose that a Banach space X admits a bump function which is upper Fréchet differentiable, Lipschitz on bounded sets, and asymptotically smooth with modulus controlled by $t^n \log^{n-1}(1/t)$. Let Y be a Banach space of dimension not exceeding n and $f : G \longrightarrow Y$ a locally Lipschitz function defined on a nonempty open subset G of X. Then f has points of Fréchet differentiability and, moreover, the multidimensional mean value estimate holds for Fréchet derivatives of f.*

The mean value estimates are discussed in Section 2.4.

As we have pointed out in Section 8.2, for $n = 1$ the assumptions of Theorem 13.1.1 hold for any space X with separable dual, since such a space admits an equivalent Fréchet smooth norm (see, e.g., Theorem 3.2.1). Hence Theorem 13.1.1 includes the result of [39] that real-valued Lipschitz functions on such spaces have points of Fréchet differentiability. It also includes the corresponding mean value estimate. However, it does not include the stronger result on Fréchet differentiability of cone-monotone functions that we proved in Chapter 12.

When $n \geq 2$, Theorem 13.1.1 is new, even when one asks only about existence of points of Fréchet differentiability. The most interesting spaces satisfying its assumptions with $n = 2$ are Hilbert spaces, as mentioned in Section 8.2. The following immediate corollary is worth pointing out.

Corollary 13.1.2. *Every pair of real-valued Lipschitz functions f, g defined on a non-empty open subset G of a Hilbert space has a common point of Fréchet differentiability. Moreover, for any $u, v \in X$ and $c \in \mathbb{R}$, the following three statements are equivalent.*

(i) *There is a point $x \in G$ at which both f, g are differentiable in the direction of $\text{span}\{u, v\}$ and $f'(x; u) + g'(x; v) > c$.*

(ii) *There is a common point $x \in G$ of Gâteaux differentiability of f, g such that $f'(x; u) + g'(x; v) > c$.*

(iii) *There is a common point $x \in G$ of Fréchet differentiability of f, g such that $f'(x; u) + g'(x; v) > c$.*

In Chapter 8 we have also noticed for which p the space ℓ^p satisfies the assumptions of Theorem 13.1.1. We again point out what this theorem says about ℓ^p spaces.

Corollary 13.1.3. *Suppose that $1 < p < \infty$, $n \in \mathbb{N}$, and $p \geq n$. Then every family of n real-valued Lipschitz functions $f_1, \ldots, f_n$ defined on a nonempty open subset G of ℓ^p possesses a common point of Fréchet differentiability. Moreover, for any $u_1, \ldots, u_n \in X$, and $c \in \mathbb{R}$, the following three statements are equivalent.*

(i) *There is a point $x \in G$ at which all f_i are differentiable in the direction of $\text{span}\{u_1, \ldots, u_n\}$ and $\sum_{k=1}^{n} f_k'(x; u_k) > c$.*

(ii) *There is a common point $x \in G$ of Gâteaux differentiability of all f_i such that $\sum_{k=1}^{n} f_k'(x; u_k) > c$.*

(iii) *There is a common point $x \in G$ of Fréchet differentiability of all f_i such that $\sum_{k=1}^{n} f_k'(x; u_k) > c$.*

Chapter 14 shows that the "moreover" part of this corollary is false once $p < n$. In particular, the "moreover" part of Corollary 13.1.2 fails for three functions. However, for what we know, the first part, that is, the existence of common points of Fréchet differentiability, may hold in all spaces with separable dual and for any number of functions (even for countably many). In Section 10.6, building on results of [28], we have identified a class of spaces in which this holds. But it is unknown, for example, for two functions on ℓ_p, $1 < p < 2$, and for three functions on a Hilbert space.

In this connection we should recall that in Chapter 4 we have seen that points of ε-Fréchet differentiability exist for Lipschitz maps from asymptotically uniformly smooth spaces to finite dimensional spaces. The problem in making $\varepsilon = 0$ is that, unlike in the proof of the results of this chapter, in the proof of Theorem 4.3.3 (as well as in the proof of this result in [22] and in the earlier, more sophisticated, proof in the special case of superreflexive spaces in [26]) there is no relation between points of ε-Fréchet differentiability for different values of ε.

13.2 REGULARITY PARAMETER

The main results of this section are Corollary 13.2.4 and Lemma 13.2.5, which estimate the following "regularity parameter." This continues and refines results of Section 9.5 in a form suitable for applications in this chapter. In particular, the target of our mappings will be a Euclidean space E of unspecified dimension (or a more general space as discussed in Section 9.5).

Definition 13.2.1. Let $g\colon \mathbb{R}^p \longrightarrow E$ be a Lipschitz function, $1 \le n \le p$, and $u \in \mathbb{R}^p$. The number

$$\operatorname{reg}_n g(u) = \sup_{\substack{|v|+|w|>0 \\ v-w \in \mathbb{R}^n}} \frac{|g(u+v) - g(u+w)|}{|v| + |w|} \tag{13.1}$$

is called the *defect of regularity* of g at the point u.

More generally, if $\Omega \subset \mathbb{R}^p$ we define $\operatorname{reg}_n g(u, \Omega)$, the *defect of restricted regularity*, by the same formula (13.1) but with the additional requirement that the open straight segment $(u + v, u + w)$ lies in the given set Ω.

Notice that $\operatorname{reg}_n g(u, \Omega)$ depends only on n and the behavior of g on $u + \Omega$. The ambient space $\mathbb{R}^p$ plays no role, and so we may, whenever we wish, change it to any $\mathbb{R}^j$ as long as $u + \Omega \subset \mathbb{R}^j$. Although this may sound obvious, in our integral estimates of the defect of regularity this remark will find an important application to restricting the number of variables over which we have to integrate.

Our goal here is to refine the results of Section 9.5 to estimates of the integral of a suitable power of $\operatorname{reg}_n g$ with the help of the integral of $\|g_n'\|^2$. The natural technique for this is to use estimates of maximal operators of the Hardy-Littlewood type, but we need these operators acting only on some coordinates. The operator $\mathbf{M}_n$, where $0 \le n \le p$, will assign to every measurable (possibly vector-valued) function h on $\mathbb{R}^p$ the function

$$\mathbf{M}_n h(u) = \sup_{r>0} \fint_{B_n(u,r)} \|h\| \, d\mathscr{L}^n.$$

By the maximal operator inequality of (9.7) and Fubini's theorem, for every measurable function h defined on $\mathbb{R}^p$, $p \ge n$,

$$\int_{\mathbb{R}^p} (\mathbf{M}_n h)^2 \, d\mathscr{L}^p \le \Lambda_n \int_{\mathbb{R}^p} \|h\|^2 \, d\mathscr{L}^p. \tag{13.2}$$

The constant Λ_n is the same as in (9.7). Although it is evident that Λ_n does not depend on p, this simple observation is of paramount importance for us. In the case when $n = p$, $\mathbf{M}_p$ is just another notation for the Hardy-Littlewood maximal operator $\mathbf{M}$.

By analogy with results of Section 9.5, we may expect that the required estimate of $\operatorname{reg}_n g$ will involve the derivative and Lipschitz constant of g with respect to the first n variables. We therefore simplify the notation for them.

Definition 13.2.2. Let $\Omega \subset \mathbb{R}^p$ be open, $g\colon \Omega \longrightarrow E$, and $0 \le n \le p$. We denote by g_n' the derivative of the map g with respect to the first n variables. In other words, $g_n'(x) = g_{\mathbb{R}^n}'(x) \in L(\mathbb{R}^n, E)$. We also denote

$$\operatorname{Lip}_n(g) = \sup_{\substack{u,v \in \Omega, \\ u-v \in \mathbb{R}^n \setminus \{0\}}} \frac{|g(u) - g(v)|}{|u - v|}.$$

The basic estimate of the defect of regularity follows from Lemma 9.5.4.

Lemma 13.2.3. *Let $n \leq j \leq p$. Assume that $g \colon \mathbb{R}^p \longrightarrow E$ is Lipschitz, $\Omega \subset \mathbb{R}^p$ is a bounded convex set, $u \in \overline{\Omega}$, and $\Omega - u \subset \mathbb{R}^j$. Suppose further that $\Omega - u$, when considered as a subset of $\mathbb{R}^j$, is open with eccentricity at most ρ. Then*

$$\big(\mathrm{reg}_n\, g(u,\Omega)\big)^j \leq \overline{K}_j (\rho \, \mathrm{Lip}_j(g))^{j-1} \mathbf{M}_j (1_\Omega\, g_n')(u), \tag{13.3}$$

where $\overline{K}_j = 3 \cdot 2^{j-1} \alpha_j K_j = 3 \cdot 24^{j-1} \alpha_j / \alpha_{j-1}$ and K_j are from Lemma 9.5.4.

Proof. Since g may be equivalently replaced by its multiple by a positive constant, we may assume that $\mathrm{Lip}(g) = 1$. Noticing that the problem is translationally invariant, we assume that $u = 0$. Finally, recalling the independence of $\mathrm{reg}_n\, g(0, \Omega)$ on the ambient space, we also assume that $p = j$.

Suppose that $v, w \in \mathbb{R}^j$ are such that $v, w \in \Omega$ and $v - w \in \mathbb{R}^n$ and denote $R = \max\{|v|, |w|\}$. By Corollary 9.5.3, $\rho(\Omega \cap B(0, R)) \leq 2\rho$. This and an application of Lemma 9.5.4 with $s = 1$ and $\Omega \cap B(0, R)$ instead of Ω give

$$|g(v) - g(w)|^j \leq 3K_j \big(\rho(\Omega \cap B(0, R))\big)^{j-1} \int_{\Omega \cap B(0,R)} \|g_n'\|\, d\mathscr{L}^j$$
$$\leq 3K_j \alpha_j R^j (2\rho)^{j-1} \mathbf{M}_j (1_\Omega\, g_n')(0).$$

Hence

$$\left(\frac{|g(v) - g(w)|}{|v| + |w|}\right)^j \leq 3 \cdot 2^{j-1} \alpha_j K_j \rho^{j-1} \mathbf{M}_j (1_\Omega\, g_n')(0)$$

and the statement follows. $\qquad\square$

The first main result of this section is a simple corollary of Lemma 13.2.3 and the maximal operator inequality. It will be used in the final stages of the proof of Fréchet differentiability.

Corollary 13.2.4. *Let $\Omega \subset \mathbb{R}^p$ be a bounded measurable set, $g \colon \mathbb{R}^p \longrightarrow E$ be Lipschitz, $n \leq j \leq p$, and $s, \lambda > 0$. Then*

$$\mathscr{L}^p\Big\{u \in \mathbb{R}^p \;\Big|\; B_j(u, s) \subset \Omega,\; \mathrm{reg}_n\, g(u, B_j(u, s)) > \lambda\Big\}$$
$$\leq \frac{\Lambda_n \overline{K}_j^2 (\mathrm{Lip}_j(g))^{2j-2}}{\lambda^{2j}} \int_\Omega \|g_n'\|^2 d\mathscr{L}^p.$$

Proof. Let $u \in \mathbb{R}^p$ be such that $B_j(u, s) \subset \Omega$. By Lemma 13.2.3 with $B_j(u, s)$ instead of Ω, we obtain

$$\big(\mathrm{reg}_n\, g(u, B_j(u, s))\big)^j \leq \overline{K}_j (\mathrm{Lip}_j(g))^{j-1} \mathbf{M}_j (1_{B_j(x,s)}\, g_n')(u)$$
$$\leq \overline{K}_j (\mathrm{Lip}_j(g))^{j-1} \mathbf{M}_j (1_\Omega\, g_n')(u).$$

Squaring, and using the maximal operator and Chebyshev's inequalities give

$$\mathscr{L}^p\Big\{u \in \mathbb{R}^p \;\Big|\; B_j(u,s) \subset \Omega,\; \mathrm{reg}_n\, g(u, B_j(u,s)) > \lambda\Big\}$$

$$\leq \frac{\overline{K}_j^2(\mathrm{Lip}_j(g))^{2j-2}}{\lambda^{2j}} \int_{\mathbb{R}^p} \mathbf{M}_j^2(1_\Omega\, g_n')\, d\mathscr{L}^p$$

$$\leq \frac{\Lambda_n \overline{K}_j^2(\mathrm{Lip}_j(g))^{2j-2}}{\lambda^{2j}} \int_\Omega \|g_n'\|^2 d\mathscr{L}^p. \qquad \square$$

The following lemma is the key to the proof of Fréchet regularity, which is the trickiest part of the proof of Fréchet differentiability. The main point of the lemma is to get rid of part of the condition $B_j(u,s) \subset \Omega$. We will in fact need to get rid of it completely, but what remains after the following lemma is more or less straightforward and so will be handled only when the lemma is applied.

Lemma 13.2.5. *There are constants* $0 < c_n < 1 < \widetilde{K}_j < \infty$ *with the following property. Let* $n, k \in \mathbb{N}$, $p = n + k$, *where* $n \geq 1$. *Let* $\Omega = B_n(0, r)$, *and let* $g \colon \mathbb{R}^p \longrightarrow E$ *be a Lipschitz function for which there are* $\varepsilon > 0$ *and* $v \in \overline{\Omega}$ *with*

$$|g(v) - g(0)| > \mathrm{Lip}(g)\varepsilon|v|.$$

Then there are $0 < t_1, \ldots, t_k \leq r \log_2\big(r^n/(c_n \varepsilon^{n+1}|v|^n)\big)$ *such that, denoting*

$$Q = \Omega \times \prod_{i=1}^k [-t_i, t_i],$$

we have

$$\fint_Q \|g_n'\|^2\, d\mathscr{L}^p > c_n \varepsilon^{n+1}(\mathrm{Lip}(g))^2 \left(\frac{|v|}{r}\right)^n,$$

and for each $n \leq j \leq p$, $\lambda > 0$ *and* $\varepsilon|v| \leq s \leq r$,

$$\mathscr{L}^p\Big\{u \in Q \;\Big|\; \mathrm{reg}_n\, g\big(u, B_j(u,s) \cap (\Omega \times \mathbb{R}^k)\big) > \lambda\Big\}$$

$$\leq \frac{\widetilde{K}_j(\mathrm{Lip}(g))^{2j-2} s^j}{\lambda^{2j} \varepsilon^j |v|^j} \int_Q \|g_n'\|^2 d\mathscr{L}^p. \qquad (13.4)$$

Since the nature of some of these estimates is crucial for validity of our main arguments, we first comment on their relative importance. Notice that the assumptions $v \in \overline{\Omega}$ and $|g(v) - g(0)| > \mathrm{Lip}(g)\varepsilon|v|$ imply that $0 < |v| \leq r$ (so allowing the use of $|v|$ in the denominator) and $\mathrm{Lip}(g)\varepsilon|v| < \mathrm{Lip}(g)|v|$. In particular, $\varepsilon < 1$, $\varepsilon|v| < r$, and so the range of s for which (13.4) applies is nonempty. For us, the most important features of the estimates given in Lemma 13.2.5 are the following.

- Lower estimate of the p-dimensional integral $\fint_Q \|g_n'\|^2\, d\mathscr{L}^p/(\mathrm{Lip}(g))^2$ by a multiple of the nth power of $|v|/r$ (no higher power would do!). Also, the multiple depends on n and ε only. So, for example, the power of ε cannot be p.

- Upper estimate (13.4) of the relative measure

$$\frac{\mathscr{L}^p\{u \in Q \mid \mathrm{reg}_n\, g(u, B_j(u, s) \cap (\Omega \times \mathbb{R}^k)) > \lambda\}}{\mathscr{L}^p Q}$$

by a multiple of the same integral $\fint_Q \|g_n'\|^2\, d\mathscr{L}^p/(\mathrm{Lip}(g))^2$. The multiple depends on

$$j, \varepsilon, \quad \frac{s}{\varepsilon|v|} \quad \text{and} \quad \frac{\mathrm{Lip}(g)}{\lambda}$$

only. So, for example, the power of $s/(\varepsilon|v|)$ again cannot be p. It is also important that this estimate can be used for the whole range of values of s; we will use it with different values of s for different j.

- The upper estimate of t_i is by a constant multiple of r, provided an upper bound for $r/|v|$ and a lower bound for ε are fixed. So, unlike in the previous estimates, an estimate considerably worse than the one actually obtained would still be sufficient for our arguments: here the constant could depend on p.

Proof. We show that the statement holds with

$$c_n = \frac{1}{9\alpha_n K_n 2^{n+2}} \quad \text{and} \quad \widetilde{K}_j = 2^{(j+2)^2} 4^{j-1} \Lambda_j \overline{K}_j^2,$$

where K_n, Λ_j and $\overline{K}_j$ are the constants from Lemma 9.5.4, (13.2), and Lemma 13.2.3, respectively. We may clearly assume that $K_n, \Lambda_j, \overline{K}_j \geq 1$, so $c_n < 1 < \widetilde{K}_j$. Since replacing g by cg and λ by $c\lambda$, where $c > 0$ leads to an equivalent statement, we will assume that $\mathrm{Lip}(g) = 1$.

For $w \in \mathbb{R}^k$ denote $\kappa(w) = \sum_{i=1}^k w_i e_{n+i} \in \mathbb{R}^p$. Let $t = \frac{1}{4}\varepsilon|v|$ and observe that for every $w \in \mathbb{R}^k$ with $|w| \leq t$,

$$|g(v + \kappa(w)) - g(\kappa(w))| \geq |g(v) - g(0)| - 2|w| > \frac{\varepsilon|v|}{2}.$$

Lemma 9.5.4 used with $n = j$ and $s = 2$ implies that for every $w \in \mathbb{R}^k$,

$$\left(\frac{\varepsilon|v|}{2}\right)^{n+1} < |g(v + \kappa(w)) - g(\kappa(w))|^{n+1} \leq 9K_n|v| \int_\Omega \|g_n'(z + \kappa(w))\|^2\, d\mathscr{L}^n(z).$$

Let $\tau_i = 2^{-i/2}t$ for $i \geq 1$. Since $\sum_{i=1}^k \tau_i^2 \leq t^2$, Fubini's theorem gives

$$\fint_{\Omega \times \prod_{i=1}^k [-\tau_i, \tau_i]} \|g_n'\|^2 d\mathscr{L}^p$$

$$= \frac{1}{2^k \tau_1 \cdots \tau_k\, \mathscr{L}^n \Omega} \int_{\prod_{i=1}^k [-\tau_i, \tau_i]} \int_\Omega \|g_n'(z + \kappa(w))\|^2\, d\mathscr{L}^n(z)\, d\mathscr{L}^k(w)$$

$$> \frac{\varepsilon^{n+1}|v|^n}{9K_n 2^{n+1} \mathscr{L}^n \Omega} = 2c_n \varepsilon^{n+1} \left(\frac{|v|}{r}\right)^n. \tag{13.5}$$

For $s_1 \geq \tau_1, \ldots, s_k \geq \tau_k$ denote

$$\Omega[s_1, \ldots, s_k] = \Omega \times \prod_{i=1}^{k} [-s_i, s_i],$$

and

$$\varphi(s_1, \ldots, s_k) = 2^{-\max\{s_1 - \tau_1, \ldots, s_k - \tau_k\}/r} \int_{\Omega[s_1, \ldots, s_k]} \|g_n'\|^2 \, d\mathscr{L}^p.$$

Since the function φ is continuous and

$$\lim_{\max\{s_1, \ldots, s_k\} \to \infty} \varphi(s_1, \ldots, s_k) = 0,$$

it attains its maximum at some point $(t_1, \ldots, t_k)$. We show that $(t_1, \ldots, t_k)$ has the required properties.

Whenever $\max\{s_1 - \tau_1, \ldots, s_k - \tau_k\} > -t + r \log_2\left(r^n/(c_n \varepsilon^{n+1} |v|^n)\right)$, that is,

$$2^{-\max\{s_1 - \tau_1, \ldots, s_k - \tau_k\}/r} < 2^{t/r} c_n \varepsilon^{n+1} \left(\frac{|v|}{r}\right)^n,$$

the inequalities $\int_{\Omega[s_1, \ldots, s_k]} \|g_n'\|^2 \, d\mathscr{L}^p \leq (\mathrm{Lip}(g))^2 = 1$, $t \leq r$, and (13.5) show that

$$\varphi(s_1, \ldots, s_k) < 2^{t/r} c_n \varepsilon^{n+1} \left(\frac{|v|}{r}\right)^n$$

$$\leq \int_{\Omega \times \prod_{i=1}^{k} [-\tau_i, \tau_i]} \|g_n'\|^2 \, d\mathscr{L}^p = \varphi(\tau_1, \ldots, \tau_k).$$

Hence at such a point $(s_1, \ldots, s_k)$ the function φ cannot attain its maximum. Consequently, $0 < \tau_i \leq t_i \leq r \log_2\left(r^n/(c_n \varepsilon^{n+1} |v|^n)\right)$, as required.

The estimate of the integral of $\|g_n'\|^2$ follows easily from (13.5),

$$\int_Q \|g_n'\|^2 \, d\mathscr{L}^p = 2^{\max\{t_1 - \tau_1, \ldots, t_k - \tau_k\}/r} \varphi(t_1, \ldots, t_k)$$

$$\geq 2^{\max\{t_1 - \tau_1, \ldots, t_k - \tau_k\}/r} \varphi(\tau_1, \ldots, \tau_k)$$

$$\geq \int_{\Omega \times \prod_{i=1}^{k} [-\tau_i, \tau_i]} \|g_n'\|^2 d\mathscr{L}^p > c_n \varepsilon^{n+1} \left(\frac{|v|}{r}\right)^n.$$

To show the remaining statement, we assume that $n \leq j \leq p$, $\lambda > 0$, and $\varepsilon |v| \leq s \leq r$. Denote $i = j - n$ and

$$\widetilde{Q} = \Omega[t_1 + s, \ldots, t_i + s, t_{i+1}, \ldots, t_k].$$

We will need an estimate of the ratio of the measures of $\widetilde{Q}$ and Q. The constant factors and powers are not important so long as they depend only on j. So we use that $s \geq \varepsilon |v|$

and $t_m \geq \tau_m = 2^{-m/2-2}\varepsilon|v|$, $m \geq 1$, to get a rough estimate

$$
\frac{\mathscr{L}^p \widetilde{Q}}{\mathscr{L}^p Q} = \prod_{m=1}^{i} \left(1 + \frac{s}{t_m}\right) \leq \prod_{m=1}^{i} \left(1 + \frac{s}{\tau_m}\right) = \prod_{m=1}^{i} \left(1 + 2^{m/2}\frac{4s}{\varepsilon|v|}\right)
$$

$$
\leq \prod_{m=1}^{i} \left(2^{m/2}\frac{8s}{\varepsilon|v|}\right) \leq \frac{2^{(i+2)^2}s^i}{\varepsilon^i|v|^i} \leq \frac{2^{(j+2)^2}s^j}{2\varepsilon^j|v|^j}.
$$

Using this and the inequality $\varphi(t_1 + s, \ldots, t_i + s, t_{i+1} \ldots, t_k) \leq \varphi(t_1, \ldots, t_k)$, and observing that the ratio of the exponential factors in the definition of these two values of φ is at most $2^{s/r} \leq 2$, we get

$$
\int_{\widetilde{Q}} \|g_n'\|^2 \, d\mathscr{L}^p \leq 2\frac{\mathscr{L}^p \widetilde{Q}}{\mathscr{L}^p Q} \int_Q \|g_n'\|^2 \, d\mathscr{L}^p \leq \frac{2^{(j+2)^2}s^j}{\varepsilon^j|v|^j} \int_Q \|g_n'\|^2 \, d\mathscr{L}^p. \tag{13.6}
$$

We plan to apply Lemma 13.2.3 with $B_j(u) := B_j(u, s) \cap (\Omega \times \mathbb{R}^k)$ instead of Ω. To this aim notice that for $u \in \Omega \times \mathbb{R}^k$ the assumption $s \leq r$ implies that the set $B_j(u) - u \subset \mathbb{R}^j$ contains a (j-dimensional) ball of radius $s/2$ (e.g., the ball with the center $-\frac{s}{2}\frac{\pi_n u}{|\pi_n u|}$ and radius $s/2$). So $B_j(u) - u$, considered as a subset of $\mathbb{R}^j$, is open with eccentricity at most 2. By Lemma 13.2.3 together with the fact that $B_j(u) \subset \widetilde{Q}$ for every $u \in Q$, one can estimate

$$
\left(\mathrm{reg}_n\, g(u, B_j(u))\right)^j \leq \overline{K}_j(2\,\mathrm{Lip}_j(g))^{j-1}\mathbf{M}_j(1_{B_j(u)}\, g_n')(u)
$$

$$
\leq \overline{K}_j(2\,\mathrm{Lip}_j(g))^{j-1}\mathbf{M}_j(1_{\widetilde{Q}}\, g_n')(u).
$$

Squaring, integrating, and using the maximal operator and Chebyshev's inequalities, we obtain from (13.6),

$$
\mathscr{L}^p\left\{u \in Q \;\middle|\; \mathrm{reg}_n g(u, B_j(u)) > \lambda\right\}
$$

$$
\leq \frac{\overline{K}_j^2(2\,\mathrm{Lip}_j(g))^{2j-2}}{\lambda^{2j}} \int_{\mathbb{R}^p} \mathbf{M}_j^2(1_{\widetilde{Q}}\, g_n') \, d\mathscr{L}^p
$$

$$
\leq \frac{\Lambda_j\overline{K}_j^2(2\,\mathrm{Lip}_j(g))^{2j-2}}{\lambda^{2j}} \int_{\widetilde{Q}} \|g_n'\|^2 d\mathscr{L}^p
$$

$$
\leq \frac{2^{(j+2)^2}\Lambda_j\overline{K}_j^2(2\,\mathrm{Lip}_j(g))^{2j-2}s^j}{\lambda^{2j}\varepsilon^j|v|^j} \int_Q \|g_n'\|^2 \, d\mathscr{L}^p
$$

$$
= \frac{\widetilde{K}_j(\mathrm{Lip}_j(g))^{2j-2}s^j}{\lambda^{2j}\varepsilon^j|v|^j} \int_Q \|g_n'\|^2 \, d\mathscr{L}^p. \qquad \square
$$

13.3 REDUCTION TO A SPECIAL CASE

We now describe a somewhat technical statement to which we intend to reduce the proof of Theorem 13.1.1. We fix $n \geq 1$ and assume that X is an infinite dimensional separable Banach space for which there is a function $\Theta\colon L(\mathbb{R}^n, X) \longrightarrow [0,1]$ satisfying the following conditions

(S1) Θ is Lipschitz and upper Fréchet smooth;

(S2) $\inf_{\|S\|>s} \Theta(S) > \Theta(0) = 0$ for every $s > 0$;

(S3) for every convergent series $\sum_{k=0}^{\infty} \lambda_k$ of positive numbers and every convergent sequence $(T_k) \subset L(\mathbb{R}^n, X)$, the function

$$\widetilde{\Theta}(T) = \sum_{k=0}^{\infty} \lambda_k \Theta(T - T_k)$$

satisfies for every $T \in L(\mathbb{R}^n, X)$,

$$\bar{\rho}_{\widetilde{\Theta}}(T; t) = o(\omega_n(t)) \text{ as } t \searrow 0.$$

The function ω_n has been defined in Section 9.3. We are interested in its behavior as $t \searrow 0$ and so recall that $\omega_n(t) = t^n \log^{n-1}(1/t)$ for small t.

We shall also assume that $f\colon X \longrightarrow \mathbb{R}^n$ is a bounded Lipschitz function and that $0 < \eta_0 < \frac{1}{2}$, $x_0 \in X$, and $T_0 \in L(\mathbb{R}^n, X)$ are such that $\|T_0\| = \frac{1}{2}$, $f'(x_0; T_0)$ exists, and

$$\| \operatorname{Id} - f'(x; T)\| \le \frac{1}{4} \tag{13.7}$$

whenever $\|T - T_0\| \le \eta_0$ and $f'(x; T)$ exists.

Proposition 13.3.1. *Under the above assumptions there are a point $x \in X$ and an operator $T \in L(\mathbb{R}^n, X)$ such that $\|x - x_0\| \le \eta_0$, $\|T - T_0\| \le \eta_0$, f is Fréchet differentiable at x, and $|f'(x; T)|_{\mathrm{H}} \ge |f'(x_0; T_0)|_{\mathrm{H}}$.*

The remaining part of this chapter will be devoted to proving this statement. But before embarking on it, we show that the differentiability results announced in this chapter follow from Proposition 13.3.1.

Proof of Theorem 13.1.1 from Proposition 13.3.1. By separable reduction arguments of Section 3.6 we may assume that X is separable. By Observation 8.2.4 the space $L(X, \mathbb{R}^n)$ admits a bump function that is upper Fréchet differentiable, Lipschitz on bounded sets, and asymptotically smooth with modulus controlled by $\omega_n(t)$. Hence Lemma 8.2.5 says precisely that there is a function Θ satisfying (S1)–(S3).

Let Y be a Banach space of dimension not exceeding n and $f\colon G \longrightarrow Y$ a locally Lipschitz function defined on a nonempty open subset G of X. Since f has points of Gâteaux differentiability, it suffices to prove only the additional statement. So suppose that $x_0 \in G$ and $T_0 \in L(\mathbb{R}^n, X)$ are such that $f'(x_0; T_0)$ exists, and let $L \in L(\mathbb{R}^n, \mathbb{R}^n)$ and $\varepsilon > 0$ be given. We have to find a point $x \in B(x_0, \varepsilon)$ at which f is Fréchet differentiable and $\langle L, f'(x; T_0) \rangle_{\mathrm{H}} > \langle L, f'(x_0; T_0) \rangle_{\mathrm{H}} - \varepsilon$.

We may and will assume that $Y = \mathbb{R}^n$, $\|T_0\| = \frac{1}{2}$, $\operatorname{rank} T_0 = n$, $\operatorname{rank} L = n$, $B(x_0, \varepsilon) \subset G$, and f is Lipschitz on $B(x_0, \varepsilon)$. Let g be an extension of f from $B(x_0, \varepsilon)$ to a bounded Lipschitz function on X. Let $R_0 \in L(X, \mathbb{R}^n)$ be such that $R_0 T_0 = \operatorname{Id}$ on $\mathbb{R}^n$. Fix a $0 < t < 1$ to satisfy the following conditions:

- $t^2 < \varepsilon$;
- $\big(\|R_0\| + 2\|L\| \operatorname{Lip}(g)\big)t < \frac{1}{4}$;
- $\big(\|R_0\| + 2\|L\| \operatorname{Lip}(g)\big)^2 nt < \varepsilon$; and
- $2n\|R_0\|t < \varepsilon$.

We will apply Proposition 13.3.1 to the function

$$h = R_0 + tL^*g \quad \text{and } \eta_0 = t^2.$$

To verify its remaining assumptions, we assume that $\|T - T_0\| \le \eta_0 = t^2$ and estimate

$$
\begin{aligned}
\|\operatorname{Id} - h'(x;T)\| &= \|R_0(T_0 - T) - tL^*g'(x;T)\| \\
&\le \|R_0\| \, \|T - T_0\| + t\|L^*\| \operatorname{Lip}(g) \, \|T\| \\
&\le \|R_0\| \, \|T - T_0\| + t\|L\| \operatorname{Lip}(g) \, (\|T_0\| + \|T - T_0\|) \\
&\le \big(\|R_0\| + 2\|L\| \operatorname{Lip}(g)\big)t \\
&< \min\left\{\frac{1}{4}, \sqrt{\frac{\varepsilon t}{n}}\right\}
\end{aligned}
\tag{13.8}
$$

by the second and third requirements in the choice of t.

Hence Proposition 13.3.1 provides a point $x \in X$ and an operator $T \in L(\mathbb{R}^n, X)$ such that $\|x - x_0\| \le \eta_0$, $\|T - T_0\| \le \eta_0$, h is Fréchet differentiable at x, and

$$|h'(x;T)|_{\mathrm{H}} \ge |h'(x_0;T_0)|_{\mathrm{H}}.
\tag{13.9}$$

Clearly, $\|x_0 - x\| \le \eta_0$ implies that $x \in B(x_0, \varepsilon)$ by the first condition in the choice of t. Using also that L^* is invertible, we see that Fréchet differentiability of h at x is equivalent to Fréchet differentiability of f at x and

$$f'(x) = \frac{1}{t}(L^*)^{-1}(h'(x) - R_0).$$

Finally, the inequality (13.9) between the Hilbert-Schmidt norms together with (13.8) imply

$$
\begin{aligned}
2\big\langle \operatorname{Id}, h'(x;T) - h'(x_0;T_0)\big\rangle_{\mathrm{H}} &= |\operatorname{Id} - h'(x_0;T_0)|_{\mathrm{H}}^2 - |\operatorname{Id} - h'(x;T)|_{\mathrm{H}}^2 \\
&\quad + |h'(x;T)|_{\mathrm{H}}^2 - |h'(x_0;T_0)|_{\mathrm{H}}^2 \\
&\ge -|\operatorname{Id} - h'(x;T)|_{\mathrm{H}}^2 \\
&\ge -n\|\operatorname{Id} - h'(x;T)\|^2 > -\varepsilon t.
\end{aligned}
$$

Using this and the last inequality in the choice of t, we obtain the remaining conclusion,

$$
\begin{aligned}
\big\langle L, f'(x;T)\big\rangle_{\mathrm{H}} - \big\langle L, f'(x_0;T_0)\big\rangle_{\mathrm{H}} &= \big\langle \operatorname{Id}, L^*f'(x;T) - L^*f'(x_0;T_0)\big\rangle_{\mathrm{H}} \\
&= \frac{1}{t}\Big(\big\langle \operatorname{Id}, h'(x;T) - h'(x_0;T_0)\big\rangle_{\mathrm{H}} - \big\langle \operatorname{Id}, R_0(T - T_0)\big\rangle_{\mathrm{H}}\Big) \\
&\ge -\frac{\varepsilon}{2} - \frac{n}{t}\|R_0\|\|T - T_0\| \ge -\frac{\varepsilon}{2} - n\|R_0\|t > -\varepsilon. \qquad \square
\end{aligned}
$$

We finish this section by several simple observations arising from the assumption (13.7) that we may sometimes use without a reference. First, it implies

$$|f(x+Tu) - f(x)| \leq \frac{5}{4}|u| \tag{13.10}$$

for every $x \in X$, $\|T - T_0\| \leq \eta_0$ and $u \in \mathbb{R}^n$. Indeed, the map $u \mapsto f(x + Tu)$ is differentiable almost everywhere by Rademacher's theorem, and the mean value estimate yields

$$|f(x+Tu) - f(x)| \leq \sup\{\|f'(z;T)\| \mid z \in X, \ f'(z;T) \text{ exists}\} |u| \leq \frac{5}{4}|u|.$$

Second, if $\|T - T_0\| \leq \eta_0$ and $f'(x;T)$ exists, then (13.7) implies that T is a linear isomorphism of $\mathbb{R}^n$ onto its image and that $f'(x;T)$ is a linear isomorphism of $\mathbb{R}^n$ onto itself. Third, we notice that $\mathrm{Lip}(f) \geq 1$: for any x such that $f'(x, T_0)$ exists and any unit vector $u \in \mathbb{R}^n$ we have $2\|T_0u\| \leq 1$. So

$$\mathrm{Lip}(f) \geq 2|f'(x; T_0u)| \geq 2(|u| - |u - f'(x; T_0u)|) \geq 1.$$

Of course, this estimate of $\mathrm{Lip}(f)$ has no serious effect, but it helps us slightly to simplify various expressions and estimates.

Setup

From now on, we will assume that f satisfies the assumptions of Proposition 13.3.1. Let $(Z_k)_{k=1}^\infty$ be an increasing sequence of subspaces of X with $\dim Z_k = k$ and with union dense in X. We define invertible linear operators $S_k \in L(\mathbb{R}^k, Z_k)$ such that S_{k+1} extends S_k, $\|S_k\| \leq 1$, and $\|S_k^{-1}\| \geq 1$. Such operators are easy to construct, for example by letting

$$S_k e_i = z_i, \ i = 1, \ldots, k,$$

where $z_i \in X$ are such that $\sum_{i=1}^\infty \|z_i\|^2 \leq 1$ and Z_k is the linear span of $\{z_1, \ldots, z_k\}$. Notice that the bounds on $\|S_k\|$ and $\|S_k^{-1}\|$ have no deep meaning but save us adding 1 in some estimates. Clearly, $\|S_k\| \leq \|S_{k+1}\|$ and $\|S_k^{-1}\| \leq \|S_{k+1}^{-1}\|$. Although it is not necessary, it will be convenient to define $Z_0 = \{0\}$ and S_0 as the (only) operator from $\mathbb{R}^0 = \{0\}$ to Z_0. The norms of S_0 and of its inverse we willbe understood as $\|S_0\| = \|S_0^{-1}\| = 1$.

We will work in the space

$$D = \left\{(x, T) \in X \times L(\mathbb{R}^n, X) \ \middle| \ \begin{array}{l} f \text{ is regularly Gâteaux differentiable} \\ \text{at } x \text{ in the direction of } T \end{array}\right\}.$$

To be sure that $D \neq \emptyset$ recall that f is Gâteaux differentiable at all $x \in X$ except a Gauss null set (Theorem 2.3.1). By Observation 9.2.2 (ii) the function f is regularly Gâteaux differentiable at all such points in the direction of every $T \in L(\mathbb{R}^n, X)$. We will metrize D by adding countably many pseudometrics to the distance inherited from $X \times L(\mathbb{R}^n, X)$.

We begin by defining pseudonorms $\|\cdot\|_k$, where $k = 0, 1, 2, \ldots$, on the space $\mathrm{Lip}(X \times \mathbb{R}^n, \mathbb{R}^n)$ of Lipschitz $\mathbb{R}^n$-valued functions on $X \times \mathbb{R}^n$ by

$$\|h\|_k = \sup_{\substack{z \in Z_k, u \in \mathbb{R}^n, \\ \|z\| + |u| > 0}} \frac{|h(z, u) - h(z, 0)|}{\|z\| + |u|}.$$

Notice that these pseudonorms are lower semicontinuous in the topology of pointwise convergence, $\|h\|_k \leq \|h\|_{k+1}$ and that

$$\|h\|_k \leq \frac{\mathrm{Lip}(h)|u|}{\|z\| + |u|} \leq \mathrm{Lip}(h).$$

For $x \in X$ and $T \in L(\mathbb{R}^n, X)$ we define $f_{x,T} \colon X \times \mathbb{R}^n \longrightarrow \mathbb{R}^n$ by

$$f_{x,T}(z, u) = f(x + z + Tu).$$

The transformation

$$(x, T) \to f_{x,T}$$

maps $X \times L(\mathbb{R}^n, X)$ to $\mathrm{Lip}(X \times \mathbb{R}^n, \mathbb{R}^n)$. In general, there is no good estimate of $\|f_{x,T} - f_{y,S}\|_k$ in terms of the difference $\|y - x\|$, even in the case $S = T$. We have only

$$\|f_{x,T} - f_{y,S}\|_k \leq \mathrm{Lip}(f)(\|T\| + \|S\|). \tag{13.11}$$

Just in the case $y = x$ we have a much better estimate,

$$\|f_{x,T} - f_{x,S}\|_k \leq \mathrm{Lip}(f)\|T - S\|. \tag{13.12}$$

We will consider D in two different topologies. Its standard topology comes from its embedding into the Banach space $X \times L(\mathbb{R}^n, X)$. For us, this topology, or any of the standard metrics inducing it, are only of minor importance. The metric that we will use is defined by

$$d\big((x, T), (y, S)\big) = \|x - y\| + \|T - S\|$$
$$+ |f'(x; T) - f'(y; S)|_{\mathrm{H}} + \sum_{k=0}^{\infty} 2^{-k} \|f_{x,T} - f_{y,S}\|_k.$$

It is rather obvious that d is a metric on D. Also, once $f'(x; T)$ is not continuous in the standard topology of D, the topology generated by d is strictly finer than the standard one. We point out some rather straightforward properties of these topologies.

Lemma 13.3.2. *Let (D, d) be the metric space introduced above.*

(i) *The function $(x, T) \to f'(x; T)$ is d-continuous on D.*

(ii) *For each $k \geq 1$, the function $\big((x, T), (y, S)\big) \to \|f_{x,T} - f_{y,S}\|_k$ is lower semicontinuous in the standard topology of $D \times D$.*

(iii) *For each $k \geq 1$, the function $\big((x,T),(y,S)\big) \to \|f_{x,T} - f_{y,T}\|_k$ (notice that this is a different function from that in (ii); for example, it does not depend on S) is lower semicontinuous in the standard topology of $D \times D$, and so also in the topology of $(D,d) \times (D,d)$.*

(iv) *The space (D,d) is separable.*

Proof. (i) is obvious. For (ii) fix for a while points $z \in Z_k$ and $u \in \mathbb{R}^n$ such that $\|z\| + |u| > 0$. Then the function

$$\big((x,T),(y,S)\big) \to \frac{|f_{x,T}(z,u) - f_{y,S}(z,u) - f_{x,T}(z,0) + f_{y,S}(z,0)|}{\|z\| + |u|}$$

is continuous in the standard topology of $D \times D$. It follows that the function

$$\big((x,T),(y,S)\big) \to \|f_{x,T} - f_{y,S}\|_k$$

is a supremum of a family of continuous functions and hence it is lower semicontinuous.

The function from (iii) is the composition of the function from (ii) with the map

$$\big((x,T),(y,S)\big) \to \big((x,T),(y,T)\big),$$

which is a continuous map of $D \times D$ equipped with the standard topology into itself. Hence it is lower semicontinuous in the standard topology of $D \times D$, and so also in the finer topology of $(D,d) \times (D,d)$.

(iv) Let H_0 be the space of all functions $\varphi \colon X \times \mathbb{R}^n \longrightarrow \mathbb{R}^n$ that have a continuous restriction to any $Z_k \times \mathbb{R}^n$. We equip H_0 with a countable family of pseudonorms

$$\|\varphi\|_k = \sup_{\substack{z \in Z_k,\, u \in \mathbb{R}^n \\ \|z\| + |u| < k}} |\varphi(z,u)|, \quad k \geq 0$$

and observe that H_0 is metrizable and separable. Indeed, the metric ϱ_0 on H_0 can be given, for example, by the formula

$$\varrho_0(\varphi,\psi) = \sum_{k=1}^{\infty} 2^{-k} \min\{1, \|\varphi - \psi\|_k\},$$

and a countable dense subset of H_0 may be obtained by defining, for each k and each polynomial g in $k + n$ variables with rational coefficients, $h \in H_0$ as a continuous extension of $h(x,u) = g(S_k^{-1}(x),u)$ from $Z_k \times \mathbb{R}^n$ to $X \times \mathbb{R}^n$.

Consequently, the space

$$H = X \times L(X,\mathbb{R}^n) \times L(\mathbb{R}^n,\mathbb{R}^n) \times H_0$$

is metrizable and separable as well. We establish (D,d) as a topological subspace of H.

For $(x, T) \in D$ define $\psi_{x,T} \colon X \times \mathbb{R}^n \to \mathbb{R}^n$ by

$$\psi_{x,T}(z, u) = \frac{f_{x,T}(z, u) - f_{x,T}(z, 0) - f'(x; Tu)}{\|z\| + |u|}, \quad \psi_{x,T}(0, 0) = 0.$$

Since the function f is regularly Gâteaux differentiable at x in the direction of T and $\mathrm{span}\big(Z_k \cup T(\mathbb{R}^n)\big)$ is finite dimensional, Observation 9.2.2 (iv) yields that the restriction of f onto $\mathrm{span}\big(Z_k \cup T(\mathbb{R}^n)\big)$ is regularly Fréchet differentiable at x in the direction of T. In terms of $\psi_{x,T}$ it means that the function $\psi_{x,T}$ belongs to H_0. Hence the map $\eta \colon D \longrightarrow H$,

$$\eta(x, T) := \big(x, T, f'(x; T), \psi_{x,T}\big),$$

is a bijection of D into H. We show that η is in fact a homeomorphism. The first three components of the map η are clearly d-continuous. The continuity of the last one follows from

$$\|\psi_{x,T} - \psi_{y,S}\|_k \leq \|f_{x,T} - f_{y,S}\|_k + \|f'(x; T) - f'(y; S)\|.$$

The continuity of η^{-1} follows similarly easily from the estimate

$$\|f_{x,T} - f_{y,S}\|_k \leq \sup_{\substack{z \in Z_k,\, u \in \mathbb{R}^n \\ \|z\| + |u| \geq k}} \frac{|f_{x,T}(z, u) - f_{x,T}(z, 0) - f_{y,S}(z, u) + f_{y,S}(z, 0)|}{\|z\| + |u|}$$

$$+ \|\psi_{x,T} - \psi_{y,S}\|_k + \|f'(x; T) - f'(y; S)\|$$

$$\leq \frac{4\|f\|_\infty}{k} + \|\psi_{x,T} - \psi_{y,S}\|_k + \|f'(x; T) - f'(y; S)\|$$

for every $k \geq 1$. $\qquad\square$

Lemma 13.3.3. (D, d) *is a complete metric space.*

Proof. Suppose that $\varepsilon_i \searrow 0$ and $(x_i, T_i) \in D$ are such that

$$d((x_j, T_j), (x_i, T_i)) < \varepsilon_i \quad \text{whenever } j \geq i.$$

Then (x_i, T_i) converge to some (x, T) in $X \times L(\mathbb{R}^n, X)$, and $L_i := f'(x_i; T_i)$ converge to some $L \in L(\mathbb{R}^n, \mathbb{R}^n)$. In particular, $\|L - L_i\| \leq \varepsilon_i$. Lemma 13.3.2 (ii) also gives that

$$\|f_{x,T} - f_{x_i,T_i}\|_k \leq \liminf_{j \to \infty} \|f_{x_j, T_j} - f_{x_i, T_i}\|_k$$

$$\leq \liminf_{j \to \infty} 2^k d((x_j, T_j), (x_i, T_i)) \leq 2^k \varepsilon_i. \tag{13.13}$$

We have to show that $(x, T) \in D$ and (x_i, T_i) converge to (x, T) in the metric d. To prove that $f'(x; T) = L$, let $\varepsilon > 0$. First find i such that $\varepsilon_i < \frac{1}{3}\varepsilon$, and then $\delta > 0$ such that $|f(x_i + T_i u) - f(x_i) - L_i u| \leq \varepsilon_i |u|$ for $|u| < \delta$. Using (13.13) with $k = 0$, we get

$$|f(x + Tu) - f(x) - Lu|$$

$$\leq |f(x_i + T_i u) - f(x_i) - L_i u| + \|L - L_i\| |u| + \|f_{x,T} - f_{x_i, T_i}\|_0 |u|$$

$$\leq \varepsilon_i |u| + \varepsilon_i |u| + \varepsilon_i |u| < \varepsilon |u|$$

for $|u| < \delta$. Hence $f'(x; T)$ exists and the sequence $f'(x_i; T_i)$ converges to $f'(x; T)$.

Similarly we prove that f is regularly Gâteaux differentiable at x in the direction of T. Let $z \in X$, $z \neq 0$, and $\varepsilon > 0$. Find k such that Z_k contains some $\tilde{z}$ with $\|z - \tilde{z}\| < \varepsilon/(1 + 6\operatorname{Lip}(f))$. Let $i \geq k$ be such that

$$2^k \varepsilon_i(1 + \|\tilde{z}\|) < \frac{\varepsilon}{3}.$$

Since f is regularly Gâteaux differentiable at x_i in the direction of T_i, there is $\delta > 0$ such that

$$|f(x_i + t\tilde{z} + T_i u) - f(x_i + t\tilde{z}) - L_i u| \leq \frac{\varepsilon}{3}(|t| + |u|)$$

whenever $|t| + |u| < \delta$. Then, for these t and u, we infer from (13.13) that

$$\begin{aligned}
|f(x + tz + Tu) - f(x + tz) - Lu| &\leq 2\operatorname{Lip}(f)|t| \, \|z - \tilde{z}\| \\
&\quad + |L_i u - Lu| + |f(x + t\tilde{z} + Tu) - f(x + t\tilde{z}) - L_i u| \\
&\leq \frac{\varepsilon}{3}|t| + \varepsilon_i|u| + \|f_{x,T} - f_{x_i,T_i}\|_k(\|t\tilde{z}\| + |u|) \\
&\quad + |f(x_i + t\tilde{z} + T_i u) - f(x_i + t\tilde{z}) - L_i u| \\
&\leq \frac{\varepsilon}{3}(|t| + |u|) + 2^k \varepsilon_i(\|t\tilde{z}\| + |u|) + \frac{\varepsilon}{3}(|t| + |u|) \\
&\leq \varepsilon(|t| + |u|).
\end{aligned}$$

This shows that the pair (x, T) belongs to D. The convergence of (x_i, T_i) to (x, T) in the metric d follows from the fact that (x_i, T_i, L_i) converge to (x, T, L) in the space $X \times L(\mathbb{R}^n, X) \times L(\mathbb{R}^n, \mathbb{R}^n)$ and from a variant of (13.13): by Lemma 13.3.2 (ii) we have for each $p \in \mathbb{N}$

$$\begin{aligned}
\sum_{k=0}^{p} 2^{-k} \|f_{x,T} - f_{x_i,T_i}\|_k &\leq \liminf_{j \to \infty} \sum_{k=0}^{p} 2^{-k} \|f_{x_j,T_j} - f_{x_i,T_i}\|_k \\
&\leq \liminf_{j \to \infty} d((x_j, T_j), (x_i, T_i)) \leq \varepsilon_i.
\end{aligned}$$

Taking the limit as $p \to \infty$ finishes the argument. $\square$

Since (D, d) is a complete separable metric space and the identity of (D, d) to $X \times L(\mathbb{R}^n, X)$ with its product topology is (one-to-one and) continuous, by classical results of descriptive set theory (e.g., [24]) a subset of D is Borel in (D, d) if and only if it is Borel in $X \times L(\mathbb{R}^n, X)$. This remark easily implies that all functions we will integrate in the sequel are measurable. For example, $(x, T) \to f'(x; T)$, being continuous on (D, d), is Borel measurable in the standard topology. (Of course, here the use of descriptive set theory could be easily avoided by slightly generalizing the results of Section 3.5.)

Use of the variational principle: plan

Here we introduce notation to be used in the rest of this chapter, explain some of the reasons behind various choices of parameters, and prove those facts that do not need technical choices of parameters. The precise choice of parameters and technical consequences will be described in the following section.

We will work in the subspace (D_0, d) of (D, d), where

$$D_0 = \{(x, T) \in D \mid \|x - x_0\| \leq \eta_0, \|T - T_0\| \leq \eta_0\}.$$

We intend to apply the variational principle of Corollary 7.2.4 on the complete metric space (D_0, d) to the function $h_0 : D_0 \longrightarrow \mathbb{R}$,

$$h_0(x, T) = -|f'(x; T)|_{\mathrm{H}}^2,$$

and with the perturbation functions $F_i \colon D_0 \times D_0 \longrightarrow [0, \infty)$, $i \geq 0$, defined by

$$F_i((x, T), (y, S)) = \lambda_i \|y - x\| + \beta_i \Theta(S - T)$$
$$+ \gamma_i |f'(y; S) - f'(x; T)|_{\mathrm{H}}^2 + \sigma_i \Delta_i((x, T), (y, S)),$$

where

$$\Delta_i((x, T), (y, S)) = \max\{0, \min\{1, \|f_{y,T} - f_{x,T}\|_i\} - s_i\}.$$

The four terms in the definition of F_i will play a role similar to that played by analogous terms in the variational proof of differentiability of one function on Asplund spaces (Chapter 12 or [29]). The key new point is that the last term involves behavior along all directions in the space and controls Gâteaux regularity. We will show in Lemma 13.3.4 that this choice makes the family (F_i) into a valid perturbation scheme on the space D_0. However, we will pay for this relatively simple way of guaranteeing Gâteaux regularity by having to prove rather delicate estimates of the Δ-term.

The perturbed function h_0 will attain its minimum at a point (x_∞, T_∞). We will show first in Section 13.4 that f is regularly Fréchet differentiable at x_∞ in the direction of T_∞; in Section 13.5 we use this information to deduce that f is, in fact, Fréchet differentiable at x_∞ and that $|f'(x_\infty; T_\infty)|_{\mathrm{H}} \geq |f'(x_0; T_0)|_{\mathrm{H}}$.

Notice that, similarly to what we have seen in Chapter 12, the peculiarity in the definition of Δ_i is not a misprint: Δ_i really does not depend on S. This will cause slight technical difficulties in the arguments of Section 13.4, but will become crucial in the final proof of Fréchet differentiability in Section 13.5: to find the value of the derivative of f at x_∞, we have to be able to differentiate $F_i((x, T), (y, S))$ with respect to S. Our choice of Δ_i guarantees that the otherwise most difficult to differentiate component of F_i, the function $\Delta_i((x, T), (y, S))$, is now trivially differentiable with respect to S. As usual, we again have to pay for this advantage. The geometric meaning of the functions Δ_i is not so clear, and their use to control the behavior of f at x_∞ is less straightforward. For the latter, we would certainly prefer to involve in the definition of Δ_i the term $\|f_{y,S} - f_{x,T}\|$. However, the difference of T and S is easily estimated using the Θ-term, and in this way we recover all the information we have lost by making Δ_i independent of the last variable.

Notice also that we do not work in the whole space (D, d). The reason behind this is simple. To apply the variational principle, we need that the function h_0 be bounded from below. As an additional bonus it immediately implies the statements

$\|x_\infty - x_0\| \leq \eta_0$ and $\|T_\infty - T_0\| \leq \eta_0$ of Proposition 13.3.1. However, this is not a genuine bonus: for our arguments to work, we will in fact have to show that these norms are estimated by a better constant $\frac{1}{2}\eta_0$.

We will now discuss the general conditions governing the choice of the parameters $\lambda_i, \beta_i, \gamma_i, \sigma_i$ and s_i, and of the additional parameters ε_i that are used in the variational principle. In order to apply the variational principle, we specify the conditions guaranteeing its assumptions. We state them in a form that is close to necessary and sufficient. Of course, we are not interested in their necessity (and so we do not discuss it further) but in their sufficiency, which is proved in the following lemma.

Lemma 13.3.4. *Suppose that for all $i \geq 0$,*

$$0 < \lambda_i, \beta_i, \gamma_i, \sigma_i, \varepsilon_i < \infty, \quad 0 \leq s_i < \infty, \quad \text{and} \quad \lim_{i\to\infty} s_i = 0.$$

Then F_i are non-negative lower semicontinuous functions on $(D_0, d) \times (D_0, d)$, satisfy $F_i((x,T),(x,T)) = 0$, and there are $r_i \searrow 0$ such that

$$\inf\big\{F_i((x,T),(y,S)) \mid d((x,T),(y,S)) \geq r_i\big\} > 0. \tag{13.14}$$

If, moreover,

$$|f'(x_0; T_0)|_\mathrm{H}^2 > \inf_{(x,T)\in D_0} |f'(x;T)|_\mathrm{H}^2 - \varepsilon_0, \tag{13.15}$$

then the function h_0 and the perturbation scheme (F_i) satisfy the assumptions of the variational principle of Corollary 7.2.4 on the metric space (D_0, d).

Proof. Clearly, $F_i \geq 0$ and $F_i((x,T),(x,T)) = 0$. Lower semicontinuity of the term Δ_i was proved in Lemma 13.3.2 (iii). Since the remaining functions of which F_i consists are continuous, this shows lower semicontinuity of F_i.

We let $t_i = 2^{-i+1}\operatorname{Lip}(f)(1+\eta_0)$ and show that $r_i = 2s_i + (6+2\operatorname{Lip}(f))t_i$ satisfy

$$\inf_{d((x,T),(y,S))\geq r_i} F_i((x,T),(y,S)) \geq \min\big\{\lambda_i t_i, \gamma_i t_i^2, \sigma_i, \sigma_i t_i, \inf_{\|L\|\geq t_i} \beta_i \Theta(L)\big\}.$$

This is obvious if $\|x - y\| \geq t_i$ or $\|T - S\| \geq t_i$ or $|f'(x;T) - f'(y;S)|_\mathrm{H} \geq t_i$. If none of these inequalities hold and $d((x,T),(y,S)) \geq r_i$, then

$$\sum_{k=0}^{\infty} 2^{-k}\|f_{x,T} - f_{y,S}\|_k \geq d((x,T),(y,S)) - 3t_i \geq 2s_i + (3 + 2\operatorname{Lip}(f))t_i.$$

Then (13.12) implies that

$$\sum_{k=0}^{\infty} 2^{-k}\|f_{x,T} - f_{y,T}\|_k \geq \sum_{k=0}^{\infty} 2^{-k}\|f_{x,T} - f_{y,S}\|_k - \sum_{k=0}^{\infty} 2^{-k}\|f_{y,S} - f_{y,T}\|_k$$

$$\geq 2s_i + (3 + 2\operatorname{Lip}(f))t_i - \sum_{k=0}^{\infty} 2^{-k}\operatorname{Lip}(f)\|S - T\|$$

$$\geq 2s_i + 3t_i.$$

Recalling that the norms $\| \cdot \|_k$ are increasing in k, we use (13.11) to infer

$$2s_i + 3t_i \leq \sum_{k=0}^{\infty} 2^{-k} \|f_{x,T} - f_{y,T}\|_k$$

$$\leq 2\|f_{x,T} - f_{y,T}\|_i + \sum_{k=i+1}^{\infty} 2^{-k} \|f_{x,T} - f_{y,T}\|_k$$

$$\leq 2\|f_{x,T} - f_{y,T}\|_i + \sum_{k=i+1}^{\infty} 2^{-k} \cdot 2\operatorname{Lip}(f)\,\|T\|$$

$$\leq 2\|f_{x,T} - f_{y,T}\|_i + 2^{-i+1}\operatorname{Lip}(f)\,(\|T_0\| + \|T - T_0\|)$$

$$\leq 2\|f_{x,T} - f_{y,T}\|_i + t_i.$$

Hence $\|f_{x,T} - f_{y,T}\|_i \geq s_i + t_i$ and $\sigma_i \Delta_i((x,T),(y,S)) \geq \sigma_i \min\{1, t_i\}$. This shows that (13.14) holds. Of course the r_i may not be decreasing, but we may always replace r_i by $\max_{j \geq i} r_j$.

The additional statement is obvious since the condition (13.15) is exactly the same as the only remaining assumption of the variational principle. $\qquad\square$

From now on we will assume that the parameters $\lambda_i, \beta_i, \gamma_i, \sigma_i, s_i, \varepsilon_i$ satisfy the assumptions of Lemma 13.3.4 and that ε_0 satisfies (13.15). Then the variational principle shows that (x_0, T_0) is the starting term of a sequence $(x_j, T_j) \in D_0$ which d-converges to some $(x_\infty, T_\infty) \in D_0$ and has the property that, denoting $\varepsilon_\infty = 0$ and

$$h_j(x, T) = -|f'(x; T)|_{\mathbb{H}}^2 + \sum_{i=0}^{j-1} F_i((x_i, T_i), (x, T)), \tag{13.16}$$

we have

$$h_\infty(x_\infty, T_\infty) \leq \min\Big\{ h_i(x_i, T_i),\ \varepsilon_i + \inf_{(x,T)\in D_0} h_i(x, T) \Big\} \tag{13.17}$$

for $0 \leq i \leq \infty$. Notice that for $j = 0$ the definition of h_0 in (13.16) agrees with the one given earlier and that for $i = \infty$ the inequality (13.17) is just a complicated way of saying that h_∞ attains its minimum on D_0 at (x_∞, T_∞). Recall also that $(x_j, T_j) \in D$, and so the derivatives

$$L_j := f'(x_j; T_j)$$

exist for all $0 \leq j \leq \infty$.

What further assumptions do we have to impose on our parameters? First, although it is not required by the variational principle, we need h_∞ to be finite. This will be needed at several points, most manifestly when we differentiate it with respect to T at the beginning of Section 13.5. Since the functions involved in the definition of F_i are uniformly bounded on D_0 and since Θ is Lipschitz, we will guarantee finiteness of h_∞ and even its upper differentiability with respect to T by imposing the condition

$$\sum_{i=0}^{\infty} (\lambda_i + \beta_i + \gamma_i + \sigma_i) < \infty. \tag{13.18}$$

The reason we mention this simple condition is that interesting information about the definition of our parameters emerges when it is combined with the following

Observation 13.3.5. *For every* $i \geq 0$,

$$\|x_\infty - x_i\| \leq \frac{\varepsilon_i}{\lambda_i}, \quad \Theta(T_\infty - T_i) \leq \frac{\varepsilon_i}{\beta_i}, \quad |L_\infty - L_i|_{\mathrm{H}} \leq \sqrt{\frac{\varepsilon_i}{\gamma_i}}.$$

Proof. It follows directly from the definition of the function h_i in (13.16) and from (13.17) that

$$h_i(x_\infty, T_\infty) + \lambda_i \|x_\infty - x_i\| + \beta_i \Theta(T_\infty - T_i) + \gamma_i |L_\infty - L_i|_{\mathrm{H}}^2$$
$$\leq h_\infty(x_\infty, T_\infty) \leq \varepsilon_i + h_i(x_\infty, T_\infty).$$

This implies all three estimates. $\qquad\square$

The need for having the estimates arising from Observation 13.3.5 small suggests that ε_i should be defined only after $\lambda_i, \beta_i, \gamma_i$. There is an exception to this rule: ε_0 has to satisfy (13.15), and so $\lambda_0, \beta_0, \gamma_0, \sigma_0$ should be chosen after ε_0. However, there is an exception to the exception: later considerations will show that we cannot take γ_0 too large. Fortunately, the estimate of $|L_\infty - L_0|_{\mathrm{H}}$ will never be used and, in fact, this norm cannot be made too small in principle. So γ_0 may be chosen independently of ε_0. Finally, it is clear that in whatever way we choose to handle the case $i = 0$, it does not invalidate the convergence requirement (13.18).

To understand the role of our parameters better, we have to discuss basic strategy of the proof of Fréchet differentiability of f at x_∞. Rather obviously, we wish to assume the opposite and show that in such a situation h_∞ does not attain its minimum on D_0 at (x_∞, T_∞). Recalling the definition of the functions F_i, we write

$$h_\infty(x_\infty, T_\infty) - h_\infty(x, T) = \left(|f'(x;T)|_{\mathrm{H}}^2 - |f'(x_\infty;T_\infty)|_{\mathrm{H}}^2\right)$$
$$- \Phi(x) - \Psi(T) - \Upsilon(f'(x;T)) - \Delta(x), \qquad (13.19)$$

where, for $x \in X$, $T \in L(\mathbb{R}^n, X)$ and $L \in L(\mathbb{R}^n, \mathbb{R}^n)$,

$$\Phi(x) = \sum_{i=0}^{\infty} \lambda_i \left(\|x - x_i\| - \|x_\infty - x_i\|\right),$$

$$\Psi(T) = \sum_{i=0}^{\infty} \beta_i \left(\Theta(T - T_i) - \Theta(T_\infty - T_i)\right),$$

$$\Upsilon(L) = \sum_{i=0}^{\infty} \gamma_i \left(|L - L_i|_{\mathrm{H}}^2 - |L_\infty - L_i|_{\mathrm{H}}^2\right),$$

and

$$\Delta(x) = \sum_{i=0}^{\infty} \sigma_i \left(\Delta_i((x_i, T_i), (x, T)) - \Delta_i((x_i, T_i), (x_\infty, T_\infty))\right).$$

Observe that all these functions are well defined and that the function Δ really does not depend on T. Also notice that the function

$$\Theta_\infty(T) := \sum_{i=0}^{\infty} \beta_i \Theta(T - T_i)$$

is positive, finite, and upper Fréchet smooth by (**S1**) and Lemma 8.1.9 (iii). Moreover, an appeal to (**S3**) yields that

$$\bar{\rho}_{\Theta_\infty}(T_\infty; t) = o(\omega_n(t)) \text{ as } t \searrow 0.$$

In the remaining part of this section we will give more detailed descriptions of various difficulties encountered in the proof and the approaches that we will use in the next section to solve them. Nothing from this description is actually needed in the proof itself, so this part may be skipped by readers who want to go straight to the technical arguments. Here we will try to avoid less important technical difficulties, especially relations between constants for which we may indicate only the main dependence. So, for example, constants denoted by a_k or b_k may depend on other parameters than k, and their value may change from one occurrence to another. In reality the relation between various occurrences may also be of importance, and it is often of paramount importance that (some of) these constants were known long before they were used, often even before we defined the perturbation functions. We will also ignore various absolute constants that have no significant influence on the flow of the arguments, such as $\mathrm{Lip}(f)$, $\|T_\infty\|$, etc.

In the first approximation, our plan is to use the assumption that f is not Fréchet differentiable at x_∞ to produce, for suitably small $r > 0$, a surface $\varphi \colon B_n(0, r) \longrightarrow X$, which is a small deformation of the affine surface $\{x_\infty + T_\infty u \mid u \in B_n(0, r)\}$ and passes through one of the points witnessing Fréchet nondifferentiability. Then we would like to use the integral estimates of the function

$$u \to h_\infty(x_\infty, T_\infty) - h_\infty(\varphi(u), \varphi'_n(u))$$

over $B_n(0, r)$ based on Lemma 13.2.5 to show that there are u with

$$h_\infty(\varphi(u), \varphi'_n(u)) < h_\infty(x_\infty, T_\infty).$$

For this to lead to a contradiction, we need that $(\varphi(u), \varphi'_n(u)) \in D_0$ for almost every u. This will be relatively easy to achieve: the distance of $(\varphi(u), \varphi'_n(u))$ from (x_0, T_0) will be close to the distance of (x_∞, T_∞) from (x_0, T_0), which we can bound by $\frac{1}{2}\eta_0$ by Lemma 13.3.5 provided the parameters are chosen well. We may still deform φ slightly to get in addition that f is Gâteaux differentiable at $\varphi(u)$ for almost every u. For this we have the simple Lemma 2.3.2 and the slightly more sophisticated set N in Corollary 9.3.3.

The surface φ that we imagine at the moment should come from the smoothness assumption via Corollary 9.3.3. We find a point x close to x_∞ witnessing "nondifferentiability behavior of f with error $\bar{\varepsilon}$" (notice that at the moment it is not clear what exactly we mean by this; although the reader can probably guess that we will eventually use the $\bar{\varepsilon}$ from Lemma 9.2.3). Let $y = x - x_\infty$ and find the function $\psi\colon \mathbb{R}^n \longrightarrow \mathbb{R}$ from Corollary 9.3.3. (The role of the Lebesgue null set $N \subset \mathbb{R}^n \times \mathbb{R}$ was explained above. However, we are not yet ready to specify the r with which Corollary 9.3.3 is used.) The deformation φ then could be of the form

$$\varphi(u) = x_\infty + \gamma(u) + T_\infty u,$$

where $\gamma(u) = \psi(u/\|y\|)y$. It satisfies that natural conditions, passing through the "bad point" ($\varphi(0) = x$), and the boundary condition $\varphi(u) = x_\infty + T_\infty u$ on $\partial B(0, \|y\|e^{1/\kappa})$. We denote $Q = B_n(0, r)$ and discuss ways of proving that the integral

$$\int_Q \left(h_\infty(x_\infty, T_\infty) - h_\infty(\varphi, \varphi')\right) d\mathcal{L}^n \tag{13.20}$$

is positive. For this, we estimate the integral of each of the terms on the right side of (13.19). Out of the five terms obtained in this way, only the first difference,

$$\int_Q \left(|f'(\varphi; \varphi')|_\mathrm{H}^2 - |f'(x_\infty; T_\infty)|_\mathrm{H}^2\right) d\mathcal{L}^n, \tag{13.21}$$

can give a positive contribution. Let us first look at the remaining terms to see what the first part has to beat. The Φ-term is more or less negligible, as it may be made as small as we wish by choosing x close to x_∞ and r small. The Υ-term is also negligible provided $\sum_{i=0}^\infty \gamma_i$ is small enough; for example, smaller than $\frac{1}{4}$ will do. (This is the promised reason why we cannot take γ_0 large.) Then this term will be majorized by a small multiple of the integral (13.21).

The Ψ-term can be estimated only by statement (iii) of Lemma 9.3.2 with

$$\theta(u) = \Theta_\infty\left(T_\infty + \frac{y \otimes u}{\|y\|}\right) - \Theta_\infty(T_\infty).$$

Here comes the first serious problem: the smoothness assumption is needed in the direction of the operators $y \otimes u$, so it would be available provided y belonged to a sufficiently small finite codimensional subspace of X. But there is no reason why this should be so. This is the point where the case of Hilbert spaces (where of course $n = 2$) substantially differs from the general case: in Hilbert space the smoothness condition is not asymptotic and so the required estimate is available for all y that we need. To solve this difficulty in the general case, we use the technique of "beefing up the complement." Fixing a small number $\sigma > 0$ that will measure how smooth Θ_∞ is, we find $\kappa > 0$ such that

$$\bar{\rho}_{\Theta_\infty}(T_\infty; t) < \sigma\omega_n(t) \text{ for } 0 < t \leq \kappa.$$

We will now be able to use Corollary 9.3.3 (and so Lemma 9.3.2); notice that this means, in particular, that $r = e^{1/\kappa}\|y\|$. Choosing the set M from Lemma 9.3.2 (iii) finite, we can find a finite codimensional space Y such that the smoothness estimate will be available for all $t \in M$ and for all operators with values in Y. Then we "beef up the complement of Y." That is, we use Corollary 4.2.9 to find a finite dimensional space Z such that every point $\widetilde{x} \in X$ can be written as a sum of a point $y \in Y$ and a point $z \in Z$ whose norms do not exceed $3\|\widetilde{x}\|$. We write $x - x_\infty = y + z$ in this way and then change the definition of φ to

$$\varphi(u) = x_\infty + \gamma(u) + z + T_\infty u.$$

Of course, we need that in the direction of T_∞ the function f behaves at the point $x_\infty + z$ similarly to how it behaved at the point x_∞ in the same direction. This is exactly what regular Gâteaux differentiability of f at x_∞ in the direction of T_∞ means! After all this work, we may finally estimate the Ψ-term: Lemma 9.3.2 (iii) will give an upper bound of the form $K_n \sigma \|y\|^n$.

Let us forget the Δ-term for a while, and consider the question of a lower bound for the term in (13.21). A natural approach seems to be to use Lemma 9.5.4 with $j = n$, $s = 2$, $u = 0$, $\Omega = B_n(0, r)$, where $r = |v| = e^{1/\kappa} \|y\|$. To this aim define

$$g(u) = f(\varphi(u)) - f'(x_\infty; T_\infty u),$$

because then the derivative

$$g' = f'(\varphi; \varphi') - f'(x_\infty; T_\infty)$$

gives the way to estimate the integral in (13.21). Although we still have not specified what "nondifferentiability behavior of f with error $\overline{\varepsilon}$" means, it is clear that the best we can hope for is that the value of

$$g(v) - g(0) = f(x_\infty + z + T_\infty v + \gamma(v)) - f(x_\infty + z + y) - f'(x_\infty; T_\infty v)$$

is of order $\overline{\varepsilon}\|y\|$. With the help of Lemma 9.5.4, this leads to the lower estimate of (13.21) by

$$a_n \frac{\overline{\varepsilon}^{n+1} \|y\|^{n+1}}{r} = a_n \overline{\varepsilon}^{n+1} \|y\|^n e^{-1/\kappa}.$$

So, to beat the Ψ-term, we would essentially need $a_n \overline{\varepsilon}^{n+1} e^{-1/\kappa} > \sigma$, for which, however, there is no hope: κ depends on how quickly the ratio $\bar{\rho}(T_\infty, t)/\omega_n(t)$ falls below σ, over which we have absolutely no control. The attempt to improve this estimate by using that the coefficients β_i in the Ψ-term are small is also hopeless. When we are choosing β_i at the very beginning of the proof, we have no idea how small κ may eventually be. After all, κ depends, for example, on T_∞.

The solution of the problem from the previous paragraph is not, in fact, too difficult. We start our arguments modestly and first show just regular Fréchet differentiability of f at x_∞ in the direction of T_∞. This approach has two advantages. First, we already have a better idea of what we mean by "nondifferentiability behavior of f with error $\overline{\varepsilon}$." Now it means that the error of regular Fréchet differentiability of f at x_∞ in the direction of T_∞ is at least $\overline{\varepsilon}$: for some $v \in \mathbb{R}^n$ and $x = x_\infty + y \in X$, both v and y with small norm,

$$|f(x_\infty + y + T_\infty v) - f(x_\infty + y) - L_\infty v| > \overline{\varepsilon}(\|y\| + |v|).$$

(But, as already pointed out, at the end the real meaning will be more delicate.) Second, the lower bound for the term in (13.21) via Lemma 9.5.4 with $s = 2$ will be done with the help of points $u = 0$ and v for which $\varphi(0)$ and $\varphi(v)$ witness Fréchet irregularity. This means that $|u - v| = |v|$ is of order $\|y\|$, and so the resulting lower estimate will be $a_n \bar{\varepsilon}^{n+1} \|y\|^n$ and this can easily beat $K_n \sigma \|y\|^n$ since σ was chosen after $\bar{\varepsilon}$.

Unfortunately, using the above plan we have no chance of controlling the Δ-term, since this term involves behavior in directions outside $T_\infty(\mathbb{R}^n)$ on which we have no information. But without the Δ-term the above plan would collapse. We have used that f is regularly Gâteaux differentiable at x_∞ in the direction of T_∞, and this was guaranteed only by involving the Δ-term. Our solution of this problem is to change the definition of the surface φ once more: we will leave the realm of n-dimensional surfaces and beef up φ to a much higher dimensional surface.

We will pick large enough k and denote $p = n + k$. Choosing a suitable subset B of $\mathbb{R}^k$, we change the domain of our surface to $Q = B_n(0, r) \times B$, and modify the definition of $\varphi \colon Q \longrightarrow X$ to

$$\varphi(u) = x_\infty + \gamma(\pi_n u) + z + T_\infty \pi_n u + S_k \pi^n u.$$

However, as one may expect, such a drastic change leads to new difficulties. The main one is that the upper estimate of the Ψ-term did not improve, while the integral of $\|g'_n\|^2$ became p-dimensional, and so its estimate by Lemma 9.5.4 would be much worse. More precisely, we may expect to get

$$\int_Q \Psi(\varphi'_n)\mathcal{L}^p \le a_n \sigma \|y\|^n \mathcal{L}^k B, \quad \text{while} \quad \int_Q \|g'_n\|^2 \mathcal{L}^p \ge K_p \bar{\varepsilon}^{p+1} \|y\|^p,$$

and there is no chance that the latter estimate beats the former. Fortunately, we have prepared ourselves for this situation in Lemma 13.2.5. Letting $B = \prod_{i=1}^k [-t_i, t_i]$, where t_i are defined in this lemma, we have the lower estimate

$$\int_Q \|g'_n\|^2 \mathcal{L}^p \ge a_n \varepsilon^{n+1} \|y\|^n \mathcal{L}^k B,$$

and we get a requirement similar to what we had before. But we have to notice that ε and $\bar{\varepsilon}$ are different: now our lower estimate of $g(v) - g(0)$ is by $\bar{\varepsilon}(\|y\| + |v|)$, while Lemma 13.2.5 requires an estimate by $\varepsilon |v|$; so we let

$$\varepsilon = \frac{\|y\| + |v|}{|v|} \bar{\varepsilon}.$$

Although the exact relation between ε and $\bar{\varepsilon}$ will become significant in the very final stages of our arguments, $\|y\|$ and $|v|$ are comparable enough to allow the use of the above arguments to control all the terms of (13.19) except Δ.

It still remains to estimate the Δ-term. For this we employ the so far unexplained technical points in the definition of Δ_i, in particular the parameters s_i and the multiples σ_i. In addition, we still have the free choice of k and κ at our disposal (recall that the radius r is $e^{1/\kappa}\|y\|$). We estimate the integral over Q of each of the terms of the infinite sum of which the Δ-term consists separately. So let

$$\Delta^i(u) := \Delta_i((x_i, T_i), (\varphi(u), \varphi'_n(u))) - \Delta_i((x_i, T_i), (x_\infty, T_\infty))$$
$$= \Delta_i((x_i, T_i), (\varphi(u), T_i)) - \Delta_i((x_i, T_i), (x_\infty, T_i)).$$

and consider first the easy case $i > k$. For this we use that $\Delta^i \leq 1$, and so estimate the total contribution to the integral average of all these terms by $\left(\sum_{i=k+1}^{\infty} \sigma_i\right)\mathscr{L}^p Q$. So k should be large enough to ensure that this is less than the positive contribution, $a_n \varepsilon^{n+1}\|y\|^n \mathscr{L}^k B$, which leads to the requirement that the choice of k be made to guarantee

$$\sum_{i=k+1}^{\infty} \sigma_i \leq a_n \varepsilon^{n+1}(\|y\|/r)^n = a_n \varepsilon^{n+1} e^{-n/\kappa}.$$

Although this seems to indicate that we simply choose k after we have chosen κ, we will shortly see that κ and k are related, and so the actual way of choosing them will be a little more complicated.

Let us now turn our attention to the most delicate estimate, that of the integral of Δ^i for $0 \leq i \leq k$. Here the main idea is that, since $\Delta^i(u) \leq 0$ whenever $\|f_{x_i, T_i} - f_{\varphi(u), T_i}\|_i \leq s_i$, it looks rather probable that $\Delta^i(u)$ is negative for most u, provided the s_i have been chosen decreasing to zero slowly enough. This doesn't quite work, but only because the behavior of f far from the points we are investigating, over which we have little control, could have caused the value $\|f_{x_i, T_i} - f_{\varphi(u), T_i}\|_i$ to be large. But, if $\|f_{x_i, T_i} - f_{\varphi(u), T_i}\|_i$ is realized by such points, $\Delta^i(u)$ is the difference of two terms that are close to each other. So this contribution may be treated in a cavalier way similar to how we treated the Φ term: it may be made as small as we wish by choosing x close to x_∞ and r small.

To make the above consideration more clear, recall again that $\Delta^i \leq 1$ and denote by P the set of $u \in Q$ such that $\Delta^i(u) > 0$. It follows that for such u,

$$\|f_{x_i, T_i} - f_{\varphi(u), T_i}\|_i > s_i, \tag{13.22}$$

and we may assume that $\|f_{x_i, T_i} - f_{\varphi(u), T_i}\|_i$ is realized by points $\widetilde{u}, \widetilde{v} \in \mathbb{R}^{n+i}$, $\widetilde{u} - \widetilde{v} \in \mathbb{R}^n$, with small norm. (Here we identify the space Z_i with $\mathbb{R}^i \subset \mathbb{R}^{n+i}$, so we allow the point $\widetilde{v} \in Z_i$ to be considered as an element of $\mathbb{R}^{n+i}$.) We wish to show that

$$\mathscr{L}^p P \leq a_i \int_Q \|g'_n\|^2 \, d\mathscr{L}^p.$$

Since the a_i depend on i only, their knowledge can be incorporated into the choice of σ_i, and so this would give the desired estimate of the integral of Δ^i.

To estimate $\mathscr{L}^p P$ we finally use what we have learned in Section 13.2 about quantitative measures of regularity. We choose a suitable radius s (depending on i) and denote by

$$B(u) := B_{n+i}(u, s) \cap (B_n(0, r) \times \mathbb{R}^k)$$

the sets appearing in Lemma 13.2.5. For $u \in P$ we wish to show the inequality $\mathrm{reg}_n\, g(u, B(u)) > a_i s_i$, so that the estimate (13.4) from Lemma 13.2.5 would imply the desired estimate for $\mathscr{L}^p P$. Since the multiple on the right hand-side of (13.4) contains the ratio $s/(\varepsilon|v|)$, we need this quantity to have a bound depending on i only, giving an upper bound for the radius s. However, s cannot be too small either, for example, because $\|f_{x_i, T_i} - f_{\varphi(u), T_i}\|_i$ would then depend on the behavior of f far from $B(u)$ while $\mathrm{reg}_n\, g(u, B(u))$ would not. We therefore choose suitable values r_i depending on i only and define s by $s/(\varepsilon|v|) = r_i$. In fact, the r_i will depend only on s_i and so are known in the very beginning of our arguments.

We also cannot estimate $\mathrm{reg}_n\, g(u, B(u))$ when u is too close to the boundary of $B_n(0, r) \times \mathbb{R}^k$, but this slight difficulty is solved in an obvious way. We consider $u \in P$ that is not too close to the boundary of $B_n(0, r) \times \mathbb{R}^k$. The required lower estimate of $\mathrm{reg}_n\, g(u, B(u))$ is straightforward when both $u+\widetilde{u}$ and $u+\widetilde{v}$ belong to $B(u)$, since it is provided by the information contained in (13.22). We finish the argument by showing that indeed both $u + \widetilde{u}$ and $u + \widetilde{v}$ belong to $\in B(u)$.

So we assume that one of the points $u+\widetilde{u}, u+\widetilde{v}$ does not belong to $B(u)$. It implies, in particular, that $|\widetilde{u}| + |\widetilde{v}| \geq s$. We want to show that $\|f_{x_i, T_i} - f_{\varphi(u), T_i}\|_i < s_i$. By an approximation, this reduces to showing that

$$|f(\varphi(u + \widetilde{u})) - f(\varphi(u + \widetilde{v})) - L_\infty(\widetilde{u} - \widetilde{v})| \leq \tfrac{1}{2}s_i(|\widetilde{u}| + |\widetilde{v}|). \qquad (13.23)$$

There are only two methods we can use to estimate the left side of this inequality: the Lipschitz condition or the regular Fréchet differentiability of the *restriction of f to Z* in the direction of T_∞. The first method works when $\widetilde{u} - \widetilde{v}$ is small. When $\widetilde{u} - \widetilde{v}$ is large, we have to use the second method. However, neither the increment $\varphi(u+\widetilde{u}) - \varphi(u+\widetilde{v})$ nor the points on the surface φ may belong to Z. So to treat the general case, we need to estimate two types of errors: the errors coming from $\varphi(u + \widetilde{u}) - \varphi(u + \widetilde{v})$ not being in Z and the errors arising from the distance of the image of φ from Z. Since $\mathrm{Lip}(\gamma) \leq \kappa$, the first of these errors may be estimated by a multiple of $\kappa|\widetilde{v} - \widetilde{u}|$. To cover all possible cases when this estimate may be needed, we have to have $\kappa \leq b_i s_i$, $i = 1, \ldots, k$. Assuming, as we may, that $b_i s_i$ decrease, we finally get the promised condition $\kappa \leq b_k s_k$ that relates the choice of κ and k.

We are now ready to finish the choice of k and κ. Letting $\kappa = b_k s_k$, we require that

$$\sum_{i=k+1}^{\infty} \sigma_i \leq a_n \varepsilon^{n+1} e^{-n/\kappa} = a_n \varepsilon^{n+1} e^{-n/b_k s_k}.$$

Here a_n and b_k are known before σ_i, but ε is not. So we choose σ_i and s_i such that $\sum_{i=k+1}^{\infty} \sigma_i = o(e^{-n/b_k s_k})$, and, whatever the ε may happen to be, sufficiently large k will have the required property.

The remaining errors arising from the distance of the image of φ to Z are not so easy to control. Obviously, they cannot be estimated by anything that happens in the space Z. So we are left only with the estimate by $\|y\|$, which is the maximal distance of the image of φ from Z. This would give the desired inequality when $\|y\| \leq b_i s_i |\widetilde{v} - \widetilde{u}|$; here b_i may be imagined as $\frac{1}{2}$ although it has to be smaller in order to compensate for errors coming from transferring the integral estimates to Euclidean spaces.

However, when $u + \widetilde{u}, u + \widetilde{v} \in B_n(0, r) \times \mathbb{R}^k$, $\|y\| > b_i s_i |\widetilde{u} - \widetilde{v}|$, and $|\widetilde{u} - \widetilde{v}|$ is not small enough, none of the above methods can be used. So the only estimate of the left-hand side of (13.23) that remains is by (a constant multiple of) $\overline{\varepsilon}(\|y\| + |\widetilde{v} - \widetilde{u}|)$, where $\overline{\varepsilon}$ is an upper(!) estimate of the error of regular Fréchet differentiability of f at x_∞ in the direction of T_∞. Here is where our strange reformulation in Lemma 9.2.3 of the assumption that f is at the point x_∞ regularly Gâteaux but not regularly Fréchet differentiable comes into play: the $\overline{\varepsilon}$ comes from this lemma, and the points y and v are chosen such that $\|T_\infty v\| \geq \overline{\alpha}\|y\|$, where $\overline{\alpha}$ depends only on $\overline{\varepsilon}$. (So here we finally specify what the "nondifferentiability behavior" of f with error $\overline{\varepsilon}$" is.) Only one adjustment is needed in the previous arguments: the finite codimensional space Y that we chose to control the smallness of modulus $\overline{\rho}_{\Theta_\infty}$ will be made smaller to incorporate the estimate from Lemma 9.2.3(i) for a suitably small α. This α will be exactly the one for which we know that the above information on $\widetilde{u}, \widetilde{v}$ guarantees applicability of Lemma 9.2.3(i). Now

$$\overline{\varepsilon}(\|y\| + |\widetilde{v} - \widetilde{u}|) \leq \overline{\varepsilon}\left(\|y\| + \frac{\|y\|}{b_i s_i}\right) \leq a_i \overline{\varepsilon}\|y\|.$$

Using that $\overline{\varepsilon} \leq \varepsilon |v|/\|y\|$, it remains only to require the last term to be estimated by $\frac{1}{2} s_i r_i \varepsilon |v| = \frac{1}{2} s_i s$. This is just a condition on how large the r_i should be, and so a condition on the choice of s. Since $\frac{1}{2} s_i s$ is smaller than $\frac{1}{2} s_i (|\widetilde{u}| + |\widetilde{v}|)$, we are done.

Once we know that f is regularly Fréchet differentiable at x_∞ in the direction of T_∞, its Fréchet differentiability follows just from the upper Fréchet differentiability of Θ_∞ at T_∞. In particular, the assumption of asymptotic smoothness need not be used any more. (A similar phenomenon occurred in Chapter 6 in the study of spaces where Lipschitz functions are Fréchet differentiable Γ-almost everywhere.) First we have to guess the value of the Fréchet derivative of f at x_∞, which we do by differentiating $h_\infty(x_\infty, T)$ with respect to T. The rest of the argument follows the same path as the proof of regular Fréchet differentiability, but is considerably less delicate. Finding x witnessing the supposed nonvalidity of our guess for the derivative, we use regular Fréchet differentiability to infer that points $x + T_\infty u$ witness nondifferentiability with the same error. This allows us to define the surface φ by a simpler formula than previously,

$$\varphi(u) = x_\infty + \psi(\pi_n u) x + T_\infty \pi_n u + S_k \pi^n u,$$

where ψ is a suitable linear form. The estimates of the integrals of the five terms on the right of (13.19) use slightly different arguments (for example, we have to use upper Fréchet differentiability instead of smoothness to estimate the Ψ-term) but are in fact easier. This is due to the fact that, thanks to having so many points witnessing Fréchet nondifferentiability, the positive contribution is much larger than it was in the proof of regular Fréchet differentiability. Also, thanks to the already proved regular Fréchet differentiability, the trickiest estimates of the Δ-term do not have to be attempted. They were needed only because we knew that regular Fréchet differentiability could be used only in the space Z, whereas now we can use it in the whole space. Hence for the set B we can take, for example, a small ball, and instead of Lemma 13.2.5 we can use much simpler Corollary 13.2.4.

Use of the variational principle: realization

We will now give precise definitions of the parameters λ_i, β_i, γ_i, s_i, σ_i, and ε_i. Except for σ_i and ε_i, these parameters will be defined rather simply by

$$\lambda_i = 2^{-i}\lambda_0, \quad \beta_i = 2^{-i}\beta_0, \quad \gamma_i = 2^{-i}\gamma_0, \quad s_i = 2^{-i}s_0.$$

For $i = 0$ we choose $\varepsilon_0 > 0$ such that

$$|f'(x_0; T_0)|_H^2 > \sup_{(x,T)\in D_0} |f'(x; T)|_H^2 - \varepsilon_0.$$

Let $\lambda_0 = 2\varepsilon_0/\eta_0$, $\gamma_0 = \frac{1}{8}$, and $s_0 = 4$. Then find $\beta_0 > 0$ large enough that

$$\Theta(S) > \frac{\varepsilon_0}{\beta_0} \quad \text{whenever} \quad \|S\| > \min\left\{\frac{\eta_0}{2}, \frac{s_0}{8\,\mathrm{Lip}(f)}\right\}.$$

For $i \geq 1$ we choose $\varepsilon_i > 0$ such that

$$\frac{\varepsilon_i}{\lambda_i} \leq \frac{\eta_0}{2}, \quad \frac{\varepsilon_i}{\gamma_i} \leq \frac{s_i^2}{64}$$

and

$$\Theta(S) > \frac{\varepsilon_i}{\beta_i} \quad \text{whenever} \quad \|S\| > \min\left\{\frac{\eta_0}{2}, \frac{s_i}{8\,\mathrm{Lip}(f)}\right\}.$$

Such a choice guarantees that the pairs $(x_i, T_i) \in D_0$ obtained by the use of the variational principle satisfy

Lemma 13.3.6. *For all $0 \leq i < \infty$ we have the following estimates:*

$$\|x_i - x_\infty\| \leq \frac{\eta_0}{2}, \ \|T_i - T_\infty\| \leq \min\left\{\frac{\eta_0}{2}, \frac{s_i}{8\,\mathrm{Lip}(f)}\right\}, \ \text{and} \ \|L_i - L_\infty\| \leq \frac{s_i}{8}.$$

We also have, even for all $0 \leq i \leq \infty$,

$$\|T_i\| \leq 1 \quad \text{and} \quad \|L_i\| \leq \frac{5}{4}.$$

Proof. The first two inequalities follow from the above definitions and Observation 13.3.5, and so does the third inequality for $i \geq 1$. The second inequality implies that $\|T_i\| \leq \|T_0\| + \|T_0 - T_\infty\| + \|T_i - T_\infty\| \leq \frac{1}{2} + \eta_0 \leq 1$. Using the second inequality once more we get

$$\|L_0 - L_\infty\| \leq \mathrm{Lip}(f)\|T_0 - T_\infty\| \leq \frac{s_0}{8}.$$

Finally, since $(x_i, T_i) \in D_0$ for all $0 \leq i \leq \infty$, the estimate $\|L_i\| \leq \frac{5}{4}$ is a consequence of our special assumption (13.7). $\square$

Before defining the last parameters σ_i, we recall the constants $\widetilde{K}_i$, K_i, Λ_i, and $\overline{K}_i$ from Lemma 13.2.5, Lemma 9.5.4, (13.2), and Lemma 13.2.3, respectively, and define subsidiary parameters r_i that will dominate the ratios $s/\varepsilon|v|$ discussed in the previous section by

$$r_i = \frac{80\,\mathrm{Lip}(f)}{s_i}\left(1 + \frac{16\,\mathrm{Lip}(f)}{s_i}\right). \tag{13.24}$$

We choose $\sigma_i > 0$, $i \geq 0$, satisfying

$$\sigma_i \leq \frac{2^{-i-3}s_i^{2n+2i}}{r_i^{n+i}\widetilde{K}_{n+i}(5\,\mathrm{Lip}(f))^{2n+2i-2}2^{4n+4i}\|S_i^{-1}\|^{2n+2i}},$$

$$\sigma_i \leq 2^{-i-3}\frac{K_n}{(i+1)\alpha_n}\exp\left(-\frac{4n(\|S_i^{-1}\| + 16\,\mathrm{Lip}(f))}{s_i}\right),$$

$$\sigma_i \leq 2^{-i-4},$$

$$\sigma_i \leq \frac{2^{-i-3}s_i^{2n+2i}}{\Lambda_n\overline{K}_{n+i}^2(5\,\mathrm{Lip}(f))^{2n+2i-2}2^{4n+4i}\|S_i^{-1}\|^{2n+2i}},$$

$$\sigma_i \leq \frac{2^{-i-5}}{i+1}\frac{s_i}{8\,\mathrm{Lip}(f)}.$$

We now briefly recapitulate what we have already shown. Since the parameters we have just defined satisfy the conditions of Lemma 13.3.4, (x_0, T_0) is a starting term of a sequence $(x_i, T_i) \in D_0$ which d-converges to some $(x_\infty, T_\infty) \in D_0$ such that the functions h_i defined by (13.16) satisfy (13.17). Moreover the (x_i, T_i) satisfy the inequalities from Lemma 13.3.6, where $L_i = f'(x_i; T_i)$; recall that the derivatives exist since $(x_i, T_i) \in D$.

Finally, we infer from (S1), (S3), and Lemma 8.1.9 (iii) that the function

$$\Theta_\infty(T) := \sum_{i=0}^{\infty}\beta_i\Theta(T - T_i)$$

is positive, finite, upper Fréchet smooth, and satisfies

$$\bar{\rho}_{\Theta_\infty}(T_\infty; t) = o(\omega_n(t)) \text{ as } t \searrow 0.$$

13.4 REGULAR FRÉCHET DIFFERENTIABILITY

As the first step of the proof of Fréchet differentiability of f at x_∞, we show that f is regularly Fréchet differentiable at x_∞ in the direction of T_∞. Arguing by contradiction and recalling that f is regularly Gâteaux differentiable at x_∞, we use Lemma 9.2.3 to fix $0 < \bar{\varepsilon} \leq 2\,\mathrm{Lip}(f)\|T_\infty\|$ and $\bar{\alpha} > 0$ such that the following two statements hold.

(F1) For every $\alpha > 0$ there is a finite codimensional subspace Y of X such that for every finite dimensional subspace Z of X there is $\delta > 0$ such that

$$|f(x_\infty + z + y + T_\infty u) - f(x_\infty + z + y) - L_\infty u| < 2\bar{\varepsilon}(\|y\| + |u|)$$

whenever $y \in Y$, $z \in Z$, $u \in \mathbb{R}^n$, $\alpha\|z\| < \|y\| < \delta$, and $0 < |u| < \delta$.

(F2) For every finite codimensional subspace Y of X there is a finite dimensional subspace Z of X such that for every $\delta > 0$ there are $y \in Y$, $z \in Z$, and $u \in \mathbb{R}^n$ such that $\bar{\alpha}\|z\| < \|y\| < \delta$, $0 < |u| < \delta$ and

$$|f(x_\infty + z + y + T_\infty u) - f(x_\infty + z + y) - L_\infty u| > \bar{\varepsilon}(\|y\| + |u|).$$

Denote by A a large constant depending only on $\bar{\varepsilon}$ and $\bar{\alpha}$; the particular inequalities we will use are

$$A \geq \left(\frac{4\operatorname{Lip}(f)}{\bar{\varepsilon}}\right)^2, \quad A \geq \left(\frac{5}{2\bar{\varepsilon}}\right)^2, \quad \text{and} \quad A \geq \left(1 + \frac{1}{\bar{\alpha}}\right)^2.$$

Recalling the constants K_n and c_n from Lemma 9.5.4 and Lemma 13.2.5, respectively, we put

$$\sigma = c_n \frac{\alpha_n}{K_n} \frac{\bar{\varepsilon}^{n+3}}{2^{n+6}} \frac{1}{(5\operatorname{Lip}(f))^{n+1}A^{2n+1}}.$$

Choose $k \in \mathbb{N}$ large enough that $k \geq 1/\sigma$ and the parameter

$$\kappa := \frac{s_k}{\|S_k^{-1}\| + 16\operatorname{Lip}(f)}$$

satisfies (besides the obvious inequalities $\kappa \leq \frac{s_k}{\|S_k^{-1}\|}$ and $\kappa \leq \frac{s_k}{16\operatorname{Lip}(f)}$)

$$(3A+1)e^{-1/\kappa} \leq k, \quad \kappa \leq \frac{\eta_0}{2}, \quad \kappa\operatorname{Lip}(f) \leq \frac{\bar{\varepsilon}}{2}, \quad r_k\bar{\varepsilon} \leq 10\operatorname{Lip}(f)e^{1/\kappa},$$

and

$$\bar{\rho}_{\Theta_\infty}(T_\infty; t) < \sigma\omega_n(t) \text{ whenever } 0 < t \leq \kappa.$$

Furthermore, we let $p = n + k$,

$$\tau = \sigma\frac{K_n e^{-n/\kappa}}{\alpha_n} = c_n\frac{\bar{\varepsilon}^{n+3}}{2^{n+6}}\frac{1}{(5\operatorname{Lip}(f))^{n+1}A^{2n+1}}e^{-n/\kappa},$$

and choose $q \geq 1$ such that

$$q \geq \log_2\left(\frac{Ae^{n/\kappa}(10\operatorname{Lip}(f))^{n+1}}{c_n\bar{\varepsilon}^{n+1}}\right).$$

Let $0 < \xi < \frac{1}{2}\bar{\varepsilon}$ be such that

$$\xi \leq \frac{s_k}{48\operatorname{Lip}(f)}, \quad \xi \leq \frac{s_k\tau}{64nkq}, \quad \xi \leq \frac{\tau}{10kqn^{3/2}}, \quad \text{and}$$

$$\xi kqe^{1/\kappa}A \leq \frac{s_i r_i\bar{\varepsilon}}{960A(\operatorname{Lip}(f))^2\|S_i^{-1}\|}.$$

Recall (F1) with $\alpha = \xi$ to find the corresponding finite codimensional subspace Y_1 of X.

Fix a finite set $M \subset (0, \kappa]$ such that $\kappa \in M$ and each point of $(0, \kappa]$ is within $e^{-n/\kappa}$ of M. Next find a finite codimensional subspace Y_2 of X such that

$$\rho_{\Theta_\infty, Y_2}(T_\infty; t) < \sigma \omega_n(t)$$

for $t \in M$ and denote $Y = Y_1 \cap Y_2$. The above inequality clearly holds also with Y in place of Y_2.

We apply (**F2**) to find a finite dimensional subspace Z of X such that for every $\delta > 0$ there are $y \in Y$, $z \in Z$, and $u \in U$ such that $\overline{\alpha} \|z\| < \|y\| < \delta, 0 < |u| < \delta$, and

$$\left| f(x_\infty + z + y + T_\infty u) - f(x_\infty + z + y) - L_\infty u \right| > \overline{\varepsilon}(\|y\| + \|u\|).$$

Since this property still holds if we make Z bigger, we shall use Corollary 4.2.9 to get that every $x \in X$ can be written as $x = y + z$, $y \in Y$, $z \in Z$, and $\|y\|, \|z\| \leq 3\|x\|$. Furthermore, we enlarge Z to contain

$$x_0, \ldots, x_k, x_\infty, \ Z_k, \ T_\infty(\mathbb{R}^n) \text{ and } \bigcup_{i=0}^{k} T_i(\mathbb{R}^n).$$

Going back to (**F1**), we use it with this subspace Z to choose $\delta_1 > 0$ such that

$$\left| f(x_\infty + z + y + T_\infty u) - f(x_\infty + z + y) - L_\infty u \right| < 2\overline{\varepsilon}(\|y\| + \|u\|)$$

whenever $y \in Y$, $z \in Z$, $u \in U$, $\xi \|z\| < \|y\| < \delta_1$, and $0 < |u| < \delta_1$.

Since Z is finite dimensional and f is Lipschitz and regularly Gâteaux differentiable at any x_i in the direction of T_i for each $i = 0, \ldots, k$ and for $i = \infty$, Observation 9.2.2 (iv) implies that the restriction of f to Z is regularly Fréchet differentiable at each such x_i in the direction of T_i. Hence there is $0 < \delta_2 < \delta_1$ such that $\delta_2 \sum_{i=0}^{\infty} \lambda_i \leq \tau$ and for each $i = 0, \ldots, k$ and for $i = \infty$,

$$\left| f(x_i + z + T_i u) - f(x_i + z) - L_i u \right| \leq \xi(\|z\| + |u|) \qquad (13.25)$$

whenever $u \in \mathbb{R}^n$, $z \in Z$, and $\|z\| + |u| < \delta_2$.

Let $0 < \delta_3 \leq \delta_2$ be such that

$$4kqe^{1/\kappa}\delta_3 \leq \min\left\{\frac{\eta_0}{2}, \frac{\delta_2}{2}\right\}, \quad 8\,\mathrm{Lip}(f)kqe^{1/\kappa}\delta_3 \leq \frac{\delta_2 \tau}{4}.$$

By the choice of Z there are points $y \in Y$, $z \in Z$, and $v \in \mathbb{R}^n$ such that $\overline{\alpha}\|z\| < \|y\| < \delta_3, 0 < \|v\| < \delta_3$ and

$$\left| f(x_\infty + y + z + T_\infty v) - f(x_\infty + y + z) - L_\infty v \right| > \overline{\varepsilon}(\|y\| + |v|). \qquad (13.26)$$

The set of pairs $(y, z) \in Y \times Z$ for which there is a $v \in \mathbb{R}^n$ with the above properties is nonempty and open. Hence Lemma 2.3.2 (used with the operator $T \colon \mathbb{R}^p \longrightarrow X$ being $Tu = T_\infty \pi_n u + S_k \pi^n u$) enables us to find $y \in Y$, $z \in Z$, and $v \in \mathbb{R}^n$ such that $\overline{\alpha}\|z\| < \|y\| < \delta_3$, $0 < \|v\| < \delta_3$, (13.26) holds, and the function f is Gâteaux differentiable at

$$x_\infty + z + ty + T_\infty \pi_n u + S_k \pi^n u$$

for almost every $(u, t) \in \mathbb{R}^p \times \mathbb{R}$. For the remaining part of this section we will fix such y, z, and v, and denote $x = y + z$.

We first observe that the norms of the three key vectors that we will use are comparable.

Lemma 13.4.1. *The values $\|x\|$, $|v|$, and $\|y\|$ are strictly positive and the ratio of any two of them is bounded by A.*

Proof. Recall that $|v| > 0$ and $\|y\| > \overline{\alpha}\|z\| \geq 0$. Furthermore, from (13.26) we see that

$$\overline{\varepsilon}(\|y\| + |v|) < |f(x_\infty + x + T_\infty v) - f(x_\infty + x) - L_\infty v|$$
$$\leq 2 \operatorname{Lip}(f)\|x\| + |f(x_\infty + T_\infty v) - f(x_\infty) - L_\infty v|$$
$$\leq 2 \operatorname{Lip}(f)\|x\| + \xi |v| \leq 2 \operatorname{Lip}(f)\|x\| + \frac{\overline{\varepsilon}}{2} |v|.$$

Hence $|v| \leq \sqrt{A}\|x\|$.

Estimating the left-hand side of (13.26) once more, this time using (13.10) and $\|L_\infty\| \leq \frac{5}{4}$, we get

$$\overline{\varepsilon}(\|y\| + |v|) < |f(x_\infty + x + T_\infty v) - f(x_\infty + x) - L_\infty v|$$
$$\leq \left(\frac{5}{4} |v| + \|L_\infty\||v|\right) \leq \frac{5}{2} |v|.$$

Hence $\|y\| \leq \sqrt{A}|v|$.

Finally, we notice that the inequality

$$\|x\| \leq \|y\| + \|z\| \leq \left(1 + \frac{1}{\alpha}\right)\|y\| \leq \sqrt{A}\|y\|$$

is all that is needed to finish the proof. $\square$

We find a function $\psi \colon \mathbb{R}^n \longrightarrow \mathbb{R}$ from Corollary 9.3.3 with the Lebesgue null set $N \subset \mathbb{R}^n \times \mathbb{R}$ being the complement of the set

$$\left\{ (u, t) \in \mathbb{R}^n \times \mathbb{R} \;\middle|\; \begin{array}{l} \text{for almost every } v \in \mathbb{R}^k, \, f \text{ is Gâteaux differentiable} \\ \text{at the point } x_\infty + z + ty + \|y\|T_\infty u + S_k v \end{array} \right\}.$$

Define now $\gamma \colon \mathbb{R}^n \longrightarrow X$ by

$$\gamma(u) = \psi\left(\frac{u}{\|y\|}\right) y.$$

The map γ is Lipschitz and belongs to $C^1(\mathbb{R}^n, X)$. Directly from Corollary 9.3.3 and Lemma 9.3.2 with the function

$$\theta(e) = \Theta_\infty\left(T_\infty + \frac{y \otimes e}{\|y\|}\right) - \Theta_\infty(T_\infty),$$

we see that it possesses the following properties.

(A) $\gamma(0) = y$;

(B) $\gamma(u) = 0$ whenever $|u| \geq e^{1/\kappa}\|y\|$;

(C) $\|\gamma(u)\| \leq \|y\|$ for all $u \in \mathbb{R}^n$;

(D) $\|\gamma'(u)\| \leq \kappa$ for almost all $u \in \mathbb{R}^n$;

(E) $\displaystyle\int_{\mathbb{R}^n} \left(\Theta_\infty(T_\infty + \gamma'(u)) - \Theta_\infty(T_\infty) \right) d\mathscr{L}^n(u) \leq \sigma K_n \|y\|^n$;

(F) defining $\varphi \colon \mathbb{R}^p = \mathbb{R}^n \times \mathbb{R}^k \longrightarrow X$ by

$$\varphi(u) = x_\infty + z + \gamma(\pi_n u) + T_\infty \pi_n u + S_k \pi^n u,$$

we have that for almost every $u \in B_n(0, e^{1/\kappa}\|y\|) \times \mathbb{R}^k$, f is Gâteaux differentiable at $\varphi(u)$.

Finally, we let $r = e^{1/\kappa}\|y\|$, $\Omega = B_n(0, r) \subset \mathbb{R}^n$, and define $g \colon \mathbb{R}^p \longrightarrow \mathbb{R}^n$ by

$$g(u) = f(\varphi(u)) - L_\infty \pi_n u. \tag{13.27}$$

We record several simple estimates that may sometimes be used without a reference. Notice that (13.28) immediately estimates expressions such as $\|\varphi(u) - x_\infty\|$.

Lemma 13.4.2. $\mathrm{Lip}(\varphi) \leq 3$, $\mathrm{Lip}(g) \leq 5 \mathrm{Lip}(f)$, and

$$\|z\| + \|\gamma(\pi_n u)\| + \|T_\infty \pi_n u\| + \|S_k \pi^n u\| + |\pi^n u| \leq 4kqr \tag{13.28}$$

for every $u \in \Omega \times [-qr, qr]^k$. Consequently, $(\varphi(u), \varphi_n'(u)) \in D_0$ for almost every such u.

Proof. Recalling from Lemma 13.3.6 that $\|T_\infty\| \leq 1$ we see that

$$\mathrm{Lip}(\varphi) \leq \mathrm{Lip}(\gamma) + \|T_\infty\| + \|S_k\| \leq \kappa + 1 + 1 \leq 3.$$

The estimate of $\mathrm{Lip}(g)$ follows from this, since $\|L_\infty\| \leq \frac{5}{4}$:

$$\mathrm{Lip}(g) \leq 3 \mathrm{Lip}(f) + \frac{5}{4} \leq 5 \mathrm{Lip}(f).$$

The expression on the left of (13.28) is majorized with the help of Lemma 13.4.1 by

$$3\|x\| + \|y\| + \|T_\infty\| r + \|S_k\| kqr + kqr \leq (3A + 1)\|y\| + r + kqr + kqr$$
$$\leq (3A + 1)e^{-1/\kappa} r + 3kqr \leq 4kqr.$$

Since $\|\varphi(u) - x_\infty\|$ is majorized by (13.28), we know from $4kqr \leq 4kqe^{1/\kappa}\delta_3 \leq \frac{1}{2}\eta_0$ that $\|\varphi(u) - x_0\| \leq \|\varphi(u) - x_\infty\| + \|x_0 - x_\infty\| \leq \eta_0$. Also, by Lemma 13.3.6,

$$\|\varphi_n'(u) - T_0\| = \|\gamma'(\pi_n u) + T_\infty - T_0\| \leq \kappa + \|T_\infty - T_0\| \leq \eta_0.$$

Finally, (F) gives that f is Gâteaux differentiable at $\varphi(u)$ for a.e. $u \in \Omega \times [-qr, qr]^k$, and Lemma 9.2.2 (ii) shows that for such u we have $(\varphi(u), \varphi_n'(u)) \in D_0$. $\square$

We are nearly ready to set the scene for the most delicate part of our argument, an application of Lemma 13.2.5. The actual use of its statement, however, will come only close to the end. To enable it, we estimate how the error in regularity from (13.26) is reflected in the difference $g(v) - g(0)$:

$$
\begin{aligned}
|g(v) - g(0)| &= \left| f(x_\infty + z + \gamma(v) + T_\infty v) - L_\infty v - f(x_\infty + y + z) \right| \\
&\geq |f(x_\infty + y + z + T_\infty v) - f(x_\infty + y + z) - L_\infty v| \\
&\quad - \mathrm{Lip}(f)\|\gamma(v) - y\| \\
&> \bar\varepsilon(\|y\| + |v|) - \mathrm{Lip}(f)\|\gamma(v) - \gamma(0)\| \\
&\geq \bar\varepsilon(\|y\| + |v|) - \mathrm{Lip}(f)\kappa\,|v| \geq \tfrac{1}{2}\,\bar\varepsilon(\|y\| + |v|).
\end{aligned}
\tag{13.29}
$$

Let

$$
\varepsilon = \frac{\bar\varepsilon}{10\,\mathrm{Lip}(f)}\,\frac{\|y\|}{|v|}.
$$

Then the above estimate can be rewritten with the help of Lemma 13.4.2 as

$$
|g(v) - g(0)| > \tfrac{1}{2}\bar\varepsilon\|y\| = 5\,\mathrm{Lip}(f)\varepsilon|v| \geq \mathrm{Lip}(g)\varepsilon|v|.
$$

Applying now Lemma 13.2.5, we find

$$
0 < t_1, \dots, t_k \leq r \log_2\!\big(r^n/(c_n \varepsilon^{n+1} |v|^n) \big) \leq qr
$$

such that, denoting $Q = \Omega \times \prod_{i=1}^{k}[-t_i, t_i]$, we have

$$
\fint_Q \|g_n'\|^2 \, d\mathscr{L}^p > c_n \varepsilon^{n+1} (\mathrm{Lip}(g))^2 \left(\frac{|v|}{r} \right)^n,
\tag{13.30}
$$

and for each $n \leq j \leq p$, $\lambda > 0$ and $\varepsilon|v| \leq s \leq r$,

$$
\mathscr{L}^p\Big\{ u \in Q \,\Big|\, \mathrm{reg}_n g(u, B_j(u, s) \cap (\Omega \times \mathbb{R}^k)) > \lambda \Big\}
\leq \frac{\widetilde{K}_j(\mathrm{Lip}(g))^{2j-2}}{\lambda^{2j}} \frac{s^j}{\varepsilon^j |v|^j} \int_Q \|g_n'\|^2 d\mathscr{L}^p.
\tag{13.31}
$$

The control of the ratio of $|v|$ and $\|y\|$ obtained in Lemma 13.4.1 implies that $A\|y\| \geq |v| \geq \|y\|/A$. Hence, by definition of ε, $\varepsilon \geq \bar\varepsilon/(10A\,\mathrm{Lip}(f))$. Inferring from (13.29) that $\mathrm{Lip}(g) \geq \bar\varepsilon/2$, we obtain from (13.30) and the choice of τ that

$$
\begin{aligned}
\fint_Q \|g_n'\|^2 \, d\mathscr{L}^p &> c_n \varepsilon^{n+1} \left(\frac{\bar\varepsilon}{2} \right)^2 \left(\frac{|v|}{\|y\|} \right)^n e^{-n/\kappa} \\
&\geq \frac{1}{4} c_n \varepsilon^{n+1} \bar\varepsilon^2 A^{-n} e^{-n/\kappa} \\
&\geq c_n \frac{\bar\varepsilon^{n+3}}{2^{n+3}} \frac{1}{(5\,\mathrm{Lip}(f))^{n+1} A^{2n+1}} e^{-n/\kappa} \\
&= 8\tau.
\end{aligned}
\tag{13.32}
$$

Our aim is to show that

$$\fint_Q \left(h_\infty(x_\infty, T_\infty) - h_\infty(\varphi, \varphi'_n) \right) d\mathscr{L}^p > 0. \tag{13.33}$$

Once this is done, a contradiction is immediately follows, since Lemma 13.4.2 says that $(\varphi(u), \varphi'_n(u)) \in D_0$ for almost every $u \in Q$, and so (13.33) implies that we may find u with $(\varphi(u), \varphi'_n(u)) \in D_0$ and

$$h_\infty(x_\infty, T_\infty) - h_\infty(\varphi(u), \varphi'_n(u)) > 0.$$

So h_∞ does not attain its minimum on D_0 at (x_∞, T_∞).

It remains to show (13.33). For this we recall from (13.19) the form of the difference

$$\begin{aligned}
h_\infty(x_\infty, T_\infty) - h_\infty(\varphi, \varphi'_n) &= \left(|f'(\varphi; \varphi'_n)|^2_{\mathrm{H}} - |L_\infty|^2_{\mathrm{H}} \right) \\
&\quad - \Phi(\varphi) - \Psi(\varphi'_n) - \Upsilon(f'(\varphi; \varphi'_n)) - \Delta(\varphi).
\end{aligned} \tag{13.34}$$

The next step is to estimate the integral of each of the five terms on the right side of (13.34). Notice that each of the functions we integrate is almost everywhere defined, measurable, and bounded on Q.

Estimate of $\fint_Q \left(|f'(\varphi; \varphi'_n)|^2_{\mathrm{H}} - |L_\infty|^2_{\mathrm{H}} \right) d\mathscr{L}^p$. Observe that

$$g'_n(u) = f'(\varphi(u); \varphi'_n(u)) - L_\infty$$

(in details, for any $v \in \mathbb{R}^n$, $g'_n(u; v) = f'(\varphi(u); \varphi'(u; v)) - L_\infty v$), and so

$$\begin{aligned}
\left| f'(\varphi(u); \varphi'_n(u)) \right|^2_{\mathrm{H}} - |L_\infty|^2_{\mathrm{H}} &= |L_\infty + g'_n(u)|^2_{\mathrm{H}} - |L_\infty|^2_{\mathrm{H}} \\
&= |g'_n(u)|^2_{\mathrm{H}} + 2\langle L_\infty, g'_n(u) \rangle_{\mathrm{H}}.
\end{aligned}$$

To estimate the last term, we first find a suitable bound on $\left| \int_Q g'_n \, d\mathscr{L}^p \right|_{\mathrm{H}}$. This bound is independent of the expression we are currently estimating and will be used once more later. Define an auxiliary function $\zeta \colon \mathbb{R}^p \longrightarrow \mathbb{R}^n$ by

$$\zeta(u) = f(x_\infty + z + S_k \pi^n u).$$

Clearly, $\zeta'_n = 0$. Denoting $\partial_n Q = \partial\Omega \times \prod_{i=1}^k [-t_i, t_i]$, we obtain by Corollary 9.4.2,

$$\begin{aligned}
\left| \int_Q g'_n \, d\mathscr{L}^p \right|_{\mathrm{H}} &= \left| \int_Q (g'_n - \zeta'_n) \, d\mathscr{L}^p \right|_{\mathrm{H}} \\
&\leq n\alpha_n r^{n-1} \prod_{i=1}^k (2t_i) \max\left\{ |g(u) - \zeta(u)| \;\middle|\; u \in \partial_n Q \right\} \\
&= \frac{n}{r} \mathscr{L}^p Q \max\left\{ |g(u) - \zeta(u)| \;\middle|\; u \in \partial_n Q \right\}.
\end{aligned} \tag{13.35}$$

We have to estimate $|g(u) - \zeta(u)|$ for $u \in \partial_n Q = \partial\Omega \times \prod_{i=1}^{k}[-t_i, t_i]$. Recall that $\gamma(\pi_n u) = 0$ by (B) and $\|z + S_k \pi^n u\| + |\pi_n u| \leq 4kqr$ by Lemma 13.4.2. Since $z + S_k \pi^n u \in Z$ and $4kqr \leq 4kqe^{1/\kappa}\delta_3 < \delta_2$, we may use (13.25) to infer from the choice of ξ that

$$
\begin{aligned}
|g(u) &- \zeta(u)| \\
&= \left| f(x_\infty + z + T_\infty \pi_n u + S_k \pi^n u) - f(x_\infty + z + S_k \pi^n u) - L_\infty \pi_n u \right| \\
&\leq \xi(\|z + S_k \pi^n u\| + |\pi_n u|) \leq 4\xi kqr \leq \frac{2\tau r}{5n^{3/2}}.
\end{aligned}
$$

Finishing thus the estimate of (13.35) we obtain

$$
\left| \int_Q g_n' \, d\mathscr{L}^p \right|_{\mathrm{H}} \leq \frac{2\tau}{5\sqrt{n}} \mathscr{L}^p Q. \tag{13.36}
$$

Since $|L_\infty|_{\mathrm{H}} \leq \|L_\infty\|\sqrt{n} \leq \frac{5}{4}\sqrt{n}$,

$$
\begin{aligned}
\fint_Q (|f'(\varphi; \varphi_n')|_{\mathrm{H}}^2 - |L_\infty|_{\mathrm{H}}^2) d\mathscr{L}^p &= \fint_Q |g_n'|_{\mathrm{H}}^2 \, d\mathscr{L}^p - 2\fint_Q \langle L_\infty, g_n' \rangle_{\mathrm{H}} \, d\mathscr{L}^p \\
&\geq \fint_Q |g_n'|_{\mathrm{H}}^2 \, d\mathscr{L}^p - 2|L_\infty|_{\mathrm{H}} \left| \fint_Q g_n' \, d\mathscr{L}^p \right|_{\mathrm{H}} \\
&\geq \fint_Q |g_n'|_{\mathrm{H}}^2 \, d\mathscr{L}^p - \tau. \tag{13.37}
\end{aligned}
$$

Estimate of $\fint_Q \Phi(\varphi) \, d\mathscr{L}^p$. For $u \in Q$ we use that Lemma 13.4.2 implies

$$
\left| \|x_i - \varphi(u)\| - \|x_i - x_\infty\| \right| \leq \|\varphi(u) - x_\infty\| \leq 4kqr \leq 4kqe^{1/\kappa}\delta_3 < \delta_2,
$$

and infer from the choice of δ_2 that

$$
|\Phi(\varphi(u))| \leq \sum_{i=0}^{\infty} \lambda_i \left| \|x_i - \varphi(u)\| - \|x_i - x_\infty\| \right| \leq \delta_2 \sum_{i=0}^{\infty} \lambda_i \leq \tau.
$$

Hence

$$
\fint_Q \Phi(\varphi) \, d\mathscr{L}^p \leq \tau. \tag{13.38}
$$

Estimate of $\fint_Q \Psi(\varphi_n') \, d\mathscr{L}^p$. Recalling that

$$
\fint_Q \Psi(\varphi_n') \, d\mathscr{L}^p = \fint_\Omega (\Theta_\infty(T_\infty + \gamma'(u)) - \Theta_\infty(T_\infty)) d\mathscr{L}^n(u),
$$

we see from (E) that

$$
\fint_Q \Psi(\varphi_n') \, d\mathscr{L}^p \leq \sigma \frac{K_n \|y\|^n}{\alpha_n r^n} = \sigma \frac{K_n e^{-n/\kappa}}{\alpha_n} = \tau. \tag{13.39}
$$

Estimate of $\int_Q \Upsilon(f'(\varphi; \varphi'_n)) \, d\mathscr{L}^p$. Denoting $R = 2 \sum_{i=0}^{\infty} \gamma_i (L_\infty - L_i)$ and noticing that $\sum_{i=0}^{\infty} \gamma_i = \frac{1}{4}$, we estimate

$$
\Upsilon\big(f'(\varphi(u); \varphi'_n(u))\big) = \Upsilon(g'_n(u) + L_\infty)
$$

$$
= \sum_{i=0}^{\infty} \gamma_i \big(|(L_i - L_\infty) - g'_n(u)|_H^2 - |L_i - L_\infty|_H^2\big)
$$

$$
= \sum_{i=0}^{\infty} \gamma_i \big(2\langle L_\infty - L_i, g'_n(u)\rangle_H + |g'_n(u)|_H^2\big)
$$

$$
= \langle R, g'_n(u)\rangle_H + \tfrac{1}{4}|g'_n(u)|_H^2.
$$

Lemma 13.3.6 implies that $|L_i - L_\infty|_H \le \sqrt{n}\|L_i - L_\infty\| \le \frac{1}{2}\sqrt{n}$ for all $i \ge 0$, which gives

$$
|R|_H \le 2 \sum_{i=0}^{\infty} \gamma_i |L_i - L_\infty|_H \le \sqrt{n}.
$$

Here comes our second use of (13.36), this time to estimate

$$
\int_Q \langle R, g'_n\rangle_H \, d\mathscr{L}^p \le |R|_H \left| \int_Q g'_n \, d\mathscr{L}^p \right|_H \le \sqrt{n}\,\frac{2\tau}{5\sqrt{n}} \le \frac{\tau}{2}.
$$

Hence

$$
\int_Q \Upsilon(f'(\varphi; \varphi'_n)) \, d\mathscr{L}^p \le \frac{\tau}{2} + \frac{1}{4} \int_Q |g'_n|_H^2 \, d\mathscr{L}^p. \tag{13.40}
$$

Estimate of $\int_Q \Delta(\varphi) \, d\mathscr{L}^p$. We estimate the integral of each of the terms

$$
\Delta_i\big((x_i, T_i), (\varphi(u), \varphi'_n(u))\big) - \Delta_i\big((x_i, T_i), (x_\infty, T_\infty)\big).
$$

For $i = 0$ there is little to do, since $s_0 > 1$ and so both terms are identically zero. Let $1 \le i \le k$ and denote

$$
j = n + i, \quad t = \frac{\tau r}{n}, \quad Q_0 = Q \cap (B_n(0, r - t) \times \mathbb{R}^k), \quad \text{and}
$$

$$
P = \big\{ u \in Q \mid \Delta_i\big((x_i, T_i), (\varphi(u), \varphi'_n(u))\big) - \Delta_i\big((x_i, T_i), (x_\infty, T_\infty)\big) > \tau \big\}.
$$

We will wish to apply (13.31) with the radius

$$
s = r_i \varepsilon |v|,
$$

where we recall the values of r_i from (13.24). Since $r_i \ge 1$, we have $s \ge \varepsilon|v|$. Furthermore, by the choice of κ,

$$
s = r_i \varepsilon|v| = r_i \frac{\overline{\varepsilon}}{10 \operatorname{Lip}(f)} \|y\| \le r_k \frac{\overline{\varepsilon}}{10 \operatorname{Lip}(f)} \|y\| \le e^{1/\kappa} \|y\| = r.
$$

Thus the estimate (13.31) will really be applicable when we get to it.

For $u \in \mathbb{R}^p$ we denote $B(u) = B_j(u, s) \cap (\Omega \times \mathbb{R}^k)$ and prove that for every $u \in P_0 := P \cap Q_0$,

$$\mathrm{reg}_n \, g(u, B(u)) > \frac{s_i}{4\|S_i^{-1}\|}. \tag{13.41}$$

For this, we suppose that $u \in P_0$ and choose $\widetilde{z} \in Z_i$ and $\widetilde{w} \in \mathbb{R}^n$ such that

$$\left| f_{x_i, T_i}(\widetilde{z}, \widetilde{w}) - f_{x_i, T_i}(\widetilde{z}, 0) - f_{\varphi(u), T_i}(\widetilde{z}, \widetilde{w}) + f_{\varphi(u), T_i}(\widetilde{z}, 0) \right|$$
$$> \left(\|f_{x_i, T_i} - f_{\varphi(u), T_i}\|_i - \frac{\tau}{2} \right) (\|\widetilde{z}\| + |\widetilde{w}|).$$

To control the expression $\varphi(u) + \widetilde{z}$, we write $\varphi(u) + \widetilde{z} = x_\infty + \widehat{z} + \gamma(\pi_n u)$, so $\widehat{z} = \widetilde{z} + z + T_\infty \pi_n u + S_k \pi^n u$, which together with Lemma 13.4.2 implies that

$$\widehat{z} \in Z \quad \text{and} \quad \|\widehat{z}\| \leq \|\widehat{z} - \widetilde{z}\| + \|\widetilde{z}\| \leq 4kqr + \|\widetilde{z}\|. \tag{13.42}$$

As the first step in the proof of (13.41), we write the vector $\widetilde{z}$ as $\widetilde{z} = S_i \pi^n \widetilde{v}$, where $\widetilde{v} \in \mathbb{R}^j$, $\pi_n \widetilde{v} = 0$, and show that the points $\widetilde{z} \in Z_i$, $\widetilde{w} \in \mathbb{R}^n$, $\widetilde{v} \in \mathbb{R}^j$, and $\widetilde{u} := \widetilde{w} + \widetilde{v} \in \mathbb{R}^j$ have the properties

$$\|\widetilde{z}\| + |\widetilde{w}| < \delta_2 \quad \text{and} \quad u + \widetilde{u}, u + \widetilde{v} \in B(u). \tag{13.43}$$

We cover all possibilities of failure of (13.43) by several cases and bring each of them to a contradiction. The first case, $\|\widetilde{z}\| + |\widetilde{w}| \geq \frac{1}{2}\delta_2$, covers, of course, the possibility that the first inequality in (13.43) fails. In the remaining cases we have $\|\widetilde{z}\| + |\widetilde{w}| < \frac{1}{2}\delta_2$. Since $\pi_n \widetilde{v} = 0$, we always have $u + \widetilde{v} \in \Omega \times \mathbb{R}^k$. So either $u + \widetilde{u} \notin \Omega \times \mathbb{R}^k$ or one of $u + \widetilde{u}, u + \widetilde{v}$ is not in $B(u)$, in which case $|\widetilde{u}| + |\widetilde{v}| \geq s$. Further distinction of cases will depend on various inequalities between the remaining parameters. We summarize the five cases we will consider in the following list, which, together with the preceding discussion, should make it easy to track how they cover all situations in which (13.43) fails.

1. $\|\widetilde{z}\| + |\widetilde{w}| \geq \dfrac{\delta_2}{2}$.

2. $\|\widetilde{z}\| + |\widetilde{w}| < \dfrac{\delta_2}{2}, \quad u + \widetilde{u} \notin \Omega \times \mathbb{R}^k$.

3. $\|\widetilde{z}\| + |\widetilde{w}| < \dfrac{\delta_2}{2}, \quad |\widetilde{u}| + |\widetilde{v}| \geq s, \quad \xi\|\widehat{z}\| \geq \|\gamma(\pi_n u)\|$.

4. $\|\widetilde{z}\| + |\widetilde{w}| < \dfrac{\delta_2}{2}, \quad |\widetilde{u}| + |\widetilde{v}| \geq s, \quad \xi\|\widehat{z}\| < \|\gamma(\pi_n u)\|, \quad \|y\| \leq \dfrac{s_i|\widetilde{w}|}{16\,\mathrm{Lip}(f)}$.

5. $\|\widetilde{z}\| + |\widetilde{w}| < \dfrac{\delta_2}{2}, \quad |\widetilde{u}| + |\widetilde{v}| \geq s, \quad \xi\|\widehat{z}\| < \|\gamma(\pi_n u)\|, \quad \|y\| > \dfrac{s_i|\widetilde{w}|}{16\,\mathrm{Lip}(f)}$.

Case 1. Suppose that $\|\tilde{z}\| + |\tilde{w}| \geq \frac{1}{2}\delta_2$. Then Lemma 13.4.2 and the choice of δ_3 imply that

$$2\operatorname{Lip}(f)\|\varphi(u) - x_\infty\| \leq 8\operatorname{Lip}(f)kqr = 8\operatorname{Lip}(f)kqe^{1/\kappa}\|y\|$$
$$< 8\operatorname{Lip}(f)kqe^{1/\kappa}\delta_3 \leq \frac{\delta_2\tau}{4}$$
$$\leq \frac{\tau}{2}\left(\|\tilde{z}\| + |\tilde{w}|\right).$$

Thus

$$(\|\tilde{z}\| + |\tilde{w}|)\|f_{x_i,T_i} - f_{x_\infty,T_i}\|_i$$
$$\geq |f_{x_i,T_i}(\tilde{z},\tilde{w}) - f_{x_i,T_i}(\tilde{z},0) - f_{x_\infty,T_i}(\tilde{z},\tilde{w}) + f_{x_\infty,T_i}(\tilde{z},0)|$$
$$\geq |f_{x_i,T_i}(\tilde{z},\tilde{w}) - f_{x_i,T_i}(\tilde{z},0) - f_{\varphi(u),T_i}(\tilde{z},\tilde{w}) + f_{\varphi(u),T_i}(\tilde{z},0)|$$
$$\quad - 2\operatorname{Lip}(f)\|\varphi(u) - x_\infty\|$$
$$\geq (\|\tilde{z}\| + |\tilde{w}|)\left(\|f_{x_i,T_i} - f_{\varphi(u),T_i}\|_i - \frac{\tau}{2}\right) - \frac{\tau}{2}(\|\tilde{z}\| + |\tilde{w}|)$$
$$= (\|\tilde{z}\| + |\tilde{w}|)(\|f_{x_i,T_i} - f_{\varphi(u),T_i}\|_i - \tau).$$

This leads to the inequality

$$\Delta_i\big((x_i, T_i), (\varphi(u), \varphi'_n(u))\big) - \Delta_i\big((x_i, T_i), (x_\infty, T_\infty)\big) \leq \tau,$$

contradicting the assumption that $u \in P$.

Assumption for the remaining cases. In the remaining cases we assume the negation of the assumption of the previous case, that is,

$$\|\tilde{z}\| + |\tilde{w}| < \frac{\delta_2}{2}. \tag{13.44}$$

Notice that this guarantees applicability of (13.25) with the points $\tilde{z}$ and $\tilde{w}$ as well as with the points $\hat{z}$ and $\tilde{w}$; the latter because of (13.42) and the inequality $4kqr \leq \frac{1}{2}\delta_2$.

From now on, we will estimate the difference of Δ by estimating only its first term. We will further reduce this to showing the inequality

$$J := |f_{\varphi(u),T_\infty}(\tilde{z},\tilde{w}) - f_{\varphi(u),T_\infty}(\tilde{z},0) - L_\infty\tilde{w}| \leq \frac{s_i}{4}(\|\tilde{z}\| + |\tilde{w}|), \tag{13.45}$$

that is, finding a suitable approximation of the increments of the function f at $\varphi(u) + \tilde{z}$ in the direction of T_∞. Were $\varphi(u) - x_\infty$ in Z, we could do it using (13.25) with the points $\varphi(u) - x_\infty + \tilde{z}$ and $\tilde{w}$. Since this is not true in general, we have approximated $\varphi(u) - x_\infty + \tilde{z}$ by $\hat{z} \in Z$ for which (13.25) is available, but we need a number of cases to estimate the error of this approximation.

Let us first see that (13.45) implies the required contradiction. Indeed, from (13.45) and Lemma 13.3.6 we deduce that

$$
\begin{aligned}
|f_{\varphi(u),T_i}(\widetilde{z},\widetilde{w}) &- f_{\varphi(u),T_i}(\widetilde{z},0) - L_i\widetilde{w}| \\
&\leq J + \mathrm{Lip}(f)\|T_i - T_\infty\|\,|\widetilde{w}| \\
&\quad + \|L_i - L_\infty\|\,|\widetilde{w}| \\
&\leq \frac{s_i}{4}\left(\|\widetilde{z}\| + |\widetilde{w}|\right) + \frac{s_i}{8}\,|\widetilde{w}| + \frac{s_i}{8}\,|\widetilde{w}| \\
&\leq \frac{s_i}{2}\left(\|\widetilde{z}\| + |\widetilde{w}|\right).
\end{aligned}
$$

Since $\|\widetilde{z}\| + |\widetilde{w}| < \delta_2$, we may use (13.25) to get

$$
|f_{x_i,T_i}(\widetilde{z},\widetilde{w}) - f_{x_i,T_i}(\widetilde{z},0) - L_i\widetilde{w}| \leq \xi(\|\widetilde{z}\| + |\widetilde{w}|) \leq \frac{s_i}{2}\left(\|\widetilde{z}\| + |\widetilde{w}|\right).
$$

Hence

$$
\begin{aligned}
\left(\|f_{x_i,T_i} - f_{\varphi(u),T_i}\|_i - \frac{\tau}{2}\right)&(\|\widetilde{z}\| + |\widetilde{w}|) \\
&\leq |f_{x_i,T_i}(\widetilde{z},\widetilde{w}) - f_{x_i,T_i}(\widetilde{z},0) - f_{\varphi(u),T_i}(\widetilde{z},\widetilde{w}) + f_{\varphi(u),T_i}(\widetilde{z},0)| \\
&\leq |f_{x_i,T_i}(\widetilde{z},\widetilde{w}) - f_{x_i,T_i}(\widetilde{z},0) - L_i\widetilde{w}| \\
&\quad + |f_{\varphi(u),T_i}(\widetilde{z},\widetilde{w}) - f_{\varphi(u),T_i}(\widetilde{z},0) - L_i\widetilde{w}| \\
&\leq \frac{s_i}{2}(\|\widetilde{z}\| + |\widetilde{w}|) + \frac{s_i}{2}(\|\widetilde{z}\| + |\widetilde{w}|) = s_i(\|\widetilde{z}\| + |\widetilde{w}|),
\end{aligned}
$$

implying that $\Delta_i\big((x_i,T_i),(\varphi(u),\varphi_n'(u))\big) \leq \frac{1}{2}\tau$ and so that $u \notin P$.

Case 2. In addition to (13.44), suppose also that $u + \widetilde{u} \notin \Omega \times \mathbb{R}^k$. Then the facts $\pi_n u \in B(0, r - t)$ and $\pi_n u + \widetilde{w} = \pi_n(u + \widetilde{u}) \notin \Omega$ imply

$$
|\widetilde{w}| \geq t \quad \text{and} \quad \gamma(\pi_n u + \widetilde{w}) = 0.
$$

Recalling the definition of $\widehat{z}$ we obtain with the help of (D) that

$$
\|(\varphi(u) + \widetilde{z}) - (x_\infty + \widehat{z})\| = \|\gamma(\pi_n u)\| \leq \|\gamma(\pi_n u) - \gamma(\pi_n u + \widetilde{w})\| \leq \kappa|\widetilde{w}|.
$$

Using (13.25) and (13.42) together with the choice of t, ξ, and κ, we obtain

$$
\begin{aligned}
J &\leq |f_{x_\infty,T_\infty}(\widehat{z},\widetilde{w}) - f_{x_\infty,T_\infty}(\widehat{z},0) - L_\infty\widetilde{w}| \\
&\quad + 2\,\mathrm{Lip}(f)\|(\varphi(u) + \widetilde{z}) - (x_\infty + \widehat{z})\| \\
&\leq \xi(\|\widehat{z}\| + |\widetilde{w}|) + 2\,\mathrm{Lip}(f)\kappa|\widetilde{w}| \\
&\leq 4\xi k q r + \xi(\|\widehat{z}\| + |\widetilde{w}|) + 2\,\mathrm{Lip}(f)\kappa|\widetilde{w}| \\
&\leq \frac{s_i}{16}t + \frac{s_i}{16}(\|\widehat{z}\| + |\widetilde{w}|) + 2\frac{s_i}{16}|\widetilde{w}| \\
&\leq \frac{s_i}{4}(\|\widehat{z}\| + |\widetilde{w}|).
\end{aligned}
$$

giving (13.45) and so a contradiction.

Assumptions for the remaining cases. As discussed above, if none of the two previous cases applies, we have

$$\|\widetilde{z}\| + |\widetilde{w}| < \frac{\delta_2}{2} \quad \text{and} \quad |\widetilde{u}| + |\widetilde{v}| \geq s. \tag{13.46}$$

Notice also that the second of these inequalities together with the choice of ξ imply that

$$\xi\|\widehat{z}\| \leq \xi\|\widetilde{z}\| + \xi k q r = \xi\|\widetilde{z}\| + \xi k q e^{1/\kappa}\|y\| \leq \xi\|\widetilde{z}\| + \xi k q e^{1/\kappa} A |v|$$
$$\leq \xi\|\widetilde{z}\| + \frac{s_i}{96\operatorname{Lip}(f)\|S_i^{-1}\|} r_i \varepsilon |v| = \xi\|\widetilde{z}\| + \frac{s_i}{96\operatorname{Lip}(f)\|S_i^{-1}\|} s$$
$$\leq \xi\|\widetilde{z}\| + \frac{s_i}{96\operatorname{Lip}(f)\|S_i^{-1}\|} (|\widetilde{u}| + |\widetilde{w}|)$$
$$\leq \frac{s_i}{48\operatorname{Lip}(f)}\|\widetilde{z}\| + \frac{s_i}{48\operatorname{Lip}(f)} (\|\widetilde{z}\| + |\widetilde{w}|)$$
$$\leq \frac{s_i}{24\operatorname{Lip}(f)} (\|\widetilde{z}\| + |\widetilde{w}|). \tag{13.47}$$

Case 3. In addition to (13.46) suppose that $\xi\|\widehat{z}\| \geq \|\gamma(\pi_n u)\|$. Then

$$J \leq |f_{x_\infty, T_\infty}(\widehat{z}, \widetilde{w}) - f_{x_\infty, T_\infty}(\widehat{z}, 0) - L_\infty \widetilde{w}|$$
$$+ 2\operatorname{Lip}(f)\|(\varphi(u) + \widetilde{z}) - (x_\infty + \widehat{z})\|$$
$$\leq \xi(\|\widehat{z}\| + |\widetilde{w}|) + 2\operatorname{Lip}(f)\|\gamma(\pi_n u)\|$$
$$\leq 3\xi\operatorname{Lip}(f)\|\widehat{z}\| + \xi|\widetilde{w}|$$
$$\leq \frac{s_i}{8}(\|\widehat{z}\| + |\widetilde{w}|) + \frac{s_i}{8}|\widetilde{w}| \leq \frac{s_i}{4}(\|\widehat{z}\| + |\widetilde{w}|),$$

giving (13.45) and so a contradiction.

Assumptions for the remaining cases. Once more, we add the negation of the assumption of the previous case to the assumptions of the remaining cases. So we assume that

$$\|\widetilde{z}\| + |\widetilde{w}| < \frac{\delta_2}{2}, \quad |\widetilde{u}| + |\widetilde{v}| \geq s, \quad \text{and} \quad \xi\|\widehat{z}\| < \|\gamma(\pi_n u)\|. \tag{13.48}$$

Case 4. In addition to (13.48) suppose that $\|y\| \leq \frac{s_i|\widetilde{w}|}{16\operatorname{Lip}(f)}$. Recalling again (13.42), in particular applicability of the estimate (13.25) with $\widehat{z}$ and $\widetilde{w}$ and the estimate of $\xi\|\widehat{z}\|$ in (13.47), we get

$$J \leq |f_{x_\infty, T_\infty}(\widehat{z}, \widetilde{w}) - f_{x_\infty, T_\infty}(\widehat{z}, 0) - L_\infty \widetilde{w}|$$
$$+ 2\operatorname{Lip}(f)\|(\varphi(u) + \widetilde{z}) - (x_\infty + \widehat{z})\|$$
$$\leq \xi(\|\widehat{z}\| + |\widetilde{w}|) + 2\operatorname{Lip}(f)\|\gamma(\pi_n u)\| \leq \xi\|\widehat{z}\| + \xi|\widetilde{w}| + 2\operatorname{Lip}(f)\|y\|$$
$$\leq \frac{s_i}{24}(\|\widehat{z}\| + |\widetilde{w}|) + \frac{s_i}{16}|\widetilde{w}| + \frac{s_i}{8}|\widetilde{w}|$$
$$\leq \frac{s_i}{4}(\|\widehat{z}\| + |\widetilde{w}|).$$

This gives (13.45) and so leads again to a contradiction.

Case 5. In addition to (13.48) suppose that $\|y\| > \frac{s_i|\widetilde{w}|}{16\operatorname{Lip}(f)}$. Unlike in the previous cases, in which the estimates were obtained by using that f is Lipschitz or that its restriction to Z is regularly Fréchet differentiable at x_∞ in the direction of T_∞, here we need the seemingly artificial reformulation of the facts that f is at x_∞ regularly Gâteaux differentiable but not regularly Fréchet differentiable in the direction of T_∞. To this aim we recall the condition imposed on the choice of δ_1 and observe that it applies to our remaining situation: $\gamma(\pi_n u) \in Y \subset Y_1$, $\widehat{z} \in Z$, $\widetilde{w} \in \mathbb{R}^n$, $\xi\|\widehat{z}\| < \|\gamma(\pi_n u)\| < \delta_1$, and $0 < |\widetilde{w}| < \delta_1$. Thus

$$|f(x_\infty + \widehat{z} + \gamma(\pi_n u) + T_\infty \widetilde{w}) - f(x_\infty + \widehat{z} + \gamma(\pi_n u)) - L_\infty \widetilde{w}| < 2\bar{\varepsilon}(\|\gamma(\pi_n u)\| + |\widetilde{w}|).$$

Since $x_\infty + \widehat{z} + \gamma(\pi_n u) = \varphi(u) + \widehat{z}$, we see that the expression on the left is exactly J. Hence

$$\begin{aligned}
J &< 2\bar{\varepsilon}(\|\gamma(\pi_n u)\| + |\widetilde{w}|) \\
&\leq 2\bar{\varepsilon}(\|y\| + |\widetilde{w}|) \\
&\leq 2\bar{\varepsilon}\left(1 + \frac{16\operatorname{Lip}(f)}{s_i}\right)\|y\| \\
&= \frac{s_i r_i}{4}\varepsilon|v| = \frac{s_i}{4}s \leq \frac{s_i}{4}(|\widetilde{u}| + |\widetilde{v}|),
\end{aligned}$$

which gives (13.45) and so leads to a contradiction.

Having thus finished the proof of (13.43), we show (13.41) by first estimating (with the help of Lemma 13.3.6)

$$\begin{aligned}
\|\varphi(u + \widetilde{w}) - \varphi(u) - T_i \widetilde{w}\| &\leq \left(\|T_i - T_\infty\| + \operatorname{Lip}_n(\gamma)\right)|\widetilde{w}| \\
&\leq \left(\frac{s_i}{8\operatorname{Lip}(f)} + \kappa\right)|\widetilde{w}| \leq \frac{3s_i}{16\operatorname{Lip}(f)}|\widetilde{w}|.
\end{aligned}$$

Then we use this estimate, (13.43), and (13.25) to get

$$\begin{aligned}
(|\widetilde{u}| + |\widetilde{v}|)\operatorname{reg}_n g(u, B(u)) &\geq |g(u + \widetilde{u}) - g(u + \widetilde{v})| \\
&= |f(\varphi(u + \widetilde{w}) + \widehat{z}) - f(\varphi(u) + \widehat{z}) - L_\infty \widetilde{w}| \\
&\geq |f_{\varphi(u),T_i}(\widehat{z}, \widetilde{w}) - f_{\varphi(u),T_i}(\widehat{z}, 0) - L_\infty \widetilde{w}| \\
&\quad - \operatorname{Lip}(f)\|\varphi(u + \widetilde{w}) - \varphi(u) - T_i \widetilde{w}\| \\
&\geq |f_{\varphi(u),T_i}(\widehat{z}, \widetilde{w}) - f_{\varphi(u),T_i}(\widehat{z}, 0) - f_{x_i,T_i}(\widehat{z}, \widetilde{w}) + f_{x_i,T_i}(\widehat{z}, 0)| \\
&\quad - |f_{x_i,T_i}(\widehat{z}, \widetilde{w}) - f_{x_i,T_i}(\widehat{z}, 0) - L_i \widetilde{w}| - \|L_i - L_\infty\||\widetilde{w}| - \frac{3s_i}{16}|\widetilde{w}| \\
&\geq \left(\||f_{x_i,T_i} - f_{\varphi(u),T_i}\||_i - \frac{\tau}{2}\right)(\|\widehat{z}\| + |\widetilde{w}|) - \xi(\|\widehat{z}\| + |\widetilde{w}|) - \left(\frac{s_i}{8} + \frac{3s_i}{16}\right)|\widetilde{w}|.
\end{aligned}$$

Since $u \in P$ we see, in particular, that $\|f_{x_i, T_i} - f_{\varphi(u), T_i}\|_i > \tau + s_i$. Hence the interrupted estimate can continue:

$$\geq s_i(\|\tilde{z}\| + |\tilde{w}|) - \frac{s_i}{16}(\|\tilde{z}\| + |\tilde{w}|) - \frac{5s_i}{16}|\tilde{w}|$$
$$> \left(s_i - \frac{s_i}{2}\right)(\|\tilde{z}\| + |\tilde{w}|) \geq \frac{s_i}{2\|S_i^{-1}\|}(|\tilde{v}| + |\tilde{w}|) \geq \frac{s_i}{4\|S_i^{-1}\|}(|\tilde{u}| + |\tilde{v}|).$$

In this way we have finally proved (13.41).

We are now ready to finish the estimate of the difference of the Δ_i for $1 \leq i \leq k$. By (13.41),

$$\mathscr{L}^p P_0 \leq \mathscr{L}^p \left\{ u \in Q \;\middle|\; \operatorname{reg}_n g(u, B(u)) \geq \frac{s_i}{4\|S_i^{-1}\|} \right\}.$$

Since

$$\frac{s}{\varepsilon |v|} = r_i,$$

we get from (13.31), from the estimate of $\operatorname{Lip}(g)$ in Lemma 13.4.2, and from the first inequality in the choice of σ_i,

$$\mathscr{L}^p P_0 \leq \frac{\tilde{K}_j (5 \operatorname{Lip}(f))^{2j-2} 2^{4j} \|S_i^{-1}\|^{2j}}{s_i^{2j}} r_i^j \int_Q \|g_n'\|^2 d\mathscr{L}^p$$
$$\leq \frac{1}{2^{i+3}\sigma_i} \int_Q \|g_n'\|^2 \, d\mathscr{L}^p.$$

Since $\Delta_i((x_i, T_i), (\varphi, \varphi_n')) \leq 1$ and

$$\mathscr{L}^p Q_0 = \left(1 - \frac{t}{r}\right)^n \mathscr{L}^p Q \geq (1 - \tau)\mathscr{L}^p Q,$$

we use that $2\sigma_i \leq 2^{-i-3}$ to conclude that

$$\sigma_i \int_Q \left(\Delta_i((x_i, T_i), (\varphi, \varphi_n')) - \Delta_i((x_i, T_i), (x_\infty, T_\infty))\right) d\mathscr{L}^p$$
$$\leq \sigma_i \int_P \Delta_i((x_i, T_i), (\varphi, \varphi_n')) \, d\mathscr{L}^p + \sigma_i \tau \mathscr{L}^p(Q \setminus P)$$
$$\leq \sigma_i \mathscr{L}^p P_0 + \sigma_i \mathscr{L}^p(Q \setminus Q_0) + \sigma_i \tau \mathscr{L}^p Q$$
$$\leq 2^{-i-3} \int_Q \|g_n'\|^2 \, d\mathscr{L}^p + 2\sigma_i \tau \mathscr{L}^p Q$$
$$\leq 2^{-i-3} \left(\int_Q \|g_n'\|^2 \, d\mathscr{L}^p + \tau \mathscr{L}^p Q\right).$$

To obtain a similar estimate for $i > k$ we infer from $\sigma > 1/i$ and the second condition in the choice of σ_i that

$$\sigma_i \leq \frac{2^{-i-3}}{i+1} \frac{K_n}{\alpha_n} e^{-n/\kappa} \leq 2^{-i-3} \sigma \frac{K_n}{\alpha_n} e^{-n/\kappa} = 2^{-i-3}\tau.$$

Hence

$$\sigma_i \int_Q \left(\Delta_i((x_i, T_i), (\varphi, \varphi'_n)) - \Delta_i((x_i, T_i), (x_\infty, T_\infty)) \right) d\mathscr{L}^p$$

$$\leq \sigma_i \mathscr{L}^p Q \leq 2^{-i-3} \tau \mathscr{L}^p Q.$$

Adding these inequalities together, we finally get

$$\fint_Q \Delta(\varphi) \, d\mathscr{L}^p \leq \sum_{i=0}^{k} 2^{-i-3} \left(\fint_Q \|g'_n\|^2 \, d\mathscr{L}^p + \tau \right) + \sum_{i=k+1}^{\infty} 2^{-i-3} \tau$$

$$\leq \frac{1}{4} \fint_Q \|g'_n\|^2 \, d\mathscr{L}^p + \frac{\tau}{4}, \tag{13.49}$$

which is the last of our estimates.

It remains only to subtract the inequalities (13.38), (13.39), (13.40), and (13.49) from (13.37) and use (13.32):

$$\fint_Q \left(h_\infty(x_\infty, T_\infty) - h_\infty(\varphi, \varphi'_n) \right) d\mathscr{L}^p \geq \frac{1}{2} \fint_Q \|g'_n\|^2 \, d\mathscr{L}^p - \frac{7\tau}{2} > \frac{\tau}{2},$$

which is what we required in (13.33) to finish the proof of Fréchet regularity of f at x_∞.

13.5 FRÉCHET DIFFERENTIABILITY

Once we know that f is regularly Fréchet differentiable at x_∞ in the direction of T_∞, its Fréchet differentiability follows just from the (upper) Fréchet differentiability of Θ_∞ at T_∞. In particular, as we have mentioned above, the assumption of asymptotic smoothness need not be used any more.

Let $L_\Theta \in L(\mathbb{R}^n, X)^*$ be an upper Fréchet derivative of Θ_∞ at T_∞. The first step of our argument is to "guess" the value of the derivative $f'(x_\infty)$. For that recall the form of the function h_∞,

$$h_\infty(x, T) = -|f'(x; T)|_{\mathrm{H}}^2 + \sum_{i=0}^{\infty} \lambda_i \|x - x_i\| + \sum_{i=0}^{\infty} \beta_i \Theta(T - T_i)$$

$$+ \sum_{i=0}^{\infty} \gamma_i |f'(x; T) - L_i|_{\mathrm{H}}^2$$

$$+ \sum_{i=0}^{\infty} \sigma_i \max\{0, \min\{1, \|f_{x,T_i} - f_{x_i, T_i}\|_i\} - s_i\}.$$

Once we fix $x = x_\infty$, the functions by which we are perturbing the map

$$h_0(x, T) = -|f'(x; T)|_{\mathrm{H}}^2$$

are either constant, differentiable, or in one case upper Fréchet differentiable with re-spect to T. If f were known to be Fréchet differentiable at x_∞, then the function $T \to h_\infty(x_\infty, T)$ would be upper differentiable and, therefore, since it attains its min-imum at T_∞ its upper derivative would be zero. Calculating the upper derivative in the direction of any $S \in L(\mathbb{R}^n, X)$ (still assuming that $f'(x_\infty)$ exists) would then lead to the system of equations for the linear operator $f'(x_\infty)$:

$$-2\langle L_\infty, f'(x_\infty; S)\rangle_{\mathrm{H}} + L_\Theta S + 2\sum_{i=0}^{\infty} \gamma_i\langle L_\infty - L_i, f'(x_\infty; S)\rangle_{\mathrm{H}} = 0,$$

for all $S \in L(\mathbb{R}^n, X)$. The above system makes sense as a system of equations for an unknown operator $L \in L(X, \mathbb{R}^n)$ in place of $f'(x_\infty)$ even when we do not know whether f is Fréchet differentiable or not. We find its solution. Denoting

$$R_\infty := 2L_\infty + 2\sum_{i=0}^{\infty} \gamma_i(L_i - L_\infty) \in L(\mathbb{R}^n, \mathbb{R}^n),$$

we rewrite this system as

$$\langle R_\infty, LS\rangle_{\mathrm{H}} = L_\Theta S, \quad S \in L(\mathbb{R}^n, X). \tag{13.50}$$

Since every $S \in L(\mathbb{R}^n, X)$ is a finite sum of rank one operators, we may consider this equation only for rank one operators $S \in L(\mathbb{R}^n, X)$. To solve it, we first observe that

Lemma 13.5.1. R_∞ *is invertible and* $\|R_\infty^{-1}\| \leq 1$.

Proof. By Lemma 13.3.6, $\|L_i - L_\infty\| \leq \frac{1}{2}$ for all $i \geq 0$. Since $\sum_{i=0}^{\infty} \gamma_i = \frac{1}{4}$ and $\|\operatorname{Id} - L_\infty\| \leq \frac{1}{4}$ by (13.7), we obtain

$$\|\operatorname{Id} - \tfrac{1}{2}R_\infty\| = \left\|\operatorname{Id} - L_\infty - \sum_{i=0}^{\infty} \gamma_i(L_i - L_\infty)\right\|$$

$$\leq \|\operatorname{Id} - L_\infty\| + \sum_{i=0}^{\infty} \gamma_i\|L_i - L_\infty\| \leq \frac{1}{2}.$$

Hence $\frac{1}{2}R_\infty = \operatorname{Id} - (\operatorname{Id} - \frac{1}{2}R_\infty)$ is invertible, its inverse being

$$(\tfrac{1}{2}R_\infty)^{-1} = \sum_{i=0}^{\infty} (\operatorname{Id} - \tfrac{1}{2}R_\infty)^i.$$

It follows that $\|R_\infty^{-1}\| \leq 1$. $\qquad\square$

Recall that a rank one operator $S \in L(\mathbb{R}^n, X)$ is of the form $S = x \otimes e$, where $x \in X$ and $e \in \mathbb{R}^n$; this operator acts by $(x \otimes e)(u) = \langle e, u\rangle x$. Since $L(x \otimes e) = Lx \otimes e$ and $\langle R_\infty, Lx \otimes e\rangle_{\mathrm{H}} = \langle R_\infty e, Lx\rangle$, the system (13.50) for the unknown value Lx becomes

$$\langle R_\infty e, Lx\rangle = L_\Theta(x \otimes e) \tag{13.51}$$

for all $e \in \mathbb{R}^n$. Using the invertibility of R_∞, (13.51) may be rewritten as

$$\langle e, Lx \rangle = L_\Theta\big(x \otimes (R_\infty^{-1}e)\big).$$

For a fixed x, the right side of this equation defines an element of $(\mathbb{R}^n)^*$, and the left side says that this functional is represented by the element $Lx \in \mathbb{R}^n$. Hence Lx is uniquely defined, and the linearity of the equation in x shows that L is linear. Moreover, L is bounded since $|Lx| \leq \|L_\Theta\| \|R_\infty^{-1}\| \|x\|$. We have thus shown

Lemma 13.5.2. *The system of equations* (13.50), *where* $S \in L(\mathbb{R}^n, X)$, *or equivalently the system of equations* (13.51), *where* $e \in \mathbb{R}^n$ *and* $x \in X$, *defines a unique operator* $L \in L(X, \mathbb{R}^n)$. *Moreover,* $\|L\| \leq \|L_\Theta\| \|R_\infty^{-1}\| \leq \|L_\Theta\|$.

Although we do not need it, let us describe L using the standard basis for the space $\mathbb{R}^n$. For that, represent L_Θ by $(x_1^*, \ldots, x_n^*) \in (X^*)^n$ as

$$L_\Theta S = \sum_{i=1}^n x_i^*(Se_i).$$

Then

$$\langle e, Lx \rangle = L_\Theta(x \otimes (R_\infty^{-1}e)) = \sum_{i=1}^n \langle R_\infty^{-1}e, e_i \rangle x_i^*(x) = \Big\langle e, \sum_{i=1}^n x_i^*(x)(R_\infty^{-1})^* e_i \Big\rangle.$$

Hence $Lx = \sum_{i=1}^n x_i^*(x)(R_\infty^{-1})^* e_i$.

From now on, L will denote the operator from Lemma 13.5.2. The rest of this section will be devoted to showing that

Proposition 13.5.3. f *is Fréchet differentiable at* x_∞ *and* $f'(x_\infty) = L$.

Revisiting the argument we used to show $f'(x_\infty) = L$, provided $f'(x_\infty)$ exists, we notice that it also shows $f'(x_\infty; T) = LT$, provided $f'(x_\infty; T)$ exists. For that we just consider (13.50) for S whose range lies in $T(\mathbb{R}^n)$. In particular, since $f'(x_\infty; T_\infty)$ exists, we have

Lemma 13.5.4. $f'(x_\infty; T_\infty) = LT_\infty$.

We will argue by contradiction, and so for the rest of this section we make the following

Assumption. L *is not the Fréchet derivative of* f *at* x_∞.

Since L is defined so indirectly, we restate the assumption that the Fréchet derivative of f at x_∞ is not L in a way which is easier to handle.

Lemma 13.5.5. *There is* $\varepsilon > 0$ *such that for every* $\delta > 0$ *and* $k \in \mathbb{N}$ *one may find* $x \in X$ *with* $\|x\| < \delta$ *such that*

- $Lx = 0$,

- $|f(x_\infty + x) - f(x_\infty)| > \varepsilon\|x\|$, and

- f is Gâteaux differentiable at $x_\infty + tx + T_\infty \pi_n u + S_k \pi^n u$ for almost every $(u, t) \in \mathbb{R}^p \times \mathbb{R}$, $p = n + k$.

Proof. Let $c = 1 + \|T_\infty L_\infty^{-1} L\| + \|L_\infty^{-1} L\|$. If L is not the Fréchet derivative of f at x_∞, then there is $\varepsilon > 0$ such that for every $\delta > 0$ one may find $\widetilde{x} \in X$ with $\|\widetilde{x}\| < \delta$ and

$$|f(x_\infty + \widetilde{x}) - f(x_\infty) - L\widetilde{x}| > 2\varepsilon c\|\widetilde{x}\|. \tag{13.52}$$

Suppose that $\delta > 0$ and $k \in \mathbb{N}$ are given. Since f is regularly Fréchet differentiable at x_∞ in the direction T_∞, we may find $\eta > 0$ such that

$$|f(x_\infty + z + T_\infty u) - f(x_\infty + z) - L_\infty u| \le \varepsilon(\|z\| + |u|) \tag{13.53}$$

whenever $z \in X$, $u \in \mathbb{R}^n$, and $\|z\| + |u| < \eta$.

Let $\widetilde{x} \in X$ be such that $c\|\widetilde{x}\| < \min\{\delta, \eta\}$ and (13.52) holds. Using Lemma 2.3.2, we move $\widetilde{x}$ slightly so that (13.52) still holds and for almost every $(u, t) \in \mathbb{R}^p \times \mathbb{R}$, f is Gâteaux differentiable at $x_\infty + t\widetilde{x} + T_\infty \pi_n u + S_k \pi^n u$.

Let $x := \widetilde{x} - T_\infty L_\infty^{-1} L\widetilde{x}$. Since $LT_\infty = L_\infty$ by Lemma 13.5.4, we get $Lx = 0$. Also,

$$\|x\| + \|L_\infty^{-1} L\widetilde{x}\| \le \|\widetilde{x}\| + \|T_\infty L_\infty^{-1} L\widetilde{x}\| + \|L_\infty^{-1} L\widetilde{x}\| \le c\|\widetilde{x}\| < \min\{\delta, \eta\}.$$

Hence both $\|x\| < \delta$ and (13.53) is applicable with $z = x$ and $u = L_\infty^{-1} L\widetilde{x}$ giving that

$$\begin{aligned}
|f(x_\infty + \widetilde{x}) &- f(x_\infty + x) - L\widetilde{x}| \\
&= |f(x_\infty + x + T_\infty L_\infty^{-1} L\widetilde{x}) - f(x_\infty + x) - L\widetilde{x}| \\
&\le \varepsilon(\|x\| + \|L_\infty^{-1} L\widetilde{x}\|) \le \varepsilon c\|\widetilde{x}\|.
\end{aligned}$$

Together with (13.52) we get the second statement by estimating

$$\begin{aligned}
|f(x_\infty + x) &- f(x_\infty)| \\
&\ge |f(x_\infty + \widetilde{x}) - f(x_\infty) - L\widetilde{x}| - |f(x_\infty + \widetilde{x}) - f(x_\infty + x) - L\widetilde{x}| \\
&> \varepsilon c\|\widetilde{x}\| \ge \varepsilon\|x\|.
\end{aligned}$$

Finally, the Gâteaux differentiability condition holds since

$$x_\infty + tx + T_\infty \pi_n u + S_k \pi^n u = x_\infty + t\widetilde{x} + T_\infty \pi_n(u - L_\infty^{-1} L\widetilde{x}) + S_k \pi^n(u - L_\infty^{-1} L\widetilde{x}),$$

and so the exceptional sets for x and $\widetilde{x}$ differ only by a shift. $\qquad\square$

From now on, we fix $0 < \varepsilon < 1$ with the property from Lemma 13.5.5. By upper Fréchet differentiability of Θ_∞ we find $k \ge 1/\varepsilon$ such that the parameter

$$\kappa := \frac{s_k}{8\operatorname{Lip}(f)}$$

satisfies $\kappa \leq \frac{1}{2}\eta_0$ and

$$\Theta_\infty(T_\infty + S) - \Theta_\infty(T_\infty) \leq L_\Theta S + \frac{\varepsilon}{4}\|S\| \tag{13.54}$$

whenever $\|S\| \leq \kappa$. Let $p = n + k$, $\tau = \frac{1}{4}\varepsilon\kappa$,

$$t = \frac{\tau}{11(n+1)\operatorname{Lip}(f)(1+|R_\infty|_H)},$$

and choose $0 < \xi \leq \frac{1}{5}t$ such that

$$\xi \leq \frac{\tau t s_k}{24\|S_k^{-1}\|((n+1)t+k)}.$$

Since f is regularly Fréchet differentiable at each x_i in the direction of T_i, there is $\delta_1 > 0$ such that for all $i = 0, \ldots, k$ and $i = \infty$,

$$|f(x_i + z + T_i v) - f(x_i + z) - L_i v| \leq \xi(\|z\| + |v|) \tag{13.55}$$

whenever $v \in \mathbb{R}^n$, $z \in X$ and $\|z\| + |v| < \delta_1$.

Let $\delta_2 > 0$ be such that

$$\delta_2 \leq \frac{\kappa\delta_1}{5}, \quad \delta_2 \leq \frac{\kappa\eta_0}{6}, \quad \delta_2 \leq \frac{\kappa\tau}{3\sum_{i=0}^\infty \lambda_i}, \quad \delta_2 \leq \frac{\delta_1\kappa\tau}{24\operatorname{Lip}(f)}.$$

Use Lemma 13.5.5 to find $\widetilde{x} \in X$ such that $\|\widetilde{x}\| < \delta_2$, $L\widetilde{x} = 0$,

$$|f(x_\infty + \widetilde{x}) - f(x_\infty)| > \varepsilon\|\widetilde{x}\|, \tag{13.56}$$

and the set H of $(u, s) \in \mathbb{R}^p \times \mathbb{R}$ for which f is Gâteaux differentiable at

$$x_\infty + s\widetilde{x} + T_\infty\pi_n u + S_k\pi^n u$$

has full $\mathscr{L}^{p+1}$-measure. Since for almost every $(u, s) \in \mathbb{R}^p \times \mathbb{R}$ the point

$$A(u, s) := \left(u, s + \frac{\kappa\langle e, u\rangle}{\|\widetilde{x}\|}\right)$$

belongs to H, we may find $\widetilde{s}$ so close to 1 that the point $x := \widetilde{s}\,\widetilde{x}$ satisfies that $\|x\| < \delta_2$,

$$|f(x_\infty + x) - f(x_\infty)| > \varepsilon\|x\|, \tag{13.57}$$

and $A(u, \widetilde{s}) \in H$ for almost every $u \in \mathbb{R}^n$. It follows that for almost every $u \in \mathbb{R}^n$, the function f is Gâteaux differentiable at the points

$$x_\infty + x + \kappa\langle e, u\rangle\frac{x}{\|x\|} + T_\infty\pi_n u + S_k\pi^n u$$

and, obviously, that $Lx = 0$.

Denote

$$r = \frac{\|x\|}{\kappa}, \quad w = \frac{1}{r}\big(f(x_\infty + x) - f(x_\infty)\big), \quad e = \frac{R_\infty^{-1} w}{|R_\infty^{-1} w|},$$

and

$$\Omega = \big\{ u \in \mathbb{R}^n \;\big|\; -r < \langle e, u \rangle < 0, |u - \langle e, u \rangle e| < r \big\}.$$

Observe that, since $\|R_\infty^{-1}\| \leq 1$,

$$\langle R_\infty e, w \rangle = \frac{|w|^2}{|R_\infty^{-1} w|} \geq |w| > \varepsilon \kappa = 4\tau. \tag{13.58}$$

Define a Lipschitz surface $\gamma \colon \mathbb{R}^n \longrightarrow X$ with $\mathrm{Lip}(\gamma) \leq \kappa$ by

$$\gamma(u) = \begin{cases} 0 & \text{if } \langle e, u \rangle \leq -r, \\ \dfrac{x}{r}(r + \langle e, u \rangle) & \text{if } -r < \langle e, u \rangle < 0, \\ x & \text{if } \langle e, u \rangle \geq 0, \end{cases}$$

and let $\varphi \colon \mathbb{R}^p \longrightarrow X$ be

$$\varphi(u) = x_\infty + \gamma(\pi_n u) + T_\infty \pi_n u + S_k \pi^n u.$$

Notice that for $u \in \mathbb{R}^p$ with $-r < \langle e, u \rangle < 0$,

$$\varphi(u) = x_\infty + x + \kappa \langle e, u \rangle \frac{x}{\|x\|} + T_\infty \pi_n u + S_k \pi^n u.$$

Hence f is Gâteaux differentiable at $\varphi(u)$ for a.e. $u \in \mathbb{R}^p$ such that $-r < \langle e, u \rangle < 0$. Notice also that

$$\varphi_n'(u) = \kappa \frac{x \otimes e}{\|x\|} + T_\infty \tag{13.59}$$

for $u \in \Omega \times \mathbb{R}^k$. Let $Q = \Omega \times B_k(0, tr)$ and define $g \colon \mathbb{R}^p \longrightarrow \mathbb{R}^n$ by

$$g(u) = f(\varphi(u)) - L_\infty \pi_n u. \tag{13.60}$$

In order to estimate various expressions needed in the sequel we observe the following lemma.

Lemma 13.5.6. *Let* φ *and* g *be the functions defined above. Then* $\mathrm{Lip}(\varphi) \leq 3$, $\mathrm{Lip}(g) \leq 5\,\mathrm{Lip}(f)$,

- $\|\varphi(u) - x_\infty\| \leq 3r$ *for all* $u \in Q$, *and*
- *for every* $u, v \in \overline{Q}$ *with* $\langle u - v, e \rangle = 0$,

$$|g(u) - g(v)| \leq \frac{\tau r}{(n+1)(1 + |R_\infty|_{\mathrm{H}})}.$$

Proof. To show the first three inequalities we invoke Lemma 13.3.6 and the choice of the parameters involved. For the first inequality we estimate

$$\operatorname{Lip}(\varphi) \leq \kappa + \|T_\infty\| + \|S_k\| \leq \frac{\eta_0}{2} + 1 + 1 \leq 3.$$

Since $\|L_\infty\| \leq \frac{5}{4}$ we immediately obtain $\operatorname{Lip}(g) \leq \frac{5}{4} + 3\operatorname{Lip}(f) \leq 5\operatorname{Lip}(f)$. The third inequality follows from

$$\|\varphi(u) - x_\infty\| \leq \|x\| + \kappa r + 2\|T_\infty\|r + \|S_k\|tr \leq r(2\kappa + 2 + t) \leq 3r.$$

The last estimate of the lemma is different from the previous ones in that it needs (13.55), which we may use only because we have already shown its Fréchet regularity.[*] Consider first the special case $u, v \in \overline{\Omega}$. Then we have $\gamma(\pi_n u) = \gamma(\pi_n v)$, and hence $\varphi(u) - \varphi(v) = T_\infty(u - v)$. We intend to use (13.55) for the vector $z = \gamma(v) + T_\infty v$ and for $u - v$ instead of u. To this aim we estimate

$$\|\gamma(v) + T_\infty v\| + |u - v| = \|\varphi(v) - x_\infty\| + |u - v| \leq 3r + 2r = 5r < \frac{5\delta_2}{\kappa} \leq \delta_1.$$

Thus (13.55) gives

$$
\begin{aligned}
&|g(u) - g(v)| \\
&= \left| f(x_\infty + \gamma(u) + T_\infty u) - f(x_\infty + \gamma(v) + T_\infty v) - L_\infty(u - v) \right| \\
&= \left| f(x_\infty + \gamma(v) + T_\infty v + T_\infty(u - v)) - f(x_\infty + \gamma(v) + T_\infty v) - L_\infty(u - v) \right| \\
&\leq \xi(\|\gamma(v) + T_\infty v\| + |u - v|) \leq 5\xi r.
\end{aligned}
$$

In the general case, we apply the above estimate to $\pi_n u$ and $\pi_n v$:

$$
\begin{aligned}
|g(u) - g(v)| &\leq |g(u) - g(\pi_n u)| + |g(v) - g(\pi_n v)| + |g(\pi_n u) - g(\pi_n v)| \\
&\leq 2\operatorname{Lip}(g)\,tr + 5\xi r \leq 10\operatorname{Lip}(f)\,tr + tr \\
&= 11\operatorname{Lip}(f)\,tr = \frac{\tau r}{(n+1)(1 + |R_\infty|_{\mathrm{H}})}. \qquad \square
\end{aligned}
$$

Our aim is to show that

$$\fint_Q \left(h_\infty(x_\infty, T_\infty) - h_\infty(\varphi, \varphi'_n) \right) d\mathscr{L}^p > 0. \tag{13.61}$$

Once this is done, a contradiction immediately follows: since for almost every $u \in Q$, f is Gâteaux differentiable at $\varphi(u)$, there is $u \in Q$ such that f is Gâteaux differentiable at $\varphi(u)$ and $h_\infty(x_\infty, T_\infty) - h_\infty(\varphi(u), \varphi'_n(u)) > 0$. But

$$\|\varphi(u) - x_0\| \leq 3r + \|x_\infty - x_0\| \leq \frac{3\delta_2}{\kappa} + \frac{\eta_0}{2} \leq \eta_0,$$

---

[*]In fact, this seemingly simple inequality is our only significant use of Fréchet regularity.

and, by (13.59)

$$\|\varphi_n'(u) - T_0\| \leq \kappa + \|T_0 - T_\infty\| \leq \frac{\eta_0}{2} + \frac{\eta_0}{2} = \eta_0.$$

This implies that $(\varphi(u), \varphi_n'(u)) \in D_0$, and so h_∞ does not attain its minimum on D_0 at (x_∞, T_∞).

It remains to show (13.61). For this we recall from (13.19) that

$$h_\infty(x_\infty, T_\infty) - h_\infty(\varphi, \varphi_n') = \left(|f'(\varphi; \varphi_n')|_{\mathrm{H}}^2 - |L_\infty|_{\mathrm{H}}^2\right) - \Phi(\varphi) \\ - \Psi(\varphi_n') - \Upsilon\left(f'(\varphi; \varphi_n')\right) - \Delta(\varphi) \tag{13.62}$$

and we estimate the integral of each of the five terms on the right side. Notice that each of the functions we integrate is almost everywhere defined, measurable, and bounded on Q.

Estimate of $\displaystyle\fint_Q \Phi(\varphi)\, d\mathscr{L}^p$. Since by Lemma 13.5.6 for $u \in Q$,

$$\big|\|x_i - \varphi(u)\| - \|x_i - x_\infty\|\big| \leq \|\varphi(u) - x_\infty\| \leq 3r \leq \frac{3\delta_2}{\kappa},$$

we get by the choice of δ_2 that

$$|\Phi(\varphi(u))| \leq \sum_{i=0}^{\infty} \lambda_i \|\varphi(u) - x_\infty\| \leq \frac{3\delta_2}{\kappa} \sum_{i=0}^{\infty} \lambda_i \leq \tau.$$

Hence

$$\fint_Q \Phi(\varphi)\, d\mathscr{L}^p \leq \tau. \tag{13.63}$$

Estimate of $\fint_Q \Psi(\varphi_n')\, d\mathscr{L}^p$. Since $Lx = 0$, (13.51) shows that $L_\Theta(x \otimes e) = 0$. Hence (13.54) gives

$$\fint_Q \Psi(\varphi_n')\, d\mathscr{L}^p = \fint_Q \left(\Theta_\infty\left(T_\infty + \kappa \frac{x \otimes e}{\|x\|}\right) - \Theta_\infty(T_\infty)\right) d\mathscr{L}^p \\ \leq \frac{\varepsilon\kappa}{4} = \tau. \tag{13.64}$$

Estimate of $\fint_Q \Upsilon(f'(\varphi; \varphi_n'))\, d\mathscr{L}^p$. We use the definition of the function g to conclude that $f'(\varphi; \varphi_n') = g_n' + L_\infty$. Since $\sum_{i=0}^{\infty} \gamma_i = \frac{1}{4}$, we have

$$\Upsilon\left(f'(\varphi(u); \varphi_n'(u))\right) = \sum_{i=0}^{\infty} \gamma_i\left(|(L_i - L_\infty) - g_n'(u)|_{\mathrm{H}}^2 - |L_i - L_\infty|_{\mathrm{H}}^2\right) \\ = \sum_{i=0}^{\infty} \gamma_i\left(2\langle L_\infty - L_i, g_n'(u)\rangle_{\mathrm{H}} + |g_n'(u)|_{\mathrm{H}}^2\right) \\ = \left\langle 2\sum_{i=0}^{\infty} \gamma_i(L_\infty - L_i), g_n'(u)\right\rangle_{\mathrm{H}} + \frac{1}{4}|g_n'(u)|_{\mathrm{H}}^2.$$

Hence

$$\fint_Q \Upsilon(f'(\varphi; \varphi_n')) \, d\mathscr{L}^p$$

$$= \fint_Q \left\langle 2 \sum_{i=0}^{\infty} \gamma_i (L_\infty - L_i), g_n' \right\rangle_{\mathrm{H}} d\mathscr{L}^p + \frac{1}{4} \fint_Q |g_n'|_{\mathrm{H}}^2 \, d\mathscr{L}^p. \tag{13.65}$$

Estimate of $\fint_Q \Delta(\varphi) \, d\mathscr{L}^p$. Consider first $0 \le i \le k$ and denote

$$j = n + i,$$

$$s = \frac{\tau t r}{(n+1)t + k}, \quad \text{and}$$

$$P = \left\{ u \in Q \ \middle| \ \Delta_i\big((x_i, T_i), (\varphi(u), \varphi_n'(u))\big) - \Delta\big((x_i, T_i), (x_\infty, T_\infty)\big) > \tau \right\}.$$

We prove that for every $u \in P$,

$$\operatorname{reg}_n g(u, B_j(u, s)) > \frac{s_i}{4\|S_i^{-1}\|}. \tag{13.66}$$

For this, we choose $\widetilde{z} \in Z_i$ and $\widetilde{w} \in \mathbb{R}^n$ such that

$$|f_{x_i, T_i}(\widetilde{z}, \widetilde{w}) - f_{x_i, T_i}(\widetilde{z}, 0) - f_{\varphi(u), T_i}(\widetilde{z}, \widetilde{w}) + f_{\varphi(u), T_i}(\widetilde{z}, 0)|$$

$$> \left(\|f_{x_i, T_i} - f_{\varphi(u), T_i}\|_i - \frac{\tau}{2} \right) (\|\widetilde{z}\| + |\widetilde{w}|).$$

We show first that the points $\widetilde{z} \in Z_i$, $\widetilde{w} \in \mathbb{R}^n$, $\widetilde{v} = S_k^{-1}\widetilde{z} = S_i^{-1}\widetilde{z} \in \mathbb{R}^i \subset \mathbb{R}^j$, and $\widetilde{u} = \widetilde{v} + \widetilde{w} \in \mathbb{R}^j$ satisfy

$$\|\widetilde{z}\| + |\widetilde{w}| < \delta_1 \quad \text{and} \quad u + \widetilde{u}, u + \widetilde{v} \in B_j(u, s). \tag{13.67}$$

The following two cases cover all possibilities of failure of (13.67); we bring both of them to a contradiction.

Case 1. Suppose that $\|\widetilde{z}\| + |\widetilde{w}| \ge \frac{1}{2}\delta_1$. Then we infer from Lemma 13.5.6 and the choice of δ_2 that

$$2\operatorname{Lip}(f)\|\varphi(u) - x_\infty\| \le 6\operatorname{Lip}(f)\, r \le 6\operatorname{Lip}(f)\frac{\delta_2}{\kappa} \le \frac{\tau}{4}\delta_1 \le \frac{\tau}{2}(\|\widetilde{z}\| + |\widetilde{w}|).$$

It follows that

$$(\|\widetilde{z}\| + |\widetilde{w}|)\|f_{x_i, T_i} - f_{x_\infty, T_i}\|_i$$

$$\ge |f_{x_i, T_i}(\widetilde{z}, \widetilde{w}) - f_{x_i, T_i}(\widetilde{z}, 0) - f_{x_\infty, T_i}(\widetilde{z}, \widetilde{w}) - f_{x_\infty, T_i}(\widetilde{z}, 0)|$$

$$\ge |f_{x_i, T_i}(\widetilde{z}, \widetilde{w}) - f_{x_i, T_i}(\widetilde{z}, 0) - f_{\varphi(u), T_i}(\widetilde{z}, \widetilde{w}) - f_{\varphi(u), T_i}(\widetilde{z}, 0)|$$

$$\quad - 2\operatorname{Lip}(f)\|\varphi(u) - x_\infty\|$$

$$\ge \left(\|f_{x_i, T_i} - f_{\varphi(u), T_i}\| - \frac{\tau}{2} \right)(\|\widetilde{z}\| + |\widetilde{w}|) - \frac{\tau}{2}(\|\widetilde{z}\| + |\widetilde{w}|)$$

$$\ge (\|\widetilde{z}\| + |\widetilde{w}|)(\|f_{x_i, T_i} - f_{\varphi(u), T_i}\|_i - \tau).$$

Hence $\Delta_i\big((x_i, T_i), (\varphi(u), \varphi_n'(u))\big) - \Delta\big((x_i, T_i), (x_\infty, T_\infty)\big) \leq \tau$, contradicting the assumption that $u \in P$. ♡

Case 2. Suppose that $\|\widetilde{z}\| + |\widetilde{w}| < \frac{1}{2}\delta_1$ and $s \leq |\widetilde{u}| + |\widetilde{v}|$. Notice that the second inequality implies that $s \leq 2\|S_k^{-1}\|(\|\widetilde{z}\| + |\widetilde{w}|)$. Define $\widehat{z} \in X$ by the requirement $\varphi(u) + \widetilde{z} = x_\infty + \widehat{z}$, that is,

$$\widehat{z} = \widetilde{z} + T_\infty \pi_n u + \gamma(\pi_n u) + S_k \pi^n u.$$

Then by Lemma 13.5.6 and by the choice of δ_2,

$$\|\widehat{z}\| + |\widetilde{w}| \leq \|\varphi(u) - x_\infty\| + \|\widetilde{z}\| + |\widetilde{w}| \leq 3r + \|\widetilde{z}\| + |\widetilde{w}| < \delta_1. \tag{13.68}$$

Thus we may apply (13.55) and use our choice of ξ and s to conclude that

$$\begin{aligned}
|f_{\varphi(u),T_i}(\widehat{z},\widetilde{w}) &- f_{\varphi(u),T_i}(\widehat{z},0) - L_i\widetilde{w}| \\
&= |f_{x_\infty,T_i}(\widehat{z}, \widetilde{w}) - f_{x_\infty,T_i}(\widehat{z},0) - L_i\widetilde{w}| \\
&\leq \xi(\|\widehat{z}\| + |\widetilde{w}|) \leq \xi(3r + \|\widetilde{z}\| + |\widetilde{w}|) \\
&\leq \frac{ss_i}{8\|S_k^{-1}\|} + \frac{s_i}{4}(\|\widetilde{z}\| + |\widetilde{w}|) \leq \frac{s_i}{4}(\|\widetilde{z}\| + |\widetilde{w}|) + \frac{s_i}{4}(\|\widetilde{z}\| + |\widetilde{w}|) \\
&= \frac{s_i}{2}(\|\widetilde{z}\| + |\widetilde{w}|). \tag{13.69}
\end{aligned}$$

Since $\|\widetilde{z}\| + |\widetilde{w}| < \delta_1$, another application of (13.55) gives

$$|f_{x_i,T_i}(\widetilde{z}, \widetilde{w}) - f_{x_i,T_i}(\widetilde{z},0) - L_i\widetilde{w}| \leq \xi(\|\widetilde{z}\| + |\widetilde{w}|) \leq \frac{s_i}{2}(\|\widetilde{z}\| + |\widetilde{w}|).$$

Hence

$$\begin{aligned}
\Big(\|f_{x_i,T_i} &- f_{\varphi(u),T_i}\|_i - \frac{\tau}{2}\Big)(\|\widetilde{z}\| + |\widetilde{w}|) \\
&\leq |f_{x_i,T_i}(\widetilde{z}, \widetilde{w}) - f_{x_i,T_i}(\widetilde{z},0) - f_{\varphi(u),T_i}(\widetilde{z}, \widetilde{w}) - f_{\varphi(u),T_i}(\widetilde{z},0)| \\
&\leq |f_{x_i,T_i}(\widetilde{z}, \widetilde{w}) - f_{x_i,T_i}(\widetilde{z},0) - L_i\widetilde{w}| \\
&\quad + |f_{\varphi(u),T_i}(\widetilde{z}, \widetilde{w}) - f_{\varphi(u),T_i}(\widetilde{z},0) - L_i\widetilde{w}| \\
&\leq s_i(\|\widetilde{z}\| + |\widetilde{w}|),
\end{aligned}$$

implying that $\Delta_i\big((x_i, T_i), (\varphi(u), \varphi_n'(u))\big) \leq \frac{1}{2}\tau$, and so $u \notin P$. ♡

Having thus finished the proof of (13.67), we are going to show (13.66). By Lemma 13.3.6 we first estimate

$$\begin{aligned}
\|\varphi(u + \widetilde{w}) - \varphi(u) - T_i\widetilde{w}\| &\leq \big(\|T_i - T_\infty\| + \mathrm{Lip}_n(\gamma)\big)|\widetilde{w}| \\
&\leq \Big(\frac{s_i}{8\,\mathrm{Lip}(f)} + \kappa\Big)|\widetilde{w}| = \frac{s_i}{4\,\mathrm{Lip}(f)}\,|\widetilde{w}|.
\end{aligned}$$

Using this estimate and (13.67) we get

$$
\begin{aligned}
(|\widetilde{u}| + |\widetilde{v}|)\,\mathrm{reg}_n\, g(u, B_j(u, s)) &\geq |g(u + \widetilde{u}) - g(u + \widetilde{v})| \\
&= |f(\varphi(u + \widetilde{w}) + \widetilde{z}) - f(\varphi(u) + \widetilde{z}) - L_\infty(\widetilde{w})| \\
&\geq |f_{\varphi(u),T_i}(\widetilde{z}, \widetilde{w}) - f_{\varphi(u),T_i}(\widetilde{z}, 0) - L_\infty(\widetilde{w})| \\
&\quad - \mathrm{Lip}(f)\|\varphi(u + \widetilde{w}) - \varphi(u) - T_i\widetilde{w}\| \\
&\geq |f_{\varphi(u),T_i}(\widetilde{z}, \widetilde{w}) - f_{\varphi(u),T_i}(\widetilde{z}, 0) - f_{x_i,T_i}(\widetilde{z}, \widetilde{w}) + f_{x_i,T_i}(\widetilde{z}, 0)| \\
&\quad - |f_{x_i,T_i}(\widetilde{z}, \widetilde{w}) - f_{x_i,T_i}(\widetilde{z}, 0) - L_i\widetilde{w}| - \|L_i - L_\infty\||\widetilde{w}| - \frac{s_i}{4}|\widetilde{w}| \\
&\geq \left(\|f_{x_i,T_i} - f_{\varphi(u),T_i}\|_i - \frac{\tau}{2}\right)(\|\widetilde{z}\| + |\widetilde{w}|) \\
&\quad - |f_{x_i,T_i}(\widetilde{z}, \widetilde{w}) - f_{x_i,T_i}(\widetilde{z}, 0) - L_i\widetilde{w}| - \|L_i - L_\infty\||\widetilde{w}| - \frac{s_i}{4}|\widetilde{w}|.
\end{aligned}
$$

The second term is estimated by (13.55), and the difference $\|L_i - L_\infty\|$ is, according to Lemma 13.3.6, bounded by $\frac{1}{8}s_i$. Also, since $u \in P$ we have, in particular, that $\|f_{x_i,T_i} - f_{\varphi(u),T_i}\|_i > \tau + s_i$. With this information at hand we continue:

$$
\begin{aligned}
&\geq \left(\|f_{x_i,T_i} - f_{\varphi(u),T_i}\|_i - \frac{\tau}{2}\right)(\|\widetilde{z}\| + |\widetilde{w}|) \\
&\quad - \xi(\|\widetilde{z}\| + |\widetilde{w}|) - \left(\frac{s_i}{8} + \frac{s_i}{4}\right)|\widetilde{w}| \\
&\geq s_i(\|\widetilde{z}\| + |\widetilde{w}|) - \frac{s_i}{8}(\|\widetilde{z}\| + |\widetilde{w}|) - \frac{3s_i}{8}(\|\widetilde{z}\| + |\widetilde{w}|) \\
&\geq \frac{s_i}{2\|S_i^{-1}\|}(|\widetilde{v}| + |\widetilde{w}|) \geq \frac{s_i}{4\|S_i^{-1}\|}(|\widetilde{u}| + |\widetilde{w}|).
\end{aligned}
$$

We are now ready to complete the first part of our estimate of the Δ-term. We let $Q_0 = \{u \in Q \mid B(u, s) \subset Q\}$ and infer from Corollary 13.2.4 that

$$
\mathscr{L}^p(P \cap Q_0) \leq \frac{4^{2j}\|S_i^{-1}\|^{2j}\Lambda_n \overline{K}_j^2(5\,\mathrm{Lip}(f))^{2j-2}}{s_i^{2j}} \int_Q \|g_n'\|^2\, d\mathscr{L}^p.
$$

Since, using the choice of s,

$$
\begin{aligned}
\mathscr{L}^p Q_0 &\geq \left(1 - \frac{2s}{r}\right)\left(1 - \frac{s}{r}\right)^{n-1}\left(1 - \frac{s}{tr}\right)^k \mathscr{L}^p Q \\
&\geq \left(1 - \frac{(n+1)s}{r} - \frac{ks}{tr}\right)\mathscr{L}^p Q = (1 - \tau)\mathscr{L}^p Q,
\end{aligned}
$$

and $\Delta_i\big((x_i, T_i), (\varphi, \varphi'_n)\big) \leq 1$, we conclude that

$$
\sigma_i \int_Q \big(\Delta_i((x_i, T_i), (\varphi, \varphi'_n)) - \Delta_i((x_i, T_i), (x_\infty, T_k))\big)\, d\mathscr{L}^p
$$

$$
\leq \sigma_i \int_P \Delta_i((x_i, T_i), (\varphi, \varphi'_n))\, d\mathscr{L}^p + \sigma_i \tau \mathscr{L}^p(Q \setminus P)
$$

$$
\leq \sigma_i \mathscr{L}^p(P \cap Q_0) + \sigma_i \mathscr{L}^p(Q \setminus Q_0) + \sigma_i \tau \mathscr{L}^p Q
$$

$$
\leq \sigma_i \frac{4^{2j}\|S_i^{-1}\|^{2j}\Lambda_n \overline{K}_j^2 (5\operatorname{Lip}(f))^{2j-2}}{s_i^{2j}} \int_Q \|g'_n\|^2\, d\mathscr{L}^p
$$

$$
+ 2\sigma_i \tau \mathscr{L}^p Q
$$

$$
\leq 2^{-i-3}\Big(\int_Q \|g'_n\|^2\, d\mathscr{L}^p + \tau \mathscr{L}^p Q\Big),
$$

where the last inequality follows from the choice of σ_i. To obtain a similar inequality for $i > k$ we infer from $\varepsilon > 1/i$ and the choice of σ_i, κ, and τ that

$$
\sigma_i \leq \frac{2^{-i-5}}{i+1}\frac{s_i}{8\operatorname{Lip}(f)} \leq 2^{-i-3}\frac{\varepsilon\kappa}{4} = 2^{-i-3}\tau.
$$

Hence

$$
\sigma_i \int_Q \big(\Delta_i((x_i, T_i), (\varphi, \varphi'_n)) - \Delta_i((x_i, T_i), (x_\infty, T_k))\big)\, d\mathscr{L}^p \leq 2^{-i-3}\tau \mathscr{L}^p Q.
$$

Adding all these inequalities together, we finally get

$$
\fint_Q \Delta(\varphi)\, d\mathscr{L}^p \leq \frac{1}{4}\fint_Q \|g'_n\|^2\, d\mathscr{L}^p + \frac{\tau}{4}, \tag{13.70}
$$

which is our desired estimate. $\qquad\square$

It remains to have a closer look at the value of the integral of the first term in (13.62), $\fint_Q(|f'(\varphi; \varphi'_n)|_{\mathrm{H}}^2 - |L_\infty|_{\mathrm{H}}^2)\, d\mathscr{L}^p$. As in the regularity part of the proof, we use that the definition of g gives $f'(\varphi; \varphi'_n) = g'_n + L_\infty$ to infer that

$$
\fint_Q (|f'(\varphi; \varphi'_n)|_{\mathrm{H}}^2 - |L_\infty|_{\mathrm{H}}^2)\, d\mathscr{L}^p
$$

$$
= \fint_Q |g'_n|_{\mathrm{H}}^2\, d\mathscr{L}^p + \fint_Q \langle 2L_\infty, g'_n\rangle_{\mathrm{H}}\, d\mathscr{L}^p. \tag{13.71}
$$

Let us try to put these estimates together. For that, we just need to recall that $R_\infty = 2L_\infty + 2\sum_{i=0}^\infty \gamma_i(L_i - L_\infty)$, subtract the inequalities (13.63), (13.64), (13.65), and (13.70) from (13.71), and use (13.58) to infer that

$$
\fint_Q \left(h_\infty(x_\infty, T_\infty) - h_\infty(\varphi, \varphi'_n) \right) d\mathscr{L}^p
$$

$$
\geq \fint_Q |g'_n|_{\mathrm{H}}^2 \, d\mathscr{L}^p + \fint_Q \langle 2L_\infty, g'_n \rangle_{\mathrm{H}} \, d\mathscr{L}^p - \frac{1}{4} \fint_Q |g'_n|_{\mathrm{H}}^2 \, d\mathscr{L}^p
$$

$$
+ \fint_Q \left\langle 2\sum_{i=0}^{\infty} \gamma_i (L_i - L_\infty), g'_n \right\rangle_{\mathrm{H}} \, d\mathscr{L}^p - \frac{1}{4} \fint_Q |g'_n|_{\mathrm{H}}^2 \, d\mathscr{L}^p - \frac{9\tau}{4}
$$

$$
\geq \fint_Q \langle R_\infty, g'_n \rangle_{\mathrm{H}} \, d\mathscr{L}^p - \frac{9\tau}{4}
$$

$$
= \langle R_\infty e, w \rangle + \fint_Q \langle R_\infty, g'_n - w \otimes e \rangle_{\mathrm{H}} \, d\mathscr{L}^p - \frac{9\tau}{4}
$$

$$
\geq \frac{7\tau}{4} - |R_\infty|_{\mathrm{H}} \left| \fint_Q (g'_n - w \otimes e) \, d\mathscr{L}^p \right|_{\mathrm{H}}.
$$

Hence we will complete the proof of (13.61) once we show that

$$
|R_\infty|_{\mathrm{H}} \left| \fint_Q (g'_n - w \otimes e) \, d\mathscr{L}^p \right|_{\mathrm{H}} < \frac{7\tau}{4}.
$$

Estimate of $\left| \fint_Q (g'_n - w \otimes e) \, d\mathscr{L}^p \right|_{\mathrm{H}}$. It is only here that we use the last inequality from Lemma 13.5.6; hence this is the only point in which we make a significant use of regular Fréchet differentiability of f in the direction of T_∞.

Define auxiliary functions $\zeta, \tilde{\zeta} \colon \mathbb{R}^p \longrightarrow \mathbb{R}^n$ by

$$
\zeta(u) = g(\langle u, e \rangle e) \quad \text{and} \quad \tilde{\zeta}(u) = \zeta(u) - g(0) - \langle u, e \rangle w.
$$

Since by Lemma 13.5.6 for every $u \in \overline{Q}$,

$$
|g(u) - \zeta(u)| \leq \frac{\tau r}{(n+1)(1 + |R_\infty|_{\mathrm{H}})},
$$

Corollary 9.4.2 yields that

$$
\left| \int_Q (g'_n - \zeta'_n) \, d\mathscr{L}^p \right|_{\mathrm{H}}
$$

$$
\leq \max\{ |g(u) - \zeta(u)| \mid u \in \partial\Omega \times B_k(0, tr) \} \, \mathscr{H}^{p-1}(\partial\Omega \times B_k(0, tr))
$$

$$
\leq \frac{\tau r \, \alpha_k(tr)^k \left(2\alpha_{n-1} r^{n-1} + (n-1)\alpha_{n-1} r^{n-1} \right)}{(n+1)(1 + |R_\infty|_{\mathrm{H}})}
$$

$$
= \frac{\tau}{1 + |R_\infty|_{\mathrm{H}}} \mathscr{L}^p Q. \tag{13.72}
$$

The function $\widetilde{\zeta}$ depends only on the projection onto $V = \mathbb{R}e$, which shows that $\widetilde{\zeta}'_n = \widetilde{\zeta}'_V \circ \pi_V$. Denote $S_0 = \{u \in \overline{Q} \mid \langle u, e \rangle = 0\}$. Fubini's theorem gives

$$
\left| \int_Q \widetilde{\zeta}'_n \, d\mathscr{L}^p \right|_{\mathrm{H}} = \left| \int_Q \widetilde{\zeta}'_V \, d\mathscr{L}^p \circ \pi_V \right|_{\mathrm{H}}
$$

$$
= \left| \int_{S_0} (\widetilde{\zeta}(-re) - \widetilde{\zeta}(0)) \otimes e \, d\mathscr{L}^{p-1} \right|_{\mathrm{H}}
$$

$$
= |\widetilde{\zeta}(-re)| \, \mathscr{L}^{p-1} S_0
$$

$$
= |g(0) - g(-re) - wr| \mathscr{L}^{p-1} S_0
$$

$$
= |f(x_\infty) - f(x - T_\infty re) + L_\infty re| \mathscr{H}^{p-1} Q_0.
$$

Since $\|T_\infty re\| + |re| < \delta_1$, we may apply (13.55) to get the estimate

$$
\left| \int_Q \widetilde{\zeta}'_n \, d\mathscr{L}^p \right|_{\mathrm{H}} \leq \xi r (\|T_\infty\| + 1) \mathscr{L}^{p-1} S_0 \leq 2\xi \mathscr{L}^p Q \leq \frac{\tau \mathscr{L}^p Q}{2(1 + |R_\infty|_{\mathrm{H}})}.
$$

Noticing that $\widetilde{\zeta}'_n = g'_n - w \otimes e$, we obtain by adding this inequality to (13.72),

$$
|R_\infty|_{\mathrm{H}} \left| \fint_Q (g'_n - w \otimes e) \, d\mathscr{L}^p \right|_{\mathrm{H}}
$$

$$
\leq |R_\infty|_{\mathrm{H}} \left| \fint_Q (g'_n - \zeta'_n) \, d\mathscr{L}^p \right|_{\mathrm{H}} + |R_\infty|_{\mathrm{H}} \left| \fint_Q (\zeta'_n - w \otimes e) \, d\mathscr{L}^p \right|_{\mathrm{H}}
$$

$$
\leq \tau + \frac{\tau}{2} < \frac{7\tau}{4},
$$

which is exactly what we needed to finish the proof.

13.6   SIMPLER SPECIAL CASES

Finally, we briefly comment on simplifications of the arguments of the proof of the main result in some special situations.

When f is everywhere Gâteaux differentiable, we have $D = X \times L(\mathbb{R}^n, X)$ since for every $(x, T) \in D$, f is regularly Gâteaux differentiable at x in the direction of T. Hence by Observation 9.2.2 (ii), Section 13.4 is redundant. In this case one can even substantially simplify the arguments of Section 13.5, in a way similar to the illustrative example of Section 12.2.

When f is real-valued, Section 13.4 is redundant as well. However, the proof in Section 13.5 has to be modified; here the argument making Δ negligible used regularity. We have essentially seen one such modification in Chapter 12, although the arguments there were somewhat more complicated than necessary because we were proving a stronger statement. For a variational proof of Theorem 13.1.1 for real-valued functions along the lines suggested here see [29].

When X is uniformly smooth with modulus $o(t^2 \log(1/t))$ (in particular, when X is a Hilbert space), we still have to prove Fréchet regularity first, but we can completely avoid the notion of regular Gâteaux differentiability (and hence the trickiest parts of the proof). The key to this is that, instead of choosing Z_k whose union is dense, we just take $Z_k = T_\infty(\mathbb{R}^2)$ for all k. This idea is used in Chapter 16 to give a self-contained proof in the Hilbert space situation and in Chapter 15 to obtain a new result on asymptotic Fréchet differentiability, which also includes the case of Hilbert spaces.

Chapter Fourteen

Unavoidable porous sets and nondifferentiable maps

In Chapters 10 and 13 we have established conditions on a Banach space X under which porous sets in X are Γ_n-null and/or the the multidimensional mean value estimates for Fréchet derivatives of Lipschitz maps into n-dimensional spaces hold. Here we show in what sense these results are close to being optimal. Under conditions that can be argued to be close to complementary to those from the previous chapters, we find a σ-porous set whose complement is null on all n-dimensional surfaces and the multidimensional mean value estimates fail even for ε-Fréchet derivatives. Particular situations in which these negative results are shown include maps of Hilbert spaces to $\mathbb{R}^3$ and, more generally, maps of ℓ^p to $\mathbb{R}^n$ for $n > p$.

14.1 INTRODUCTION AND MAIN RESULTS

Here we investigate in what sense the assumptions of the main results from Chapters 10 and 13 are close to being optimal. An optimal complement to these results would of course be that they fail whenever the space in question fails to be asymptotically uniformly smooth with modulus of smoothness $o(t^n \log^{n-1}(1/t))$. We need to assume more than this optimal assumption: while the natural strengthening of the complementary assumption would require asymptotic uniform convexity with modulus $t^n \log^{n-1}(1/t)$, for our strongest results we assume slightly more than asymptotic uniform convexity with modulus t^p for some $p < n$, namely, asymptotic uniform smoothness of the dual space with modulus of smoothness t^q, where $q > n/(n-1)$. (Recall that these two formulations are equivalent when the space is reflexive, but the dual formulation is stronger in general.) For some results, such as construction of a porous set whose complement is null on all n-dimensional surfaces, we can improve the dual modulus to $o(t^{n/(n-1)}/\log^\beta(1/t))$, where $\beta > 1$, and our methods should even lead to the optimal $o(t^{n/(n-1)}/\log(1/t))$ with only a log log error. However, for the main results such as the decomposition of a porous set into a union of a Γ_n-null G_δ set and a Haar null set proved in Theorem 10.4.5 (which was basic for ε-Fréchet differentiability results of Chapter 11) or the multidimensional mean value estimates for Fréchet derivatives of Lipschitz mappings proved in Theorem 13.1.1, our methods do not seem to lead to such an improvement of the complementary assumptions.

We will, in fact, prove our results under a considerably weaker assumption on X^* than asymptotic smoothness. To state it quickly, we recall that a normalized sequence of elements $x_i \in X$ is said to satisfy the *upper q estimate* if there is a constant A such that for any sequence (c_i) of real numbers and any $k \in \mathbb{N}$,

$$\left\| \sum_{i=1}^{k} c_i x_i \right\| \leq A \left(\sum_{i=1}^{k} |c_i|^q \right)^{1/q}.$$

It is not difficult to see, and we will prove it shortly in greater generality, that a space admitting an asymptotically uniformly smooth norm with modulus of smoothness $o(t^q)$, where $q > 1$, contains a normalized sequence satisfying the upper q estimate. (The converse is obviously false.)

We are now ready to state the first main result of this chapter, which will be proved in the following sections.

Theorem 14.1.1. *Let X be a separable Banach space, and let $n > 1$. Suppose that, for some $q > n/(n-1)$, the dual space X^* contains a normalized sequence satisfying the upper q estimate. Then we may find a Lipschitz map $f \colon X \longrightarrow \mathbb{R}^n$ and a continuous linear form $\mathscr{T}$ on $L(X, \mathbb{R}^n)$ such that*

- $\mathscr{T}(f'(x)) \geq 0$ *whenever f is Gâteaux differentiable at x,*

- *there are x at which f is Gâteaux differentiable and $\mathscr{T}(f'(x)) > 0$,*

- $\mathscr{T}(Q) \leq \varepsilon$ *whenever Q is the ε-Fréchet derivative of f at some x.*

In particular, for every sufficiently small $\varepsilon > 0$ the set

$$\Big\{ f'(x) \,\Big|\, f \text{ is Gâteaux differentiable at } x \text{ and } \mathscr{T}(f'(x)) > \varepsilon \Big\}$$

is a nonempty slice of the set of all Gâteaux derivatives of f containing no ε-Fréchet derivative.

An immediate consequence on existence of large porous sets follows directly from Theorem 11.3.6.

Corollary 14.1.2. *If X^* is separable, $n > 1$ and, for some $q > n/(n-1)$, X^* contains a normalized sequence satisfying the upper q estimate, then X contains a porous set that cannot be covered by a union of a Haar null set and a Γ_n-null G_δ set.*

To state a more natural result on existence of big porous sets, we describe a more general approach to the notion of upper q estimates, upper ω estimates. Although we need only ω behaving like $t^q \log^r(1/t)$ for t small, we believe that working with more general Orlicz functions reveals the ideas of the proof better. Recall that an *Orlicz function* is a convex increasing function $\omega \colon [0, \infty) \longrightarrow [0, \infty)$ such that $\omega(0) = 0$, $\omega(t) > 0$ for $t > 0$, and $\lim_{t \to \infty} \omega(t) = \infty$. (In all we do, only behavior of $\omega(t)$ for t close to zero matters, so ω need only be defined on an interval $[0, \varepsilon)$ for some $\varepsilon > 0$.) We list here some basic notions and facts related to Orlicz functions, which can be found, for example, in [30, Chapter 4].

The complementary (or dual) function ω^* to ω is defined as

$$\omega^*(t) = \max\big\{ st - \omega(s) \,\big|\, 0 \leq s < \infty \big\}.$$

There are two norms $\| \cdot \|_\omega$ and $| \cdot |_\omega$ on the space of sequences of numbers derived from the Orlicz function ω:

$$\|(c_i)\|_\omega = \inf\left\{ \lambda > 0 \mid \sum_{i=1}^\infty \omega(|c_i|/\lambda) \leq 1 \right\},$$

$$|(c_i)|_\omega = \sup\left\{ \sum_{i=1}^\infty c_i d_i \mid \sum_{i=1}^\infty \omega^*(|d_i|) \leq 1 \right\}.$$

The former is called the Luxemburg norm, the latter Orlicz norm, and the relation between them is

$$\| \cdot \|_\omega \leq | \cdot |_\omega \leq 2\| \cdot \|_\omega.$$

For later use we note the easily established fact

$$|(c_i)|_\omega \leq 1 + \sum_{i=1}^\infty \omega(|c_i|). \tag{14.1}$$

A normalized sequence of elements $x_i \in X$ is said to satisfy the *upper ω estimate* if there is a constant A such that for any sequence (c_i) of real numbers and any $k \in \mathbb{N}$

$$\left\| \sum_{i=1}^k c_i x_i \right\| \leq A \left\| (c_i)_{i=1}^k \right\|_\omega.$$

The notion of upper q estimate defined above is, of course, obtained with $\omega(t) = t^q$. It is useful for us to notice certain flexibility in this notion. First, slight perturbation of the vectors x_i preserve the upper ω estimate: Let $\delta_i > 0$ be such that $\sum_{i=1}^\infty \omega^*(\delta_i) \leq 1$. Then for any vectors u_i with $\|u_i\| \leq \delta_i$ and any $k \in \mathbb{N}$ we have

$$\left\| \sum_{i=1}^k c_i(x_i + u_i) \right\| \leq \left\| \sum_{i=1}^k c_i x_i \right\| + \sum_{i=1}^k c_i |\delta_i| \leq A \left\| (c_i)_{i=1}^k \right\|_\omega + \left| (c_i)_{i=1}^k \right|_\omega$$

$$\leq (A+2) \left\| (c_i)_{i=1}^k \right\|_\omega.$$

Thus the sequence $(x_i + u_i)_{i \in \mathbb{N}}$ satisfies the upper ω estimate as well. Second, if two Orlicz functions ω and σ are related by

$$Q^{-1}\omega(q^{-1}t) \leq \sigma(t) \leq Q\omega(qt)$$

for some constants $q, Q > 0$, then the corresponding norms $\| \cdot \|_\omega$ and $\| \cdot \|_\sigma$ are equivalent. The functions ω and σ are in that case also called equivalent. Consequently, we may always assume that $A = 1$ in the inequality from the definition of the upper ω estimate, passing from ω to an equivalent Orlicz function.

Notice that a space asymptotically uniformly smooth with modulus ω necessarily contains a normalized sequence satisfying the upper ω estimate. Indeed, let X be a Banach space with a modulus of asymptotic uniform smoothness ω. Without loss of generality we assume that $\omega(1) = 1$. Let $\varepsilon_j > 0$ be such that $\sum_{j=1}^\infty j\varepsilon_j \leq 1$.

The unit vectors (x_i) will be selected recursively. Let x_1 be any unit vector, and assume that $x_1, \ldots, x_k$ have been already chosen. By definition of asymptotic uniform smoothness, for each unit vector $x \in X$ and $t \geq 0$,

$$\inf_{\substack{Y \subset X \\ \dim X/Y < \infty}} \sup_{\substack{y \in Y \\ \|y\| \leq t}} \|x + y\| - 1 \leq \omega(t).$$

Let U_k be the unit sphere in $\operatorname{span}\{x_1, \ldots, x_k\}$ and let $M_k \subset U_k \times [0,1]$ be a finite set such that for every $(x, u) \in U_k \times [0,1]$ there is $(z, v) \in M_k$ such that

$$\|z - x\| + |v - u| < \tfrac{1}{2}\varepsilon_{k+1}.$$

Then there is a finite codimensional subspace Y_k of X such that for every $(x, t) \in M_k$,

$$\sup_{\substack{y \in Y_k \\ \|y\| \leq t}} \|x + y\| - 1 < \tfrac{1}{2}\varepsilon_{k+1} + \omega(t).$$

Since the left side of this inequality is Lipschitz with constant one in both x and t, we infer that

$$\sup_{\substack{y \in Y_k \\ \|y\| \leq t}} \|x + y\| - 1 < \varepsilon_{k+1} + \omega(t) \tag{14.2}$$

for every $x \in U_k$ and $t \in [0,1]$.

Take now x_{k+1} to be any unit vector in Y_k.

If $x \in \operatorname{span}\{x_1, \ldots, x_k\}$, $\|x\| \geq 1$, and $|c| \leq 1$, then (14.2) used with $x/\|x\|$ and $y = cx_{k+1}/\|x\|$, and the convexity of ω imply that

$$\|x + cx_{k+1}\| - \|x\| < \varepsilon_{k+1}\|x\| + \|x\|\omega(|c|/\|x\|) \leq \varepsilon_{k+1}\|x\| + \omega(|c|). \tag{14.3}$$

Let (c_i) be a sequence of real numbers such that $\sum_{i=1}^{\infty} \omega(|c_i|) \leq 1$. Denoting $S_k = \sum_{i=1}^{k} c_i x_i$, it suffices to show that $\|S_k\| \leq 4$ for each k, which will yield the desired inequality with $A = 4$.

Before estimating $\|S_k\|$, notice that $|c_i| \leq 1$ for each i and that the sequence $\widetilde{c}_i = c_i/\|(c_i)\|_\omega$ satisfies $\sum_{i=1}^{\infty} \omega(\widetilde{c}_i) = 1$. Given k, find the largest $0 \leq l \leq k$ such that $\|S_l\| \leq 1$. (Since empty sums are zero, $S_0 = 0$, and so such an l exists.) Notice that $|c_i| \leq 1$ for each i, giving

$$\|S_j\| \leq \|S_l\| + \sum_{i=l+1}^{j} |c_i| \leq 1 + j - l$$

for every j with $l \leq j \leq k$. In particular, we may assume that $l + 1 < k$. Since

$\|S_{l+1}\| \leq 2$ and $\|S_j\| > 1$ for $j = l+1, \ldots, k$, we infer from (14.3) that

$$\|S_k\| = \|S_{k-1} + c_k x_k\| \leq \|S_{k-1}\| + \varepsilon_k \|S_{k-1}\| + \omega(|c_k|)$$

$$\vdots$$

$$\leq \|S_{l+1}\| + \sum_{j=l+1}^{k-1} \left(\varepsilon_{j+1} \|S_j\| + \omega(|c_{j+1}|) \right)$$

$$\leq 2 + \sum_{j=l+1}^{k-1} \varepsilon_{j+1}(j+1-l) + \sum_{j=l+1}^{k-1} \omega(|c_{j+1}|) \leq 4.$$

We have already indicated that existence of a sequence with the upper ω estimate is weaker than existence of an equivalent norm with modulus of asymptotic uniform smoothness equal to ω. If a particular example is required, the space $C[0,1]$ provides it. Let (f_i) be a normalized sequence of functions in $C[0,1]$ having disjoint support. Then, clearly,

$$\left\| \sum_{i=1}^{k} c_i f_i \right\| = \max\{ |c_i| \mid i = 1, \ldots, k \} \leq A \|(c_i)_{i=1}^{k}\|_\omega,$$

where A is such that $\omega(A) = 1$, for *any* (nondegenerate) Orlicz function ω. However, $C[0,1]$ is not isomorphic to an asymptotically uniformly smooth Banach space, since every such space must be an Asplund space; see Proposition 4.2.7.

We can now state the second main results of this Chapter.

Theorem 14.1.3. *Let X be a separable Banach space, and let $n > 1$. Suppose that, for some $\beta > 1$, the dual space X^* contains a normalized sequence satisfying the upper $t^{n/(n-1)}/\log^\beta(1/t)$ estimate. Then X contains a σ-porous subset whose complement meets every n-dimensional Lipschitz surface in a set of n-dimensional Hausdorff measure zero.*

It follows that the space X from this theorem contains a σ-porous subset that is neither Haar null nor Γ_n-null. Of course, the existence of a σ-porous set that is not Haar null is not new. In fact, every infinite dimensional separable Banach space contains a σ-porous set whose complement is null on every line (see [40] or [4, Theorem 6.39]). Such a set obviously cannot be Haar null.

It may be also interesting to remark that the real condition on ω for Theorem 14.1.3, respectively Theorem 14.1.1, to hold can be given as a requirement for existence of a solution of a certain partial differential inequality. This can be seen in the statement of Lemma 14.3.2.

The above results treat only the case $n > 1$. For the case $n = 1$ differentiability problems are well understood, even for more general functions than Lipschitz (see Chapter 12). The relation between porosity and Γ_1-nullness has been clarified in Theorems 10.5.1 and 10.5.2. Nevertheless, as an illustration of our techniques, we will start by constructing a large σ-porous subset of ℓ_1. This example is in fact stronger than that provided by the results referred to above: the complement of the σ-porous set that we construct is not only Γ_1-null but also null (in the sense of one-dimensional Hausdorff measure) on every Lipschitz curve. It is not known whether such a set exists in every separable Banach space with nonseparable dual. The construction we present here can be clearly extended beyond ℓ_1, at least to spaces containing ℓ_1. We do not do this, since it is outside the aims of this text, and since it seems that all spaces to which the method presented here would apply are covered by an elegant result of Maleva [32] saying that such a set exists in any separable Banach space that admits ℓ_1 as a Lipschitz quotient. Recall that in [22] it is proved that any separable Banach space containing a copy of ℓ_1 has ℓ_1 as a Lipschitz quotient. Notice also that, as pointed out in [22], the question whether any non-Asplund space has ℓ_1 as a Lipschitz quotient was left open. A positive answer to this question would (via Maleva's result) also give a positive answer to the problem above.

The relation between porosity and differentiability is, in spite of many results in the previous chapters and examples in this chapter, still not well understood. One of many open problems is whether every Lipschitz map of an Asplund space X (or just of a Hilbert space) to an RNP space is Gâteaux differentiable at a point outside any given σ-porous subset S of X. If true, one could try to prove a Fréchet differentiability result by maximizing a suitable functional of the Gâteaux derivatives $f'(x)$ over $x \in D(f) \backslash S$, where S is a suitable σ-porous subset of X, for example, the set of points at which f behaves irregularly. (Recall that $D(f)$ denotes the set of points at which f is Gâteaux differentiable.) The weakest form of this question is whether, given $n \in \mathbb{N}$, one can decompose

$$X = S \cup \bigcup_{i=1}^{n} (X \setminus D(f_i)),$$

where S is σ-porous and f_i are real-valued Lipschitz function on X. Provided X has separable dual, this is impossible with $n = 1$ since the results of Chapter 12 show that one can find a point of Fréchet differentiability of any real-valued Lipschitz function outside any given σ-porous set. It looks plausible that a similar refinement of the results of Chapter 13 would show that such a decomposition does not exist when $X = \ell_p$ and $2 \le n \le p$. But we do not see a method capable of treating the case when X is the Hilbert space and $n = 3$.

In connection to the above problem recall that every separable Banach space can be decomposed into the union of a σ-porous set and a set that is null on all straight lines. In particular, a point of Gâteaux differentiability outside this σ-porous set cannot be found just by referring to the Gâteaux differentiability results using Gauss null sets or Haar null sets. It is unknown whether it can be found by using stronger results on Gâteaux differentiability almost everywhere. In particular, it is not known whether the Hilbert space can be decomposed into the union of a σ-porous set and a set belonging to the class $\widetilde{\mathcal{A}}$ defined in Section 2.3.

Another group of problems arises by observing details of the constructions in this chapter. Namely, the function we construct in Theorem 14.1.1 has points of Gâteaux nondifferentiability, and, more interestingly, the set of points for which $f'(x)$ belongs to the "bad" slice is σ-porous. We do not know whether these are just features of our construction or general phenomena. More precisely, we do not know the answer to the following questions about a Lipschitz function f from a space with separable dual into a finite dimensional space. If f is everywhere Gâteaux differentiable, does every slice of the Gâteaux derivatives of f contain an ε-Fréchet derivative? If for some $\varepsilon > 0$, a slice of the Gâteaux derivatives of f contains no ε-Fréchet derivative, is the set of points for which $f'(x)$ belongs to this slice necessarily σ-porous?

14.2   AN UNAVOIDABLE POROUS SET IN ℓ_1

This section is purely illustrative: its purpose is to show a simplified version of the construction of a "big" porous set whose more sophisticated form will be used in the following sections. Readers can compare the relatively easy steps here with more involved similar arguments in what follows, which could help them isolate the idea of the construction from the technical calculations.

Theorem 14.2.1. *There is a σ-porous subset A of ℓ_1 whose complement meets every C^1 curve in a set of length zero, i.e.*

$$\mathscr{L}\big\{t \in [0,1] \ \big| \ \varphi'(t) \neq 0, \ \varphi(t) \notin A\big\} = 0$$

for any C^1 curve $\varphi \colon [0,1] \longrightarrow \ell_1$.

Since ℓ_1 has the RNP, one can show by standard methods that the same set A meets any Lipschitz curve in a set of length zero.

The set A provides a particular example of a σ-porous set with Γ_1-null complement (which can be found by Theorem 10.5.1 in every separable Banach space with nonseparable dual) and of a σ-porous set with Gauss null complement (which can be found by [40] in every separable Banach space).

As we have already noticed, porous sets and Fréchet differentiability of Lipschitz maps are closely related. Indeed, the last time we saw this was in Corollary 14.1.2. It should therefore be no surprise that the proofs of the main results of this chapter, Theorems 14.1.1 and 14.1.3, are deeply related to each other. To illustrate this point in ℓ_1 is not so straightforward: while Theorem 14.2.1 is a clear analogy of Theorem 14.1.3, an ℓ_1 analogy of Theorem 14.1.1 is, in view of the nowhere Fréchet differentiability of its norm, not obvious. It can be found, in fact, only in the proof of the main results. Basically, we construct a function $f \colon X \longrightarrow \mathbb{R}^n$ with $\mathrm{Lip}(f) \leq 1$ and $\|f\|_\infty$ arbitrarily small that is (to all practical purposes) constant on a porous set and has large positive divergence (i.e., the value of the form $\mathscr{T}$ from Theorem 14.1.1) outside. The porous set is already one of those of which the σ-porous set required in Theorem 14.1.3 will be composed. To prove Theorem 14.1.1, we still have to go through an additional iteration which is reflected on the difference in assumptions.

The arguments alluded to in the previous paragraph are relatively involved mainly due to the need for the careful control of the behavior of the divergence of the maps from which the final function is constructed. An analogy in ℓ_1 is much simpler, since here we construct a real-valued function and positivity of the divergence is replaced by positivity of the derivative in a particular direction, which is much easier to imagine and handle. So the key observation (which is essentially what we prove in Lemma 14.2.5) can be stated rather simply as follows.

Observation 14.2.2. *For every $\varepsilon > 0$ there are a porous set $A \subset \ell_1$ and a 1-Lipschitz function $f \colon \ell_1 \longrightarrow [0, \varepsilon]$ such that f is increasing in the direction of the basis vector e_0 and $f'(z; e_0) = 1$ whenever $z \notin A$.*

It follows that every curve φ with tangent vector $\varphi'(t)$ close to e_0 must intersect the complement of A only in a set of small measure, since the increment of the composition $f \circ \varphi$ is small along the curve. With this fact in hand it is not difficult to complete the proof of Theorem 14.2.1.

Perhaps it is also worth pointing out another, purely technical reason why the arguments of this section are simpler than those of the following sections. To guarantee that a function $f \colon \ell_1 \longrightarrow \mathbb{R}$ be Lipschitz with constant, say, L, it is sufficient to show that f is L-Lipschitz in the direction of every vector e_j from the standard basis of ℓ_1.

Construction

To start the construction of the set A, let $e_j \in \ell_1$, $j \geq 0$, denote the standard basis of ℓ_1. We decompose the set $\mathbb{N}$ of all positive integers into countably many disjoint infinite sets N_m, $\mathbb{N} = \bigcup_{m=1}^{\infty} N_m$.

For $m \geq 1$ let X_m be the closed subspace of ℓ_1 generated by vectors e_j with $j \in N_m$,

$$X_m = \overline{\mathrm{span}}\{e_j \mid j \in N_m\},$$

and $X_0 = \mathbb{R}e_0$. Then

$$\ell_1 = \Big(\sum_{m=0}^{\infty} \oplus X_m \Big)_{\ell_1}.$$

We let (d_i) be a countable dense set in X_0. Let $\pi_m \colon \ell_1 \longrightarrow X_m$, $m \geq 0$, denote the canonical projections of ℓ_1 onto X_m.

We want to define suitable families of open cylinders in ℓ_1 whose unions will represent the complements of the desired porous sets. To this aim let $r_m = 2^{-m}$ and denote, for $m, k \geq 1$, by $E_{m,k}$ a maximal $9kr_m$ discrete subset of X_m, that is, $E_{m,k}$ is a maximal subset of X_m such that any two different elements in $E_{m,k}$ are at least $9kr_m$ apart. For each $x \in E_{m,k}$ we choose points denoted by $x(i) \in X_m$, $i \in \mathbb{N}$, such that

(i) $x(1) = x$,

(ii) $\|x(i) - x\| = 3kr_m$ for $i > 1$,

(iii) $\|x(i) - x(j)\| \geq 3kr_m$ for $i \neq j$.

This implies, in particular, that

$$\|x(i) - \widetilde{x}(j)\| \geq 3kr_m \tag{14.4}$$

if either $i \neq j$ or $x \neq \widetilde{x}$ for $x, \widetilde{x} \in E_{m,k}$.

Let now $x \in E_{m,k}$ and $i \geq 1$. We put

$$V_{m,k}(x,i) = \left\{ z \in \ell_1 \mid \frac{|\pi_0 z - d_i|}{r_m} + \frac{\|\pi_m z - x(i)\|}{kr_m} < 1 \right\}.$$

Observe that $V_{m,k}(x,i)$ is a cylinder with the basis in the space $X_0 \oplus X_m$. The basis is an "ellipsoid" having the length r_m of the semiprincipal axis in the direction of e_0 and the length kr_m of remaining semiprincipal axes in the directions of e_j, $j \in N_m$. Since the definition of $V_{m,k}(x,i)$ does not depend on the coordinates with indices outside the set $\{0\} \cup N_m$, we have

$$V_{m,k}(x,i) = \mathbb{R}e_j + V_{m,k}(x,i) \quad \text{for all } e_j \notin X_0 \oplus X_m. \tag{14.5}$$

From condition (14.4) it also follows that any two different cylinders $V_{m,k}(x,i)$ and $V_{m,k}(\widetilde{x},j)$, where $x, \widetilde{x} \in E_{m,k}$, have distance at least kr_m. Indeed, consider any $z \in V_{m,k}(x,i)$ and $\widetilde{z} \in V_{m,k}(\widetilde{x},j)$. Then

$$\|z - \widetilde{z}\| \geq \|\pi_m z - \pi_m \widetilde{z}\|$$
$$\geq \|x(i) - \widetilde{x}(j)\| - \|\pi_m z - x(i)\| - \|\pi_m \widetilde{z} - \widetilde{x}\|$$
$$\geq 3kr_m - kr_m - kr_m = kr_m.$$

Since for $y \in \ell_1$ the projection $\pi_m(y + X_0) = \pi_m y$ is a singleton, another consequence of (14.4) is that the line $y + X_0$ intersects at most one of the cylinders $V_{m,k}(x,i)$ for fixed $m, k \geq 1$.

We also denote

$$W_{m,k} = \bigcup_{i \in \mathbb{N}} \bigcup_{x \in E_{m,k}} V_{m,k}(x,i).$$

Lemma 14.2.3. *For every $k \geq 1$ the set*

$$A_k = \ell_1 \setminus \bigcup_{m \geq k} W_{m,k}$$

is closed and c_k-porous, where $c_k = 1/(12k + 1)$.

Proof. The sets A_k are clearly closed.

Let $y \in A_k$ and let $\varepsilon > 0$. We find $m \geq k$ such that $(12k + 1)r_m \leq \varepsilon$. By the maximality of $E_{m,k}$ in X_m there is $x \in E_{m,k}$ such that

$$\|\pi_m y - x\| \leq 9kr_m.$$

We also choose $d_i \in X_0$ to be at distance at most r_m from $\pi_0 y$. Then the open ball $B(w, r_m)$ with center

$$w = d_i + x(i) + (y - (\pi_0 + \pi_m)y)$$

is contained in $V_{m,k}(x,i)$. Indeed, since $\pi_0 w = d_i$ and $\pi_m w = x(i)$, we have for $z \in B(w, r_m)$ the inequality

$$r_m > \|w - z\| \geq |d_i - \pi_0 z| + \|x(i) - \pi_m z\|.$$

This shows that $B(w, r_m) \subset V_{m,k}(x, i)$. Also, the center w is at distance at most $(12k + 1)r_m$ from y:

$$\|w - y\| = |d_i - \pi_0 y| + \|x(i) - \pi_m y\|$$
$$\leq |d_i - \pi_0 y| + \|x(i) - x\| + \|x - \pi_m y\|$$
$$\leq r_m + 3kr_m + 9kr_m = (12k + 1)r_m.$$

Consequently, the ball $B(w, r_m)$ does not meet A_k and its center is at distance at most $r_m/c_k = (12k + 1)r_m \leq \varepsilon$ from y. $\qquad\square$

We define, for each $k \geq 1$, the function $f_k \colon \ell_1 \longrightarrow \mathbb{R}$ as follows. Let $z \in \ell_1$ be written as $z = te_0 + y$, where $t \in \mathbb{R}$ and $y \in \left(\sum_{m=1}^{\infty} \oplus X_m \right)_{\ell_1}$. We put

$$f_k(te_0 + y) = \mathscr{L}\{s \in \mathbb{R} \mid s \leq t, \, se_0 + y \notin A_k\}$$
$$= \mathscr{L}\left\{s \in \mathbb{R} \,\Big|\, s \leq t, \, se_0 + y \in \bigcup_{m \geq k} W_{m,k}\right\}.$$

For each $m, k \geq 1$, the half line $\{se_0 + y \mid s \leq t\}$ meets at most one of the sets $V_{m,k}(x, i)$. This implies that

$$f_k(te_0 + y) \leq \sum_{m \geq k} \mathscr{L}\left\{s \in \mathbb{R} \,\Big|\, s \leq t, se_0 + y \in W_{m,k}\right\} \leq \sum_{m \geq k} 2r_m.$$

Hence

$$\lim_{k \to \infty} f_k(z) = 0 \quad \text{for all } z \in \ell_1. \tag{14.6}$$

It also follows from (14.5) that for any $t \in \mathbb{R}$ and $m, k \geq 1$,

$$\mathscr{L}\left\{s \in \mathbb{R} \,\Big|\, s \leq t, \, se_0 + y \in W_{m,k}\right\}$$
$$= \mathscr{L}\left\{s \in \mathbb{R} \,\Big|\, s \leq t, \, se_0 + y + \mathbb{R}e_j \subset W_{m,k}\right\},$$

whenever $e_j \notin X_0 \oplus X_m$. In particular,

$$f_k(z) = f_k(z + w) \tag{14.7}$$

whenever $w \in X_1 \oplus \cdots \oplus X_{k-1}$.

Before stating the main lemma we make the following simple observation.

Observation 14.2.4. *Let $h_1, h_2 \colon \mathbb{R} \longrightarrow \mathbb{R}$ be two Lipschitz functions with Lipschitz constant L, and let $h_1 \leq h_2$. Then for any Borel set $M \subset \mathbb{R}$ the function*

$$\theta(\tau) = \mathscr{L}\{s \in \mathbb{R} \mid s \in [h_1(\tau), h_2(\tau)] \setminus M\}$$

is also Lipschitz with Lipschitz constant at most $2L$.

Proof. Since the increment of both h_1 and h_2 between two points $\tau, \tau' \in \mathbb{R}$ is at most $L|\tau' - \tau|$, the increment of the function θ between these points is dominated by

$$|\theta(\tau') - \theta(\tau)| \le |h_1(\tau') - h_1(\tau)| + |h_2(\tau') - h_2(\tau)|$$
$$\le 2L|\tau' - \tau|. \qquad \square$$

We will denote by $\overline{D}f(x; v)$ and $\underline{D}f(x; v)$ the upper and lower derivatives of f at the point x in the direction v. For $X = \mathbb{R}$ they are defined in Section 12.3, and in general they are are defined as upper and lower derivatives of the function $t \in \mathbb{R} \to f(x + tv)$ at $t = 0$.

Lemma 14.2.5. *Let* $k \in \mathbb{N}$ *and let* $z \in \ell_1$. *Then*

(i) $0 \le \underline{D}f_k(z; e_0) \le \overline{D}f_k(z; e_0) \le 1$, *and* $f_k'(z; e_0) = 1$ *if* $z \notin A_k$;

(ii) f_k *is constant along* $\left(\sum_{m=1}^{k-1} \oplus X_m\right)_{\ell_1}$, *so* $f_k'(z; e_j) = 0$ *for all* $e_j \in X_m$ *and* $m = 1, \ldots, k-1$;

(iii) f_k *is* $2/k$-*Lipschitz on* $\left(\sum_{m \ge k} \oplus X_m\right)_{\ell_1}$.

Proof. (i) Since f_k is 1-Lipschitz and increasing in the direction of e_0, the three inequalities are clear. The last part follows from the fact that the complement of A_k is an open set.

(ii) was proved in (14.7).

(iii) Let $e_j \in \left(\sum_{m \ge k} \oplus X_m\right)_{\ell_1}$ be fixed. We show that the function

$$\tau \to f_k(z + \tau e_j)$$

is $2/k$-Lipschitz for any $z \in \ell_1$. Let $z \in \ell_1$ and write $z = te_0 + y$ with $\pi_0 y = 0$. Let $e_j \in X_n$ for some $n \ge k$ and let $\tau \in \mathbb{R}$. Then $f_k(z + \tau e_j)$ is the sum of two terms,

$$\mathscr{L}\left\{ s \in \mathbb{R} \;\middle|\; s \le t, se_0 + y + \tau e_j \in \bigcup_{\substack{m \ge k \\ m \ne n}} W_{m,k} \right\} \tag{14.8}$$

and

$$\left\{ s \in \mathbb{R} \;\middle|\; s \le t, se_0 + y + \tau e_j \in W_{n,k} \setminus \bigcup_{\substack{m \ge k \\ m \ne n}} W_{m,k} \right\}. \tag{14.9}$$

Since $\tau e_j \in X_n$, (14.8) does not depend on τ by (14.5). Therefore it remains to show only that (14.9) is $2/k$-Lipschitz in τ.

Consider the segment $I(z) = (z - \frac{1}{2}kr_n e_j, z + \frac{1}{2}kr_n e_j)$. Its length is kr_n so it cannot intersect more than one of the cylinders $V_{n,k}(x, i)$ with $x \in E_{n,k}$, $i \ge 1$. If the segment $I(z)$ does not meet any $V_{n,k}(x, i)$, then f_k is constant on $I(z)$ and (iii) holds true on $I(z)$. If $V_{n,k}(x, i)$ is the only set meeting $I(z)$, it suffices to show that f_k is $2/k$-Lipschitz on the open subinterval

$$I(z) \cap V_{n,k}(x, i).$$

To this aim we apply Observation 14.2.4. Fix τ and look at those $s \le t$ for which $se_0 + y + \tau e_j \in V_{n,k}(x, i)$. It means

$$\frac{|s - d_i|}{r_n} + \frac{\|\pi_n y + x(i) - \tau e_j\|}{k r_n} < 1,$$

which is, by rearrangement,

$$|s - d_i| < r_n - \frac{1}{k} \|\pi_n y + x(i) - \tau e_j\|.$$

Denoting

$$h_1(\tau) = d_i - \max\left\{0, r_n - \frac{1}{k}\|\pi_n y + x(i) - \tau e_j\|\right\},$$

$$h_2(\tau) = d_i + \max\left\{0, r_n - \frac{1}{k}\|\pi_n y + x(i) - \tau e_j\|\right\}, \quad \text{and}$$

$$M = \left\{s \in \mathbb{R} \ \Big| \ se_0 + y \in \bigcup_{\substack{m \ge k \\ m \ne n}} W_{m,k}\right\} \cup \{s \in \mathbb{R} \mid s > t\},$$

we express (14.9) in the form

$$\mathscr{L}\big\{s \in \mathbb{R} \mid s \in [h_1(\tau), h_2(\tau)] \setminus M\big\}.$$

Since h_1, h_2 are obviously $1/k$-Lipschitz functions, Observation 14.2.4 gives that

$$\tau \longrightarrow f_k(z + \tau e_j)$$

is $2/k$-Lipschitz on $I(z)$ for any $z \in \ell_1$. Hence f_k is $2/k$-Lipschitz along all directions $e_j \in \left(\sum_{m \ge k} \oplus X_m\right)_{\ell_1}$. Because the domain space is ℓ_1, this implies (iii). $\square$

Lemma 14.2.6. *Let $\varphi \colon [0, 1] \longrightarrow \ell_1$ be C^1 curve such that $\pi_0 \varphi'(t) \ne 0$ for all $t \in [0, 1]$. Then*

$$\mathscr{L}\left\{t \in [0, 1] \ \Big| \ \varphi(t) \notin \bigcup_{i=1}^{\infty} A_i\right\} = 0.$$

Proof. Without loss of generality we may assume that $\pi_0 \varphi'(t)$ is a positive multiple of e_0 for all $t \in [0, 1]$. Let $k \in \mathbb{N}$ and denote by $\varphi_0(t)$ the projection of $\varphi(t)$ onto the subspace $\left(\sum_{m=0}^{k-1} \oplus X_m\right)_{\ell_1}$, that is,

$$\varphi_0(t) = \left(\sum_{m=0}^{k-1} \pi_m\right)\varphi(t).$$

Let $\varphi_1(t) = \varphi(t) - \varphi_0(t)$. We estimate the lower derivative of the composition $f_k \circ \varphi$:

$$\underline{D}(f_k \circ \varphi)(t) = \liminf_{h \to 0} \frac{f_k(\varphi(t + h)) - f_k(\varphi(t))}{h}$$

$$\ge \liminf_{h \to 0} \frac{f_k(\varphi(t + h)) - f_k(\varphi(t) + h\,\varphi'(t))}{h}$$

$$+ \liminf_{h \to 0} \frac{f_k(\varphi(t) + h\,\varphi'(t)) - f_k(\varphi(t))}{h}.$$

Since f_k is a Lipschitz function, the first term in the above sum is zero and we can continue:

$$\geq \liminf_{h \to 0} \frac{f_k\big(\varphi(t) + h\,\varphi_0'(t) + h\,\varphi_1'(t)\big) - f_k(\varphi(t) + h\,\varphi_0'(t))}{h}$$

$$+ \liminf_{h \to 0} \frac{f_k(\varphi(t) + h\,\varphi_0'(t)) - f_k(\varphi(t))}{h}$$

$$\geq -\frac{2}{k}\,\|\varphi_1'(t)\| + \liminf_{h \to 0} \frac{f_k(\varphi(t) + h\,\varphi_0'(t)) - f_k(\varphi(t))}{h},$$

where in the last step we used Lemma 14.2.5 (iii). Notice that

$$\varphi_0'(t) = \pi_0 \varphi'(t) + \Big(\sum_{m=1}^{k-1} \pi_m \Big) \varphi'(t).$$

Thus, using (14.7), we finally get

$$\underline{D}(f_k \circ \varphi)(t) \geq -\frac{2}{k}\,\|\varphi_1'(t)\| + \liminf_{h \to 0} \frac{f_k(\varphi(t) + h\,\pi_0\varphi'(t)) - f_k(\varphi(t))}{h}$$

$$= -\frac{2}{k}\,\|\varphi_1'(t)\| + \underline{D}f_k(\varphi(t); \pi_0\varphi'(t)).$$

Let $\varepsilon = \min\{\|\pi_0\varphi'(t)\| \mid t \in [0,1]\} > 0$. Then, by Lemma 14.2.5 (i), we have that $\underline{D}f_k(\varphi(t); \pi_0\varphi'(t)) \geq 0$ for every $t \in [0,1]$ and $\underline{D}f_k(\varphi(t); \pi_0\varphi'(t)) \geq \varepsilon$ when $\varphi(t) \notin \bigcup_i A_i$. Using also that the composition $f_k \circ \varphi$ is Lipschitz, hence differentiable a.e., and the above inequality, we get

$$f_k(\varphi(1)) - f_k(\varphi(0)) = \int_0^1 \frac{d}{dt} f_k(\varphi(t)\,dt$$

$$\geq \int_0^1 \Big(\underline{D}f_k(\varphi(t); \pi_0\varphi'(t)) - \frac{2}{k}\,\|\varphi_1'(t)\| \Big)\,dt$$

$$\geq \varepsilon\,\mathscr{L}\Big\{ t \in [0,1] \,\Big|\, \varphi(t) \notin \bigcup_{i=1}^{\infty} A_i \Big\} - \frac{2}{k} \int_0^1 \|\varphi_1'(t)\|\,dt$$

$$\geq \varepsilon\,\mathscr{L}\Big\{ t \in [0,1] \,\Big|\, \varphi(t) \notin \bigcup_{i=1}^{\infty} A_i \Big\} - \frac{2}{k}\,\text{length}(\varphi).$$

Hence

$$\varepsilon\,\mathscr{L}\Big\{ t \in [0,1] \,\Big|\, \varphi(t) \notin \bigcup_{i=1}^{\infty} A_i \Big\} \leq f_k(\varphi(1)) - f_k(\varphi(0)) + \frac{2}{k}\,\text{length}(\varphi).$$

Letting $k \to \infty$, we obtain using (14.6) that

$$\mathscr{L}\Big\{ t \in [0,1] \,\Big|\, \varphi(t) \notin \bigcup_{i=1}^{\infty} A_i \Big\} = 0$$

for any C^1 curve with $\pi_0\varphi' > 0$. $\qquad\square$

We define for $n \geq 1$ an isometry $Q_n \colon \ell_1 \longrightarrow \ell_1$ by

$$Q_n \Big(\sum_{i=0}^{\infty} z_i e_i \Big) = z_n e_0 + z_0 e_n + \sum_{\substack{i=1 \\ i \neq n}}^{\infty} z_i e_i.$$

Proof of Theorem 14.2.1. Let A_k be the porous sets from Lemma 14.2.3 and Q_n the isometry of ℓ_1 introduced above. Then, for every $n \geq 1$, the image $Q_n (\bigcup_k A_k)$ is also a σ-porous set, and consequently so is

$$A := \bigcup_{n=1}^{\infty} Q_n \Big(\bigcup_{k=1}^{\infty} A_k \Big).$$

Let $\varphi \colon [0,1] \longrightarrow \ell_1$ be a C^1 curve. If $\varphi'(t_0) \neq 0$ for some $t_0 \in (0,1)$, there are a basis functional $e_j^* \in \ell_\infty$ and $\delta > 0$ such that

$$e_j^*(\varphi'(t)) \neq 0 \quad \text{for } t \in [t_0 - \delta, t_0 + \delta] \subset [0,1].$$

Lemma 14.2.6 applied to the set $Q_j(\bigcup_k A_k)$ and to the restriction of φ on $[t_0 - \delta, t_0 + \delta]$ gives that for almost all $t \in [t_0 - \delta, t_0 + \delta]$ the point $\varphi(t)$ belongs to $Q_j(\bigcup_k A_k) \subset A$. Since the set $\{t \in [0,1] \mid \varphi'(t) \neq 0\}$ can be covered by at most countably many such intervals, we get the statement. $\qquad\square$

14.3 PRELIMINARIES TO PROOFS OF MAIN RESULTS

For the rest of this chapter, we fix $n \geq 2$ and work with n-dimensional surfaces for porosity questions and the target space $\mathbb{R}^n$ for differentiability questions.

Since our notion of surfaces is defined with the help of cubes, and since it seems that some readers find the use of the divergence theorem simpler for cubes than for Euclidean balls, it will be convenient to work with the maximum norm in the space $\mathbb{R}^n$, which we will denote by $\| \cdot \|$. Our application of the divergence theorem will be, in fact, limited to the inequality

$$\Big| \int_{\|x-a\| \leq r} \operatorname{div} H \, d\mathscr{L}^n \Big| \leq 2^n n r^{n-1} \max_{\|x-a\|=r} \|H(x)\|, \qquad (14.10)$$

where $H \in C^1(\mathbb{R}^n, \mathbb{R}^n)$.

The Euclidean norm in $\mathbb{R}^n$ will be used only in the proof of Lemma 14.3.2, where it will be denoted by $| \cdot |$.

In Lemmas 14.4.1–14.5.1 we will use the notion of *upper metric derivative* of a mapping ψ from a Banach space X to a Banach space Y. The upper (one-sided) metric derivative of ψ at x in the direction u is defined by

$$\overline{D}_u \psi(x) = \limsup_{t \searrow 0} \frac{\|\psi(x + tu) - \psi(x)\|}{t}.$$

We will need a suitable candidate for the set of centers of balls that will play the role of holes of the porous set we wish to construct. The idea is that this set should be b-dense in X, but yet its points should be sufficiently well separated by linear functionals that surfaces have a hard time trying to pass close to many of them.

Lemma 14.3.1. *Suppose that* $0 < a < b < \infty$ *and that* (x_i^*) *is a normalized sequence in* X^*, w^*-*converging to* 0. *Then there is* $E \subset X$ *such that every* $x \in X$ *is at distance less than* b *from* E *and* $\sup_{i \in \mathbb{N}} |x_i^*(x - y)| > a$ *whenever* $x, y \in E$ *are different.*

Proof. Let $0 < \varepsilon < 1$ be such that $b(1 - \varepsilon)^2 - 2\varepsilon > a$. Let (y_i) be a countable dense set in X. We define the elements x_i of E recursively as follows.

Put $x_1 = y_1$. Whenever $x_1, \ldots, x_{k-1}$ have been already defined, find the index $m = m_k$ large enough that

$$|x_m^*(y_k)| < \varepsilon \quad \text{and} \quad |x_m^*(x_i)| < \varepsilon \text{ for } 1 \leq i < k.$$

This can be clearly done since $x_i^* \xrightarrow{w^*} 0$. Let $u \in X$ with $\|u\| \leq 1$ be such that $x_m^*(u) > 1 - \varepsilon$. We put $x_k = y_k + b(1 - \varepsilon)u$.

When this construction is finished, it is obvious that every $x \in X$ is at distance less than b from E. To prove the last statement, we observe that

$$x_m^*(x_k) = x_m^*(y_k) + b(1 - \varepsilon)\, x_m^*(u) > -\varepsilon + b(1 - \varepsilon)^2.$$

Hence for x_i, $x_k \in E$, $i < k$, we conclude that

$$|x_{m_k}^*(x_k - x_i)| > b(1 - \varepsilon)^2 - \varepsilon - x_{m_k}^*(x_i) > b(1 - \varepsilon)^2 - 2\varepsilon > a. \qquad \square$$

We now state general assumptions, which could be understood as assumptions on the given Orlicz function ω, under which we can prove our main results. Should some readers be interested, for example, in the possibility of replacing the $\log^\beta(1/t)$ term in Theorem 14.1.3 by a term of the form $\log(1/t) \log^\beta \log(1/t)$, their first port of call should be checking whether these assumptions remain valid for such a choice. After stating them, we show that they hold for the Orlicz functions used in our main results, and from then on all we do will be based only on these assumptions on ω.

Assumptions. *We will assume that we are given an Orlicz function* ω, *a function* $g \in C^1(\mathbb{R}^n)$, *a vector field* $\psi \in C^1(\mathbb{R}^n, \mathbb{R}^n)$, *and a constant* $C_0 \geq 1$ *such that*

(α) $\omega(t) = o(t)$ *as* $t \searrow 0$;

(β) $\|\psi(x)\| \leq C_0$ *and* $0 \leq \operatorname{div} \psi(x) \leq C_0$ *for* $x \in \mathbb{R}^n$, $\operatorname{div} \psi(x) = 1$ *if* $\|x\| \leq 1$, *and* $\lim_{\|x\| \to \infty} \|\psi'(x)\| = 0$;

(γ) $0 \leq g(t) \leq 1$, $|g'(t)| \leq C_0$ *for* $t \in \mathbb{R}$, *and*

$$g(t) = \begin{cases} 1 & |t| \leq 1, \\ 0 & |t| \geq 2; \end{cases}$$

(δ) $\left\| g'(t)\psi(x) \right\| \le C_0 \omega^{-1}\big(g(t)\operatorname{div}\psi(x)\big)$ *for every* $t \in \mathbb{R}$ *and* $x \in \mathbb{R}^n$.

In the proof of Theorem 14.1.1 we also need ψ *to satisfy*

(ε) $\left\| \psi'(x) \right\| \le C_0 \operatorname{div}\psi(x)$.

In order to treat both proofs together, we do not require ψ *to satisfy* (ε) *throughout; instead we always give the additional statements in case this property holds.*

Notice that the limit requirement in (β) implies that ψ has bounded derivative, $\lim_{\|x\|\to\infty} \operatorname{div}\psi(x) = 0$, and $\operatorname{div}\psi$ is uniformly continuous on $\mathbb{R}^n$. We also notice two simple but useful consequences of (α). First we observe that any normalized sequence (x_i) having an upper ω estimate weakly converges to zero. Indeed, otherwise, passing to a subsequence if necessary (denoted again by (x_i)), we may assume that $x^*(x_i) \ge c$ for some $x^* \in X^*$, $\|x^*\| = 1$, and $c > 0$. But then, with $k \in \mathbb{N}$ and with $c_i = 1/kc$, $i = 1,\dots,k$, we have

$$\big\|(c_i)_{i=i}^k\big\|_\omega = \inf\Big\{\lambda > 0 \ \Big| \ \sum_{i=1}^k \omega\big(|c_i|/\lambda\big) \le 1\Big\}$$

$$= \inf\Big\{\lambda > 0 \ \Big| \ k\,\omega\big(1/\lambda kc\big) \le 1\Big\} \overset{k\to\infty}{\longrightarrow} 0.$$

On the other hand,

$$\Big\|\sum_{i=1}^k c_i x_i \Big\| \ge \sum_{i=1}^k \frac{1}{kc} x^*(x_i) \ge 1.$$

Second, based on this and on the fact that slight perturbations preserve the upper ω estimate, we obtain the following. If X^* contains a normalized sequence (x_i^*) with an upper ω estimate, and if (V_i) is any increasing sequence of finite dimensional subspaces of X, then X^* contains (another) normalized sequence (z_i^*) also having an upper ω estimate and such that for all x belonging to $\bigcup_i V_i$ the set $\{i \mid z_i^*(x) \neq 0\}$ is finite. To prove this, let (δ_i) be a sequence of positive numbers such that $\sum_k \omega^*(\delta_i) \le 1$ and choose $0 < \alpha_i < 1$ such that $2\alpha_i/(1 - \alpha_i) < \delta_i$. This choice is in order to use the following fact. If $\|x\| = 1$ and $\|y\| < \alpha_i$, then

$$\big\|(x - y) - \|x - y\|\,x\big\| \le \big|1 - \|x - y\|\big| + \|y\| \le 2\|y\|,$$

and so

$$\Big\|\frac{(x - y)}{\|x - y\|} - x\Big\| \le \frac{2\|y\|}{\|x - y\|} < \frac{2\alpha_i}{1 - \alpha_i} < \delta_i.$$

For any $i \in \mathbb{N}$, we use that x_k^* weakly converge to zero to find $k_i > k_{i-1}$ ($k_1 \ge 1$) such that the functional $x_{k_i}^*$ satisfies

$$\big|x_{k_i}^*(x)\big| < \alpha_i \|x\| \quad \text{for } x \in V_i.$$

Extend the restriction $x_{k_i}^*|_{V_i}$ by the Hahn-Banach theorem to a functional u_i^* of norm less than α_i and define the functional $z_i^* = (x_{k_i}^* - u_i^*)/\|x_{k_i}^* - u_i^*\|$. Then clearly

$\|z_i^* - x_{k_i}^*\| < \delta_i$ and by the observation mentioned at the beginning of this section the sequence (z_i^*) satisfies the upper ω estimate. Moreover, if $x \in \bigcup_k V_k$, then $x \in V_k$ for some k and therefore $z_i^*(x) = x_{k_i}^*(x) - u_i^*(x) = 0$ for all $i \geq k$.

We finish this section by showing that the above assumptions hold for the Orlicz functions we are interested in. The key to this is in the following lemma, which constructs the required function ψ. Adding the cut-off function g is, as we will see later, not so delicate.

Lemma 14.3.2. *Suppose that for some $\beta > 1$,*

$$\omega(t) = O(t^{n/(n-1)} \log^\beta(1/t)) \text{ as } t \searrow 0.$$

Then there are a constant $C > 0$ and a bounded C^1 vector field $\psi \colon \mathbb{R}^n \longrightarrow \mathbb{R}^n$ with bounded derivative such that

(i) $\operatorname{div} \psi(x) = 1$ *for* $\|x\| \leq 1$;

(ii) $\lim_{\|x\| \to \infty} \|\psi'(x)\| = 0$;

(iii) $\omega(\|\psi(x)\|) \leq C \operatorname{div} \psi(x)$.

If, for some $q > n/(n-1)$, $\omega(t) = O(t^q)$ as $t \searrow 0$, then we may also require that

(iv) $\|\psi'(x)\| \leq C \operatorname{div} \psi(x)$.

Proof. In this proof we will use the Euclidean norm $|\cdot|$ and we will actually prove the statement in this norm. Note that this is without loss of generality since it just changes the value of the constant C in (iii) and (iv).

We will look for the vector field ψ of the form $\psi(x) = h(|x|)\, x$, where the function $h \colon [0, \infty) \longrightarrow (0, \infty)$ is a C^1 function such that

- $h(s) = 1/n$ for $0 \leq s \leq 1$;

- $\lim_{s \to \infty}(s|h'(s)| + h(s)) = 0$;

- $sh(s)$ is bounded; and

- $s^n h(s)$ has strictly positive derivative on $(0, \infty)$.

Clearly, $\psi \colon \mathbb{R}^n \longrightarrow \mathbb{R}^n$ so defined is a C^1 vector field. Since the derivative of ψ in the direction u is

$$\psi'(x; u) = h'(|x|)\langle x, u\rangle \frac{x}{|x|} + h(|x|)u,$$

we have

$$\operatorname{div} \psi(x) = h'(|x|)|x| + nh(|x|) \quad \text{and} \quad |\psi'(x)| \leq |h'(|x|)||x| + h(|x|).$$

Hence (i) and (ii) are satisfied.

We are therefore left with finding h satisfying our requirements and such that, for $s \geq 0$,

$$\omega(sh(s)) \leq C(sh'(s) + nh(s)), \tag{14.11}$$

and, under the stronger assumption on ω, also

$$s|h'(s)| + h(s) \le C(sh'(s) + nh(s)). \tag{14.12}$$

Notice that both sides of these inequalities are continuous functions of s. Their right side, being constant on $[0, 1]$ and $Cs^{-n+1}\frac{d}{ds}(s^n h(s))$ on $(0, \infty)$, is strictly positive. Hence both inequalities hold provided s is restricted to a compact subinterval of $[0, \infty)$ with a constant possibly depending on the subinterval. It therefore suffices to show that they hold for s large enough.

Assume first that $\beta > 1$ and

$$\omega(t) \le \frac{Kt^{n/(n-1)}}{\log^\beta(1/t)} \tag{14.13}$$

for $0 < t \le t_0 < 1$. Let $r > 0$ be such that

$$\beta - 1 > \frac{rn}{n-1},$$

and let $h_0 \colon [0, \infty) \longrightarrow [0, \infty)$ be a C^1 function with strictly positive derivative and satisfying

$$h_0(s) = \begin{cases} \dfrac{s^n}{n} & \text{if } s \in [0, 1], \\[2mm] \log^r s & \text{if } s \in [e, \infty). \end{cases}$$

If we put $h(s) = s^{-n}h_0(s)$, then the function h satisfies all requirements needed for (i) and (ii). We check now the inequality (14.11). Let $s_1 > e$ be such that for all $s \ge s_1$,

$$sh(s) = s^{-n+1}\log^r s \le t_0$$

and

$$\frac{\log^{rn/(n-1)} s}{\big((n-1)\log s - r\log\log s\big)^\beta} \le \frac{1}{\log s}.$$

Using the growth estimate (14.13) we obtain

$$\omega(sh(s)) \le K\frac{(sh(s))^{n/(n-1)}}{\log^\beta(1/sh(s))}$$

$$= Ks^{-n}\frac{\log^{rn/(n-1)} s}{\big((n-1)\log s - r\log\log s\big)^\beta}$$

$$\le Ks^{-n}\frac{1}{\log s} \le Ks^{-n}\log^{r-1} s.$$

Notice, finally, that $sh'(s) + nh(s) = rs^{-n}\log^{r-1} s$ for $s \ge e$. The inequality (14.11) thus holds with $C = K/r$ if we restrict s to $[s_1, \infty)$ and so, as already pointed out, it holds for all $s \ge 0$ (possibly with a different C).

Assume now that $q > n/(n-1)$ and $\omega(t) \le Kt^q$ for $0 < t \le t_0$. We denote by $p = q/(q-1)$ the dual power and proceed in the same way as in the previous case but replacing now h_0 by a C^1 function with strictly positive derivative and satisfying

$$
h_0(s) = \begin{cases} \dfrac{s^n}{n} & \text{if } s \in [0,1], \\[2mm] s^{n-p} & \text{if } s \in [s_0, \infty). \end{cases}
$$

for some $s_0 > 1$. As above, we put $h(s) = s^{-n}h_0(s)$ and find $s_1 > s_0$ such that $sh(s) \le t_0$ for $s \ge s_1$. For such s we obtain

$$
\omega(sh(s)) \le K(sh(s))^q = Ks^{-p}
$$

and

$$
s|h'(s)| + h(s) = (1+p)s^{-p}.
$$

Since $sh'(s) + nh(s) = (n-p)s^{-p}$, it follows that both inequalities (14.11) and (14.12) hold with $C = (K+1+p)/(n-p)$ and for all s large. As already pointed out, they hold for all $s \ge 0$ (possibly with a different C). $\qquad\square$

The following simple cut-off functions will complete the proof that the assumptions above hold in the situation we need to prove our main results. The power p we will use will be obtained from the growth assumption on ω in Observation 14.3.4.

Lemma 14.3.3. *For each $p > 1$ and $0 < a < b < \infty$, there are $g \in C^1(\mathbb{R})$ and a constant $C > 0$ such that $0 \le g(t) \le 1$,*

$$
g(t) = \begin{cases} 1 & \text{if } |t| \le a, \\ 0 & \text{if } |t| \ge b, \end{cases}
$$

and $|g'(t)|^p \le Cg(t)$ for every $t \in \mathbb{R}$.

Proof. Choose any number c with $\max\{a, b-1\} < c < b$, and find $g \in C^1(\mathbb{R})$ such that $0 < g(t) \le 1$ for $|t| < b$, and

$$
g(t) = \begin{cases} 1 & \text{if } |t| \le a, \\ (b - |t|)^{p/(p-1)} & \text{if } c \le |t| \le b, \\ 0 & \text{if } |t| \ge b. \end{cases}
$$

Only the inequality $|g'(t)|^p \le C\,g(t)$ needs an argument. Since

$$
|g'(t)|^p = (p/(p-1))^p g(t) \quad \text{for } c \le |t| \le b,
$$

for these t it holds with $C = (p/(p-1))^p$. On the compact interval $[-c, c]$ the inequality holds with some (possibly bigger) C since both sides are continuous and the right side is strictly positive. Finally, for $|t| > b$ it holds with any C since both sides are zero. $\qquad\square$

Observation 14.3.4. *Suppose that ω satisfies the growth condition*

$$\liminf_{t \searrow 0} \frac{\omega(2t)}{\omega(t)} > 2,$$

and that ψ and C have the properties (β) and $\omega(\|\psi'(x)\|) \le C \operatorname{div} \psi(x)$ for $x \in \mathbb{R}^n$. Then there are C_0 and g such that (α)–(δ) hold.

Proof. We first notice the following consequence of the growth condition. There are $p > 1$, $b_0 > 0$, and $K > 0$ such that

$$\omega(ab) \le K a^p \omega(b) \text{ whenever } a \in [0, 1] \text{ and } b \in [0, b_0]. \tag{14.14}$$

Indeed, let $b_0 > 0$ and $p > 1$ be such that $\omega(t) \ge 2^p \omega(t/2)$ for $t \in [0, b_0]$. By induction, we see that

$$\omega\left(2^{-m}t\right) \le 2^{-pm}\omega(t), \ m \in \mathbb{N}.$$

Put $K = 2^p$. Given any $a \in (0, 1]$ we find $m \in \mathbb{N}$ with $2^{-m} < a \le 2^{-m+1}$. Then for every $b \in [0, b_0]$ we have

$$\omega(ab) \le \omega\left(2^{-m+1}b\right) \le 2^{(-m+1)p}\omega(b) = K\, 2^{-mp}\omega(b) \le K\, a^p \omega(b).$$

It is obvious that the growth condition also implies that $\omega(t) = o(t)$ as $t \searrow 0$, that is, (α). It remains to notice that the function g obtained by the use of Lemma 14.3.3 with the above p, $a = 1$, and $b = 2$ has the property (γ). As for (δ), recall that ψ is a bounded vector field satisfying (iii) of Lemma 14.3.2, so we can find $0 < c \le 1$ such that

$$\|\psi(x)\| \le \frac{b_0}{c}, \ \ \omega(\|\psi(x)\|) \le \frac{\operatorname{div}\psi(x)}{c}, \ \ |g'(t)|^p \le \frac{|g(t)|}{Kc^{p-1}}, \ \text{ and } \ |g'(t)| \le \frac{1}{c^2}.$$

Then

$$\begin{aligned}
\omega\left(c^2\|g'(t)\psi'(x)\|\right) &= \omega\left((c|g'(t)|)(c\|\psi'(x)\|)\right) \\
&\le Kc^p|g'(t)|^p\omega\left(c\|\psi'(x)\|\right) \\
&\le cg(t)\,\omega\left(\|\psi'(x)\|\right) \le g(t)\operatorname{div}\psi(x).
\end{aligned}$$

Hence, letting $C_0 = c^{-2}$ we get $|g'(t)| \le C_0$ and

$$\|g'(t)\psi(x)\| \le C_0\,\omega^{-1}\left(g(t)\operatorname{div}\psi(x)\right). \qquad \square$$

Corollary 14.3.5. *Assumptions (α)–(δ) hold for Orlicz functions satisfying*

$$\omega(t) = t^{n/(n-1)}\log^{\beta}(1/t)\,\text{for small }t > 0,$$

provided $\beta > 1$, and all assumptions (α)–(ε) hold for $\omega(t) = t^q$ when $q > n/(n-1)$.

14.4   THE MAIN CONSTRUCTION, PART I

The main point of this part is to transfer the vector field ψ on $\mathbb{R}^n$ whose existence we assume to an appropriate map of our Banach space to $\mathbb{R}^n$. This will be achieved in Lemma 14.5.1.

For maps from X to $\mathbb{R}^n$, the role of the divergence will be played by the following. If $T \in L(\mathbb{R}^n, X)$ and $\varphi \colon X \longrightarrow \mathbb{R}^n$ is differentiable at a point $x \in X$ in the direction of $T(\mathbb{R}^n)$, we denote by $\mathrm{div}_T\, \varphi(x)$ the divergence of the function $u \mapsto \varphi(x + Tu)$ at $u = 0$. In other words,

$$\mathrm{div}_T\, \varphi(x) = \sum_{i=1}^{n} \varphi_i'(x; Te_i) = \mathrm{Trace}(\varphi'(x)T),$$

where (e_i) is the standard basis for $\mathbb{R}^n$.

Except in the final part of the proof of Theorem 14.1.3, we will work with fixed $T \in L(\mathbb{R}^n, X)$ and $P \in L(X, \mathbb{R}^n)$ such that PT is the identity on $\mathbb{R}^n$. Observe that T is an isomorphism onto $T(\mathbb{R}^n)$ and X is the direct sum $X = T(\mathbb{R}^n) \oplus \mathrm{Ker}\, P$, with the corresponding projections TP and $\mathrm{Id} - TP$.

We also assume that (x_i^*) is a normalized sequence in X^* satisfying the upper ω estimate such that $x_i^* T = 0$ for $i \in \mathbb{N}$. This is possible since we have assumed that $\omega(t) = o(t)$ for $t \searrow 0$, and, as we noticed above, all x_i^* can be zero on the fixed finite dimensional subspace $T(\mathbb{R}^n)$. To minimize the number of unnecessary constants, we will also assume that the constant A from the definition of the upper ω estimate is equal to one.

In this section only we denote by $\|x\|_\mathbb{N}$ the pseudonorm

$$\|x\|_\mathbb{N} = \sup_{i \in \mathbb{N}} |x_i^*(x)|,$$

and fix a set $E = \{x_1, x_2, \dots\} \subset X$ from Lemma 14.3.1 with $a = 5$ and $b = 6$. So every $x \in X$ is in distance less than 6 from E and $\|x_i - x_j\|_\mathbb{N} > 5$ whenever $i \neq j$. The results of this part will be used for several subsequences of (x_i^*), and so with several different sets E. It is therefore important that the constants in the following estimates do not depend on the x_i^* or on E, which is the reason we try to express them with the help of C_0 and norms of T and P.

Lemma 14.4.1. *For every $\kappa > 1 + \|TP\|$ there is a Lipschitz function $\varphi \colon X \longrightarrow \mathbb{R}^n$ such that*

(i) $\|\varphi(x)\| \leq C_0$ *and* $\varphi(x) = 0$ *whenever* $\|x\|_\mathbb{N} \geq 2$ *or* $\|x - TPx\| \geq 2\kappa$;

(ii) φ *is differentiable in the direction* $T(\mathbb{R}^n)$, *the divergence* $\mathrm{div}_T\, \varphi$ *is uniformly continuous on* X, $0 \leq \mathrm{div}_T\, \varphi \leq C_0$, *and*

$$\lim_{\|Px\| \to \infty} \mathrm{div}_T\, \varphi(x) = 0;$$

(iii) $\mathrm{div}_T\, \varphi(x) = 1$ *for* $\|x\| \leq \dfrac{1}{1 + \|P\|};$

(iv) $\overline{D}_y\varphi(x) \leq C_0\omega^{-1}\big(\mathrm{div}_T\,\varphi(x)\big)\,\|y\|_{\mathbb{N}} + \dfrac{C_0^2}{\kappa}\,\|y\|$ *for every* $x \in X$ *and every* $y \in \mathrm{Ker}\,P$;

(v) *if* (ε) *holds, then at every* $x \in X$, *the function* φ *is differentiable in the direction of* $T(\mathbb{R}^n)$ *and*

$$\|\varphi'(x; Tu)\| \leq C_0\,\mathrm{div}_T\,\varphi(x)\|u\| \quad \text{for every } u \in \mathbb{R}^n.$$

Proof. We show that the statement holds with

$$\varphi(x) := g\big(\|x\|_{\mathbb{N}}\big)\,g\Big(\frac{\|x - TPx\|}{\kappa}\Big)\,\psi(Px).$$

Conditions (β) and (γ) immediately imply that φ satisfies (i). The same conditions also imply that the functions

$$x \mapsto g\big(\|x\|_{\mathbb{N}}\big)\,g\Big(\frac{\|x - TPx\|}{\kappa}\Big) \quad \text{and} \quad x \mapsto \psi(Px)$$

are bounded and Lipschitz. Moreover, since $x_i^* T = 0$ for all $i \in \mathbb{N}$, the first of these functions is constant in the direction of $T(\mathbb{R}^n)$ and the second in the direction of $\mathrm{Ker}\,P$. In particular, we have that for $u \in \mathbb{R}^n$,

$$\varphi'(x; Tu) = g\big(\|x\|_{\mathbb{N}}\big)\,g\Big(\frac{\|x - TPx\|}{\kappa}\Big)\,\psi'(Px; u). \tag{14.15}$$

Hence

$$\mathrm{div}_T\,\varphi(x) = \sum_{i=1}^{n} \varphi_i'(x; Te_i)$$

$$= g\big(\|x\|_{\mathbb{N}}\big)\,g\Big(\frac{\|x - TPx\|}{\kappa}\Big) \sum_{i=1}^{n} \psi_i'(Px; e_i)$$

$$= g\big(\|x\|_{\mathbb{N}}\big)\,g\Big(\frac{\|x - TPx\|}{\kappa}\Big)\,\mathrm{div}\,\psi(Px). \tag{14.16}$$

The above observation together with (β), (γ) immediately implies all statements of (ii).

Let $\|x\| \leq 1/(1 + \|P\|)$. Then we have

$$\frac{\|x - TPx\|}{\kappa} \leq \frac{1 + \|TP\|}{\kappa(1 + \|P\|)} \leq 1, \quad \|Px\| \leq 1, \quad \text{and} \quad \|x\|_{\mathbb{N}} \leq 1.$$

By condition (γ), $g\big(\|x\|_{\mathbb{N}}\big)\,g\big(\|x - TPx\|/\kappa\big) = 1$, and by (β), we have $\mathrm{div}_T\,\varphi(x) = 1$. Hence (iii) holds true.

Let $x \in X$ and $y \in \operatorname{Ker} P$. Since the mapping $\psi(Px)$ is constant in the direction of $\operatorname{Ker} P$, a direct calculation gives

$$
\begin{aligned}
\overline{D}_y \varphi(x) &= \overline{D}_y \left[g(\|x\|_{\mathbb{N}}) \, g\left(\frac{\|x - TPx\|}{\kappa} \right) \right] \|\psi(Px)\| \\
&\leq \left| g'(\|x\|_{\mathbb{N}}) \right| \|y\|_{\mathbb{N}} \, g\left(\frac{\|x - TPx\|}{\kappa} \right) \|\psi(Px)\| \\
&\quad + g(\|x\|_{\mathbb{N}}) \left| g'\left(\frac{\|x - TPx\|}{\kappa} \right) \right| \frac{\|y\|}{\kappa} \|\psi(Px)\|.
\end{aligned}
$$

By (δ) and the concavity of ω^{-1} the first term on the right is bounded by

$$
\begin{aligned}
C_0 \, \omega^{-1} &\left(g\left(\frac{\|x - TPx\|}{\kappa} \right) g(\|x\|_{\mathbb{N}}) \operatorname{div} \psi(Px) \right) \|y\|_{\mathbb{N}} \\
&= C_0 \, \omega^{-1} \big(\operatorname{div}_T \varphi(x) \big) \|y\|_{\mathbb{N}}.
\end{aligned}
$$

Because of the estimates of $\|\psi\|$ and $|g'|$ in (β) and (γ), the second term is bounded by $C_0^2 \|y\|/\kappa$. Hence

$$
\overline{D}_y \varphi(x) \leq C_0 \omega^{-1} \big(\operatorname{div}_T \varphi(x) \big) \|y\|_{\mathbb{N}} + \frac{C_0^2}{\kappa} \|y\|,
$$

as required by (iv).

Statement (v) follows immediately from (14.15), (14.16), and (ε):

$$
\begin{aligned}
\|\varphi'(x; Tu)\| &\leq g(\|x\|_{\mathbb{N}}) \, g\left(\frac{\|x - TPx\|}{\kappa} \right) C_0 \operatorname{div} \psi(Px) \|u\| \\
&= C_0 \operatorname{div}_T \varphi(x) \|u\|. \qquad \square
\end{aligned}
$$

Lemma 14.4.2. *Let $\delta > 0$ and let $\eta \colon X \longrightarrow [0, \infty)$ be a uniformly continuous function such that $\sup_{x \in X} \eta(x) < 1$. Then for every sufficiently small $r > 0$ there is a Lipschitz function $\Phi \colon X \longrightarrow \mathbb{R}^n$ such that*

(i) $\|\Phi(x)\| \leq r$;

(ii) Φ *is differentiable in the direction of* $T(\mathbb{R}^n)$, $\operatorname{div}_T \Phi$ *is uniformly continuous on X, $\operatorname{div}_T \Phi \geq 0$, and $\sup_{x \in X} \big(\eta(x) + \operatorname{div}_T \Phi(x) \big) < 1$;*

(iii) $\operatorname{div}_T \Phi(x) \geq \dfrac{1}{2C_0} (1 - \eta(x))$ *for every* $x \in \displaystyle\bigcup_{z \in E} B\left(rz, \dfrac{r}{1 + \|P\|} \right)$;

(iv) $\overline{D}_y \Phi(x) \leq C_0 \omega^{-1} \big(\operatorname{div}_T \Phi(x) \big) \|y\|_{\mathbb{N}} + \delta \|y\|$ *for all $x \in X$ and $y \in \operatorname{Ker} P$; and*

(v) *if ψ has the property (ε), then at every $x \in X$, Φ is differentiable in the direction of $T(\mathbb{R}^n)$ and*

$$
\|\Phi'(x; Tu)\| \leq C_0 \operatorname{div}_T \Phi(x) \|u\| \quad \text{for every } u \in \mathbb{R}^n.
$$

Proof. We assume that $0 < \delta \leq 1$ and denote

$$\varepsilon = \frac{1}{6}\Big(1 - \sup_{x \in X} \eta(x)\Big).$$

Next we find κ such that

$$\kappa > 1 + \|TP\|, \ \kappa \geq \frac{C_0}{\delta} \quad \text{and} \quad |\eta(x) - \eta(y)| < \varepsilon \text{ whenever } \|x - y\| < \frac{1}{\kappa}.$$

Let φ be the map defined in Lemma 14.4.1. By the last statement of Lemma 14.4.1 (ii) there is $s \geq 1$ such that $\mathrm{div}_T\, \varphi(x) < \varepsilon$ whenever $\|Px\| > s$.

Now let $r > 0$ be small enough to verify

$$r < \frac{\delta}{\kappa(2\kappa + s\|T\|)}.$$

Recall that the elements of the set E are denoted by $(x_j)_{j \in \mathbb{N}}$. We show that the map Φ with the required properties may be defined by

$$\Phi(x) := r \sum_{j=1}^{\infty} \alpha_j \varphi\Big(\frac{x - rx_j}{r}\Big), \tag{14.17}$$

where $\alpha_j = (1 - \eta(rx_j) - 2\varepsilon)/C_0$. Notice that $0 \leq \alpha_j \leq 1/C_0$.

For the proof, the key observation is that the support of the jth summand is contained in the set

$$W_j = \Big\{x \in X \ \Big| \ \|x - rx_j\|_{\mathbb{N}} \leq 2r, \ \|(x - rx_j) - TP(x - rx_j)\| \leq 2\kappa r\Big\},$$

and that these sets are r separated in $\|\cdot\|_{\mathbb{N}}$. The former is an easy consequence of Lemma 14.4.1 (i). For the latter, let $x \in W_i$ and $y \in W_j$, $i \neq j$. The set E is 5 discrete in the pseudonorm $\|\cdot\|_{\mathbb{N}}$. Hence $\|rx_i - rx_j\|_{\mathbb{N}} > 5r$, and so

$$\|x - y\|_{\mathbb{N}} \geq \|rx_i - rx_j\|_{\mathbb{N}} - \|x - rx_i\|_{\mathbb{N}} - \|y - rx_j\|_{\mathbb{N}}$$
$$> 5r - 2r - 2r = r.$$

This key observation now implies that the summands in (14.17) are disjointly supported. Hence Φ is Lipschitz and statements (i) and (ii) of Lemma 14.4.1 imply that $\|\Phi\| \leq r$, Φ is differentiable in the direction of $T(\mathbb{R}^n)$, and $\mathrm{div}_T\, \Phi$ is uniformly continuous and non-negative. To finish the proof of (ii) we show that $\eta(x) + \mathrm{div}_T\, \Phi(x) \leq 1 - \varepsilon$. This is obvious if $\mathrm{div}_T\, \Phi(x) = 0$. Assume that $\mathrm{div}_T\, \Phi(x) \neq 0$. We find j such that $x \in W_j$ and calculate

$$\mathrm{div}_T\, \Phi(x) = \alpha_j\, \mathrm{div}_T\, \varphi\Big(\frac{x - rx_j}{r}\Big). \tag{14.18}$$

We distinguish two cases, depending on the size of $\|P(x - rx_j)\|$. First, if

$$\|P(x - rx_j)\| > rs \quad \text{i.e.} \quad \Big\|P\Big(\frac{x - rx_j}{r}\Big)\Big\| > s,$$

then by the choice of the value s, $\operatorname{div}_T \varphi\big((x - rx_j)/r\big) < \varepsilon$. Thus

$$\eta(x) + \operatorname{div}_T \Phi(x) \leq (1 - 2\varepsilon) + \varepsilon = 1 - \varepsilon.$$

Second, if $\|P(x - rx_j)\| \leq rs$, we have

$$\|x - rx_j\| \leq \|(x - rx_j) - TP(x - rx_j)\| + \|TP(x - rx_j)\|$$
$$\leq 2\kappa r + rs\|T\|$$
$$< (2\kappa + \|T\|s) \frac{\delta}{\kappa(2\kappa + \|T\|s)} \leq \frac{1}{\kappa}.$$

This time we obtain the estimate

$$\eta(x) + \operatorname{div}_T \Phi(x) = \eta(x) + \alpha_j \operatorname{div}_T \varphi\left(\frac{x - rx_j}{r}\right)$$
$$\leq \eta(rx_j) + \varepsilon + \alpha_j \operatorname{div}_T \varphi\left(\frac{x - rx_j}{r}\right)$$
$$= \eta(rx_j) + \varepsilon + \frac{1 - \eta(rx_j) - 2\varepsilon}{C_0} \operatorname{div}_T \varphi\left(\frac{x - rx_j}{r}\right).$$

Since $\operatorname{div}_T \varphi \leq C_0$ by Lemma 14.4.1 (ii), we can continue

$$\leq \eta(rx_j) + \varepsilon + \frac{1 - \eta(rx_j) - 2\varepsilon}{C_0} C_0 = 1 - \varepsilon.$$

This finishes the proof of (ii).

A simpler argument using the uniform continuity of η is also used to get (iii). If $z \in E$ and $x \in B(rz, r/(1 + \|P\|))$, then $z = x_j$ for some j, and

$$\|x - rx_j\| \leq \frac{r}{1 + \|P\|}, \quad \text{i.e.,} \quad \left\|\frac{x - rx_j}{r}\right\| \leq \frac{1}{1 + \|P\|}.$$

Using Lemma 14.4.1 (iii) we get

$$\operatorname{div}_T \Phi(x) = \alpha_j \operatorname{div}_T \varphi\left(\frac{x - rx_j}{r}\right) = \alpha_j = \frac{1 - \eta(rx_j) - 2\varepsilon}{C_0}.$$

Moreover, since $r < 1/\kappa$, we also know that $\|x - rx_j\| < 1/\kappa$. By the uniform continuity of $\eta(x)$ we may continue

$$\geq \frac{1 - \eta(x) - 3\varepsilon}{C_0} \geq \frac{1 - \eta(x)}{2C_0}.$$

Statements (iv) and (v) follow from the key observation and Lemma 14.4.1 (iv) and (v), respectively. If $x \notin \bigcup_j W_j$, then $\Phi = 0$ on some neighborhood of x and both (iv) and (v) are trivially satisfied. If $x \in W_j$, then by Lemma 14.4.1 (iv), (14.18), and the concavity of ω^{-1},

$$
\begin{aligned}
\overline{D}_y \Phi(x) &= \alpha_j \overline{D}_y \varphi \Big(\frac{x - rx_j}{r} \Big) \\
&\leq C_0 \alpha_j \omega^{-1} \Big(\operatorname{div}_T \varphi \Big(\frac{x - rx_j}{r} \Big) \Big) \|y\|_{\mathbb{N}} + \frac{C_0}{\kappa} \|y\| \\
&\leq C_0 \omega^{-1} \Big(\alpha_j \operatorname{div}_T \varphi \Big(\frac{x - rx_j}{r} \Big) \Big) \|y\|_{\mathbb{N}} + \delta \|y\| \\
&= C_0 \omega^{-1} \big(\operatorname{div}_T \Phi(x) \big) \|y\|_{\mathbb{N}} + \delta \|y\|.
\end{aligned}
$$

Similarly, according to Lemma 14.4.1 (v), we obtain

$$
\begin{aligned}
\|\Phi'(x; Tu)\| &= \Big\| \alpha_j \varphi' \Big(\frac{x - rx_j}{r}; Tu \Big) \Big\| \\
&\leq C_0 \alpha_j \operatorname{div}_T \varphi \Big(\frac{x - rx_j}{r} \Big) \|u\| = C_0 \operatorname{div}_T \Phi(x) \|u\|. \qquad \square
\end{aligned}
$$

14.5 THE MAIN CONSTRUCTION, PART II

In the second step of the construction we add, in a rather straightforward way, the functions found in Lemma 14.4.2 to define mappings G_k. They will, in the proof of Theorem 14.1.1, converge to an affine transform of the desired map and, in the proof of Theorem 14.1.3, they allow the use of the divergence theorem to estimate the measure of the intersection of surfaces with a simply constructed σ-porous set.

Despite its simplicity, it is the following lemma where upper ω estimates are essential to infer that different estimates provided by Lemma 14.4.2 (iv) can be added together.

Lemma 14.5.1. *There are sets $E_k \subset X$, Lipschitz maps $G_k \colon X \longrightarrow \mathbb{R}^n$, and numbers $r_k \searrow 0$ such that $G_0 = 0$, $r_0 = 1$, and for each $k \geq 1$,*

(i) *every point of X is within distance less than 6 from E_k;*

(ii) $\|G_k(x)\| \leq 1$, $\|G_k(x) - G_{k-1}(x)\| \leq r_k$ *and* $r_k \leq 2^{-k} r_{k-1}$;

(iii) G_k *is differentiable in the direction of $T(\mathbb{R}^n)$, $\operatorname{div}_T G_k$ is uniformly continuous on X, $\operatorname{div}_T G_k \geq \operatorname{div}_T G_{k-1} \geq 0$, and $\sup_{x \in X} \operatorname{div}_T G_k(x) < 1$;*

(iv) *For every* $x \in \bigcup_{z \in E_k} B\Big(r_k z, \dfrac{r_k}{1 + \|P\|} \Big)$,

$$
\operatorname{div}_T G_k(x) \geq \operatorname{div}_T G_{k-1}(x) + \frac{1}{2C_0} \big(1 - \operatorname{div}_T G_{k-1}(x) \big);
$$

(v) $| \operatorname{div}_T G_{k-1}(y) - \operatorname{div}_T G_{k-1}(z) | \leq 2^{-k}$ *for* $\|y - z\| \leq 8 r_k$;

(vi) $\overline{D}_y G_k(x) \leq (4C_0 + 1) \|y\|$ *for all $x \in X$ and $y \in \operatorname{Ker} P$;*

(vii) *if (ε) holds, then at every $x \in X$, G_k is differentiable in the direction of $T(\mathbb{R}^n)$, and*

$$\|G_k'(x; Tu)\| \le C_0\|u\| \quad \text{for every } u \in \mathbb{R}^n.$$

Proof. We choose infinite mutually disjoint subsets $N_k \subset \mathbb{N}$, $k \in \mathbb{N}$, and define the pseudonorms

$$\|x\|_k = \sup_{j \in N_k} |x_j^*(x)|.$$

By Lemma 14.3.1 one can find sets $E_k \subset X$ such that every point of X has distance at most 6 from E_k (so (i) holds) and $\|x - y\|_k > 5$ whenever x, y are distinct points of E_k.

We will recursively construct the mappings G_k and numbers r_k, using the construction from Part I in the kth step with the pseudonorm $\| \cdot \|_k$ in the place of $\| \cdot \|_\mathbb{N}$. The initial values G_0 and r_0 are given in the statement of the lemma.

Assume that $G_0, \ldots, G_{k-1}$, and $r_0, \ldots, r_{k-1}$ have been already defined and that they verify the requirements (i)–(vii). To find G_k and r_k we proceed as follows. Apply Lemma 14.4.2 with the pseudonorm $\| \cdot \|_\mathbb{N}$ replaced by $\| \cdot \|_k$, with $E = E_k$, $\delta = 2^{-k}r_{k-1}$, and the function $\eta(x) = \operatorname{div}_T G_{k-1}(x)$. Notice that by the induction hypothesis (iii), we match the assumptions of this lemma. Hence we may find $0 < r_k \le 2^{-k}r_{k-1}$ and Lipschitz maps $\Phi_k \colon X \longrightarrow \mathbb{R}^n$ with all properties from Lemma 14.4.2 (i)–(iv). We can choose r_k even smaller to guarantee that, at the same time, (v) holds true. The recursive construction is finished by letting

$$G_k = G_{k-1} + \Phi_k.$$

Clearly, G_k is Lipschitz, $r_k \searrow 0$, and $\|G_k(x) - G_{k-1}(x)\| = \|\Phi_k(x)\| \le r_k$ by Lemma 14.4.2 (i). Moreover, it is immediate that

$$\|G_k(x)\| \le \sum_{j=1}^{k} \|G_j(x) - G_{j-1}(x)\| \le \sum_{j=1}^{k} r_j \le \sum_{j=1}^{k} 2^{-j}r_{j-1} \le r_0 \sum_{j=1}^{k} 2^{-j} \le 1,$$

and (ii) holds true. The statements (iii) follow directly from $\operatorname{div}_T G_{k-1} \ge 0$ and Lemma 14.4.2 (ii).

The statement (iv) is just rewritten (iii) of Lemma 14.4.2.

To prove (vi), consider any $x \in X$ and $y \in \operatorname{Ker} P$. By the statement (iv) of Lemma 14.4.2,

$$\overline{D}_y \Phi_i(x) \le C_0 \omega^{-1}(\operatorname{div}_T \Phi_i(x))\|y\|_i + 2^{-i}r_{i-1}\|y\|, \quad i = 1, \ldots, k. \quad (14.19)$$

We choose $m_i \in N_i$ such that $|x_{m_i}^*(y)| \ge \frac{1}{2}\|y\|_i$, $i = 1, \ldots, k$, and abbreviate $c_i = \operatorname{sign}(x_{m_i}^*(y)) \, \omega^{-1}(\operatorname{div}_T \Phi_i(x))$. Since the m_i are mutually different and the whole sequence $(x_j^*)_{j \in \mathbb{N}}$ satisfies the upper ω estimate, we see that

$$\left\| \sum_{i=1}^{k} \operatorname{sign}\big(x_{m_i}^*(y)\big)\, \omega^{-1}\big(\operatorname{div}_T \Phi_i(x)\big) \cdot x_{m_i}^* \right\|$$

$$= \left\| \sum_{i=1}^{k} c_i x_{m_i}^* \right\|$$

$$\leq \left\| (c_i)_{i=1}^{k} \right\|_\omega \leq 1 + \sum_{i=1}^{k} \omega(|c_i|)$$

$$\leq 1 + \sum_{i=1}^{k} \omega\big(\big|\operatorname{sign}\big(x_{n_i}^*(y)\big)\omega^{-1}(\operatorname{div}_T \Phi_i(x))\big|\big)$$

$$= 1 + \sum_{i=1}^{k} \operatorname{div}_T \Phi_i(x) = 1 + \operatorname{div}_T G_k \leq 2. \qquad (14.20)$$

Hence, combining (14.19) and the just finished estimate (14.20), we obtain

$$\overline{D}_y G_k(x) \leq \sum_{i=1}^{k} \overline{D}_y \Phi_i(x)$$

$$\leq \sum_{i=1}^{k} \Big(C_0 \omega^{-1}\big(\operatorname{div}_T \Phi_i(x)\big)\, \|y\|_i + 2^{-i} r_{i-1} \|y\| \Big)$$

$$\leq \sum_{i=1}^{k} 2 C_0 \omega^{-1}\big(\operatorname{div}_T \Phi_i(x)\big)\, |x_{m_i}^*(y)| + \|y\|$$

$$= 2C_0 \sum_{i=1}^{k} \operatorname{sign}\big(x_{m_i}^*(y)\big)\omega^{-1}\big(\operatorname{div}_T \Phi_i(x)\big)\, x_{m_i}^*(y) + \|y\|$$

$$\leq 2C_0 \left\| \sum_{i=1}^{k} \operatorname{sign}\big(x_{m_i}^*(y)\big)\, \omega^{-1}\big(\operatorname{div}_T \Phi_i(x)\big) \cdot x_{m_i}^* \right\| \|y\| + \|y\|$$

$$\leq (4C_0 + 1)\|y\|.$$

In the case when (ε) holds, then in view of Lemma 14.4.2 (v),

$$\|G_k'(x; Tu)\| \leq \sum_{i=1}^{k} \|\Phi_i'(x; Tu)\|$$

$$\leq \sum_{i=1}^{k} C_0 \operatorname{div}_T \Phi_i(x)\|u\|$$

$$= C_0 \operatorname{div}_T G_k(x)\|u\| \leq C_0\|u\|,$$

as required by (vii). $\qquad\qquad\qquad\qquad\qquad\qquad\qquad\qquad\qquad\qquad\qquad\square$

14.6 PROOF OF THEOREM 14.1.3

Lemma 14.6.1. *Let $T \in L(\mathbb{R}^n, X)$ be of rank n. Then there are a σ-porous set $A \subset X$ and $\varepsilon > 0$ such that for every $\gamma \in \Gamma_n(X)$,*

$$\mathscr{L}^n \Big\{ u \in [0,1]^n \;\Big|\; \gamma(u) \notin A,\ \|\gamma'(u) - T\| < \varepsilon \Big\} = 0.$$

Proof. We continue using the notation from the previous sections; in particular, the operator $P \in L(X, \mathbb{R}^n)$ is such that $PT = \mathrm{Id}$ on $\mathbb{R}^n$. We will also assume that the sequence (x_i^*) with the upper ω estimate also satisfies $x_i^* T = 0$ for all i. Note that this causes no loss of generality as explained at the beginning of the previous section.

Lemma 14.5.1 provides the sets $E_k \subset X$, Lipschitz maps $G_k \colon X \longrightarrow \mathbb{R}^n$, and a sequence $r_k \searrow 0$ with the properties (i)–(vi). Let $c = 1/(1 + \|T\|)$. We put

$$M = \bigcup_{k=1}^{\infty} \bigcup_{z \in E_k} B(r_k z, c r_k).$$

Then $X \setminus M$ is a porous set: To check it, let $x \in X \setminus M$ and $k \in \mathbb{N}$. Since $r_k E_k$ is $6r_k$-dense, there exists $z \in E_k$ such that $\|x - r_k z\| \le 6r_k$. We see that at the distance at most $6r_k$ from x there is a ball $B(r_k z, c r_k)$ disjoint from $X \setminus M$. Consequently, $X \setminus M$ is a porous set with the porosity constant $c/6$. The desired σ-porous set A will be

$$A = \bigcup_{m=1}^{\infty} \Big(X \setminus \frac{1}{m} M \Big) = X \setminus \bigcap_{m=1}^{\infty} \frac{1}{m} M.$$

We let $\alpha = 1/(8nC_0)$ and choose $\varepsilon > 0$ such that

$$\varepsilon \|P\| < 1, \quad \varepsilon \, \frac{(4C_0 + 1)(1 + \|TP\|)}{1 - \varepsilon \|P\|} \le \alpha.$$

Assume, in order to obtain a contradiction, that there is $\gamma_0 \in \Gamma_n(X)$ for which the statement is false. Then there is a density point v_0 of the set

$$\Big\{ v \in (0,1)^n \;\Big|\; \gamma_0(v) \notin A,\ \|\gamma_0'(v) - T\| < \varepsilon \Big\}.$$

Consider the auxiliary mapping $I \colon \mathbb{R}^n \times \mathbb{R}^n \longrightarrow \mathbb{R}^n$ given by

$$I(v, u) = P\big(\gamma_0(v) - \gamma_0(v_0) - Tu\big).$$

Notice that the point $(v_0, 0) \in \mathbb{R}^n \times \mathbb{R}^n$ is a solution of the equation

$$I(v, u) = 0,$$

and that the partial differential with respect to the v-coordinate at $(v_0, 0)$ has full rank: Indeed, the partial differential is equal to $P\gamma_0'(v_0)$ and

$$\|P\gamma_0'(v_0) - \mathrm{Id}\| = \|P(\gamma_0'(v_0) - T)\| \le \|P\| \, \varepsilon < 1.$$

The implicit function theorem provides us with a C^1 diffeomorphism η of some ball $B(0, \rho_0) \subset \mathbb{R}^n$ onto a neighborhood of $v_0 = \eta(0) \in \mathbb{R}^n$ such that

$$I\big(\eta(u), u\big) = 0, \quad \text{i.e.,} \quad P\big(\gamma_0(\eta(u)) - \gamma_0(v_0) - Tu\big) = 0, \quad u \in B(0, \rho_0).$$

The density point v_0 is mapped by means of η^{-1} to 0, so 0 is a density point of the set

$$\Omega_0 = \Big\{ u \in B(0, \rho_0) \ \Big| \ \gamma_0(\eta(u)) \notin A, \ \|\gamma_0'(\eta(u)) - T\| < \varepsilon \Big\}.$$

For $u \in B(0, \rho_0)$ define

$$\xi(u) = \gamma_0(\eta(u)) - \gamma_0(v_0) - Tu.$$

Clearly, $\xi(u) \in \operatorname{Ker} P$. In order to estimate the norm $\|\xi'(u)\|$ for $u \in \Omega_0$ we notice the following. First, differentiating the identity

$$P\big(\gamma_0(\eta(u)) - \gamma_0(v_0) - Tu\big) = 0,$$

we obtain

$$P\big(\gamma_0'(\eta(u))\big)\, \eta'(u) - PT = 0, \quad \text{i.e.,} \quad P\big(\gamma_0'(\eta(u))\, \eta'(u) = \operatorname{Id}.$$

In other words, $\eta'(u)$ is the inverse operator to $P\gamma_0'(\eta(u))$, and so, since

$$\|P\gamma_0'(\eta(u)) - \operatorname{Id}\| \le \|P\|\,\|\gamma_0'(\eta(u)) - T\| \le \varepsilon \|P\| < 1,$$

the standard estimate of the norm of the inverse gives

$$\|\eta'(u)\| \le \sum_{j=0}^{\infty} \|\operatorname{Id} - P\gamma_0'(\eta(u))\|^j \le \frac{1}{1 - \varepsilon \|P\|}.$$

Hence

$$\begin{aligned}
\|\xi'(u)\| &= \|\gamma_0'(\eta(u))\, \eta'(u) - T\| \\
&\le \|\gamma_0'(\eta(u))\, \eta'(u) - T\eta'(u)\| + \|T\eta'(u) - T\| \\
&\le \|\gamma_0'(\eta(u)) - T\|\,\|\eta'(u)\| + \|T - TP\gamma_0'(\eta(u))\|\,\|\eta'(u)\| \\
&\le \varepsilon \|\eta'(u)\| + \|TP\|\,\|T - \gamma_0'(\eta(u))\|\,\|\eta'(u)\| \\
&\le \varepsilon \|\eta'(u)\| + \varepsilon \|TP\|\,\|\eta'(u)\| = \varepsilon \|\eta'(u)\|\,(1 + \|TP\|). \\
&\le \varepsilon \frac{1 + \|TP\|}{1 - \varepsilon \|P\|}. \quad\quad\quad (14.21)
\end{aligned}$$

Pick $0 < \rho < \rho_0$ small enough that the set $B(0, \rho) \cap \Omega_0$ has $\mathscr{L}^n$-measure greater than $2^{n-1}\rho^n$. (Recall that the balls in $\mathbb{R}^n$ are considered with respect to the maximum norm.) Let $m \in \mathbb{N}$ be such that $m > 8nC_0/\rho$. The set

$$\big\{ u \in B(0, \rho) \mid \gamma(\eta(u)) \in M/m \big\}$$

contains $B(0, \rho) \cap \Omega_0$, so its measure is greater than $2^{n-1}\rho^n$. Since

$$\frac{M}{m} = \bigcup_{j=1}^{\infty} \bigcup_{z \in E_j} B\Big(\frac{r_j z}{m}, \frac{cr_j}{m}\Big),$$

we can find $k \in \mathbb{N}$ for which the set

$$\Omega := \Big\{ u \in B(0, \rho) \ \Big| \ \gamma(\eta(u)) \in \bigcup_{j=1}^{k} \bigcup_{z \in E_j} B\Big(\frac{r_j z}{m}, \frac{cr_j}{m}\Big) \Big\}$$

has measure still greater than $2^{n-1}\rho^n$.

Consider now the vector field G_k. Since for all $u, v \in \mathbb{R}^n$, the directional derivative $\xi'(u; v)$ belongs to $\operatorname{Ker} P$, we may apply Lemma 14.5.1 (vi) together with (14.21) to get

$$\overline{D}_{\xi'(u;v)} G_k(x) \leq (4C_0 + 1)\|\xi'(u; v)\|$$
$$\leq (4C_0 + 1) \frac{(1 + \|T\|)\varepsilon\|v\|}{1 - \varepsilon\|P\|} \leq \alpha\|v\|. \tag{14.22}$$

Hence for every $\widetilde{x} \in X$, the function

$$v \mapsto G_k(\widetilde{x} + m\xi(v))$$

has Lipschitz constant at most αm. We use this information to show that the vector field $H \colon B(0, \rho_0) \longrightarrow \mathbb{R}^n$ defined by

$$H(u) = G_k\big(m\gamma_0(\eta(u))\big) = G_k\big(m(\xi(u) + Tu + \gamma_0(t_0))\big)$$

satisfies

$$\operatorname{div} H(u) \geq m \, (\operatorname{div}_T G_k)\big(m\gamma_0(\eta(u))\big) - nm\alpha. \tag{14.23}$$

For this it clearly suffices to have the following estimate of the difference of directional derivatives:

$$\|H'(u; v) - G_k'(m\gamma_0(\eta(u)); mTv)\| \leq \alpha m\|u\|, \ \ v \in \mathbb{R}^n.$$

But, denoting for a moment $\widetilde{x} = m(T(u + tv) + \gamma_0(v_0))$, we see that

$$\big\|H(u + tv) - H(u) - \big(G_k(m\gamma_0(\eta(u))) + tmTv - G_k(m\gamma_0(\eta(u)))\big)\big\|$$
$$= \|G_k(\widetilde{x} + m\xi(u + tv)) - G_k(\widetilde{x} + m\xi(u))\| \leq m\alpha|t| \, \|v\|.$$

Since by Lemma 14.5.1 (iii) $\operatorname{div}_T G_k \geq 0$, we obtain from (14.23) that

$$\operatorname{div} H(u) \geq -mn\alpha = -\frac{m}{8C_0} \ \ u \in B(0, \rho_0). \tag{14.24}$$

For $u \in \Omega$ we will need a more precise estimate of div $H(u)$. In this case the point $x = m\gamma_0(\eta(u)) \in \bigcup_{j=1}^{k} \bigcup_{z \in E_j} B(r_j z_j, c r_j)$ and so by Lemma 14.5.1 (iv),

$$\text{div}_T\, G_k(x) \geq \text{div}_T\, G_{k-1}(x) + \frac{1}{2C_0} - \frac{1}{2C_0}\, \text{div}_T\, G_{k-1}(x) \geq \frac{1}{2C_0}.$$

Consequently, (14.23) gives

$$\text{div}\, H(u) \geq m\, \frac{1}{2C_0} - mn\alpha = \frac{3m}{8C_0}, \quad u \in \Omega.$$

This, (14.24), and (14.10) yield the required contradiction,

$$
\begin{aligned}
2n\, 2^{n-1} \rho^{n-1} &\geq \int_{\partial B(0,\rho)} \|H\|\, d\mathscr{H} \\
&\geq \int_{B(0,\rho)} \text{div}\, H\, d\mathscr{L}^n \\
&= \int_{\Omega} \text{div}\, H\, d\mathscr{L}^n + \int_{B(0,\rho)\backslash\Omega} \text{div}\, H\, d\mathscr{L}^n \\
&\geq \frac{3m}{8C_0}\, \mathscr{L}^n \Omega - \frac{m}{8C_0}\, \mathscr{L}^n \big(B(0,\rho)\backslash\Omega\big) \\
&> \frac{3m}{8C_0}\, 2^{n-1} \rho^n - \frac{m}{8C_0}\, 2^{n-1} \rho^n = \frac{m}{4C_0}\, 2^{n-1} \rho^n \\
&> 2n\, 2^{n-1} \rho^{n-1}. \qquad\qquad\qquad\qquad\qquad\qquad\qquad\Box
\end{aligned}
$$

Proof of Theorem 14.1.3. In view of Lemma 14.6.1 we find for each $T \in L(\mathbb{R}^n, X)$ with rank n a σ-porous set A_T and an $\varepsilon_T > 0$. The set of rank n operators from $L(\mathbb{R}^n, X)$ is covered by the family of open sets

$$\{Q \in L(\mathbb{R}^n, X) \mid \|Q - T\| < \varepsilon_T\}, \quad T \in L(\mathbb{R}^n, X).$$

Since $L(\mathbb{R}^n, X)$ is separable, the Lindelöf theorem says that this cover has a countable subcover. Hence there is a sequence (T_j) of operators having rank n such that for any rank n operator T one can find an index j with $\|T - T_j\| < \varepsilon_{T_j}$.

Letting $A = \bigcup_j A_{T_j}$, Lemma 14.6.1 implies that

$$
\begin{aligned}
\mathscr{L}^n &\Big\{ t \in [0,1]^n \;\Big|\; \gamma(t) \notin A,\; \text{rank}\, \gamma'(t) = n \Big\} \\
&\leq \mathscr{L}^n \bigcup_{j=1}^{\infty} \Big\{ t \in [0,1]^n \;\Big|\; \gamma(t) \notin A_{T_j},\; \|\gamma'(t) - T_j\| < \varepsilon_{T_j} \Big\} \\
&\leq \sum_{j=1}^{\infty} \mathscr{L}^n \Big\{ t \in [0,1]^n \;\Big|\; \gamma(t) \notin A_{T_j},\; |\gamma'(t) - T_j| < \varepsilon_{T_j} \Big\} \\
&= 0. \qquad\qquad\qquad\qquad\qquad\qquad\qquad\qquad\qquad\qquad\qquad\qquad\Box
\end{aligned}
$$

14.7   PROOF OF THEOREM 14.1.1

We still continue using the notation from Sections 14.3–14.5, assuming of course that (ε) also holds. To simplify the constants, we will also assume that $\|T\| = 1$.

Use Lemma 14.5.1 to find sets $E_k \subset X$, Lipschitz maps $G_k \colon X \longrightarrow \mathbb{R}^n$, and a sequence $r_k \searrow 0$ with the properties (i)–(vii). By (ii), G_k converge uniformly to some $G \colon X \longrightarrow \mathbb{R}^n$ with $\|G\| \leq 1$. To see that G is Lipschitz, we use both (vi) and (vii). Let $y \in \operatorname{Ker} P$. Then for any $k \in \mathbb{N}$

$$\|G_k(x+y) - G_k(x)\| \leq \int_0^1 \overline{D}_y G_k(x+ty)\,dt \leq (4C_0 + 1)\|y\|.$$

Letting $k \to \infty$ we conclude that G is Lipschitz in the direction of $\operatorname{Ker} P$. A similar argument with (vii) instead of (vi) implies that G is Lipschitz in the direction of $T(\mathbb{R}^n)$. Since $X = \operatorname{Ker} P \oplus T(\mathbb{R}^n)$, we infer that G is Lipschitz.

We first notice a simple estimate of the divergence of G_k in terms of G.

Lemma 14.7.1. *Suppose that $x \in X$, $k \in \mathbb{N}$, $s > 0$, $\varepsilon > 0$, and $Q \in L(X, \mathbb{R}^n)$ are such that*

$$\|G(x+y) - G(x) - Qy\| \leq \varepsilon s \text{ whenever } y \in T(\mathbb{R}^n) \text{ and } \|y\| \leq s.$$

Then there is $z \in B(x, s)$ such that

$$|\operatorname{div}_T G_k(z) - \operatorname{Trace} QT| \leq n\left(\varepsilon + \frac{2^{-k+1} r_k}{s}\right).$$

Proof. For a fixed $x \in X$ we define an auxiliary vector field $H \colon \mathbb{R}^n \longrightarrow \mathbb{R}^n$ by

$$H(u) = G_k(x+Tu) - G_k(x) - Q(Tu).$$

Notice that Lemma 14.5.1 (ii) implies that $\|G(y) - G_k(y)\| \leq 2^{-k} r_k$ for every $y \in X$. Hence,

$$\begin{aligned}
\|H(u)\| &\leq \|G_k(x+Tu) - G(x+Tu)\| + \|G(x) - G_k(x)\| \\
&\quad + \|G(x+Tu) - G(x) - QT(u)\| \\
&\leq 2^{-k+1} r_k + \varepsilon s,
\end{aligned}$$

whenever $\|u\| \leq s$. By (14.10),

$$\left| \int_{\|u\| \leq s} \operatorname{div} H \, d\mathscr{L}^n \right| \leq 2n(\varepsilon s + 2^{-k+1} r_k)\, 2^{n-1} s^{n-1}.$$

Since $\operatorname{div} H$ is continuous, there is $u \in \mathbb{R}^n$ with $\|u\| \leq s$ and

$$|\operatorname{div} H(u)| \leq n\left(\varepsilon + \frac{2^{-k+1} r_k}{s}\right).$$

Noting that $\operatorname{div} H(u) = \operatorname{div}_T G_k(x+Tu) - \operatorname{Trace} QT$, we see that the statement holds with $z = x + Tu$. $\qquad\square$

Our final lemma shows that G has, up to a simple transformation, the properties required in the statement of Theorem 14.1.1. Recall that $D(G)$ denotes the set of $x \in X$ at which G is Gâteaux differentiable.

Lemma 14.7.2. *The Lipschitz map $G\colon X \longrightarrow \mathbb{R}^n$ has the following properties.*

(i) $0 \le \operatorname{div}_T G(x) \le 1$ *at every $x \in D(G)$.*

(ii) *For any $\tau > 0$ there is $x \in D(G)$ at which $\operatorname{div}_T G(x) \le \tau$.*

(iii) *There is a constant $\kappa > 0$ such that, whenever Q is an ε-Fréchet derivative of G at some $x \in X$, then $|1 - \operatorname{Trace} QT| \le \kappa\varepsilon$.*

Proof. To prove (i), fix an $x \in D(G)$. Let $Q = G'(x)$ and find, for any given $\varepsilon > 0$, an index k such that $2^{-k+1} \le \varepsilon$ and

$$\|G(x + y) - G(x) - Qy\| \le \varepsilon r_k \text{ for all } y \in T(\mathbb{R}^n) \text{ with } \|y\| \le r_k.$$

By Lemma 14.7.1 there is $z \in B(x, r_k)$ such that

$$|\operatorname{div}_T G_k(z) - \operatorname{Trace} QT| \le n\Big(\varepsilon + \frac{2^{-k+1} r_k}{r_k}\Big) \le 2n\varepsilon.$$

Since $\operatorname{div}_T G(x) = \operatorname{Trace} QT$ and $0 \le \operatorname{div}_T G_k(z) \le 1$ for all z, we infer that

$$-2n\varepsilon \le \operatorname{div}_T G(x) \le 1 + 2n\varepsilon$$

for each $\varepsilon > 0$. Hence $0 \le \operatorname{div}_T G(x) \le 1$.

To prove (ii), we recall that G is Gâteaux differentiable except for a Gauss null set (Theorem 2.3.1), so there exists $y \in X$ such that $y + Tu \in D(G)$ for $\mathscr{L}^n$ almost all $u \in \mathbb{R}^n$. Define $H\colon \mathbb{R}^n \longrightarrow \mathbb{R}^n$ by

$$H(u) = G\Big(y + \frac{2n}{\tau} Tu\Big).$$

Then

$$\operatorname{div} H(u) = \frac{2n}{\tau} \operatorname{div}_T G\Big(y + \frac{2n}{\tau} Tu\Big) \ge 0$$

for almost all $u \in \mathbb{R}^n$. The divergence theorem says that

$$\int_{[0,1]^n} \operatorname{div} H \, d\mathscr{L}^n \le 2n,$$

and so $\operatorname{div} H \le 2n$ on a set of positive measure. Hence there is $u \in (0,1)^n$ such that $\operatorname{div} H(u) \le 2n$ and G is Gâteaux differentiable at $y + (2n/\tau)Tu$. It follows that $x = y + (2n/\tau)Tu$ is the point we want.

Finally, we show that (iii) holds with $\kappa = 2C_0 n(17 + 14\|P\|)$. Let $\varepsilon > 0$ and let Q be an ε-Fréchet derivative of G at x. By definition, there is $\delta > 0$ such that

$$\|G(y) - G(x) - Q(y - x)\| \le \varepsilon\|y - x\|$$

whenever $\|y - x\| < \delta$. Let $k \in \mathbb{N}$ be such that $2^{-k+2} < \varepsilon$ and $7r_k < \delta$. Applying Lemma 14.7.1 for the given x and ε with k replaced by $k - 1$ and s replaced by r_k, we find a point $z \in B(x, r_k)$ such that

$$|\operatorname{div}_T G_{k-1}(z) - \operatorname{Trace} QT| \leq n\Big(\varepsilon + \frac{2^{-k+2}r_k}{r_k}\Big) < 2n\varepsilon. \tag{14.25}$$

Let $x_0 \in r_k E_k$ be such that $\|x - x_0\| \leq 6r_k$. If $\|y - x_0\| \leq r_k$, then $\|y - x\| \leq 7r_k < \delta$ and we may estimate

$$
\begin{aligned}
\|G(y) &- G(x_0) - Q(y - x_0)\| \\
&\leq \|G(y) - G(x) - Q(y - x)\| + \|G(x_0) - G(x) - Q(x_0 - x)\| \\
&\leq \varepsilon\|y - x\| + \varepsilon\|x_0 - x\| \leq 13\varepsilon r_k.
\end{aligned}
$$

Thus Lemma 14.7.1 can be used once more with x_0, k, $s = r_k/(1 + \|P\|)$, and $13\varepsilon(1 + \|P\|)$. We obtain a point $z_0 \in B(x_0, r_k/(1 + \|P\|))$ such that

$$
\begin{aligned}
|\operatorname{div}_T G_k(z_0) - \operatorname{Trace} QT| &\leq n\Big(13\varepsilon(1 + \|P\|) + \frac{2^{-k+1}r_k}{r_k}(1 + \|P\|)\Big) \\
&\leq 14n\varepsilon(1 + \|P\|).
\end{aligned}
\tag{14.26}
$$

The two estimates (14.25) and (14.26) provide us with the needed control over $\operatorname{Trace} QT$. On one side,

$$\operatorname{Trace} QT \leq \operatorname{div}_T G_{k-1}(z) + 2n\varepsilon \leq 1 + \kappa\varepsilon.$$

On the other side,

$$
\begin{aligned}
\operatorname{Trace} QT = {}&2C_0 \operatorname{div}_T G_k(z_0) - (2C_0 - 1)\operatorname{div}_T G_{k-1}(z) \\
&- 2C_0\big(\operatorname{div}_T G_k(z_0) - \operatorname{Trace} QT\big) \\
&+ (2C_0 - 1)\big(\operatorname{div}_T G_{k-1}(z) - \operatorname{Trace} QT\big) \\
\geq {}&2C_0 \operatorname{div}_T G_k(z_0) - (2C_0 - 1)\operatorname{div}_T G_{k-1}(z) \\
&- 2C_0|\operatorname{div}_T G_k(z_0) - \operatorname{Trace} QT| \\
&- (2C_0 - 1)|\operatorname{div}_T G_{k-1}(z) - \operatorname{Trace} QT|.
\end{aligned}
$$

Applying Lemma 14.5.1 (iv) to the first term we continue:

$$
\begin{aligned}
\geq {}&1 - (2C_0 - 1)\big(\operatorname{div}_T G_{k-1}(z) - \operatorname{div}_T G_{k-1}(z_0)\big) \\
&- 2C_0|\operatorname{div}_T G_k(z_0) - \operatorname{Trace} QT| \\
&- (2C_0 - 1)|\operatorname{div}_T G_{k-1}(z) - \operatorname{Trace} QT|.
\end{aligned}
$$

Since also $\|z_0 - z\| \leq \|z - x\| + \|x - x_0\| + \|x_0 - z_0\| \leq 8r_k$, we can use the estimate (v) of Lemma 14.5.1 for the difference of divergences. Together with (14.25) and (14.26) we finally obtain

$$\geq 1 - (2C_0 - 1)2^{-k} - 2C_0 \, 14n\varepsilon(1 + \|P\|) - (2C_0 - 1)2n\varepsilon$$
$$\geq 1 - 2C_0\big(1 + 14n(1 + \|P\|) + 2n\big)\varepsilon$$
$$\geq 1 - 2C_0 n(17 + 14\|P\|)\varepsilon = 1 - \kappa\varepsilon.$$

So $|1 - \operatorname{Trace} QT| \leq \kappa\varepsilon$ as required. $\qquad\square$

Proof of Theorem 14.1.1. Let

$$F(x) = \frac{1}{n} P(x) - G(x) \ \text{ and } \ \mathscr{T}(Q) = \frac{1}{\kappa} \operatorname{Trace} QT,$$

where κ is the constant from Lemma 14.7.2 (iii).

Clearly, $F \colon X \longrightarrow \mathbb{R}^n$ is Lipschitz and $\mathscr{T}$ is a bounded linear form on $L(X, \mathbb{R}^n)$. By Lemma 14.7.2 (i),

$$\mathscr{T}(F'(x)) = \frac{1}{\kappa}(1 - \operatorname{div}_T G(x)) \geq 0$$

for every $x \in D(F) = D(G)$.

By Lemma 14.7.2 (ii), for any $\tau > 0$ there is $x \in D(F)$ such that $\operatorname{div}_T G(x) \leq \tau$. Hence

$$\mathscr{T}(F'(x)) = \frac{1}{\kappa}\big(1 - \operatorname{div}_T G(x)\big) \geq \frac{1 - \tau}{\kappa},$$

and so the slice of Gâteaux derivatives

$$\Big\{ F'(x) \ \Big| \ x \in D(F), \ \mathscr{T}(F'(x)) > \varepsilon \Big\}$$

is nonempty whenever $0 < \varepsilon < 1/\kappa$.

Consider now any $x \in X$ at which F is ε-Fréchet differentiable with Q as an ε-derivative. Then $G = P/n - F$ is ε-Fréchet differentiable at x with ε-derivative $P/n - Q$. Hence, by Lemma 14.7.2 (iii),

$$|\operatorname{Trace} QT| = \big|1 - \operatorname{Trace}\big((P/n - Q)T\big)\big| \leq \kappa\varepsilon,$$

which implies that $\mathscr{T}(Q) \leq \varepsilon$.

The above facts imply all statements of Theorem 14.1.1. $\qquad\square$

Chapter Fifteen

Asymptotic Fréchet differentiability

This chapter should be considered slightly experimental. We return to nonvariational arguments for proving differentiability, and try to construct the sequence converging to a point of differentiability by a less straightforward algorithm. The reason for this is the hope that a different algorithm can avoid the pitfalls indicated in Chapter 14 and prove, at least, that Lipschitz mappings of Hilbert spaces to finite dimensional spaces have points of Fréchet differentiability. From this point of view the results of this chapter are negative, although we provide a new proof of Corollary 13.1.2 on Fréchet differentiability of Lipschitz maps of Hilbert spaces to $\mathbb{R}^2$, and we prove a new result on asymptotic Fréchet differentiability of functions on spaces with appropriate control of moduli of smoothness. The negative aspect mentioned above means that in all results of this chapter an appropriate version of the multidimensional mean value estimate holds, and so Theorem 14.1.1 implies that the methods used here cannot find points of Fréchet differentiability of Lipschitz maps of Hilbert spaces to $\mathbb{R}^3$.

15.1 INTRODUCTION

We prove a result which, at least in the most interesting case of asymptotic smoothness, is contained in Theorem 13.1.1. It also implies that Lipschitz maps of Hilbert spaces into $\mathbb{R}^2$ have points of Fréchet differentiability. However, this is not the genuine reason for including this chapter, since readers interested in the Hilbert space case only will find a more accessible proof in Chapter 16. But there are still several reasons we believe this material should be of interest.

First, this was actually our first proof of existence of points of Fréchet differentiability for maps of Hilbert spaces into $\mathbb{R}^2$, which then developed into the variational approach presented in the previous chapters. In this sense, even the extension of the differentiability results for real-valued functions from Lipschitz functions to cone-monotone ones has its roots in what we do here. We therefore believe that readers interested in further developments will find the material from this chapter helpful in developing their ideas.

The second reason for trying the different approach of this chapter is that, while the key idea of the previous constructions of points of Fréchet differentiability could be roughly expressed as maximizing an additive perturbation of the norm of the derivative in direction T, that is,

$$(x, T) \in X \times L(\mathbb{R}^n, X) \to |f'(x; T)|_{\mathrm{H}} + \Theta(x, T),$$

the perturbation that we try to maximize here is

$$(x, T) \in X \times L(\mathbb{R}^n, X) \to \frac{|f'(x; T)|_{\mathrm{H}}}{\Theta(x, T)}.$$

One can heuristically argue that this perturbation has a better chance of finding points of Fréchet differentiability without showing the validity of the multidimensional mean value estimate. In view of Theorem 14.1.1, this is necessary for any attempt to find points of Fréchet differentiability for Lipschitz maps of Hilbert spaces into $\mathbb{R}^3$. We will not describe the heuristic arguments since the opposite turned out to be true: the method presented here also shows the multidimensional mean value estimate, and, checking the proof carefully, one can convince oneself that this should hold for other similar attempts as well.

Finally, we actually prove a new result, which illustrates the interesting point that we have already met (and that caused us serious difficulties in Chapter 13): it is sometimes easier to control the behavior of the function "at infinity" than to control its behavior in the direction of finite dimensional subspaces. The idea here is to skip the control of the behavior on finite dimensional subspaces and prove differentiability "at infinity" only. Before giving the precise notion of differentiability we have in mind here, we discuss what led us to it. Seen from the point of view of (nonasymptotic) Fréchet differentiability, our aim is, in particular, to show that in separable spaces whose modulus of smoothness satisfies

$$\rho_X(t) = o(t^n \log^{n-1}(1/t)) \text{ as } t \searrow 0,$$

every slice of the set of Gâteaux derivatives of any Lipschitz map to a space of dimension not exceeding n contains a Fréchet derivative.

As pointed out after the definition of $\rho_X(t)$ (Definition 4.2.1), spaces whose modulus is controlled in this way exist for $n = 1, 2$ only. We therefore turn our attention to spaces admitting a suitable bump smooth in the direction of a family of subspaces $\mathcal{Y}$ with modulus controlled by $t^n \log^{n-1}(1/t)$; see Definition 8.2.3. The introduction of $\mathcal{Y}$ (as opposed to using only the case when $\mathcal{Y}$ is the family of finite codimensional subspaces of the given space X, as we have done before) allows us to treat, for example, the Hilbert space with $n = 2$ and $\mathcal{Y} = \{X\}$, and so obtain the full differentiability result for $\mathbb{R}^2$-valued maps. In the general case we even find a way of stating our result in a form that includes a mean value estimate. As in Chapter 8, we will always assume that $\mathcal{Y}$ is a nonempty, downward directed family of subspaces of the given Banach space X. Since the general statement, which includes a version of the mean value theorem, is technically complicated we first give a simpler but considerably more special result. To state it, we introduce the definition of "differentiability at infinity" alluded to above.

Definition 15.1.1. Let $\mathcal{Y}$ be a nonempty downward directed family of subspaces of X. A function $f\colon X \longrightarrow Z$ is called *asymptotically Fréchet differentiable at a point* $x \in X$ *with respect to* $\mathcal{Y}$ if there is $Q \in L(X, Z)$ such that, for every $\varepsilon > 0$, there are $Y \in \mathcal{Y}$ and $\delta > 0$ such that

$$\|f(x + y) - f(x) - Qy\| \le \varepsilon \|y\|$$

whenever $y \in Y$ and $\|y\| \leq \delta$.

Theorem 15.1.2. *Suppose a Banach space X admits a bump function which is upper Fréchet differentiable, Lipschitz on bounded sets, and asymptotically smooth in the direction of $\mathcal{Y}$ with modulus controlled by $t^n \log^{n-1}(1/t)$. Then every Lipschitz map f from a nonempty open subset G of X to a space of dimension not exceeding n is asymptotically Fréchet differentiable with respect to $\mathcal{Y}$ at some point of G.*

This result clearly follows from statement (i) of the following theorem, which includes a variant of the multidimensional mean value estimate appropriate in our context, and which is the main result of this chapter.

Theorem 15.1.3. *Suppose a Banach space X admits a bump function which is upper Fréchet differentiable, Lipschitz on bounded sets, and asymptotically smooth in the direction of $\mathcal{Y}$ with modulus controlled by $t^n \log^{n-1}(1/t)$. Let f be a Lipschitz map of an open subset G of X to a space V of dimension not exceeding n. Suppose further that $x_0 \in G$, $v_1^*, \ldots, v_n^* \in V^*$ and $z_1, \ldots, z_n \in X$ are such that f is differentiable at x_0 in the direction of the linear span of $z_1, \ldots, z_n$. Then for any $\eta > 0$ there are $x \in G$, $y_1, \ldots, y_n \in X$ and $Q \in L(X, V)$ such that*

(i) *for every $\varepsilon > 0$ there are $\delta > 0$ and $Y \in \mathcal{Y}$ such that*

$$\|f(x+y) - f(x) - Qy\| \leq \varepsilon\|y\|$$

for every y in the linear span of $Y \cup \{y_1, \ldots, y_n\}$ with $\|y\| < \delta$;

(ii) $\|y_i - z_i\| \leq \eta$;

(iii) *f is differentiable at x in the direction of the linear span of $y_1, \ldots, y_n$;*

(iv) $\displaystyle\sum_{i=1}^{n} v_i^*(f'(x; y_i)) > \sum_{i=1}^{n} v_i^*(f'(x_0; z_i)) - \eta.$

As we have already pointed out, statement (i) implies that, at the point x, f is asymptotically Fréchet differentiable with respect to $\mathcal{Y}$. A version of the mean value estimate is obtained in (iv). The complicated approach to the mean value estimate is caused by the fact that in general we are unable to add the original points $z_1, \ldots, z_n$ instead of $y_1, \ldots, y_n$ to Y in (i) or to show Gâteaux differentiability of f at x. If, for example, f happens to be Gâteaux differentiable at x, we easily deduce the "correct" form of the mean value estimate instead of (iv) (ignoring an unimportant change of η): $\sum_{i=1}^{n} v_i^*(f'(x; z_i)) > \sum_{i=1}^{n} v_i^*(f'(x_0; z_i)) - \eta$.

We will not discuss easy corollaries of Theorem 15.1.3 that follow, often in a stronger form, from Theorem 13.1.1. We mention only one, because it was alluded to in the explanation of motivation, although by Theorem 13.1.1 it holds even with asymptotic smoothness instead of uniform smoothness (where, unlike here, it is a nonempty statement also for $n \geq 3$).

Corollary 15.1.4. *Suppose that X is a Banach space whose modulus of smoothness satisfies $\rho_X(t) = o(t^n \log^{n-1}(1/t))$ as $t \searrow 0$. Let f be a Lipschitz map of a nonempty open subset G of X to a space of dimension not exceeding n. Then every w^*-slice of the set of Gâteaux derivatives of f contains a Fréchet derivative.*

Proof. Let S be a weak*-slice of the set of all Gâteaux derivatives of $f \colon X \longrightarrow V$ where $\dim V \leq n$. Choose $x_0 \in G$ such that $f'(x_0) \in S$ and find $v_1^*, \ldots, v_n^* \in V^*$, $z_1, \ldots, z_n \in X$ and $\varepsilon > 0$ such that $f'(x) \in S$ whenever f is Gâteaux differentiable at x and

$$\sum_{j=1}^n v_j^*\big(f'(x; z_j)\big) > \sum_{j=1}^n v_j^*\big(f'(x_0; z_j)\big) - \varepsilon.$$

We let $\mathcal{Y} = \{X\}$ and recall Corollary 8.2.8 to see that the assumptions of Corollary 15.1.4 allow us to use Theorem 15.1.3. We will do so with $\eta > 0$ such that $\eta(1 + \sum_{i=1}^n \|v_i^*\| \operatorname{Lip}(f)) < \varepsilon$. Because of the choice of $\mathcal{Y}$, Theorem 15.1.3 (i) means that f is Fréchet differentiable at x. Moreover, (iv) and (ii) imply

$$\sum_{i=1}^n v_i^*(f'(x; z_i)) > \sum_{i=1}^n v_i^*(f'(x; y_i)) - \sum_{i=1}^n \|v_i^*\| \operatorname{Lip}(f)\|y_i - z_i\|$$

$$> \sum_{i=1}^n v_i^*(f'(x_0; z_i)) - \eta - \sum_{i=1}^n \|v_i^*\| \operatorname{Lip}(f)\|y_i - z_i\|$$

$$> \sum_{i=1}^n v_i^*(f'(x_0; z_i)) - \varepsilon.$$

Hence $f'(x) \in S$ and we are done. □

Before turning our attention to the proof of Theorem 15.1.3, we remark that our arguments do not use any a priori knowledge of existence of points of Gâteaux differentiability, and so we do not need to assume that the space X is separable. Unusually, it is not obvious how the nonseparable case can be proved by the methods of separable reduction described in Section 3.6. Also, this remark may be useful if one wants to replace in our arguments Fréchet differentiability by Gâteaux differentiability. Since in separable spaces we know that real-valued Lipschitz functions are Gâteaux differentiable almost everywhere (in various meanings), such results would be new only in the nonseparable situation.

The rest of this chapter is devoted to the proof of Theorem 15.1.3. It consists of five sections. The main point of the first is to rewrite some integral estimates that we have proved in Section 9.5 in a way suitable for the current application.

The next three sections prove the differentiability statement of Theorem 15.1.3 and a variant of its mean value estimate for a somewhat special class of functions f. In particular, here we will treat only the case $V = \mathbb{R}^n$ and the dual basis v_i^* to the standard basis e_i. The exact requirements are given at the beginning of Section 15.3 where we construct, starting from x_0, a sequence of points x_k converging to the required point x_∞. We also construct a sequence of operators $T_k \in L(\mathbb{R}^n, X)$ converging to

$T_\infty \in L(\mathbb{R}^n, X)$. These operators are related to the statement of Theorem 15.1.3 by $z_i = T_0 e_i$ and $y_i = T_\infty e_i$. Various properties we show in this section include that f is differentiable in direction $T_\infty(\mathbb{R}^n)$ as well as the required form of the mean value estimate.

The proof of (asymptotic) differentiability is divided into two parts. In Section 15.4 we show that at the point x_∞, f behaves regularly in the direction of $T_\infty(\mathbb{R}^n)$. This means that, under the restriction that $z \in T_\infty(\mathbb{R}^n)$, the difference $f(y + z) - f(y)$ is approximated by $f'(x_\infty; z)$ with error $o(\|y\| + \|z\|)$ as $\|y\| + \|z\| \to 0$. It is in this section where we use the full power of the assumption on control of the modulus of smoothness of our bump (which is the only one that depends on n).

In the following Section 15.5 we use the upper differentiability assumption of our theorem and the regularity statement of the previous section to show the linear approximation result required in Theorem 15.1.3 (i).

Finally, in the short Section 15.6 we transform the general problem to the special case treated in the previous sections.

15.2 AUXILIARY AND FINITE DIMENSIONAL LEMMAS

We will need special cases of some results proved in the previous chapters, most important Deformation Lemma 9.3.2 and integral estimates from Section 9.5. From the former we will just need the function $\psi(u) := c\psi_\kappa(u/c)$ whose properties we list in the following lemma. Recall also the function ω_n defined in Definition 9.3.1 (of whose properties we will use mainly that $\omega_n(t) = t^n \log^{n-1}(1/t)$ for small $t > 0$).

Lemma 15.2.1. *For each $0 < \kappa \le 1$ and $c > 0$ there is $\psi \in C^1(\mathbb{R}^n)$ such that*

(i) $\psi(0) = c$, $\psi(u) = 0$ for $|u| \ge e^{1/\kappa} c$, and $|\psi(u)| \le c$ for $u \in \mathbb{R}^n$;

(ii) $|\psi'(u)| \le \kappa$ for all $u \in \mathbb{R}^n$; and

(iii) *if $\theta \colon \mathbb{R}^n \longrightarrow \mathbb{R}$ is bounded, Borel measurable, $\theta(0) = 0$, and $M \subset (0, \kappa]$ is such that $\kappa \in M$ and each point of $(0, \kappa]$ is within $e^{-n/\kappa}$ of M, then*

$$\int_{B(0,e^{1/\kappa}c)} \theta(\psi'(u)) \, d\mathscr{L}^n(u) \le K_n c^n \sup_{t \in M} \sup_{|z| < t} \frac{\theta(z) + \theta(-z)}{\omega_n(t)},$$

where K_n is a constant depending only on n.

Since we will use the integral estimates from Section 9.5 for special domains only, we will agree from now on that $\Omega \subset \mathbb{R}^n$ is either a ball or a right circular cylinder with equal height and radius. We also recall that for $u \in \Omega$ the Lipschitz constant of a function g at $u \in \Omega$ is the least number $\operatorname{Lip}_u(g) = \operatorname{Lip}_{u,\Omega}(g) \in [0, \infty)$ such that for every $v \in \Omega$,

$$|g(v) - g(u)| \le \operatorname{Lip}_{u,\Omega}(g)|v - u|.$$

Since the eccentricity of Ω is at most $\sqrt{5}$, Lemma 9.5.4 with $s = 2$ and its Corollary 9.5.5 immediately imply the following

Lemma 15.2.2. *There is a constant K_n with the following property. Let $g \colon \Omega \longrightarrow \mathbb{R}^n$ be a Lipschitz function. Then*

$$\int_\Omega \left(\mathrm{Lip}_{u,\Omega}(g)\right)^{2n} d\mathscr{L}^n(u) \le 5^{n-1} K_n \left(\mathrm{Lip}(g)\right)^{2(n-1)} \int_\Omega \|g'\|^2 \, d\mathscr{L}^n,$$

and for every $u, v \in \Omega$,

$$|g(v) - g(u)|^{n+1} \le K_n (\mathrm{Lip}(g))^{n-1} |v - u| \int_\Omega \|g'\|^2 \, d\mathscr{L}^n.$$

Throughout the rest of this chapter, we will use K_n to denote a constant for which Lemma 15.2.1 (iii) and both statements of Lemma 15.2.2 hold.

It will be convenient to rewrite the first statement of the previous lemma in a rather complicated technical way which will be suitable for later use. The apparent abundance in the notation is meant to strengthen the similarity with the situation where it will be applied.

Lemma 15.2.3. *Let $g \colon \Omega \longrightarrow \mathbb{R}^n$ be a Lipschitz map, $\vartheta \colon L(\mathbb{R}^n, \mathbb{R}^n) \longrightarrow \mathbb{R}$ a continuous function, $\theta \colon \Omega \to (0, \infty)$ a measurable function, $\tau, A > 0$, and $L \in L(\mathbb{R}^n, \mathbb{R}^n)$ with $\vartheta(L) \ge 0$. Suppose further that*

$$
\begin{aligned}
2\tau + \frac{\vartheta(L)}{A} \fint_\Omega (\theta(u) - A)) \, d\mathscr{L}^n(u) + \fint_\Omega \|g'\|^2 \, d\mathscr{L}^n \\
< \fint_\Omega \left(\vartheta(L + g'(u)) - \vartheta(L)\right) d\mathscr{L}^n(u).
\end{aligned}
\tag{15.1}
$$

If $s > 0$ is such that

$$\int_{\{u \in \Omega : B(u,s) \not\subset \Omega\}} \vartheta(L + g'(u)) \, d\mathscr{L}^n(u) \le \tau \mathscr{L}^n \Omega, \tag{15.2}$$

then there is $w \in \Omega$ such that g is differentiable at w, $B(w, s) \subset \Omega$, and

$$\frac{\vartheta(L + g'(w))}{\theta(w)} - \frac{\vartheta(L)}{A} > \frac{\tau}{\theta(w)} + \frac{\left(\mathrm{Lip}_w(g)\right)^{2n}}{K_n \theta(w) \left(\mathrm{Lip}(g)\right)^{2(n-1)}}.$$

Proof. Let $s > 0$ satisfy the assumptions of the lemma. Denote

$$\Omega_s = \{u \in \Omega \mid B(u, s) \not\subset \Omega\}, \quad D_s = \{u \in \Omega \setminus \Omega_s \mid g'(u) \text{ exists}\},$$

and

$$\zeta(u) = \begin{cases} \vartheta(L + g'(u)) - \dfrac{\vartheta(L)}{A} \theta(u) - \tau & \text{if } u \in D_s, \\ 0 & \text{if } u \in \Omega \setminus D_s. \end{cases}$$

Then, using that D_s and $\Omega \setminus \Omega_s$ differ only by a set of measure zero, (15.2), (15.1), and the first statement of Lemma 15.2.2, we may estimate

$$
\begin{aligned}
\int_\Omega \zeta \, d\mathscr{L}^n &= \int_{\Omega \setminus \Omega_s} \vartheta(L + g'(u)) \, d\mathscr{L}^n(u) - \frac{\vartheta(L)}{A} \int_{\Omega \setminus \Omega_s} \theta \, d\mathscr{L}^n - \tau \mathscr{L}^n(\Omega \setminus \Omega_s) \\
&\geq \int_\Omega \vartheta(L + g'(u)) \, d\mathscr{L}^n(u) - \int_{\Omega_s} \vartheta(L + g'(u)) \, d\mathscr{L}^n(u) \\
&\quad - \frac{\vartheta(L)}{A} \int_\Omega \theta \, d\mathscr{L}^n - \tau \mathscr{L}^n \Omega \\
&\geq \int_\Omega \vartheta(L + g'(u)) \, d\mathscr{L}^n(u) - \frac{\vartheta(L)}{A} \int_\Omega \theta \, d\mathscr{L}^n - 2\tau \mathscr{L}^n \Omega \\
&= \int_\Omega \left(\vartheta(L + g'(u)) - \vartheta(L) \right) d\mathscr{L}^n(u) - \frac{\vartheta(L)}{A} \int_\Omega (\theta(u) - A) \, d\mathscr{L}^n(u) \\
&\quad - 2\tau \mathscr{L}^n \Omega \\
&> \int_\Omega \|g'^2\| \, d\mathscr{L}^n \\
&\geq \frac{1}{K_n \left(\mathrm{Lip}(g) \right)^{2(n-1)}} \int_\Omega \left(\mathrm{Lip}_u(g) \right)^{2n} d\mathscr{L}^n(u).
\end{aligned}
$$

It follows that the inequality

$$
\zeta(w) > \frac{\left(\mathrm{Lip}_w(g) \right)^{2n}}{K_n \left(\mathrm{Lip}(g) \right)^{2(n-1)}}
$$

holds for some $w \in \Omega$. In particular, $\zeta(w) > 0$, which implies $B(w, s) \subset \Omega$ as well as $w \in D_s$. Hence g is differentiable at w. If we substitute for $\zeta(w)$, we obtain

$$
\vartheta\left(L + g'(w) \right) - \frac{\vartheta(L)}{A} \theta(w) - \tau > \frac{\left(\mathrm{Lip}_u g \right)^{2n}}{K_n \left(\mathrm{Lip}\, g \right)^{2(n-1)}},
$$

from which the statement follows by adding τ and dividing by $\theta(w)$. $\qquad\square$

In the applications of Lemma 15.2.3, the function ϑ will have a special form, and the following lemma will be used to estimate the right hand side of (15.1).

Lemma 15.2.4. *Let $\Omega \subset \mathbb{R}^n$ be a bounded open set and $g \colon \Omega \longrightarrow \mathbb{R}^n$ a Lipschitz map. Suppose that $\vartheta(L) = a|L|_{\mathrm{H}}^2 + \langle H, L \rangle_{\mathrm{H}} + b$, where $2 \leq a \leq 3$, $b \in \mathbb{R}$, and $\|H\| \leq 1$. Then for every $L \in L(\mathbb{R}^n, \mathbb{R}^n)$ such that $\|L - \mathrm{Id}\| \leq \frac{1}{2}$ and every $w \in \mathbb{R}^n$,*

$$
\begin{aligned}
\fint_\Omega &\left(\vartheta(L + g'(u)) - \vartheta(L) \right) d\mathscr{L}^n(u) \\
&\geq |w|^2 + 2 \fint_\Omega \|g'\|^2 \, d\mathscr{L}^n - 10\sqrt{n} \left| \fint_\Omega (g'(u) - w \otimes w) \, d\mathscr{L}^n(u) \right|_{\mathrm{H}}.
\end{aligned}
$$

Proof. Recall that $w \otimes w$ is identified with an operator from $L(\mathbb{R}^n, \mathbb{R}^n)$ such that $\langle H_0, w \otimes w \rangle_{\mathrm{H}} = \langle H_0 w, w \rangle$ for any $H_0 \in L(\mathbb{R}^n, \mathbb{R}^n)$ and $w \in \mathbb{R}^n$. Given L and w, we use this with $H_0 = H + 2aL = 2a \, \mathrm{Id} + H + 2a(L - \mathrm{Id})$ to estimate

$$
\begin{aligned}
\langle H_0, w \otimes w \rangle_{\mathrm{H}} &= 2a|w|^2 + \langle Hw, w \rangle + 2a \langle (L - \mathrm{Id})w, w \rangle \\
&\geq 2a|w|^2 - \|H\| \|w\|^2 - 2a\|L - \mathrm{Id}\| \|w\|^2 \\
&= 2a(1 - \|L - \mathrm{Id}\|)|w|^2 - \|H\| \|w\|^2 \geq |w|^2.
\end{aligned}
$$

This estimate yields a lower bound of the expression

$$
\begin{aligned}
\vartheta(L + g'(u)) &- \vartheta(L) \\
&= a|L + g'(u)|_{\mathrm{H}}^2 + \langle H, L + g'(u) \rangle_{\mathrm{H}} - a|L|_{\mathrm{H}}^2 - \langle H, L \rangle_{\mathrm{H}} \\
&= a|g'(u)|_{\mathrm{H}}^2 + \langle H + 2aL, g'(u) \rangle_{\mathrm{H}} \\
&\geq 2\|g'(u)\|^2 + \langle H_0, g'(u) \rangle_{\mathrm{H}} \\
&= 2\|g'(u)\|^2 + \langle H_0, w \otimes w \rangle_{\mathrm{H}} + \langle H_0, g'(u) - w \otimes w \rangle_{\mathrm{H}} \\
&\geq 2\|g'(u)\|^2 + |w|^2 + \langle H_0, g'(u) - w \otimes w \rangle_{\mathrm{H}}.
\end{aligned}
$$

Integrating the above inequality over Ω and using the Cauchy-Schwarz inequality for the Hilbert-Schmidt scalar product we get

$$
\fint_\Omega \left(\vartheta(L + g'(u)) - \vartheta(L) \right) d\mathscr{L}^n(u)
$$

$$
\geq |w|^2 + 2 \fint_\Omega \|g'\|^2 \, d\mathscr{L}^n + \left\langle H_0, \fint_\Omega (g'(u) - w \otimes w) \, d\mathscr{L}^n(u) \right\rangle_{\mathrm{H}}
$$

$$
\geq |w|^2 + 2 \fint_\Omega \|g'\|^2 \, d\mathscr{L}^n - |H_0|_{\mathrm{H}} \left| \fint_\Omega (g'(u) - w \otimes w) \, d\mathscr{L}^n(u) \right|_{\mathrm{H}}.
$$

It suffices to realize that

$$
|H_0|_{\mathrm{H}} \leq \sqrt{n} \|2a \, \mathrm{Id} + H + 2a(L - \mathrm{Id})\| \leq 10\sqrt{n},
$$

and the required inequality is established. □

Our proof will actually construct the required asymptotic derivative only on some subspace of the space X, and we will need the following two simple results from linear algebra to define it on the whole space.

Lemma 15.2.5. *Let $\mathcal{Y}$ be a nonempty family of subspaces of X which is closed under finite intersections, and let V be a finite dimensional subspace of X. Then there are $0 < c < \infty$ and $Y_0 \in \mathcal{Y}$ such that for every $Y \in \mathcal{Y}$, $Y \subset Y_0$, every $x \in Y + V$ can be written as $x = y + v$, where $y \in Y$, $v \in V$, and $\|y\| + \|v\| \leq c\|x\|$.*

Proof. Since every strictly decreasing sequence of subspaces of the finite dimensional space V is finite, there is $Y_0 \in \mathcal{Y}$ such that $Y \cap V = Y_0 \cap V$ for every $Y \in \mathcal{Y}$, $Y \subset Y_0$.

Let $W \subset V$ be a complement of $Y_0 \cap V$ in V. Then $Y_0 \cap W = \{0\}$, which, since W is finite dimensional, shows that $Y_0 + V$ is the direct sum $Y_0 + V = Y_0 \oplus W$. Denote by π the projection of $Y_0 + V$ onto W along Y_0.

Suppose that $Y \in \mathcal{Y}$, $Y \subset Y_0$. Since $Y \cap V = Y_0 \cap V$, we have $Y + V = Y + W$. If $x \in Y + V$, we write $x = y + v$ where $y \in Y$ and $v \in W$. Noticing that $y \in Y_0$ and $Y_0 + V = Y_0 \oplus W$, we infer that $v = \pi x$. Hence

$$\|y\| + \|v\| = \|x - \pi x\| + \|\pi x\| \le (1 + 2\|\pi\|)\|x\|. \qquad \square$$

Lemma 15.2.6. *Let* $\Psi \in L(\mathbb{R}^n, X)^*$. *Then there is a subspace Z of X of codimension at most n such that* $\Psi(T) = 0$ *for every* $T \in L(\mathbb{R}^n, X)$ *with image $T(\mathbb{R}^n)$ contained in Z.*

Proof. It suffices to represent Ψ as $\sum_{i=1}^{n} x_i^* \otimes e_i$, $x_i^* \in X^*$, and let Z be the intersection of the kernels of the x_i^*. $\qquad \square$

15.3   THE ALGORITHM

We assume that X is a Banach space and $\mathcal{Y}$ is a nonempty family of its subspaces that is closed under finite intersections. In view of Lemma 8.2.5 and the transfer in Observation 8.2.4 we further assume that there is a bounded function $\Theta \colon L(\mathbb{R}^n, X) \longrightarrow [0, \infty)$ with the following properties.

(S.1) $\Theta(0) = 0$ and $\inf_{\|x\| > s} \Theta(x) > 0$ for every $s > 0$.

(S.2) For every convergent series $\sum_{k=0}^{\infty} \lambda_k$ of positive numbers and every convergent sequence $(T_k) \subset L(\mathbb{R}^n, X)$, the function

$$\widetilde{\Theta}(T) = \sum_{k=0}^{\infty} \lambda_k \Theta(T - T_k)$$

is Lipschitz, upper Fréchet smooth, and such that

$$\bar{\rho}_{\widetilde{\Theta}, \mathcal{Y}}(T; t) = t^n \log^{n-1}(1/t) \text{ as } t \searrow 0$$

for every $T \in L(\mathbb{R}^n, X)$.

As mentioned in the Introduction, we consider a Lipschitz function $f \colon X \longrightarrow \mathbb{R}^n$ satisfying certain special requirements. To state these requirements, we recall from Definition 9.2.1 the notion of Fréchet derivative $f'(x; T)$ of f in the direction of an operator $T \in L(\mathbb{R}^n, X)$ at a point $x \in X$: it is the derivative of the map $u \mapsto f(x + Tu)$ (from $\mathbb{R}^n$ to $\mathbb{R}^n$) at the point $u = 0$. So $f'(x; T) \in L(\mathbb{R}^n, \mathbb{R}^n)$ and

$$\|f'(x; T)\| \le \operatorname{Lip}(f)\|T\|.$$

We denote by D the set

$$D = \big\{ (x, T) \in X \times L(\mathbb{R}^n, X) \mid f'(x; T) \text{ exists} \big\}.$$

We are now ready to state the special requirement mentioned above: we will assume that we are given $0 < \eta_0 < 1$ and $(x_0, T_0) \in D$ such that

(R) $\|f'(x;T) - \mathrm{Id}\,\| \leq \frac{1}{2}$ whenever $(x,T) \in D$ and $\|T - T_0\| \leq \eta_0$.

Under this assumption, our goal is to give an algorithm recursively constructing a sequence $(x_k, T_k) \in D$ converging to $(x_\infty, T_\infty) \in D$ such that $\|x_\infty - x_0\| \leq \eta_0$, $\|T_\infty - T_0\| \leq \eta_0$, $|f'(x_\infty; T_\infty)|_\mathrm{H} \geq |f'(x_0; T_0)|_\mathrm{H}$, and f satisfies x_∞ the asymptotic differentiability requirement (i) of Theorem 15.1.3 with $y_i = T_\infty(e_i)$. Roughly, the sequence (x_k, T_k) is chosen in such a way that a certain weight of the pairs (x, T) tends to its maximum under suitable constraints. However, the weight as well as the constraints keep changing with changing k.

We first give some easy consequences of the requirement (R).

Corollary 15.3.1. *Let* $T \in L(\mathbb{R}^n, X)$ *be such that* $\|T - T_0\| \leq \eta_0$. *Then*

(i) *if* $(x,T) \in D$ *then* $\frac{1}{2} \leq \|f'(x;T)\| \leq 2$, $\frac{1}{2} \leq |f'(x;T)|_\mathrm{H} \leq 2\sqrt{n}$ *and* $f'(x;T)$ *is a linear isomorphism of* $\mathbb{R}^n$ *to its image with inverse of norm not exceeding 2;*

(ii) T *has rank* n;

(iii) $|f(x + Tu) - f(x)| \leq 2|u|$ *for every* $x \in X$ *and* $u \in \mathbb{R}^n$.

Proof. The first statement is obvious. To see the second, we observe that for some x, $f'(x;T)$ exists. By the first statement, $f'(x;T)$ has rank n, so T must have rank n as well. The last statement is an immediate corollary of the first and of the mean value estimate. Indeed, let $x \in X$ be given. The Lipschitz map $u \mapsto f(x + Tu)$ is differentiable a.e. by Rademacher's theorem; and the mean value estimate yields

$$|f(x + Tu) - f(x)| \leq \sup\{\|f'(z;T)\| \mid (z,T) \in D\}\,|u| \leq 2|u|. \qquad \square$$

We could define all parameters of our construction recursively, but it is perhaps easier to define now those that can be determined based on the data we have in our hands. We begin by choosing decreasing sequences $\eta_j \searrow 0$ and $\zeta_j \searrow 0$, $j \geq 0$, of positive numbers (η_0 has been already fixed) that will serve to estimate

$$\|T_j - T_\infty\| \leq \eta_j \quad \text{and} \quad \|f'(x_j; T_j) - f'(x_\infty; T_\infty)\| \leq \zeta_j,$$

respectively. For the η_j we require

$$\sum_{j=1}^{\infty} \eta_j < \frac{1}{4n}, \tag{15.3}$$

while for the ζ_j we only require the validity of the desired inequality for $j = 0$, which is achieved by choosing $\zeta_0 = 4$.

For each $j \geq 0$ we use (S.1) to find $c_j > 0$ with the property that

$$\Theta(T) \leq c_j \text{ implies } \|T\| \leq \eta_j.$$

Multiplying Θ by a suitable constant, we may assume that $c_0 = 3n\,\mathrm{Lip}(f)^2(\|T_0\|+1)^2$.

Next we define a large enough constant Λ and a small enough positive constant μ that will serve to obtain upper (resp. lower) bounds of certain weight functions Θ_j that we need to define in our algorithm, as well as upper bounds on Lipschitz constants of some functions and norms of some operators that we encounter in the future,

$$\Lambda = 1 + 12n + 2 \sup_{T \in L(\mathbb{R}^n, X)} \Theta(T) \quad \text{and} \quad \mu = \frac{1}{5^{2(n-1)} K_n \Lambda}. \tag{15.4}$$

Recall that K_n is the constant for which the statement of Lemma 15.2.1 and both statements of Lemma 15.2.2 hold.

Our next step is to choose a strictly decreasing sequence $\beta_j \searrow 0$ of positive numbers with $\beta_0 > \beta_1 \geq 6(12n/\mu)^{1/2n}$ that will be used (in a slightly roundabout way) to estimate the speed of approximation of the increment $f(x_j + T_j u) - f(x_j)$ by the value of $f'(x_j; T_j)$.

Finally we choose, starting from $\tau_0 = 12n$, a decreasing sequence $\tau_j \searrow 0$ of positive numbers such that $\tau_j < \frac{1}{2}$ for $j \geq 1$. These constants will control how close is our weight to its supremum at the jth step. The main requirements are that β_j and even β_{j+1} are much bigger than τ_j:

$$\tau_j \leq e^{-j/\beta_j}, \quad \tau_j \leq \mu \left(\frac{\beta_{j+1}}{6} \right)^{2n} \quad \text{for } j \geq 0,$$
$$\tau_j \leq \frac{\zeta_j^2 \eta_j}{\Lambda}, \quad \tau_j \leq \frac{c_j \eta_j}{\Lambda} \quad \text{for } j \geq 1. \tag{15.5}$$

We are now ready to start the description of our algorithm. It will recursively define and/or choose

- positive constants σ_k;

- pairs $(x_k, T_k) \in D$;

- operators $L_k \in L(\mathbb{R}^n, \mathbb{R}^n)$;

- positive constants Δ_k, δ_k;

- functions ϑ_k, Θ_k, and Υ_k defined on $L(\mathbb{R}^n, \mathbb{R}^n)$, $L(\mathbb{R}^n, X)$, and D, respectively;

and, finally,

- nonempty sets $D_k \subset D$ together with functions β_k on D_k.

The functions β_k will give a finer measure of approximation of our function by its derivative than the constants β_k defined earlier. So they are related to but different from them, but this should cause no confusion. For $k \geq 1$, all these parameters will be defined in the above order. For consistency, we will keep this order in the starting case $k = 0$ as well and define all the required objects so that analogues of the future requirements for $k \geq 1$ hold also for $k = 0$, even though some of them will never be used.

We start by letting $\sigma_0 = 1 + 12n$, recalling that the pair $(x_0, T_0) \in D$ has been given to us and denoting $L_0 = f'(x_0; T_0)$. Then we let $\Delta_0 = 1$ and $\delta_0 = \eta_0$ and define the function $\vartheta_0 \colon L(\mathbb{R}^n, \mathbb{R}^n) \longrightarrow \mathbb{R}$ by

$$\vartheta_0(L) = 1 + 3|L|_{\mathrm{H}}^2,$$

the function $\Theta_0 \colon L(\mathbb{R}^n, X) \longrightarrow \mathbb{R}$ by

$$\Theta_0(T) = 1 + \Theta(T - T_0)$$

and the function $\Upsilon_0 \colon D \longrightarrow \mathbb{R}$ by

$$\Upsilon_0(x, T) = \frac{\vartheta_0(f'(x; T))}{\Theta_0(T)}.$$

By Corollary 15.3.1, when $\|T - T_0\| \le \eta_0$, then

$$0 < \Upsilon_0(x, T) \le 1 + 3|f'(x; T)|_{\mathrm{H}}^2 \le 1 + 12n = \sigma_0. \tag{15.6}$$

Finally, we denote

$$D_0 = \left\{ (x, T) \in D \;\middle|\; \|x - x_0\| < \delta_0, \; \|T - T_0\| \le 1, \; \Upsilon_0(x, T) \ge \Upsilon_0(x_0, T_0) \right\}$$

and define the function $\beta_0(x, T) = \beta_0/2$ on D_0.

It is clear that $D_0 \subset D$ is nonempty since $(x_0, T_0) \in D_0$. However, we cannot (and do not) claim that D_0 contains any other points.

The starting choice of parameters has been made so that, in particular, the following simple observations hold.

Lemma 15.3.2. *For every* $(x, T) \in D_0$,

(i) $\|T - T_0\| \le \eta_0$;

(ii) $1 = \sigma_0 - \tau_0 < 1 + \frac{3}{4} \le \Upsilon_0(x_0, T_0) \le \Upsilon_0(x, T) \le \sigma_0$;

(iii) $\|f'(x; T) - L_0\| \le \zeta_0$;

(iv) $|f'(x; T)|_{\mathrm{H}} \ge |L_0|_{\mathrm{H}}$.

Proof. For (i) we use the condition $\Upsilon_0(x, T) \ge \Upsilon_0(x_0, T_0)$, that is,

$$\frac{1 + 3|f'(x; T)|_{\mathrm{H}}^2}{1 + \Theta(T - T_0)} \ge 1 + 3|L_0|_{\mathrm{H}}^2 \ge 1. \tag{15.7}$$

Since $\|T - T_0\| \le 1$, it implies that

$$\Theta(T - T_0) \le 3|f'(x; T)|_{\mathrm{H}}^2 \le 3n \operatorname{Lip}(f)^2 (\|T_0\| + 1)^2 = c_0,$$

which guarantees that $\|T - T_0\| \le \eta_0$.

The first and second inequalities in (ii) follow from the definitions of τ_0 and σ_0. Since $|L_0|_{\mathrm{H}} \ge \frac{1}{2}$ by Corollary 15.3.1, we have $\Upsilon_0(x_0, T_0) = 1 + 3|L_0|_{\mathrm{H}}^2 \ge 1 + \frac{3}{4}$. The next two inequalities in (ii) are immediate directly from definitions, and the last is (15.6).

The statement (iv) follows from the first inequality in (15.7):

$$1 + 3|f'(x; T)|_{\mathrm{H}}^2 \ge \big(1 + \Theta(T - T_0)\big)(1 + 3|L_0|_{\mathrm{H}}^2) \ge 1 + 3|L_0|_{\mathrm{H}}^2.$$

Finally, statement (iii) holds because ζ_0 was chosen large enough. Indeed,

$$\|f'(x; T) - L_0\| \le \|f'(x; T)\| + \|L_0\| \le 4 = \zeta_0. \qquad \square$$

We are now ready to describe the iterative part of our algorithm. Assume that $k \geq 1$ and all our objects have already been defined for $k - 1$. We define, choose, and/or denote:

(A1) $\sigma_k = \sup\{\Upsilon_{k-1}(x, T) \mid (x, T) \in D_{k-1}\}$.

(A2) $(x_k, T_k) \in D_{k-1}$ such that $\Upsilon_{k-1}(x_k, T_k) > \sigma_k - \tau_k$.

Denote $L_k = f'(x_k; T_k)$.

(A3) $\Delta_k < \frac{1}{2}\Delta_{k-1}$ such that $|f(x_k + T_k u) - f(x_k) - L_k u| \leq \frac{1}{2}\beta_k |u|$ for $u \in \mathbb{R}^n$, $|u| \leq \Delta_k$.

(A4) $0 < \delta_k < \dfrac{1}{4} \min\left\{\delta_{k-1} - \|x_k - x_{k-1}\|, \dfrac{\beta_{k-1} - \beta_{k-1}(x_k, T_k)}{\mathrm{Lip}(f)} \Delta_k\right\}$.

(A5) $\vartheta_k(L) = \vartheta_{k-1}(L) - \eta_k |L - L_k|_{\mathrm{H}}^2$, $\Theta_k(T) = \Theta_{k-1}(T) + \eta_k \Theta(T - T_k)$, and

$$\Upsilon_k(x, T) = \frac{\vartheta_k(f'(x; T))}{\Theta_k(T)} \text{ for } (x, T) \in D.$$

(A6) Finally we define the set D_k and the function $\beta_k(x, T)$ on D_k by

$$D_k = \left\{(x, T) \in D_{k-1} \left| \begin{array}{l} \|x - x_k\| < \delta_k, \ \Upsilon_k(x, T) \geq \Upsilon_k(x_k, T_k), \\ \text{there is } \beta = \beta_k(x, T) < \beta_k \text{ such that} \\ \text{if } u \in \mathbb{R}^n, |u| \leq \Delta_k \text{ there is } (y, S) \in D_{k-1} \\ \text{with } |f(x + Su) - f(x) - f'(y; Su)| \leq \beta|u|. \end{array} \right. \right\}$$

This recursive construction is possible since, as the following simple observation shows, $D_{k-1} \neq \emptyset$ and $\Upsilon_{k-1}(x, T)$ is bounded on D_{k-1}, and since $\|x_k - x_{k-1}\| < \delta_{k-1}$ and $\beta_{k-1}(x_k, T_k) < \beta_{k-1}$.

Observation 15.3.3. $(x_k, T_k) \in D_k$ and for every $k \geq 1$ and $(x, T) \in D_k$,

$$1 \leq \sigma_k - \tau_k < \Upsilon_{k-1}(x_k, T_k) = \Upsilon_k(x_k, T_k) \leq \Upsilon_k(x, T) \leq \Upsilon_{k-1}(x, T) \leq \sigma_k.$$

Proof. The only point of the statement $(x_k, T_k) \in D_k$ that may require an explanation is the existence of the function $\beta = \beta_k(x, T) < \beta_k$ such that for every $u \in \mathbb{R}^n$ with $|u| \leq \Delta_k$ there is $(y, S) \in D_{k-1}$ with

$$|f(x_k + Su) - f(x_k) - f'(y; Su)| \leq \beta|u|.$$

However, the requirement (A3) shows that the choice $\beta = \beta_k/2$ and $(y, S) = (x_k, T_k)$ works for any such u.

In the chain of inequalities all except the first one are immediate consequences of the definitions. As for the first one, it suffices to notice that

$$\sigma_k \geq \Upsilon_k(x_k, T_k) \geq \cdots \geq \Upsilon_0(x_0, T_0) \geq 1 + \frac{3}{4}$$

and that $\tau_k < \frac{1}{2}$ for all k. $\square$

Notice that, as in the case $k = 0$, we do not claim that D_k is in any sense big; in fact it can contain the element (x_k, T_k) only, in which case all the subsequent D_i are equal to D_k.

Lemma 15.3.4. *Let $(x, T) \in D_k$ and $0 \le j < k$. Then*

$$\|x - x_j\| \le \delta_j - \delta_{j+1}, \quad \|T - T_j\| \le \eta_j, \text{ and } \quad \|f'(x; T) - L_j\| \le \zeta_j.$$

Consequently, the limits

$$x_\infty := \lim_{j \to \infty} x_j \in X,$$

$$T_\infty := \lim_{j \to \infty} T_j \in L(\mathbb{R}^n, X), \text{ and}$$

$$L_\infty := \lim_{j \to \infty} L_j \in L(\mathbb{R}^n, \mathbb{R}^n)$$

all exist and for each $j \ge 0$ satisfy $\|x_\infty - x_j\| \le \delta_j - \delta_{j+1}$, $\|T_\infty - T_j\| \le \eta_j$, and $\|L_\infty - L_j\| \le \zeta_j$.

Proof. By the requirement (A4) of the iterative construction we see that

$$\|x_{j+1} - x_j\| \le \delta_j - 4\delta_{j+1} \le \delta_j - 2\delta_{j+1}, \quad j \ge 0.$$

Hence

$$\begin{aligned}
\|x - x_j\| &\le \|x - x_k\| + \|x_k - x_{k-1}\| + \cdots + \|x_{j+1} - x_j\| \\
&\le \delta_k + (\delta_{k-1} - 2\delta_k) + \cdots + (\delta_j - 2\delta_{j+1}) \\
&\le \delta_k + (\delta_{k-1} - \delta_k) + \cdots + (\delta_j - \delta_{j+1}) - \delta_{j+1} = \delta_j - \delta_{j+1}.
\end{aligned}$$

We show the estimates $\|T - T_j\| \le \eta_j$ and $\|f'(x; T) - L_j\| \le \zeta_j$ by induction. The case $j = 0$ follows from Lemma 15.3.2. Assume that $j \ge 1$ and the estimates hold for $i = 0, \ldots, j - 1$. Using Observation 15.3.3 and the inequality $\vartheta_{j-1}(f'(x; T)) \le \sigma_j \Theta_{j-1}(T)$, which follows from the definition of σ_j, we get

$$\begin{aligned}
\sigma_j - \tau_j < \Upsilon_j(x, T) &= \frac{\vartheta_{j-1}(f'(x; T)) - \eta_j |f'(x; T) - L_j|_{\mathrm{H}}^2}{\Theta_{j-1}(T) + \eta_j \Theta(T - T_j)} \\
&\le \frac{\sigma_j \Theta_{j-1}(T) - \eta_j |f'(x; T) - L_j|_{\mathrm{H}}^2}{\Theta_{j-1}(T) + \eta_j \Theta(T - T_j)}.
\end{aligned}$$

A rearrangement reveals that

$$(\sigma_j - \tau_j)\eta_j \Theta(T - T_j) + \eta_j |f'(x; T) - L_k|_{\mathrm{H}}^2 \le \tau_j \Theta_{j-1}(T).$$

Since

$$\Theta_{j-1}(T) = 1 + \Theta(T - T_0) + \sum_{i=1}^{j-1} \eta_i \Theta(T - T_i) \le 1 + \left(1 + \sum_{i=1}^{\infty} \eta_i\right) \sup_T \Theta(T) \le \Lambda,$$

we have
$$(\sigma_j - \tau_j)\eta_j\Theta(T - T_j) + \eta_j|f'(x;T) - L_k|_H^2 \le \tau_j\Lambda.$$
Hence $|f'(x;T) - L_j|_H^2 \le \tau_j\Lambda/\eta_j \le \zeta_j^2$ by the choice of τ_j in (15.5). Similarly,

$$\Theta(T - T_j) \le \frac{\tau_j\Lambda}{\eta_j(\sigma_j - \tau_j)} \le \frac{\tau_j\Lambda}{\eta_j} \le c_j,$$

giving $\|T - T_j\| \le \eta_j$ by the condition imposed upon c_j. □

Lemma 15.3.5. *The sequence of functions* $\vartheta_k(L) = 1 + 3|L|_H^2 - \sum_{i=1}^{k}\eta_i|L - L_i|_H^2$ *and* $\Theta_k(T) = 1 + \Theta(T - T_0) + \sum_{i=1}^{k}\eta_i\Theta(T - T_i)$ *converge, as* $k \to \infty$, *to the limits*

$$\vartheta_\infty(L) := 1 + 3|L|_H^2 - \sum_{i=1}^{\infty}\eta_i|L - L_i|_H^2, \quad L \in L(\mathbb{R}^n, \mathbb{R}^n), \text{ and}$$

$$\Theta_\infty(T) := 1 + \Theta(T - T_0) + \sum_{i=1}^{\infty}\eta_i\Theta(T - T_i), \quad T \in L(\mathbb{R}^n, X),$$

respectively. Moreover, $\vartheta_k \searrow \vartheta_\infty$, $\Theta_k \nearrow \Theta_\infty$, *both convergences are uniform on bounded sets, and* $0 \le \vartheta_k \le 1 + 3|L|_H^2$ *and* $1 \le \Theta_k \le \Lambda$ *for all* k.
The function ϑ_∞ *is a positive second degree polynomial,*

$$\vartheta_\infty(L) = a|L|_H^2 + \langle H, L\rangle_H + b, \tag{15.8}$$

where $2 \le a \le 3$, $|H|_H \le 1$, *and* $0 \le b \le 1$.
The function Θ_∞ *is continuous, upper Fréchet asymptotically differentiable with respect to* $\mathcal{Y}$ *and such that* $\bar{\rho}_{\Theta_\infty,\mathcal{Y}}(T;t) = o\big(t^n\log^{n-1}(1/t)\big)$, $t \searrow 0$, *for every* $T \in L(\mathbb{R}^n, X)$.

Proof. Since $\sum_i \eta_i$ converges by (15.3) and $\|L_i\| \le 2$ by Corollary 15.3.1, the series defining ϑ_∞, and so the sequence ϑ_k, converge uniformly on bounded sets. The inequality $\vartheta_k \le 1 + 3|L|_H^2$ is obvious and, using (15.3) and

$$|L_i|_H \le \sqrt{n}\|L_i\| \le 2\sqrt{n},$$

we estimate ϑ_k from below by

$$1 + 3|L|_H^2 - 2\sum_{i=1}^{k}\eta_i(|L|_H^2 + |L_i|_H^2) \ge 1 - 4n\sum_{i=1}^{k}\eta_i + \Big(3 - 2\sum_{i=1}^{k}\eta_i\Big)|L|_H^2 > 0.$$

The expression (15.8) for $\vartheta_\infty(L)$ holds with

$$a = 3 - \sum_{i=1}^{\infty}\eta_i, \quad H = 2\sum_{i=1}^{\infty}\eta_iL_i \quad \text{and} \quad b = 1 - \sum_{i=1}^{\infty}\eta_i|L_i|_H^2.$$

The estimates of a and b follow again from (15.3) and $|L_i|_{\mathrm{H}} \leq 2\sqrt{n}$, and so does

$$|H|_{\mathrm{H}} \leq 2\sum_{i=1}^{\infty} \eta_i |L_i|_{\mathrm{H}} \leq 4\sqrt{n}\sum_{i=1}^{\infty} \eta_i \leq 1.$$

The series defining Θ_{∞} and so the sequence Θ_k converge uniformly since Θ is bounded. In particular, Θ_{∞} is continuous. The inequality $\Theta_k \geq 1$ is obvious and Θ_k is bounded above by $1 + \left(1 + \sum_{i=1}^{\infty} \eta_i\right) \sup_T \Theta(T) \leq \Lambda$. Since $T_k \to T_{\infty}$ by Lemma 15.3.4, the requirement (S.2) gives that Θ_{∞} is bounded, Lipschitz, upper Fréchet smooth, and

$$\bar{\rho}_{\Theta_{\infty},\mathcal{Y}}(T;t) = o\left(t^n \log^{n-1}(1/t)\right), \quad t \searrow 0$$

for every $T \in L(\mathbb{R}^n, X)$. $\square$

Lemma 15.3.6. *Let $(x, T) \in D$ and denote*

$$\Upsilon_{\infty}(x, T) = \frac{\vartheta_{\infty}(f'(x; T))}{\Theta_{\infty}(T)}.$$

Then the sequence $\Upsilon_k(x, T)$ decreases and converges to $\Upsilon_{\infty}(x, T)$, the pair (x_{∞}, T_{∞}) belongs to D, $f'(x_{\infty}; T_{\infty}) = L_{\infty}$, and the sequence $\Upsilon_k(x_k, T_k)$ increases and converges to $\Upsilon_{\infty}(x_{\infty}, T_{\infty})$.

Proof. Obviously $\Upsilon_k(x, T) \searrow \Upsilon_{\infty}(x, T)$. To show that $f'(x_{\infty}; T_{\infty})$ exists and is equal to L_{∞}, let $\varepsilon > 0$ and choose $j \in \mathbb{N}$ such that

$$2\eta_j \operatorname{Lip}(f) + 2\zeta_j + \beta_{j+2} \leq \varepsilon.$$

Consider any $k \geq j + 2$ and $|u| \leq \Delta_{j+2}$. Since $(x_k, T_k) \in D_{j+2}$, we may find $(y, S) \in D_{j+1}$ such that

$$|f(x_k + Su) - f(x_k) - f'(y; Su)| \leq \beta_{j+2}|u|.$$

Moreover, by Lemma 15.3.4,

$$\|f'(y; S) - L_{\infty}\| \leq \|f'(y; S) - L_j\| + \|L_j - L_{\infty}\| \leq 2\zeta_j,$$

and

$$\|S - T_{\infty}\| \leq \|S - T_j\| + \|T_j - T_{\infty}\| \leq 2\eta_j.$$

Hence

$$\begin{aligned}
|f(x_{\infty} + T_{\infty}u) &- f(x_{\infty}) - L_{\infty}u| \\
&\leq \left|f(x_k + Su) - f(x_k) - f(x_{\infty} + T_{\infty}u) + f(x_{\infty})\right| \\
&\quad + \left|f(x_k + Su) - f(x_k) - f'(y; Su)\right| + |f'(y; Su) - L_{\infty}u| \\
&\leq \operatorname{Lip}(f)\left(2\|x_k - x_{\infty}\| + \|S - T_{\infty}\||u|\right) + \beta_{j+2}|u| + 2\zeta_j|u| \\
&\leq \varepsilon|u| + 2\operatorname{Lip}(f)\|x_k - x_{\infty}\|.
\end{aligned}$$

Taking limit for $k \to \infty$, we obtain

$$|f(x_\infty + T_\infty u) - f(x_\infty) - L_\infty u| \le \varepsilon |u|$$

for $|u| \le \Delta_{j+2}$.

To see that the sequence $\Upsilon_k(x_k, T_k)$ is increasing, it suffices to notice that the condition $(x_{k+1}, T_{k+1}) \in D_k$ implies

$$\Upsilon_k(x_k, T_k) \le \Upsilon_k(x_{k+1}, T_{k+1}) = \Upsilon_{k+1}(x_{k+1}, T_{k+1}).$$

The convergence $\Upsilon_k(x_k, T_k) \to \Upsilon_\infty(x_\infty, T_\infty)$ follows, since $f'(x_k; T_k) = L_k$ converge to $L_\infty = f'(x_\infty; T_\infty)$, T_k converge to T_∞, and since ϑ_k and Θ_k are equibounded on bounded sets and converge uniformly on bounded sets to ϑ_∞ and Θ_∞, respectively. □

Our next lemma handles the slight but rather significant problem that β from (A6) depends on $(x, T) \in D_k$. It shows that this dependence disappears if we restrict ourselves to the pairs $(x, T) \in D_{k+1}$. Although this is a purely technical point, it will be crucial in several key stages of our argument, such as in the proof that (x_∞, T_∞) belongs to the all the D_k in Lemma 15.3.8. The reader may notice that some technical points of our algorithm are there only in order to make this statement true (and easy to show).

Lemma 15.3.7. *For each $k \ge 1$ there is $0 < \widetilde{\beta}_k < \beta_k$ such that, for every $u \in \mathbb{R}^n$ with $|u| \le \Delta_k$ and every $(x, T) \in D_{k+1}$, there is $(y, S) \in D_{k-1}$ such that*

$$|f(x + Su) - f(x) - f'(y; Su)| \le \widetilde{\beta}_k |u|.$$

Proof. We show that the statement holds with

$$\widetilde{\beta}_k = \max\left\{\beta_{k+1}, \frac{\beta_k + \beta_k(x_{k+1}, T_{k+1})}{2}\right\}.$$

Recalling that β_k are strictly decreasing, we see that indeed $0 < \widetilde{\beta}_k < \beta_k$.

Let $(x, T) \in D_{k+1}$ and $u \in \mathbb{R}^n$. We consider two cases. If $|u| \le \Delta_{k+1}$, it follows from the requirement (A6) of the construction that there is $(y, S) \in D_k \subset D_{k-1}$ such that

$$|f(x + Su) - f(x) - f'(y; Su)| \le \beta_{k+1}(x, T)|u| \le \beta_{k+1}|u| \le \widetilde{\beta}_k |u|.$$

If, on the other hand, $\Delta_{k+1} < |u| \le \Delta_k$, then applying the requirement (A6) for $(x_{k+1}, T_{k+1}) \in D_k$ there is $(y, S) \in D_{k-1}$ such that

$$|f(x_{k+1} + Su) - f(x_{k+1}) - f'(y; Su)| \le \beta_k(x_{k+1}, T_{k+1})|u|.$$

Since $\|x - x_{k+1}\| \le \delta_{k+1}$, this implies

$$
\begin{aligned}
|f(x + Su) - f(x) - f'(y; Su)| &\le |f(x_{k+1} + Su) - f(x_{k+1}) - f'(y; Su)| \\
&\quad + |f(x + Su) - f(x_{k+1} + Su)| + |f(x) - f(x_{k+1})| \\
&\le \beta_k(x_{k+1}, T_{k+1})|u| + 2\operatorname{Lip}(f)\delta_{k+1} \\
&\le \beta_k(x_{k+1}, T_{k+1})|u| + \frac{\beta_k - \beta_k(x_{k+1}, T_{k+1})}{2}\Delta_{k+1} \\
&\le \widetilde{\beta}_k|u|,
\end{aligned}
$$

where we have used the choice of δ_{k+1} in (A6).   □

Lemma 15.3.8. (x_∞, T_∞) *belongs to* D_k *for each* $k \ge 0$.

Proof. Recall that $\|x_\infty - x_k\| < \delta_k$ and $\|T_\infty - T_k\| \le \eta_k$ by Lemma 15.3.4. Furthermore, Lemma 15.3.6 yields that $(x_\infty, T_\infty) \in D$ and

$$
\Upsilon_k(x_k, T_k) \le \Upsilon_\infty(x_\infty, T_\infty) \le \Upsilon_k(x_\infty, T_\infty).
$$

For $k = 0$ this already gives that $(x_\infty, T_\infty) \in D_0$.

Let $k \ge 1$. To show that $(x_\infty, T_\infty) \in D_k$ we need to verify the last condition in the definition of D_k in (A6). Let $u \in \mathbb{R}^n$, $0 < |u| \le \Delta_k$. Find $j \ge k + 1$ such that

$$
\operatorname{Lip}(f)\|x_j - x_\infty\| \le \frac{\beta_k - \widetilde{\beta}_k}{4}|u|.
$$

Since $(x_j, T_j) \in D_{k+1}$ we use Lemma 15.3.7 to find a pair $(y, S) \in D_{k-1}$ such that

$$
|f(x_j + Su) - f(x_j) - f'(y; Su)| \le \widetilde{\beta}_k|u|.
$$

Hence

$$
\begin{aligned}
|f(x_\infty + Su) - f(x_\infty) - f'(y; Su)| &\le |f(x_j + Su) - f(x_j) - f'(y; Su)| \\
&\quad + |f(x_\infty + Su) - f(x_j + Su)| + |f(x) - f(x_\infty)| \\
&\le \widetilde{\beta}_k|u| + 2\operatorname{Lip}(f)\|x_j - x_\infty\| \\
&\le \left(\widetilde{\beta}_k + \frac{\beta_k - \widetilde{\beta}_k}{2}\right)|u| = \frac{\beta_k + \widetilde{\beta}_k}{2}|u| < \beta_k|u|.
\end{aligned}
$$

This together with the consequences of Lemmas 15.3.4 and 15.3.6 mentioned at the beginning of the proof shows that $(x_\infty, T_\infty) \in D_k$.   □

15.4   REGULARITY OF f AT x_∞

We show that at the point x_∞, the function f behaves regularly in the direction of a suitable subspace $Y \in \mathcal{Y}$.

Lemma 15.4.1. *For every $\varepsilon > 0$ there are $\delta > 0$ and a subspace $Y \in \mathcal{Y}$ such that*

$$\left| f(x_\infty + y + T_\infty v) - f(x_\infty + y) - L_\infty v \right| \leq \varepsilon(\|y\| + |v|)$$

whenever $y \in Y$, $v \in \mathbb{R}^n$ and $\|y\| + |v| \leq \delta$.

Proof. Let $0 < \varepsilon < 1$ be given. We begin by choosing several parameters. Denote

$$\sigma = \frac{\varepsilon^{2n+1}}{2^{3(n+1)}5^{n-1}K_n^2 \Upsilon_\infty(x_\infty, T_\infty)}.$$

Then choose $j \geq 1$ large enough that $\kappa := \frac{\beta_j}{6\operatorname{Lip}(f)}$ has the following properties:

- $\kappa + \eta_0 \leq 1$, $\operatorname{Lip}(f)\kappa < \dfrac{\varepsilon}{2}$, $\dfrac{4\operatorname{Lip}(f)}{\varepsilon} \leq e^{1/\kappa}$;

- $\bar{\rho}_{\Theta_\infty, \mathcal{Y}}(T_\infty; t) < \sigma\omega_n(t)$ for $0 < t < \kappa$;

- $\tau := \dfrac{\varepsilon^{2n+1}}{2^{3(n+1)}5^{n-1}\alpha_n K_n e^{n/\kappa}} = \dfrac{\Upsilon_\infty(x_\infty, T_\infty)K_n\sigma}{\alpha_n e^{n/\kappa}} > \Lambda\tau_j.$

Notice that the last inequality holds for large j since $\tau_j \leq e^{-j/\beta_j}$.

Finally, we choose $0 < s < 1$ such that

$$(1 + 75n^2)(1 - (1-s)^n) \leq \tau,$$

and let

$$\xi = \min\left\{ \frac{\varepsilon}{2}, \frac{\tau}{10n^{3/2}}, \frac{s\beta_j}{12(1+s)} \right\}.$$

Since $f'(x_\infty; T_\infty)$ exists and is equal to L_∞ by Lemma 15.3.6, we can find $\Delta > 0$ such that for every $u \in \mathbb{R}^n$, $|u| \leq \Delta$,

$$\left| f(x_\infty + T_\infty u) - f(x_\infty) - L_\infty u \right| \leq \xi|u|. \tag{15.9}$$

Let $M \subset (0, \kappa]$ be a finite set such that $\kappa \in M$ and every point of $(0, \kappa]$ is within $e^{-n/\kappa}$ of an element of M. Using that M is finite, $\mathcal{Y}$ is closed under finite intersections and $\bar{\rho}_{\Theta_\infty, \mathcal{Y}}(T_\infty, t) < \sigma\omega_n(t)$ for $t \in M$, we find $Y \in \mathcal{Y}$ such that

$$\rho_{\Theta_\infty, Y}(T_\infty, t) < \sigma\omega_n(t) \quad \text{for all } t \in M. \tag{15.10}$$

Let $\delta > 0$ be smaller than each of the following numbers:

$$\frac{e^{-1/\kappa}\Delta}{2}, \quad \frac{\delta_{j+2}}{1 + e^{1/\kappa}\|T_\infty\|}, \quad \frac{\Delta \min_{1 \leq i \leq j}(\beta_i - \widetilde{\beta}_i)}{8\operatorname{Lip}(f)(1 + e^{1/\kappa}\|T_\infty\|)}.$$

We claim that the statement holds with Y and δ. To show that this is the right choice, suppose to the contrary that $y \in Y$ and $v \in \mathbb{R}^n$ are such that $\|y\| + |v| \leq \delta$ and

$$\left| f(x_\infty + y + T_\infty v) - f(x_\infty + y) - L_\infty v \right| > \varepsilon(\|y\| + |v|). \tag{15.11}$$

We first observe that $\|y\|$ and $|v|$ are comparable. More precisely, we show that

$$4|v| > \varepsilon\|y\| \quad \text{and} \quad \varepsilon|v| < 4\,\mathrm{Lip}(f)\|y\|. \tag{15.12}$$

To prove the first inequality, it suffices to use Corollary 15.3.1 to estimate the left-hand side of (15.11) by

$$|f(x_\infty + y + T_\infty v) - f(x_\infty + y) - L_\infty v| \le 2|v| + \|L_\infty\||v| \le 4|v|.$$

Estimating the left-hand side of (15.11) once more, now with the help of (15.9) and the choice of ξ, we obtain

$$\begin{aligned}
|f(x_\infty + y + T_\infty v) &- f(x_\infty + y) - L_\infty v| \\
&\le |f(x_\infty + T_\infty v) - f(x_\infty) - L_\infty v| + 2\,\mathrm{Lip}(f)\|y\| \\
&\le \xi|v| + 2\,\mathrm{Lip}(f)\|y\| \le \frac{\varepsilon}{2}|v| + 2\,\mathrm{Lip}(f)\|y\|,
\end{aligned}$$

implying the second inequality from (15.12).

Define now $\gamma \in C^1(\mathbb{R}^n, X)$ by

$$\gamma(u) = \psi(u)\,\frac{y}{\|y\|},$$

where ψ is the function from Lemma 15.2.1 with $c = \|y\|$. Then γ is Lipschitz and

$$\gamma'(u) = \frac{y}{\|y\|} \otimes \psi'(u),$$

where $\psi'(u)$ is considered as an element of $\mathbb{R}^n$. Directly from Lemma 15.2.1 we see that γ has the following properties; the last property follows from 15.2.1 (iii) with

$$\theta(e) = \Theta_\infty\left(T_\infty + \frac{y \otimes e}{\|y\|}\right) - \Theta_\infty(T_\infty).$$

(A) $\gamma(0) = y$.

(B) $\gamma(u) = 0$ whenever $|u| \ge e^{1/\kappa}\|y\|$.

(C) $\|\gamma(u)\| \le \|y\|$ for all $u \in \mathbb{R}^n$.

(D) $\|\gamma'(u)\| \le \kappa$ for almost all $u \in \mathbb{R}^n$.

(E) $\displaystyle\int_{\mathbb{R}^n} \left(\Theta_\infty(T_\infty + \gamma'(u)) - \Theta_\infty(T_\infty)\right) d\mathscr{L}^n(u) \le \sigma K_n\|y\|^n$.

Let $r = e^{1/\kappa}\|y\|$, $\Omega = B(0, r)$ and define $g\colon \Omega \longrightarrow \mathbb{R}^n$ by

$$g(u) = f(x_\infty + T_\infty u + \gamma(u)) - f(x_\infty) - L_\infty u.$$

Then, using Corollary 15.3.1,

$$
\begin{aligned}
|g(u) - g(v)| &\leq \left| f(x_\infty + T_\infty u + \gamma(u)) - f(x_\infty + T_\infty v + \gamma(v)) \right| \\
&\quad + \|L_\infty\| |u - v| \\
&\leq \left| f(x_\infty + T_\infty u + \gamma(u)) - f\left(x_\infty + T_\infty u + T_\infty(v - u) + \gamma(u)\right) \right| \\
&\quad + \left| f(x_\infty + T_\infty v + \gamma(u)) - f(x_\infty + T_\infty v + \gamma(v)) \right| \\
&\quad + 2|u - v| \\
&\leq 2|u - v| + \mathrm{Lip}(f)\kappa |u - v| + 2|u - v|.
\end{aligned}
$$

So $\mathrm{Lip}(g) \leq 4 + \mathrm{Lip}(f)\kappa \leq 5$. Moreover,

$$
\begin{aligned}
|g(v) - g(0)| &= \left| f(x_\infty + T_\infty v + \gamma(v)) - L_\infty v - f(x_\infty + y) \right| \\
&\geq |f(x_\infty + y + T_\infty v) - f(x_\infty + y) - L_\infty v| - \mathrm{Lip}(f)\|\gamma(v) - y\| \\
&> \varepsilon(\|y\| + |v|) - \mathrm{Lip}(f)\|\gamma(v) - \gamma(0)\| \\
&> \varepsilon|v| - \mathrm{Lip}(f)\kappa|v| \geq \varepsilon|v| - \frac{\varepsilon}{2}|v| = \frac{\varepsilon}{2}|v|.
\end{aligned}
$$

Hence, recalling from (15.12) that $|v| < 4\,\mathrm{Lip}(f)\|y\|/\varepsilon \leq e^{1/\kappa}\|y\| = r$, we infer from the second statement of Lemma 15.2.2 that

$$
\left(\frac{\varepsilon}{2}|v|\right)^{n+1} \leq |g(v) - g(0)|^{n+1} \leq 5^{n-1}K_n|v| \int_\Omega \|g'\|^2 \, d\mathscr{L}^n.
$$

This implies

$$
\begin{aligned}
\fint_\Omega \|g'\|^2 \, d\mathscr{L}^n &\geq \frac{\varepsilon^{n+1}|v|^n}{2^{n+1}5^{n-1}K_n\alpha_n r^n} > \frac{\varepsilon^{n+1}\|y\|^n\varepsilon^n/4^n}{2^{n+1}5^{n-1}K_n\alpha_n\|y\|^n e^{n/\kappa}} \\
&= \frac{\varepsilon^{2n+1}}{2^{3n+1}5^{n-1}K_n\alpha_n e^{n/\kappa}} = 4\tau.
\end{aligned}
\tag{15.13}
$$

We intend to use Lemma 15.2.3 with Ω, τ, and g defined above, the functions $\vartheta = \vartheta_\infty$ and $\theta(u) = \Theta_\infty(T_\infty + \gamma'(u))$, $A = \Theta_\infty(T_\infty)$, and $L = L_\infty$. First we check its assumption (15.1), that is,

$$
\begin{aligned}
2\tau + \Upsilon_\infty(x_\infty, T_\infty)\fint_\Omega \left(\Theta_\infty(T_\infty + \gamma'(u)) - \Theta_\infty(T_\infty)\right) d\mathscr{L}^n(u) + \fint_\Omega \|g'\|^2 \, d\mathscr{L}^n \\
< \fint_\Omega \left(\vartheta_\infty(L_\infty + g'(u)) - \vartheta_\infty(L_\infty)\right) d\mathscr{L}^n(u). \quad (15.14)
\end{aligned}
$$

To estimate of the right-hand side of inequality (15.14) we employ Lemma 15.2.4. Note that Lemma 15.3.5 guarantees that parameters a and H in the representation of ϑ_∞ match the conditions needed in this Lemma. Also, the assumption (R) gives $\|L_\infty - \mathrm{Id}\,\| = \|f'(x_\infty; T_\infty) - \mathrm{Id}\,\| \leq \frac{1}{2}$. Hence Lemma 15.2.4 with $w = 0$ implies

$$
\begin{aligned}
\fint_\Omega (\vartheta_\infty(L_\infty + g'(u)) - \vartheta_\infty(L_\infty))\, d\mathscr{L}^n(u) \\
\geq 2\fint_\Omega \|g'\|^2 d\mathscr{L}^n - 10\sqrt{n}\left|\fint_\Omega g'(u)\, d\mathscr{L}^n(u)\right|_{\mathrm{H}}.
\end{aligned}
$$

To estimate the last integral, we infer from $r \leq e^{1/\kappa}\delta \leq \Delta$ and (15.9) that for every $u \in \partial\Omega$,

$$|g(u)| \leq |f(x_\infty + T_\infty u) - f(x_\infty) - L_\infty u| \leq \xi|u| = \xi r.$$

Using Corollary 9.4.2 we conclude that

$$\left| \fint_\Omega g' \, d\mathscr{L}^n \right|_H \leq \frac{\xi r}{\alpha_n r^n} n\alpha_n r^{n-1} = \xi n \leq \frac{\tau}{10\sqrt{n}}.$$

Hence

$$\fint_\Omega (\vartheta_\infty(L_\infty + g'(u)) - \vartheta_\infty(L_\infty)) \, d\mathscr{L}^n(u) \geq 2\fint_\Omega \|g'\|^2 d\mathscr{L}^n - \tau. \qquad (15.15)$$

The left-hand side of (15.14) can be estimated as follows. Notice that using (E) and the definition of σ, we have

$$\Upsilon_\infty(x_\infty, T_\infty) \fint_{B(0,r)} \left(\Theta_\infty(T_\infty + \gamma'(u)) - \Theta_\infty(T_\infty) \right) d\mathscr{L}^n(u)$$

$$\leq \frac{\Upsilon_\infty(x_\infty, T_\infty) K_n \sigma \|y\|^n}{\alpha_n r^n} = \frac{\Upsilon_\infty(x_\infty, T_\infty) K_n \sigma e^{-n/\kappa}}{\alpha_n} = \tau.$$

This inequality and (15.13) show that the left-hand side of (15.14) is at most

$$2\tau + \Upsilon_\infty(x_\infty, T_\infty) \fint_\Omega \left(\Theta_\infty(T_\infty + \gamma'(u)) - \Theta_\infty(T_\infty) \right) d\mathscr{L}^n(u)$$

$$+ \fint_\Omega \|g'\|^2 \, d\mathscr{L}^n$$

$$\leq 3\tau + \fint_\Omega \|g'\|^2 \, d\mathscr{L}^n < 2\fint_\Omega \|g'\|^2 d\mathscr{L}^n - \tau,$$

which is the same as the lower estimate of its right-hand side in (15.15).

Noting also that by Lemma 15.3.5,

$$\vartheta_\infty(g'(u)) \leq 1 + 3|g'(u)|_H^2 \leq 1 + 3n^2 \operatorname{Lip}^2(g) \leq 1 + 75n^2,$$

we infer from the choice of the value of s that

$$\int_{\{u \in \Omega | B(u, sr) \not\subset \Omega\}} \vartheta_\infty(g'(u)) \, d\mathscr{L}^n(u) \leq (1 + 75n^2)\alpha_n \left(r^n - r^n(1-s)^n\right)$$

$$\leq \tau\alpha_n r^n = \tau\mathscr{L}^n\Omega.$$

Having verified its assumptions, we may finally use Lemma 15.2.3 to find a point $w \in \Omega$ such that $B(w, sr) \subset \Omega$, g is differentiable at w and

$$\frac{\vartheta_\infty(L_\infty + g'(w))}{\Theta_\infty(T_\infty + \gamma'(w))} - \Upsilon_\infty(x_\infty, T_\infty)$$

$$> \frac{\tau}{\Theta_\infty(T_\infty + \gamma'(w))} + \frac{(\operatorname{Lip}_w(g))^{2n}}{K_n\Theta_\infty(T_\infty + \gamma'(w))(\operatorname{Lip}(g))^{2(n-1)}}. \qquad (15.16)$$

We denote

$$x = x_\infty + T_\infty w + \gamma(w),$$
$$T = T_\infty + \gamma'(w), \text{ and}$$
$$L = L_\infty + g'(w).$$

Then $(x, T) \in D$ and $f'(x; T) = L$. Indeed, putting

$$\tilde{\gamma}(u) = \gamma(w + u) - \gamma(w) - \gamma'(w)(u),$$

we have, for $|w + u| \leq r$, the identity

$$f(x + Tu) - f(x) - Lu = \big(g(w + u) - g(w) - g'(w)(u)\big) \\ + \big(f(x + Tu) - f(x + Tu + \tilde{\gamma}(u))\big). \tag{15.17}$$

The first summand is obviously $o(u)$, $u \to 0$, while the second summand is dominated by $\text{Lip}(f)\|\tilde{\gamma}(u)\| = o(u)$, $u \to 0$.

Substituting

$$\frac{\vartheta_\infty(L_\infty + g'(w))}{\Theta_\infty(T_\infty + \gamma'(w))} = \Upsilon_\infty(x, T)$$

into (15.16) and using $\Theta_\infty(T) \leq \Lambda$ and $\text{Lip}(g) \leq 5$ leads to

$$\Upsilon_\infty(x, T) - \Upsilon_\infty(x_\infty, T_\infty) > \frac{\tau}{\Lambda} + \frac{(\text{Lip}_w(g))^{2n}}{5^{2(n-1)} K_n \Lambda}.$$

The choice of μ in (15.4) and the inequality $\tau > \Lambda \tau_j$ allows us to rewrite this estimate in the form

$$\Upsilon_\infty(x, T) - \Upsilon_\infty(x_\infty, T_\infty) > \tau_j + \mu(\text{Lip}_w(g))^{2n}. \tag{15.18}$$

In particular, we see that the left-hand side of (15.18) is positive which, together with Lemma 15.3.6, shows that for each $i \geq 0$,

$$\Upsilon_i(x, T) \geq \Upsilon_\infty(x, T) \geq \Upsilon_\infty(x_\infty, T_\infty) \geq \Upsilon_i(x_i, T_i). \tag{15.19}$$

Our aim is to show by induction that $(x, T) \in D_i$ for each $0 \leq i \leq j$. For this we need to estimate $\|x - x_i\|$. To start, recall that $x = x_\infty + T_\infty w + \gamma(w)$, where $w \in B(0, r)$. In view of (C),

$$\|x - x_\infty\| = \|T_\infty w + \gamma(w)\| \leq \|T_\infty\| r + \|y\|$$
$$= \|y\|\left(1 + e^{1/\kappa}\|T_\infty\|\right) \leq \delta\left(1 + e^{1/\kappa}\|T_\infty\|\right). \tag{15.20}$$

Since the choice of δ gives that the right side is $\leq \delta_{j+2}$, we see that for $i \leq j$,

$$\|x - x_i\| \leq \|x - x_\infty\| + \|x_i - x_\infty\| \leq \delta_{j+2} + (\delta_i - \delta_{i+1}) < \delta_i.$$

Also

$$\|T - T_0\| \leq \|T_\infty - T_0\| + \|\gamma'(w)\| \leq \eta_0 + \kappa \leq 1.$$

Hence $(x, T) \in D_0$.

To continue the induction, suppose $1 \leq i \leq j$ and $(x, T) \in D_{i-1}$. In particular, since both (x, T) and (x_{i-1}, T_{i-1}) belong to D_{i-1} we infer from Lemma 15.3.2 (ii) in case $i = 1$ that

$$\Upsilon_0(x_0, T_0) \geq \Upsilon_0(x, T) - \tau_0.$$

For $i > 1$ we use that σ_i decrease and (A2) of the algorithm to get

$$\Upsilon_{i-1}(x_{i-1}, T_{i-1}) = \Upsilon_{i-2}(x_{i-1}, T_{i-1}) > \sigma_{i-1} - \tau_{i-1}$$
$$\geq \sigma_i - \tau_{i-1} \geq \Upsilon_{i-1}(x, T) - \tau_{i-1}.$$

Together with (15.19) and (15.18) this implies

$$\tau_{i-1} \geq \Upsilon_\infty(x, T) - \Upsilon_\infty(x_\infty, T_\infty) \geq \mu(\text{Lip}_w(g))^{2n}.$$

Recalling the condition $\tau_{i-1} \leq \mu(\beta_i/6)^{2n}$ from (15.5), we infer that for every u such that $w + u \in \Omega$,

$$|g(w + u) - g(w)| \leq \text{Lip}_w(g)|u| \leq \left(\frac{\tau_{i-1}}{\mu}\right)^{1/2n} |u| \leq \frac{\beta_i}{6}|u|. \tag{15.21}$$

Since we already know that $\|x - x_i\| < \delta_i$ and $\Upsilon_i(x, T) \geq \Upsilon_i(x_i, T_i)$, the proof of $(x, T) \in D_i$ will be finished by finding, for each $|u| \leq \Delta_i$, a suitable $(y, S) \in D_{i-1}$ for which we can estimate $|f(x + Su) - f(x) - f'(y; S)(u)|$. To do this, we consider the following three cases.

Case 1. $|w + u| \leq r$. Here we choose $(y, S) = (x_\infty, T_\infty)$ and write

$$f(x + T_\infty u) - f(x) - L_\infty u$$
$$= \big(g(w + u) - g(w)\big)$$
$$- \big(f\big(x + T_\infty u + \gamma(w + u) - \gamma(w)\big) - f(x + T_\infty u)\big).$$

Using that $w + u \in \Omega$ we estimate the first term by (15.21). Since $\text{Lip}(\gamma) \leq \kappa$, the second term is dominated by $\text{Lip}(f)\kappa|u|$. So we get that

$$|f(x + T_\infty u) - f(x) - L_\infty u| \leq \frac{\beta_i}{6}|u| + \text{Lip}(f)\kappa|u| = \frac{\beta_i}{3}|u|. \tag{15.22}$$

Case 2. $r < |w + u| \leq \Delta$. Our choice of (y, S) is again (x_∞, T_∞). Recall that $B(w, sr) \subset \Omega = B(0, r)$, so $|w| \leq (1 - s)r$. We find a positive multiple of u, say $\tilde{u}$, such that $|w + \tilde{u}| = r$ and $|u| \geq |\tilde{u}| \geq sr \geq s|w|$. Using that $\gamma(w + \tilde{u}) = \gamma(w + u) = 0$, we write

$$f(x + T_\infty u) - f(x) - L_\infty u$$
$$= \big(g(w + \tilde{u}) - g(w)\big)$$
$$+ \Big(f\big(x_\infty + T_\infty(w + u)\big) - f\big(x_\infty + T_\infty(w + \tilde{u})\big) - L_\infty(u - \tilde{u})\Big)$$
$$+ \Big(f\big(x_\infty + T_\infty(w + u) + \gamma(w)\big) - f\big(x_\infty + T_\infty(w + u)\big)\Big).$$

We estimate all three summands above. The norm of the first term is dominated by $\frac{1}{6}\beta_i|\widetilde{u}| \leq \frac{1}{6}\beta_i|u|$ according to (15.21). Using $|w+\widetilde{u}| = r \leq \Delta$ and (15.9), we estimate the second term by

$$
\begin{aligned}
&\left|f\left(x_\infty + T_\infty(w+u)\right) - f(x_\infty) - L_\infty(w+u)\right| \\
&\quad + \left|f\left(x_\infty + T_\infty(w+\widetilde{u})\right) - f(x_\infty) - L_\infty(w+\widetilde{u})\right| \\
&\quad \leq \xi\left(|u+w| + |\widetilde{u}+w|\right) \leq \xi(2|w| + |u| + |\widetilde{u}|) \\
&\quad \leq 2\xi\left(1 + \frac{1}{s}\right)|u| \leq \frac{\beta_i}{6}|u|.
\end{aligned}
$$

Finally, the third term is estimated by

$$
\mathrm{Lip}(f)\|\gamma(w)\| = \mathrm{Lip}(f)\|\gamma(w) - \gamma(w+\widetilde{u})\| \leq \mathrm{Lip}(f)\kappa|\widetilde{u}| \leq \frac{\beta_i}{6}|u|.
$$

Hence

$$
|f(x + T_\infty u) - f(x) - L_\infty u| \leq \frac{\beta_i}{2}|u|. \tag{15.23}
$$

Case 3. $\Delta < |w+u|$. Since $|u| \leq \Delta_i$, Lemma 15.3.7 applied to $(x_\infty, T_\infty) \in D_{i+1}$, gives us $(y, S) \in D_{i-1}$ such that

$$
|f(x_\infty + Su) - f(x_\infty) - f'(y; Su)| \leq \widetilde{\beta}_i|u|.
$$

This choice of (y, S) works also in our remaining case. Indeed, we first estimate

$$
\begin{aligned}
&|f(x + Su) - f(x) - f'(y; Su)| \\
&\quad \leq |f(x_\infty + Su) - f(x_\infty) - f'(y; Su)| \\
&\qquad + |f(x + Su) - f(x_\infty + Su)| + |f(x) - f(x_\infty)| \\
&\quad \leq \widetilde{\beta}_i|u| + 2\,\mathrm{Lip}(f)\|x_\infty - x\|. \tag{15.24}
\end{aligned}
$$

By the choice of δ we have $|w| \leq r \leq e^{1/\kappa}\delta \leq \Delta/2$. Hence $\Delta \leq 2|u|$. Applying the estimate (15.20) and using the choice of δ once more, we see that

$$
\begin{aligned}
2\,\mathrm{Lip}(f)\|x_\infty - x\| &\leq 2\,\mathrm{Lip}(f)\,\delta\left(1 + e^{1/\kappa}\|T_\infty\|\right) \\
&\leq \frac{\beta_i - \widetilde{\beta}_i}{4}\Delta \leq \frac{\beta_i - \widetilde{\beta}_i}{2}|u|.
\end{aligned}
$$

Substituting this inequality into (15.24), we obtain

$$
|f(x + Su) - f(x) - f'(y; Su)| \leq \widetilde{\beta}_i|u| + \frac{\beta_i - \widetilde{\beta}_i}{2}|u| = \frac{\beta_i + \widetilde{\beta}_i}{2}|u|.
$$

In all three cases we managed to find the required pair (y, S), thus showing that $(x, T) \in D_i$. Hence, by induction, we conclude that $(x, T) \in D_j$. But this, (15.19), and (15.18) imply that

$$
\tau_j \geq \Upsilon_j(x, T) - \Upsilon_j(x_j, T_j) \geq \Upsilon_\infty(x, T) - \Upsilon_\infty(x_\infty, T_\infty) > \tau_j,
$$

and we have found our desired contradiction. $\qquad\square$

15.5 LINEAR APPROXIMATION OF f AT x_∞

Once we know that f behaves regularly at x_∞ (in the sense of Lemma 15.4.1), we use (upper) Fréchet differentiability of Θ_∞ at T_∞ to finish the proof. We first list some simple facts following from the assumptions made at the beginning of Section 15.3.

Lemma 15.5.1. (i) T_∞ *is a linear isomorphism of* $\mathbb{R}^n$ *onto* $T_\infty(\mathbb{R}^n)$.

(ii) *There are* $0 < c < \infty$ *and* $Y_0 \in \mathcal{Y}$ *such that for every* $Y \subset Y_0$, $Y \in \mathcal{Y}$, *every* $z \in Y + T_\infty(\mathbb{R}^n)$ *can be written as* $z = x + T_\infty u$, *where* $x \in Y$, $u \in \mathbb{R}^n$ *and* $\|x\| + |u| \le c\|z\|$.

Proof. Part (i) was proved in Corollary 15.3.1 and part (ii) follows from it and from Lemma 15.2.5. $\square$

Let $\Psi \in L(\mathbb{R}^n, X)^*$ be an upper Fréchet derivative of $\Theta_\infty(T)$ at T_∞. Let $Z \subset X$ be a subspace of codimension at most n such that $\Psi(T) = 0$ for every $T \in L(\mathbb{R}^n, X)$ with image contained in Z; see Lemma 15.2.6.

The following lemma gives the required differentiability result in the direction of Z, which we will show later to be the kernel of the asymptotic derivative we are looking for. This is the last of the main steps in the proof of the results of this chapter. The rest of the argument will consist of a much simpler use of Lemma 15.4.1 to extend differentiability to the remaining directions (Proposition 15.5.4) and, in the following section, of a simple modification of the original function to satisfy the requirements introduced at the beginning of Section 15.3.

Although there are significant similarities between the proofs of Lemmas 15.5.2 and 15.4.1, there are also substantial differences, the main ones stemming from the use of the case $w \ne 0$ of Lemma 15.2.3.

Lemma 15.5.2. *For every* $\varepsilon > 0$ *there are* $\delta > 0$ *and a subspace* $Y \in \mathcal{Y}$ *such that for every* $z \in (Y + T_\infty(\mathbb{R}^n)) \cap Z$ *with* $\|z\| < \delta$,

$$|f(x_\infty + z) - f(x_\infty)| \le \varepsilon\|z\|.$$

Proof. We may assume that $\varepsilon \le 1$ and find $c \ge 1$ and $Y_0 \in \mathcal{Y}$ as in Lemma 15.5.1 (ii). Since Ψ is an upper Fréchet derivative of Θ_∞ at T_∞, there is $0 < \eta \le 1 - \eta_0$ such that $\mathrm{Lip}(f)\eta \le 1$ and for every T with $\|T - T_\infty\| \le \eta$

$$\Theta_\infty(T) - \Theta_\infty(T_\infty) - \Psi(T - T_\infty) \le \varepsilon\,\frac{\|T - T_\infty\|}{4\Lambda}.$$

Let $\tau = \varepsilon\eta/4$. We choose $j \in \mathbb{N}$ such that $\tau_j < \tau/\Lambda$, and $0 < s < 1$ satisfying $(1 + 75n^2)\big(1 - (1 - s)^{n-1}(1 - 2s)\big) \le \tau$.

Recalling further that c denotes the constant from Lemma 15.5.1 (ii), let

$$\xi = \min\left\{\frac{\tau}{20(c + 1)(n + 1)\sqrt{n}}, \frac{s\beta_j}{16c(1 + \|T_\infty\|)}\right\}.$$

We apply Lemma 15.4.1 with $\varepsilon = \xi$ to find $Y \subset Y_0$ and $\Delta > 0$ such that

$$|f(x_\infty + y + T_\infty u) - f(x_\infty + y) - L_\infty u| \leq \xi(\|y\| + |u|) \tag{15.25}$$

whenever $y \in Y$, $u \in \mathbb{R}^n$ and $\|y\| + |u| \leq 2\Delta$.

We show that the statement of the lemma holds with this Y and any $\delta > 0$ smaller than

$$\frac{4\tau\Delta}{2c\tau + \varepsilon}, \quad \frac{2\tau\delta_{j+2}}{\varepsilon(1 + \|T_\infty\|)}, \quad \frac{4\tau\Delta}{2c\varepsilon(1 + \|T_\infty\|)}, \quad \frac{\tau\Delta \min_{1 \leq i \leq j}(\beta_i - \widetilde{\beta}_i)}{2\varepsilon(1 + \|T_\infty\|)\,\mathrm{Lip}(f)}.$$

For that, suppose to the contrary that $z \in (Y + T_\infty(\mathbb{R}^n)) \cap Z$ is such that $\|z\| < \delta$ and still

$$|f(x_\infty + z) - f(x_\infty)| > \varepsilon\|z\|.$$

Let $w = f(x_\infty + z) - f(x_\infty)$, so $\varepsilon\|z\| < |w|$. Also put $\widehat{w} = w/|w|$ and $a = \varepsilon\|z\|/(4\tau)$. Notice for future use that $\|z\| = 4\tau a/\varepsilon \leq a$. Define

$$\Omega = \{u + r\widehat{w} \in \mathbb{R}^n \mid 0 < r < a, \langle u, w \rangle = 0, |u| < a\},$$

$$Tu = T_\infty u + \frac{1}{a}\langle u, \widehat{w}\rangle z, \quad \text{and}$$

$$g(u) = f(x_\infty + Tu) - f(x_\infty) - L_\infty u.$$

Then

$$\|T - T_\infty\| = \frac{\|z\|}{a} = \frac{4\tau}{\varepsilon} = \eta,$$

and, writing the function $g \colon \Omega \longrightarrow \mathbb{R}^n$ for a moment in the form

$$g(u) = \big(f(x + Tu) - f(x + T_\infty u)\big) + \big(f(x + T_\infty u) - f(x_\infty)\big) - L_\infty u$$

we obtain that g is a Lipschitz map with

$$\mathrm{Lip}(g) \leq \mathrm{Lip}(f)\frac{\|z\|}{a} + 2 + 2 = 4 + \mathrm{Lip}(f)\,\eta \leq 5.$$

We intend to use Lemma 15.2.3 with Ω, g, and τ defined above, $\theta(u) = \Theta_\infty(T)$, $\vartheta = \vartheta_\infty$, $A = \Theta_\infty(T_\infty)$, and $L = L_\infty$. For this we have to check assumption (15.1) in Lemma 15.2.3, that is,

$$2\tau + \Upsilon_\infty(x_\infty, T_\infty)\fint_\Omega (\Theta_\infty(T) - \Theta_\infty(T_\infty))\,d\mathscr{L}^n + \fint_\Omega \|g'\|^2\,d\mathscr{L}^n$$
$$< \fint_\Omega (\vartheta_\infty(L_\infty + g'(u)) - \vartheta_\infty(L_\infty))\,d\mathscr{L}^n(u). \tag{15.26}$$

The right-hand side will be estimated with the help of Lemma 15.2.4 with w replaced by

$$\widetilde{w} = \widehat{w}\sqrt{\frac{|w|}{a}}.$$

The function $\vartheta = \vartheta_\infty$ satisfies the assumptions required in Lemma 15.2.4 due to Lemma 15.3.5. To apply Lemma 15.2.4, we also need to know the size of

$$\int_\Omega (g'(u) - \widetilde{w} \otimes \widetilde{w}) \, d\mathscr{L}^n(u).$$

To estimate the norm of this integral we use Divergence Theorem 9.4.1. Let ν_Ω denote the outer unit normal vector to $\partial\Omega$. The boundary of Ω is the union of two disks,

$$\Omega_0 := \{u \mid \langle u, \widehat{w} \rangle = 0, |u| \le a\} \quad \text{and} \quad \Omega_a := \{u \mid \langle u, \widehat{w} \rangle = a, |u| \le a\sqrt{2}\},$$

and of the cylinder

$$\Omega_C := \{u + r\widehat{w} \mid 0 \le r \le a, |u| = a, \langle u, \widehat{w} \rangle = 0\}.$$

The outer normal $\nu_\Omega = -\widehat{w}$ on Ω_0, $\nu_\Omega = \widehat{w}$ on Ω_a, and ν_Ω is orthogonal to $\widehat{w}$ on Ω_C. Hence, defining an auxiliary function $\varphi \colon \mathbb{R}^n \longrightarrow \mathbb{R}^n$,

$$\varphi(u) := f\big(x_\infty + \tfrac{1}{a} \langle u, \widehat{w} \rangle \, z\big) - f(x_\infty) - \frac{1}{a} \langle u, \widehat{w} \rangle \, w,$$

we observe that the product $\varphi(u) \langle \widehat{w}, \nu_\Omega \rangle = 0$ on $\partial\Omega$, since $\varphi(u) = 0$ on $\Omega_0 \cup \Omega_a$ and $\langle \widehat{w}, \nu_\Omega \rangle = 0$ on Ω_C. It follows that

$$\int_\Omega \varphi'(u)(\widehat{w}) \, d\mathscr{L}^n(u) = \int_{\partial\Omega} \varphi(u) \langle \widehat{w}, \nu_\Omega \rangle \, d\mathscr{H}^{n-1}(u) = 0.$$

Noting that $\varphi'(u)(v) = 0$ if v is orthogonal to $\widehat{w}$, we conclude that

$$\int_\Omega \varphi' \, d\mathscr{L}^n = 0.$$

Thus, letting $h \colon \mathbb{R}^n \longrightarrow \mathbb{R}^n$,

$$h(u) = g(u) - \langle u, \widetilde{w} \rangle \widetilde{w} - \varphi(u),$$

we observe that

$$h'(u) = g'(u) - \langle \cdot, \widetilde{w} \rangle \widetilde{w} - \varphi'(u) = g'(u) - \widetilde{w} \otimes \widetilde{w} - \varphi'(u).$$

So by Corollary 9.4.2,

$$\left| \int_\Omega (g'(u) - \widetilde{w} \otimes \widetilde{w}) \, d\mathscr{L}^n(u) \right|_H = \left| \int_\Omega h'(u) \, d\mathscr{L}^n(u) \right|_H \qquad (15.27)$$
$$\le \mathscr{H}^{n-1}(\partial\Omega) \max_{u \in \partial\Omega} |h(u)|,$$

and all that remains is to estimate the norm of h on the boundary of Ω.

For $u \in \Omega_0$ we have $|u| \le a \le 2\Delta$, and so (15.25) implies

$$|h(u)| = |g(u)| = |f(x_\infty + Tu) - f(x_\infty) - L_\infty u| \le \xi|u| \le \xi a \le 2\xi a(c+1).$$

Similarly, for $u \in \Omega_a$ we first notice that any $z \in Y + T_\infty(\mathbb{R}^n)$ can be written as $z = y + T_\infty v$, where $y \in Y$, $v \in \mathbb{R}^n$ and $\|y\| + |v| \le c\|z\|$ by Lemma 15.5.1 (ii). It follows that $z + T_\infty u = y + T_\infty(v + u)$ and

$$\|y\| + |v + u| \le c\|z\| + |u| \le c\delta + 2a \le c\delta + \frac{\varepsilon\delta}{2\tau} \le 2\Delta.$$

This allows us to again use (15.25) to get

$$
\begin{aligned}
|h(u)| &= |g(u) - \langle u, \widetilde{w}\rangle \widetilde{w}| \\
&= |g(u) - w| \\
&= |f(x_\infty + z + T_\infty u) - f(x_\infty) - L_\infty u - w| \\
&= |f(x_\infty + z + T_\infty u) - f(x_\infty + z) - L_\infty u| \\
&\le \big|f(x_\infty + y + T_\infty(v + u)) - f(x_\infty + y) - L_\infty(v + u)\big| \\
&\quad + |f(x_\infty + y + T_\infty v) - f(x_\infty + y) - L_\infty v| \\
&\le \xi(2\|y\| + 2|v| + |u|) \le \xi(2c\|z\| + |u|) \le 2\xi a(c + 1).
\end{aligned}
$$

Finally, for $\widetilde{u} := u + r\widehat{w} \in \Omega_C$ we have

$$
\begin{aligned}
|h(\widetilde{u})| &= |g(\widetilde{u}) - \langle \widetilde{u}, \widetilde{w}\rangle \widetilde{w} - \varphi(\widetilde{u})| = \big|g(\widetilde{u}) - \tfrac{1}{a}\langle \widetilde{u}, \widehat{w}\rangle\, w - \varphi(\widetilde{u})\big| \\
&= \big|f(x_\infty + T\widetilde{u}) - f\big(x_\infty + \tfrac{1}{a}\langle \widetilde{u}, \widehat{w}\rangle\, z\big) - L_\infty \widetilde{u}\big| \\
&= \big|f\big(x_\infty + \tfrac{1}{a}\langle \widetilde{u}, \widehat{w}\rangle\, z + T_\infty \widetilde{u}\big) - f\big(x_\infty + \tfrac{1}{a}\langle \widetilde{u}, \widehat{w}\rangle\, z\big) - L_\infty \widetilde{u}\big| \\
&= \big|f\big(x_\infty + \tfrac{1}{a}rz + T_\infty \widetilde{u}\big) - f\big(x_\infty + \tfrac{1}{a}rz\big) - L_\infty \widetilde{u}\big| \\
&\le \big|f\big(x_\infty + \tfrac{1}{a}ry + T_\infty\big(\widetilde{u} + \tfrac{1}{a}rv\big)\big) - f\big(x_\infty + \tfrac{1}{a}ry\big) - L_\infty\big(\widetilde{u} + \tfrac{1}{a}rv\big)\big| \\
&\quad + \big|f\big(x_\infty + \tfrac{1}{a}ry + T_\infty\big(\tfrac{1}{a}rv\big)\big) - f\big(x_\infty + \tfrac{1}{a}rv\big) - L_\infty\big(\tfrac{1}{a}rv\big)\big|.
\end{aligned}
$$

Similarly as above $\big\|\tfrac{1}{a}ry\big\| + \big|\widetilde{u} + \tfrac{1}{a}rv\big| \le c\|z\| + |\widetilde{u}| \le c\delta + 2a \le 2\Delta$, we use (15.25) to continue

$$\le \xi(2\|y\| + 2|v| + |\widetilde{u}|) \le \xi(2c\|z\| + 2a) \le 2\xi a(c + 1).$$

Hence (15.27) gives

$$
\begin{aligned}
\left| \int_\Omega (g'(u) - \widetilde{w} \otimes \widetilde{w})\, d\mathscr{L}^n(u) \right|_{\mathrm{H}} \\
\le \frac{2\xi a(c + 1)\big(\boldsymbol{\alpha}_{n-1}a^{n-1} + \boldsymbol{\alpha}_{n-1}a^{n-1} + (n - 1)\boldsymbol{\alpha}_{n-1}a^{n-1}\big)}{\boldsymbol{\alpha}_{n-1}a^n} \\
= 2\xi(c + 1)(n + 1) \le \frac{\tau}{10\sqrt{n}},
\end{aligned}
$$

and we infer from Lemma 15.2.4 that

$$\fint_\Omega (\vartheta_\infty(L_\infty + g'(u)) - \vartheta_\infty(L_\infty)) \, d\mathscr{L}^n(u)$$

$$\geq \frac{|w|}{a} + 2\fint_\Omega \|g'\|^2 d\mathscr{L}^n - 10\sqrt{n} \left| \int_\Omega (g'(u) - \widetilde{w} \otimes \widetilde{w}) \, d\mathscr{L}^n(u) \right|_{\mathrm{H}}$$

$$> \frac{\varepsilon \|z\|}{a} + \fint_\Omega \|g'\|^2 d\mathscr{L}^n - \tau$$

$$= 3\tau + \fint_\Omega \|g'\|^2 d\mathscr{L}^n. \tag{15.28}$$

In order to verify the inequality (15.26), we observe that the image of

$$T - T_\infty = \frac{1}{a} \langle \cdot, \widehat{w} \rangle z$$

lies in Z, and so $\Psi(T - T_\infty) = 0$. Hence

$$\Theta_\infty(T) - \Theta_\infty(T_\infty) \leq \frac{\varepsilon \|T - T_\infty\|}{4\Lambda} \leq \frac{\varepsilon \eta}{4\Lambda} = \frac{\tau}{\Lambda}. \tag{15.29}$$

Noticing that by Corollary 15.3.1,

$$\Upsilon_\infty(x_\infty, T_\infty) \leq 1 + 3|L_\infty|_{\mathrm{H}}^2 \leq 1 + 12n \leq \Lambda,$$

we see that the left-hand side of (15.26) is

$$2\tau + \Upsilon_\infty(x_\infty, T_\infty) \fint_\Omega (\Theta_\infty(T) - \Theta_\infty(T_\infty)) \, d\mathscr{L}^n(u) + \fint_\Omega \|g'\|^2 \, d\mathscr{L}^n$$

$$\leq 2\tau + \Upsilon_\infty(x_\infty, T_\infty) \frac{\tau}{\Lambda} + \fint_\Omega \|g'\|^2 \, d\mathscr{L}^n$$

$$\leq 3\tau + \fint_\Omega \|g'\|^2 \, d\mathscr{L}^n,$$

which is the same as the lower estimate of its right-hand side in (15.28). Moreover, since $|g'(u)|_{\mathrm{H}} \leq \sqrt{n} \operatorname{Lip}(g) \leq 5\sqrt{n}$ we get

$$\vartheta_\infty(g'(u)) \leq 1 + 3|g'(u)|_{\mathrm{H}}^2 \leq 1 + 75n^2,$$

and so we infer from the choice of s that

$$\int_{\{u \in \Omega | B(u,sa) \not\subset \Omega\}} \vartheta_\infty(g'(u)) \, d\mathscr{L}^n(u) \leq (1 + 75n^2)\big(1 - (1-s)^{n-1}(1-2s)\big)\mathscr{L}^n\Omega$$

$$\leq \tau \mathscr{L}^n \Omega.$$

Having verified all its assumptions, we are finally able to use Lemma 15.2.3 to find a point $v \in \Omega$ such that $B(v, sa) \subset \Omega$, g is differentiable at v, and

$$\frac{\vartheta_\infty(L_\infty + g'(v))}{\Theta_\infty(T)} - \Upsilon_\infty(x_\infty, T_\infty) > \frac{\tau}{\Theta_\infty(T)} + \frac{(\operatorname{Lip}_v(g))^{2n}}{K_n \Theta_\infty(T)(\operatorname{Lip}(g))^{2(n-1)}}.$$

We let $x = x_\infty + Tv$ and observe that

$$g'(v) = f'(x_\infty + Tv; T) - L_\infty = f'(x; T) - L_\infty.$$

Hence

$$\frac{\vartheta_\infty(L_\infty + g'(v))}{\Theta_\infty(T)} = \Upsilon_\infty(x, T),$$

and so

$$\Upsilon_\infty(x, T) - \Upsilon_\infty(x_\infty, T_\infty) > \frac{\tau}{\Theta_\infty(T)} + \frac{(\mathrm{Lip}_v(g))^{2n}}{K_n \Theta_\infty(T)(\mathrm{Lip}(g))^{2(n-1)}}.$$

Recalling $\Theta_\infty(T) \leq \Lambda$, $\tau > \Lambda \tau_j$, the estimate $\mathrm{Lip}(g) \leq 5$ of the Lipschitz constant of g established above, and the value of μ chosen in (15.4), we get

$$\Upsilon_\infty(x, T) - \Upsilon_\infty(x_\infty, T_\infty) > \tau_j + \mu(\mathrm{Lip}_v(g))^{2n}. \tag{15.30}$$

Having deduced this key inequality, we now follow the same path as in the proof of Lemma 15.4.1. Lemma 15.3.6 and (15.30) show that for each $i \geq 0$,

$$\Upsilon_i(x, T) \geq \Upsilon_\infty(x, T) \geq \Upsilon_\infty(x_\infty, T_\infty) \geq \Upsilon_i(x_i, T_i). \tag{15.31}$$

We estimate $\|x_\infty - x\|$ by

$$\begin{aligned}\|x_\infty - x\| = \|Tv\| &\leq \|T_\infty\|\|v\| + \frac{\|z\|}{a}|v| \\ &\leq 2(1 + \|T_\infty\|)a \leq (1 + \|T_\infty\|)\frac{\varepsilon\delta}{2\tau}.\end{aligned} \tag{15.32}$$

Since the choice of δ gives that the right side is $\leq \delta_{j+2}$, we see that for $i \leq j$,

$$\|x - x_i\| \leq \|x - x_\infty\| + \|x_i - x_\infty\| \leq \delta_{j+2} + (\delta_i - \delta_{i+1}) < \delta_i.$$

Also,

$$\|T - T_0\| \leq \|T - T_\infty\| + \|T_\infty - T_0\| \leq \frac{\|z\|}{a} + \eta_0 \leq \eta + \eta_0 \leq 1.$$

In particular, $(x, T) \in D_0$.

We prove by induction that $(x, T) \in D_i$ for all $i = 0, \ldots, j$. For that, suppose that $1 \leq i \leq j$ and $(x, T) \in D_{i-1}$. Since we have already proved that $\|x - x_i\| < \delta_i$ and $\Upsilon_i(x, T) \geq \Upsilon_i(x_i, T_i)$, the proof of $(x, T) \in D_i$ will be finished by estimating, for each $|u| \leq \Delta_i$, $|f(x + Su) - f(x) - f'(y; S)(u)|$ for a suitable $(y, S) \in D_{i-1}$. For this, we consider three cases.

Case 1. $|u| < sa$. Since both (x, T) and (x_{i-1}, T_{i-1}) belong to D_{i-1}, we infer from (A2) of the algorithm in case $i > 1$ or from Lemma 15.3.2 (ii) in case $i = 1$ that

$$\Upsilon_{i-1}(x_{i-1}, T_{i-1}) \geq \Upsilon_{i-1}(x, T) - \tau_{i-1}.$$

Together with (15.30) and (15.31), we obtain

$$\tau_{i-1} \geq \Upsilon_{i-1}(x,T) - \Upsilon_{i-1}(x_{i-1}, T_{i-1}) \geq \mu(\mathrm{Lip}_v(g))^{2n}.$$

Recalling the inequality $\tau_{i-1} \leq \mu(\beta_i/6)^{2n}$, we see that

$$\mathrm{Lip}_v(g) \leq (\tau_{i-1}/\mu)^{1/2n} \leq \frac{\beta_i}{6}.$$

Since $B(v, sa) \subset \Omega$, $v + u$ lies in Ω, and we get that $(y, S) = (x, T)$ is a good choice:

$$|f(x + Tu) - f(x) - f'(x; Tu)| = |g(v + u) - g(v) - g'(v)(u)|$$

$$\leq 2\,\mathrm{Lip}_v(g)|u| \leq \frac{\beta_i}{3}|u|. \qquad (15.33)$$

Case 2. $sa \leq |u| < \Delta$. This time, our choice is $(y, S) = (x_\infty, T_\infty)$. Since $Tv = T_\infty v + \langle v, \widehat{w}\rangle z/a \in Y + T_\infty(\mathbb{R}^n)$, and since $Y \subset Y_0$, where Y_0 has been chosen by Lemma 15.5.1 (ii), we can write $Tv = y + T_\infty \widetilde{v}$, where $y \in Y$, $\widetilde{v} \in \mathbb{R}^n$, and

$$\|y\| + |\widetilde{v}| \leq c\|Tv\| \leq 2c(1 + \|T_\infty\|)a.$$

In particular, by the choice of δ this implies that

$$\|y\| + |\widetilde{v}| + |u| \leq 2c(1 + \|T_\infty\|)\frac{\varepsilon\delta}{4\tau} + \Delta \leq 2\Delta.$$

Hence we may use (15.25) to infer that

$$|f(x + T_\infty u) - f(x) - L_\infty u|$$
$$= |f(x_\infty + Tv + T_\infty u) - f(x_\infty + Tv) - L_\infty u|$$
$$\leq |f(x_\infty + y + T_\infty(\widetilde{v} + u)) - f(x_\infty + y) - L_\infty(\widetilde{v} + u)|$$
$$\quad + |f(x_\infty + y + T_\infty\widetilde{v}) - f(x_\infty + y) - L_\infty\widetilde{v}|$$
$$\leq \xi(2\|y\| + 2|\widetilde{v}| + |u|) \leq 4\xi c(1 + \|T_\infty\|)a + \xi|u|$$
$$\leq \frac{\beta_i}{4}sa + \frac{\beta_i}{4}|u| \leq \frac{\beta_i}{2}|u|. \qquad (15.34)$$

Case 3. $\Delta \leq |u| \leq \Delta_i$. Since $(x_\infty, T_\infty) \in D_{i+1}$, by Lemma 15.3.7, there is $(y, S) \in D_{i-1}$ such that

$$|f(x_\infty + Su) - f(x_\infty) - f'(y; Su)| \leq \widetilde{\beta}_i|u|.$$

The estimate (15.32) and the choice of δ show that

$$2\,\mathrm{Lip}(f)\|x_\infty - x\| \leq \frac{\beta_i - \widetilde{\beta}_i}{2}\Delta \leq \frac{\beta_i - \widetilde{\beta}_i}{2}|u|.$$

Now we get

$$|f(x + Su) - f(x) - f'(y; Su)|$$
$$\leq |f(x_\infty + Su) - f(x_\infty) - f'(y; Su)|$$
$$+ |f(x + Su) - f(x_\infty + Su)| + |f(x) - f(x_\infty)|$$
$$\leq \widetilde{\beta}_i |u| + 2 \operatorname{Lip}(f) \|x_\infty - x\|$$
$$\leq \widetilde{\beta}_i |u| + \frac{\beta_i - \widetilde{\beta}_i}{2} |u| = \frac{\widetilde{\beta}_i + \beta_i}{2} |u|. \tag{15.35}$$

In all three cases we found the required pair (y, S), thus showing that $(x, T) \in D_i$. Hence, by induction, we conclude that $(x, T) \in D_j$. But this, (15.31), and (15.30) imply that

$$\tau_j \geq \Upsilon_j(x, T) - \Upsilon_j(x_j, T_j) \geq \Upsilon_\infty(x, T) - \Upsilon_\infty(x_\infty, T_\infty) > \tau_j,$$

and we have found our desired contradiction. $\square$

The final point that has to be added to Lemma 15.5.2 is the construction of the (asymptotic) derivative S of f at x_∞. Since we have already said that Z should be (a subset of) its kernel, and since in the direction of $T_\infty(\mathbb{R}^n)$ the derivative can be found on the basis of $f'(x_\infty; T_\infty)$, we connect these facts together by the following simple lemma.

Lemma 15.5.3. *X is the direct sum of Z and $T_\infty(\mathbb{R}^n)$.*

Proof. Since Lemma 15.5.1 (i) implies that $T_\infty(\mathbb{R}^n)$ has dimension n, and since Z has codimension $\leq n$, it suffices to show that $Z \cap T_\infty(\mathbb{R}^n) = \{0\}$.

Let $z \in Z \cap T_\infty(\mathbb{R}^n)$ and write $z = T_\infty v$, $v \in \mathbb{R}^n$. Let $\varepsilon > 0$. If $t > 0$ is small enough, the fact that $f'(x_\infty; T_\infty) = L_\infty$ implies that

$$|f(x_\infty + tz) - f(x_\infty) - tL_\infty v| \leq \varepsilon t \|v\|.$$

On the other hand, Lemma 15.5.2 gives (independently of what Y has been produced) that

$$|f(x_\infty + tz) - f(x_\infty)| \leq \varepsilon t \|z\|.$$

Hence $|L_\infty v| \leq \varepsilon \|z\| + \varepsilon \|v\| \leq \varepsilon(1 + \|T_\infty\|) \|v\|$. Since $\varepsilon > 0$ is arbitrary, we infer that $L_\infty v = 0$. Recalling that $L_\infty \colon \mathbb{R}^n \longrightarrow \mathbb{R}^n$ has rank n, we conclude that $v = 0$ and hence $z = T_\infty v = 0$. $\square$

We are now ready to finish the "algorithm" part of our proof by showing that f has at x_∞ the appropriate version of derivative. We also add some statements that will be needed to show Theorem 15.1.3 and reformulate the increase in the Hilbert-Schmidt norm as a mean value inequality.

Proposition 15.5.4. *The pair $(x_\infty, T_\infty) \in D \times L(\mathbb{R}^n, X)$ and the operator*

$$L_\infty = f'(x_\infty; T_\infty) \in L(\mathbb{R}^n, \mathbb{R}^n)$$

have the following properties:

(a) $\|x_\infty - x_0\| \leq \delta_0 \ (= \eta_0)$ *and* $\|T_\infty - T_0\| \leq \eta_0$;

(b) $\operatorname{Trace} f'(x_\infty; T_\infty) \geq \operatorname{Trace} f'(x_0; T_0) - n\|f'(x_\infty; T_\infty) - \operatorname{Id}\|^2$; *and*

(c) *there is* $S \in L(X, \mathbb{R}^n)$ *such that* $L_\infty = ST_\infty$ *and for every* $\varepsilon > 0$ *there are* $\delta > 0$ *and* $Y \in \mathcal{Y}$ *such that*

$$|f(x_\infty + y) - f(x_\infty) - Sy| \leq \varepsilon \|y\|$$

for every $y \in Y + T_\infty(\mathbb{R}^n)$ *with* $\|y\| < \delta$.

Proof. First recall that $(x_\infty, T_\infty) \in D_0$ by Lemma 15.3.8. So (a) follows from the definition of D_0 and Lemma 15.3.2 (i).

Inferring from Lemma 15.3.2 (iv) that $|L_\infty|_{\mathrm{H}}^2 \geq |L_0|_{\mathrm{H}}^2$, we get (b) by estimating

$$
\begin{aligned}
2\operatorname{Trace}(L_\infty - L_0) &= 2\langle L_\infty - L_0, \operatorname{Id}\rangle_{\mathrm{H}} \\
&= |L_0 - \operatorname{Id}|_{\mathrm{H}}^2 - |L_\infty - \operatorname{Id}|_{\mathrm{H}}^2 + |L_\infty|_{\mathrm{H}}^2 - |L_0|_{\mathrm{H}}^2 \\
&\geq -|L_\infty - \operatorname{Id}|_{\mathrm{H}}^2 \geq -n\|L_\infty - \operatorname{Id}\|^2.
\end{aligned}
$$

It remains to prove (c). Using Lemma 15.5.3, we denote by P the projection $P\colon X \longrightarrow T_\infty(\mathbb{R}^n)$ onto $T_\infty(\mathbb{R}^n)$ along Z. We show that the statement holds with $S = L_\infty T_\infty^{-1} P$. Clearly, $L_\infty = ST_\infty$.

Use Lemma 15.5.1 and the fact that T_∞ has rank n to find $Y_0 \in \mathcal{Y}$ and a constant $c \geq 1 + \|P\| + \|T_\infty^{-1}\|\|P\|$ such that for every $Y \in \mathcal{Y}$, $Y \subset Y_0$, every $y \in Y + T_\infty(\mathbb{R}^n)$ can be written as

$$y = \widetilde{y} + T_\infty v,$$

where $\widetilde{y} \in Y$, $v \in \mathbb{R}^n$ and $\|\widetilde{y}\| + |v| \leq c\|y\|$. Notice also that the lower bound on c was chosen such that for every $y \in X$

$$\|y - Py\| + |T_\infty^{-1} Py| \leq \left(1 + \|P\| + \|T_\infty^{-1} P\|\right)\|y\| \leq c\|y\|.$$

Let $\xi = \varepsilon/(4c)$ and use Lemmas 15.4.1 and 15.5.2 to find $\Delta > 0$ and $Y \subset Y_0$ such that both

$$|f(x_\infty + z) - f(x_\infty)| \leq \xi \|z\| \tag{15.36}$$

for every $z \in (Y + T_\infty(\mathbb{R}^n)) \cap Z$ with $\|z\| \leq \Delta$, and

$$|f(x_\infty + y + T_\infty u) - f(x_\infty + y) - L_\infty u| \leq \xi(\|y\| + |u|) \tag{15.37}$$

whenever $y \in Y$, $u \in \mathbb{R}^n$ and $\|y\| + |u| \leq \Delta$.

Let $\delta = \Delta/(2c)$. Assume that $y \in Y + T_\infty(\mathbb{R}^n)$ and $\|y\| < \delta$. Write $y = \widetilde{y} + T_\infty v$, where $\widetilde{y} \in Y$, $v \in \mathbb{R}^n$, and $\|\widetilde{y}\| + |v| \leq c\|y\|$. Further let $u = T_\infty^{-1} Py$ and notice that $Sy = L_\infty u$. Hence

$$
\begin{aligned}
|f(x_\infty + y) &- f(x_\infty) - Sy| \\
&\leq |f(x_\infty + \widetilde{y} + T_\infty v) - f(x_\infty + \widetilde{y}) - L_\infty v| \\
&\quad + |f(x_\infty + \widetilde{y} + T_\infty(v - u)) - f(x_\infty + \widetilde{y}) - L_\infty(v - u)| \\
&\quad + |f(x_\infty + \widetilde{y} + T_\infty(v - u)) - f(x_\infty)|.
\end{aligned} \tag{15.38}
$$

Since $\|\widetilde{y}\| + |v| \leq c\|y\| \leq \Delta$ and

$$\|\widetilde{y}\| + |v - u| \leq (\|\widetilde{y}\| + |v|) + |u| \leq c\|y\| + \|T_\infty^{-1}\|\|P\|\|y\| \leq 2c\|y\| \leq \Delta,$$

the first two terms on the right may be estimated using (15.37) by

$$\xi(\|\widetilde{y}\| + |v|) + \xi(\|\widetilde{y}\| + |v - u|) \leq 2\xi(\|\widetilde{y}\| + |v|) + \xi|u| \leq 3c\xi\|y\|.$$

To estimate the last term we employ (15.36) with

$$z = \widetilde{y} + T_\infty(v - u) = \widetilde{y} + T_\infty v - Py = y - Py.$$

The last expression shows that $z \in Z$ and $\|z\| \leq (1 + \|P\|)\|y\| \leq c\|y\| \leq \Delta$. Hence (15.36) implies that the third term on the right-hand side of (15.38) may be estimated by $\xi\|z\| \leq c\xi\|y\|$. Putting all three estimates together, we get

$$|f(x_\infty + y) - f(x_\infty) - Sy| \leq 4c\xi\|y\| = \varepsilon\|y\|. \qquad \square$$

15.6 PROOF OF THEOREM 15.1.3

We may assume that $\eta < 1$ is small enough that $B(x_0, \eta) \subset G$, and f is Lipschitz on $B(x_0, \eta)$. We may also assume that $\dim V = n$ and, moving the v_i^* and z_i slightly if necessary, that both $v_1^*, \ldots, v_n^*$ and $z_1, \ldots, z_n$ are linearly independent. Define $T_0 \in L(\mathbb{R}^n, X)$ by $T_0(e_i) = z_i$ (where $e_1, \ldots, e_n$ is the standard basis of $\mathbb{R}^n$), and let $S_0 \in L(X, \mathbb{R}^n)$ be such that $S_0 T_0 = \text{Id}$ on $\mathbb{R}^n$. Also, define $P \colon V \longrightarrow \mathbb{R}^n$ by $Pv = \sum_{i=1}^n v_i^*(v)e_i$.

Let $h \colon B(x_0, \eta) \longrightarrow \mathbb{R}^n$ be given as

$$h(x) = P(f(x))$$

and extend h to a Lipschitz map of X to $\mathbb{R}^n$. Fix a small $0 < t < \frac{1}{2}\eta$ to guarantee that the following inequalities hold true:

$$2n\big(\|S_0\| + \text{Lip}(h)(\|T_0\| + 1)\big)^2 t < \eta \quad \text{and} \quad 2n\|S_0\|t < \eta. \qquad (15.39)$$

We intend to apply the algorithm of Section 15.3 with $\eta_0 = t^2$ to the function $g := S_0 + th$ (in the place of f). Notice that, whenever $\|T - T_0\| \leq \eta_0$, the first estimate in (15.39) implies

$$
\begin{aligned}
\|g'(x; T) - \text{Id}\|^2 &= \|S_0(T - T_0) + th'(x; T)\|^2 \\
&\leq \big(\|S_0\|\|T - T_0\| + t\,\text{Lip}(h)(\|T_0\| + \|T - T_0\|)\big)^2 \\
&\leq \big(\|S_0\|t + t\,\text{Lip}(h)(\|T_0\| + 1)\big)^2 \qquad (15.40) \\
&\leq \big(\|S_0\| + \text{Lip}(h)(\|T_0\| + 1)\big)^2 t^2 \\
&< \frac{\eta t}{2n}.
\end{aligned}
$$

The requirement (R) of Section 15.3, page 364, follows from this since $\eta t/(2n) \leq \frac{1}{4}$. Hence the algorithm provides us with $x \in X$ and $T \in L(\mathbb{R}^n, X)$ (which are the x_∞ and T_∞ of Section 15.3) such that $g'(x; T)$ exists and, by Proposition 15.5.4,

(α) $\|x - x_0\| \le \eta_0$ and $\|T - T_0\| \le \eta_0$;

(β) Trace $g'(x; T) \ge$ Trace $g'(x_0; T_0) - n\|g'(x; T) - \mathrm{Id}\,\|^2$; and

(γ) there is $S \in L(X, \mathbb{R}^n)$ such that for every $\varepsilon > 0$ there are $\delta > 0$ and $Y \in \mathcal{Y}$ such that

$$|g(x + y) - g(x) - Sy| \le \varepsilon\|y\|$$

for every $y \in Y + T(\mathbb{R}^n)$ with $\|y\| < \delta$.

We show that all the statements of Theorem 15.1.3 hold with our x,

$$y_i = T(e_i) \quad \text{and} \quad Q = \frac{1}{t}P^{-1}(S - S_0).$$

By (α), $x \in G$ and $\|y_i - z_i\| \le \|T - T_0\| \le \eta_0 < \eta$, so we have (ii). To show (i), we notice that $B(x, \eta_0) \subset B(x_0, \eta)$ and so

$$f(z) = \frac{1}{t}P^{-1}(g(z) - S_0(z)) \text{ for } z \in B(x, \eta_0).$$

Hence, obtaining δ and Y from (γ) with $\varepsilon t/\|P^{-1}\|$ instead of ε, we get that for every $y \in Y + T(\mathbb{R}^n)$ with $\|y\| < \min\{\delta, \eta_0\}$,

$$|f(x + y) - f(x) - Qy| \le \frac{1}{t}\|P^{-1}\|\,|g(x + y) - g(x) - Sy| \le \varepsilon\|y\|.$$

In addition to Theorem 15.1.3 (i), an immediate consequence of the existence of $g'(x; T)$ is that f is differentiable at x in the direction of $T(\mathbb{R}^n)$. In particular, the statement in 15.1.3 (iii) holds since $T(\mathbb{R}^n)$ is the linear span of $y_1, \ldots, y_n$. Furthermore,

$$tf'(x; y_i) = tQy_i = P^{-1}(Sy_i - S_0 y_i) = P^{-1}\big(g'(x; T)(e_i) - S_0 Te_i\big),$$

and a completely similar argument gives

$$tf'(x_0; z_i) = P^{-1}\big(g'(x_0; T_0)(e_i) - S_0 T_0 e_i\big).$$

Hence

$$t\sum_{i=1}^{n} v_i^*(f'(x; y_i)) = t\sum_{i=1}^{n}(P^*e_i^*)\big(f'(x; y_i)\big) = \sum_{i=1}^{n} e_i^*\big(g'(x; T) - S_0 T\big)e_i$$

$$= \mathrm{Trace}(g'(x; T) - S_0 T),$$

and analogically

$$t\sum_{i=1}^{n} v_i^*(f'(x_0; z_i)) = \mathrm{Trace}(g'(x_0; T_0) - S_0 T_0).$$

We are now ready to deduce Theorem 15.1.3 (iv). Using (β), (15.40), and the second condition in (15.39), we estimate

$$t \sum_{i=1}^{n} v_i^*(f'(x; y_i)) = \operatorname{Trace} g'(x; T) - \operatorname{Trace} S_0 T$$

$$\geq \operatorname{Trace} g'(x_0; T_0) - \operatorname{Trace} S_0 T - n \|g'(x; T) - \operatorname{Id}\|^2$$

$$\geq t \sum_{i=1}^{n} v_i^*(f'(x_0; z_i)) - \operatorname{Trace}(S_0(T - T_0)) - \frac{\eta t}{2}$$

$$\geq t \sum_{i=1}^{n} v_i^*(f'(x_0; z_i)) - n \|S_0\| \eta_0 - \frac{\eta t}{2}$$

$$= t \sum_{i=1}^{n} v_i^*(f'(x_0; z_i)) - n \|S_0\| t^2 - \frac{\eta t}{2}$$

$$> t \sum_{i=1}^{n} v_i^*(f'(x_0; z_i)) - \eta t.$$

This completes the proof of Theorem 15.1.3.

Chapter Sixteen

Differentiability of Lipschitz maps on Hilbert spaces

For the benefit of those readers whose main interest is in Hilbert spaces, we give here a separate proof of existence of points of Fréchet differentiability of $\mathbb{R}^2$-valued Lipschitz maps on such spaces. Although the arguments are based on ideas from the previous chapters, only two technical lemmas whose proof may be easily read independently from the previous chapters are actually used. We also use this occasion to explain several ideas for treating the differentiability problem that may not have been apparent in the generality in which we have worked so far.

16.1 INTRODUCTION

We give here an essentially self-contained proof of the following result on existence of points of Fréchet differentiability of $\mathbb{R}^2$-valued Lipschitz maps on Hilbert spaces. This result has been already proved twice, in Corollary 13.1.2 and Corollary 15.1.4. The treatment presented here uses similar ideas. However, the special structure of the Hilbert space is heavily used and readers interested predominantly in the Hilbert space case should find the arguments of this chapter easier to follow.

Theorem 16.1.1. *Every Lipschitz map of a Hilbert space to a two-dimensional space has points of Fréchet differentiability.*

Standard strengthenings of this result, for example, to locally Lipschitz maps defined on nonempty open sets or to the validity of the mean value estimates for Fréchet derivatives may be obtained, similarly to what was done in Proposition 13.3.1, by applying Theorem 16.1.1 to a suitably modified function, and so they will not be treated here.

We also use this occasion to explain yet another variant of the strategy for finding points of Fréchet differentiability. Observe that a possible algorithmic procedure for finding a sequence converging to a point of Fréchet differentiability of a real-valued Lipschitz function f on a Hilbert space H may be sketched as follows. Among the point-direction pairs (x, e), where $x \in H$, $e \in H$ with $\|e\| = 1$ and $f'(x; e)$ exists, we choose a pair (x_0, e_0) such that $f'(x_0; e_0)$ is almost largest possible. Next we choose (x_1, e_1) (from a certain restricted set of pairs, but that does not interest us at the moment) such that $f'(x_1; e_1) - \alpha_0 \|e_1 - e_0\|^2$ is almost largest possible. Noticing that

$$f'(x; e) - \alpha_0 \|e - e_0\|^2 = f'(x; e) + 2\alpha_0 \langle e_0, e \rangle + \text{constant},$$

we may understand the second step as maximizing the directional derivative of the function $x \mapsto f(x) + 2\alpha_0 \langle e_0, x \rangle$, hence of a *linear perturbation of f*. This approach was used by Doré and Maleva in their deep study of the two-dimensional version of the differentiability problem in [15]. Notice, however, that there is a subtle difference between maximizing perturbations of the function $(x, e) \mapsto f'(x; e)$ as we do in Chapter 12 and maximizing a linear perturbation as Doré and Maleva do. While the former allows the vector e to move over the whole space, the latter constrains it to unit vectors. When translated to the setting of two-dimensional ranges, we in Chapter 13 maximize a perturbation of the function $(x, T) \to |f'(x) \circ T|_H$ where T runs over $L(\mathbb{R}^2, H)$, while the approach that we will follow here will be essentially equivalent to maximizing a perturbation the function $(x, U) \to \left|f'(x)|_U\right|_H$ where U runs through two-dimensional subspaces of H.

The approach via linear perturbations also has an interesting geometric interpretation. Imagine for the sake of argument that we are treating the one-dimensional range and the pairs (x, e) are restricted to those for which f is Gâteaux differentiable at x. By adding to f a multiple of the linear functions $x \to \langle x, e_0 \rangle$, we have shifted the set of Gâteaux derivatives so that the new maximizing direction e_1 is necessarily close to e_0. This is expressed in an abstract form in Lemma 16.2.2. In fact, this property may be traced back to the uniform convexity of the (dual norm to the) Hilbertian norm. So it appears that linear perturbations may be used to show the existence of points of Fréchet differentiability of real-valued Lipschitz functions on superreflexive spaces. However, it seems that the method presented here cannot work in general Asplund spaces, even though Stegall's variational principle (see, for example, [37, Corollary 5.22]), and the connection between differentiability problems and minima attaining perturbations may suggest that there is a way of using just linear perturbations even in the general situation.

The construction of a point of Fréchet differentiability of $\mathbb{R}^2$-valued Lipschitz functions is based on the above idea, in which we replace the pairs (x, e) by pairs (x, X), where X is a two-dimensional linear subspace of H in the direction of which f is differentiable. The norm of $f'(x; e)$ is then replaced by the Hilbert-Schmidt norm of the derivative of f in the direction of X. It follows that the arguments presented here work for both $\mathbb{R}$- and $\mathbb{R}^2$-valued functions. However, they fail for $\mathbb{R}^n$-valued maps where $n \geq 3$. The reason is hidden in a problem we have not discussed yet: the need for restricting the set of pairs (x, X) in each step of the construction in a way that will assure that in the limiting point the function f has at least some differentiability properties. We handle this point by (essentially 3-dimensional) Lemma 16.2.1. The only point where the restriction to $n \leq 2$ is needed is in the part of the proof of Lemma 16.2.1 in which we show regularity. For our method this failure is irreparable. The validity of Lemma 16.2.1 for certain n implies not only Theorem 16.1.1 for Lipschitz maps from H to $\mathbb{R}^n$ but also that the mean value estimates hold for Fréchet derivatives. But the results of Chapter 14 show that they fail for $\mathbb{R}^n$-valued Lipschitz maps with $n \geq 3$.

Notation

We will denote by $G(H, n)$ the set of n-dimensional linear subspaces of the Hilbert space H. Using orthogonal projections we will consider derivatives in the direction of a subspace of H as defined on the whole of H. When f maps H to a Banach space Y and $V \in G(H, n)$, we denote by $D_V f(x)$, provided it exists, a linear operator $D_V f(x) \colon H \longrightarrow Y$ such that

$$D_V f(x)(v) = \begin{cases} f'(x; v) & \text{for } v \in V, \\ 0 & \text{for } v \in V^\perp. \end{cases}$$

In other words, the existence of $D_V f(x)$ says that f is Gâteaux differentiable at x in the direction of V, and the value $D_V f(x)$ is the composition of this derivative with the orthogonal projection π_V of H onto V.

For $f \colon X \longrightarrow Y$, $x \in X$, and $L \in L(X, Y)$, we will use the following simple measure of how close L is to being the Fréchet derivative of f at x in the direction of a given subspace Z:

$$\varepsilon_Z(f, x, L, \delta) = \sup_{\substack{z \in Z \\ 0 < \|z\| < \delta}} \frac{\|f(x+z) - f(x) - Lz\|}{\|z\|}.$$

In the case when $Z = X$ we write $\varepsilon(f, x, L, \delta)$ instead of $\varepsilon_X(f, x, L, \delta)$.

The symbol $\mathcal{D}_2(f)$ will denote the set of pairs $(x, X) \in H \times G(H, 2)$ such that $D_X f(x)$ exists.

We will use orthogonal projections also to measure the distance of two subspaces $V, W \in G(H, n)$ by the norm $\|\pi_V - \pi_W\|$. It will sometimes be convenient to estimate it by

$$\|\pi_V - \pi_W\| = \|\pi_V \pi_{W^\perp} - \pi_{V^\perp} \pi_W\| \leq \|\pi_V \pi_{W^\perp}\| + \|\pi_{V^\perp} \pi_W\| = 2\|\pi_V \pi_{W^\perp}\|.$$

Here we recalled that $\|\pi_{V^\perp} \pi_W\| = \|\pi_{W^\perp} \pi_V\| = \|\pi_V \pi_{W^\perp}\|$. Indeed, the first equality can be proved, for example, by finding orthonormal bases (v_i) and (w_i) for V and W, respectively, such that $\pi_V w_i = \lambda_i v_i$ and observing that $\pi_W v_i = \lambda_i w_i$ as well. Then

$$\|\pi_{V^\perp} \pi_W\|^2 = \sup\left\{ \sum_{i=1}^n (1 - \lambda_i^2)\alpha_i^2 \;\Big|\; \sum_{i=1}^n \alpha_i^2 \leq 1 \right\} = \|\pi_{W^\perp} \pi_V\|^2.$$

The second equality follows by taking adjoints.

Finally, recall the relation between the Hilbert-Schmidt norm $|\cdot|_H$ and the operator norm $\|\cdot\|$ on the space $L(\mathbb{R}^n, \mathbb{R}^n)$,

$$\|\cdot\| \leq |\cdot|_H \leq \sqrt{n}\|\cdot\|.$$

16.2 PRELIMINARIES

Here we state the following key lemma, which forms the main ingredient of our proof of Theorem 16.1.1. Although we state it in an arbitrary Hilbert space, one can notice that its validity in the general case follows from its three-dimensional version. Once again, we point out that this is where the assumption that the range is two-dimensional plays a major role: an analogous lemma for the three-dimensional target is false.

Lemma 16.2.1. *There is an increasing function* $\chi\colon (0,\infty) \longrightarrow (0,\infty)$ *such that* $\lim_{t \searrow 0} \chi(t) = 0$ *and for every* $\varepsilon > 0$ *there are* $c_0, C_0 \in (0,\infty)$ *for which the following statement holds.*

Suppose that $g\colon H \longrightarrow \mathbb{R}^2$, $\xi > 0$, $(x,X) \in \mathcal{D}_2(g)$, *and* $L \in L(H,\mathbb{R}^2)$ *satisfy*

$$\mathrm{Lip}(g) \le 1, \quad L = L\pi_X, \quad \varepsilon_X(g,x,L,C_0\xi) < c_0, \quad \text{and} \quad \varepsilon(g,x,L,\xi) > \varepsilon.$$

Then there are $(y_0,Y) \in \mathcal{D}_2(g)$ *and* $c > c_0$ *such that*

$$\|y_0 - x\| < C_0\xi, \quad |D_Y g(y_0)|_{\mathrm{H}} \ge |L|_{\mathrm{H}} + c$$

and for every $u \in X$,

$$\big|(g(y_0 + u) - g(y_0)) - (g(x + u) - g(x))\big| \le \chi(c)\|u\|.$$

We defer the proof of this lemma to Section 16.5.

It will be convenient to have the numbers c_0 and C_0 from Lemma 16.2.1 in the form $c_0 = \varphi(\varepsilon)$ and $C_0 = 1/\varphi(\varepsilon)$, where $\varphi\colon (0,\infty) \longrightarrow (0,\infty)$ is an increasing function. Such a function may be obtained, for example, by choosing the corresponding values c_k, C_k say, for $\varepsilon = 1/k$ and letting

$$\varphi(t) = \begin{cases} \min\{c_1, 1/C_1\} & \text{for } t > 1, \\ \min\{c_1, \ldots, c_k, 1/C_1 \ldots, 1/C_k\} & \text{for } 1/k < t \le 1/(k-1). \end{cases}$$

Clearly, we may diminish φ even further. In particular, we may assume that $\varphi(t) \le t$ and that the following statement holds.

Lemma 16.2.2. *Let* $L, P \in L(H,\mathbb{R}^n)$, $V, W \in G(H,n)$, *and* $c, \eta > 0$ *be such that* $|Lu| \ge c\|\pi_V u\|$ *for all* $u \in H$ *and*

$$|P + \eta L\pi_W|_{\mathrm{H}} > (1+\eta)(1 - \varphi(\eta))S,$$

where $S \ge \max\{|L|_{\mathrm{H}}, |P|_{\mathrm{H}}\}$. *Then*

$$\|\pi_W - \pi_V\| < \frac{\eta S}{c} \quad \text{and} \quad \|P - L\| < \eta S\sqrt{\frac{S}{c}}.$$

Proof. We show that the required inequalities hold whenever

$$\varphi(\eta) \le \frac{2\alpha^2 \eta}{1+\eta}, \quad \text{where} \quad \alpha = \frac{\eta^2}{10(1+\eta)(1+\eta^2)\sqrt{n}}.$$

The assumptions imply that $|\eta L\pi_W|_{\mathrm{H}} > (1+\eta)(1 - \varphi(\eta))S - S$, giving that

$$|L\pi_W|_{\mathrm{H}} > \Big(1 - \frac{1+\eta}{\eta}\varphi(\eta)\Big)S \ge (1 - 2\alpha^2)S.$$

Since the right side of this inequality is positive, we get

$$S^2 \geq |L|_H^2$$
$$= |L\pi_W|_H^2 + |L\pi_{W^\perp}|_H^2 \geq \left(1 - 2\alpha^2\right)^2 S^2 + |L\pi_{W^\perp}|_H^2$$
$$\geq \left(1 - 4\alpha^2\right)S^2 + |L\pi_{W^\perp}|_H^2.$$

Hence $|L\pi_{W^\perp}|_H^2 \leq 4\alpha^2 S^2$ and, using the assumption $|Lu| \geq c\|\pi_V u\|$, we obtain that

$$\|\pi_V - \pi_W\| \leq 2\|\pi_V \pi_{W^\perp}\| \leq \frac{2}{c}|L\pi_{W^\perp}|_H \leq \frac{4\alpha S}{c}. \tag{16.1}$$

Since $4\alpha < \eta$, the first statement of the lemma is proved.

We now turn our attention to the second statement. Since $\|P - L\| \leq 2S$, we may assume that $\eta^2 S/c \leq 4$. We also notice that $L = L\pi_V$: the assumption $|Lu| \geq c\|\pi_V u\|$ implies that $\ker L \subset \ker \pi_V$ and so, since $\ker L$ has codimension at most n, we see that $\ker L = \ker \pi_V$. Then, in particular, $S \geq \|L\| = \|L\pi_V\| \geq c$ and $\eta \leq 2\sqrt{c/S} \leq 2$.

Using (16.1) and the inequality $(1 + \eta)\varphi(\eta) \leq \eta\alpha \leq \eta\alpha S\sqrt{n}/c$, we calculate

$$|P + \eta L|_H = |P + \eta L\pi_V|_H \geq |P + \eta L\pi_W|_H - \eta\|L\|\|\pi_V - \pi_W|_H$$
$$> \left((1 + \eta)(1 - \varphi(\eta)) - \eta \frac{4\alpha S\sqrt{n}}{c}\right) S$$
$$\geq \left(1 + \eta - 5\eta \frac{\alpha S\sqrt{n}}{c}\right) S.$$

Since $5\eta\alpha S\sqrt{n}/c \leq 10\alpha\sqrt{n}/\eta \leq 1$, the last expression is positive. By squaring we obtain

$$|P|_H^2 + 2\eta\langle P, L\rangle_H + \eta^2|L|_H^2 \geq \left(1 + \eta - 5\eta \frac{\alpha S\sqrt{n}}{c}\right)^2 S^2$$
$$\geq \left(1 + 2\eta + \eta^2 - 10(1 + \eta)\eta \frac{\alpha S\sqrt{n}}{c}\right)S^2.$$

Hence $2\langle P, L\rangle_H \geq (2 - 10(1 + \eta)\alpha S\sqrt{n}/c)S^2$ and

$$|P - L|_H^2 = |P|_H^2 + |L|_H^2 - 2\langle P, L\rangle_H$$
$$\leq 2S^2 - \left(2 - 10(1 + \eta) \frac{\alpha S\sqrt{n}}{c}\right)S^2 < \eta^2 S^3/c. \qquad \square$$

16.3   THE ALGORITHM

As in all our constructions of points of differentiability, we start with the choice of suitable parameters to measure how far we are from differentiability. We let $\varepsilon_k = 2^{-k}$ and, recalling the function φ introduced before Lemma 16.2.2, we denote $\alpha_0 = \varphi(\varepsilon_0)$ and

$$\alpha_k = \min\left\{\frac{\alpha_{k-1}}{16}, \varphi(\varepsilon_k)\right\} \quad \text{for } k \geq 1.$$

We intend recursively to construct a sequence (x_k) of points in H that will converge to a Fréchet differentiability point of f. In addition to the points x_k we will also define subspaces $X_k \in G(H, 2)$ and positive numbers η_k, s_k, δ_k, Δ_k, and ξ_k. It will in fact be more convenient to consider x_k and X_k together as a pair (x_k, X_k) that we will require to belong to $\mathcal{D}_2(f)$.

In the starting step $k = 0$, the choice of our objects will be nearly arbitrary; the main point of this step is to modify the function f so that future estimates will become easier to manage. Denote $E_0 = \mathcal{D}_2(f)$ and choose $(x_0, X_0) \in E_0$ and $\Delta_0 > 0$ arbitrarily. Let $L_0 \in L(H, \mathbb{R}^2)$ be such that $|L_0 x| = \frac{3}{4} \|\pi_{X_0} x\|$ for every $x \in H$. Hence $\|L_0\| = \frac{3}{4}$ and $|L_0|_H = \sqrt{2} \|L_0\|$.

Choose $0 < \eta_0 \le 1$ such that

$$\eta_0 < \frac{\alpha_0}{5}, \quad 4\sqrt{\varphi(\eta_0)} < \frac{\alpha_0}{10}, \quad \|L_0\| \left(1 + \frac{\varphi(\eta_0)}{2}\right) e^{\eta_0} \le 1, \qquad (16.2)$$

and put $s_0 = \frac{1}{2}\varphi(\eta_0)$. Notice that

$$s_0 \le \frac{\eta_0}{2} < \frac{\alpha_0}{10} \le \frac{1}{10}.$$

Finally, let $\sigma > 0$ be small enough that the function $\sigma\big(f - D_{X_0} f(x_0)\big)$ has Lipschitz constant less than $\|L_0\| s_0 = \frac{1}{\sqrt{2}} |L_0|_H s_0$. Define now

$$f_0(x) = L_0 x + \sigma\big(f(x) - D_{X_0} f(x_0)(x)\big).$$

We observe that $D_{X_0} f_0(x_0) = L_0$ and

$$\mathrm{Lip}(f_0) \le \|L_0\| + \|L_0\| s_0 = \|L_0\| \left(1 + \frac{\varphi(\eta_0)}{2}\right) < 1.$$

Let $S_0 := \sup_{(x,X) \in E_0} |D_X f_0(x)|_H$. Since

$$S_0 \le |L_0|_H + |L_0|_H s_0 = |L_0|_H (1 + s_0),$$

it follows that $|L_0|_H > S_0(1 - s_0)$ and hence the pair (x_0, X_0) belongs to the set

$$D_0 := \{(x, X) \in E_0 \mid |D_X f_0(x)|_H > S_0(1 - s_0)\}.$$

Although not all the following observations will be used, it is convenient to know that elements $(x, X) \in D_0$ have properties matching those of Lemma 16.3.1 for elements of D_k, $k > 0$. Some of them are obvious, since

$$\varepsilon(f_0, x, L_0, \delta) \le \mathrm{Lip}(f_0 - L_0) = \|L_0\| s_0 < 1 = \varepsilon_0$$

for every $\delta > 0$. The other two estimates need a little argument. Observe that for every $(x, X) \in \mathcal{D}_2(f)$,

$$\big||D_X f_0(x)|_H - |L_0 \pi_X|_H\big| \le |D_X f_0(x) - L_0 \pi_X|_H \le \sqrt{2}\,\mathrm{Lip}(f_0 - L_0) \le |L_0|_H s_0.$$

Hence for $(x, X) \in D_0$,

$$|L_0 \pi_X|_{\mathrm{H}} \geq |D_X f_0(x)|_{\mathrm{H}} - |L_0|_{\mathrm{H}} s_0 \geq S_0(1 - s_0) - S_0 s_0 \geq 0.$$

By squaring the first of these inequalities we obtain

$$
\begin{aligned}
|L_0 \pi_X|_{\mathrm{H}}^2 &\geq \left(|D_X f_0(x)|_{\mathrm{H}} - |L_0|_{\mathrm{H}} s_0\right)^2 \\
&\geq |D_X f_0(x)|_{\mathrm{H}}^2 - 2|D_X f_0(x)|_{\mathrm{H}} |L_0|_{\mathrm{H}} s_0 \\
&\geq S_0^2 (1 - s_0)^2 - 2 S_0^2 s_0 \\
&\geq S_0^2 (1 - 4 s_0) \geq |L_0|_{\mathrm{H}}^2 (1 - 4 s_0) = \|L_0\|^2 (2 - 8 s_0).
\end{aligned}
$$

On the other hand, using that $L_0 = L_0 \pi_{X_0}$ we have

$$
\begin{aligned}
|L_0 \pi_X|_{\mathrm{H}}^2 &\leq \|L_0\|^2 |\pi_{X_0} \pi_X|_{\mathrm{H}}^2 \\
&= \|L_0\|^2 \left(|\pi_{X_0}|_{\mathrm{H}}^2 - |\pi_{X_0} \pi_{X^\perp}|_{\mathrm{H}}^2\right) \\
&\leq \|L_0\|^2 \left(2 - \|\pi_{X_0} \pi_{X^\perp}\|^2\right) \leq \|L_0\|^2 \left(2 - \frac{\|\pi_{X_0} - \pi_X\|^2}{4}\right).
\end{aligned}
$$

Combining the last two estimates we obtain $\|\pi_X - \pi_{X_0}\|^2 \leq 32 s_0 = 16 \varphi(\eta_0)$. The second condition in (16.2) now implies that

$$\|\pi_X - \pi_{X_0}\| < \frac{\alpha_0}{10}. \tag{16.3}$$

Since $\|D_X f_0(x) - L_0 \pi_X\| \leq \mathrm{Lip}(f_0 - L_0) \leq \|L_0\| s_0 < \alpha_0/10$, we also have

$$\|D_X f_0(x) - L_0\| \leq \|D_X f_0(x) - L_0 \pi_X\| + \|L_0\| \|\pi_X - \pi_{X_0}\| < \frac{\alpha_0}{5}.$$

For future use, we also observe that for $(x, X) \in D_0$ and $u \in X$,

$$
\begin{aligned}
|D_X f_0(x)(u)| &\geq |L_0 \pi_X u| - \|L_0\| s_0 \|u\| \\
&= \|L_0\| \|\pi_{X_0} \pi_X u\| - \|L_0\| s_0 \|u\| \\
&\geq \|L_0\| \|u\| - \|L_0\| \|\pi_X - \pi_{X_0}\| \|u\| - \|L_0\| s_0 \|u\| \\
&\geq \|L_0\| \left(1 - \frac{1}{10} - \frac{1}{10}\right) \|u\| = \frac{3}{5} \|u\|. \tag{16.4}
\end{aligned}
$$

The values of δ_0 and ξ_0 can be arbitrary, but for definiteness we let

$$\delta_0 = \xi_0 = \Delta_0.$$

Having thus defined all our objects in step 0, we proceed recursively, defining the kth objects by the following lemma.

Lemma 16.3.1. *For all $k \geq 1$ there are pairs $(x_k, X_k) \in D_{k-1}$ and positive numbers η_k, s_k, δ_k, Δ_k, and ξ_k such that, denoting*

$$f_k = f_{k-1} + \eta_k D_{X_k} f_{k-1}(x_k), \tag{16.5}$$

$$L_k = D_{X_k} f_k(x_k) = (1 + \eta_k) D_{X_k} f_{k-1}(x_k), \tag{16.6}$$

$$E_k = \big\{(x, X) \in D_{k-1} \;\big|\; x \in B(x_k, \Delta_k),\; \varepsilon_{X_k}(f_k, x, L_k, \delta_k) < \alpha_k \big\}, \tag{16.7}$$

$$S_k = \sup\big\{ |D_X f_k(x)|_{\mathrm{H}} \;\big|\; (x, X) \in E_k \big\}, \tag{16.8}$$

$$D_k = \big\{(x, X) \in E_k \;\big|\; |D_X f_k(x)|_{\mathrm{H}} > S_k(1 - s_k) \big\}, \tag{16.9}$$

we have that $S_k \geq S_{k-1}$, $(x_k, X_k) \in D_k$ and for every $(x, X) \in D_k$,

$$\|D_X f_k(x) - L_k\| < \frac{\alpha_k}{5}, \quad \|\pi_X - \pi_{X_k}\| < \frac{\alpha_k}{5}, \quad \text{and} \quad \varepsilon(f_k, x, L_k, \xi_k) \leq \varepsilon_k.$$

Proof. Choose η_k to satisfy the following requirements:

$$\eta_k \leq \frac{S_0}{4\sqrt{2}} \varphi(\eta_{k-1}), \quad \eta_k < \frac{\alpha_k}{20}, \quad \chi(2\sqrt{2}\eta_k) \leq \frac{\alpha_k}{5}. \tag{16.10}$$

We notice, in particular, that the first condition also implies $\eta_k \leq \frac{1}{2}\eta_{k-1}$. Define f_k and L_k by (16.5) and (16.6), respectively. We will often use without any reference the simple fact that

$$\mathrm{Lip}(f_i) \leq \mathrm{Lip}(f_0) \prod_{j=1}^{i}(1 + \eta_j) \leq \mathrm{Lip}(f_0) e^{\eta_0} \leq 1$$

to replace $\mathrm{Lip}(f_i)$ or $\|D_X f_i(x)\|$ by one in various estimates. We will also use the corollary that for $0 \leq i < j$,

$$\mathrm{Lip}(f_j - f_i) \leq \sum_{m=i+1}^{j} \eta_m \mathrm{Lip}(f_{m-1}) \leq \sum_{m=i+1}^{\infty} \eta_m \leq 2\eta_{i+1}.$$

Moreover, observe that $f_i - f_j$ are linear maps, so in particular

$$\mathcal{D}_2(f_i) = \mathcal{D}_2(f_j) = \mathcal{D}_2(f).$$

Let $s_k = \frac{1}{2}\varphi(\eta_k)$. Noticing that in the definition of S_{k-1} the supremum can be taken over $(x, X) \in D_{k-1}$ only, we can choose a pair $(x_k, X_k) \in D_{k-1}$ such that $|D_{X_k} f_{k-1}(x_k)|_{\mathrm{H}} > S_{k-1}(1 - s_k)$. It follows that

$$|D_{X_k} f_k(x_k)|_{\mathrm{H}} = (1 + \eta_k)|D_{X_k} f_{k-1}(x_k)|_{\mathrm{H}} > (1 + \eta_k)S_{k-1}(1 - s_k). \tag{16.11}$$

Since the derivative $D_{X_k} f_k$ at x_k exists, we may find $\delta_k > 0$ such that

$$\varepsilon_{X_k}(f_k, x_k, L_k, \delta_k) < \alpha_k.$$

We will need, however, a little better estimate of the value of $\varepsilon_{X_k}(f_k, x_k, L_k, \delta_k)$. Since the sets D_i and E_i are nested,

$$E_0 \supset D_0 \supset E_1 \supset \cdots \supset E_{k-1} \supset D_{k-1},$$

the pair (x_k, X_k) belongs to E_i for each $i < k$. Hence there is $0 < \beta_k < 1$ such that

$$\varepsilon_{X_i}(f_i, x_k, L_i, \delta_i) < \alpha_i - \beta_k \quad \text{for all } 0 \le i \le k.$$

Finally, choose $0 < \Delta_k \le \frac{1}{8}\beta_k\delta_k$ such that $B(x_k, 2\Delta_k) \subset B(x_{k-1}, \Delta_{k-1})$ and put $\xi_k = \varphi(\varepsilon_k)\Delta_k$. Having chosen these parameters, we have also defined the remaining objects E_k, S_k, and D_k by (16.7), (16.8), and (16.9), respectively. The rest of the proof deals with the verification of their claimed properties.

Since $(x_k, X_k) \in E_k \subset E_{k-1}$, we have that for every $(x, X) \in E_{k-1}$,

$$|D_X f_k(x)|_H \le |D_X f_{k-1}(x)|_H + \eta_k |D_{X_k} f_{k-1}(x_k)\pi_X|_H \le (1 + \eta_k)S_{k-1},$$

and we infer that $S_k \le (1 + \eta_k)S_{k-1}$. Together with (16.11) this implies two things. First, due to the choice of $s_k = \frac{1}{2}\varphi(\eta_k) \le \frac{1}{2}\eta_k$, we have

$$S_k \ge |D_{X_k} f_k(x_k)|_H \ge (1 + \eta_k)S_{k-1}(1 - s_k) \ge S_{k-1}.$$

Second, we infer that

$$|D_{X_k} f_k(x_k)|_H > (1 + \eta_k)S_{k-1}(1 - s_k) \ge S_k(1 - s_k),$$

which implies that $(x_k, X_k) \in D_k$.

To prove the three inequalities required by the lemma, let $(x, X) \in D_k$. For showing the first inequality, it is our intention to use Lemma 16.2.2 with $L = D_{X_k} f_{k-1}(x_k)$, $P = D_X f_{k-1}(x)$, $V = X_k$, and $W = X$.

Let $u \in H$. Using (16.4) and $\eta_1 \le 1/20$ gives

$$\begin{aligned}
|D_{X_k} f_{k-1}(x_k)(u)| &= |D_{X_k} f_{k-1}(x_k)(\pi_{X_k} u)| \\
&\ge |D_{X_k} f_0(x_k)(\pi_{X_k} u)| - \|D_{X_k}(f_{k-1} - f_0)(x_k)\| \|\pi_{X_k} u\| \\
&\ge |D_{X_k} f_0(x_k)(\pi_{X_k} u)| - \mathrm{Lip}(f_{k-1} - f_0)\|\pi_{X_k} u\| \\
&\ge \frac{3}{5}\|\pi_{X_k} u\| - 2\eta_1\|\pi_{X_k} u\| \ge \frac{1}{2}\|\pi_{X_k} u\|.
\end{aligned}$$

This inequality shows that the first assumption of Lemma 16.2.2 holds with $c = \frac{1}{2}$. The remaining assumption is of the form

$$\left|D_X f_{k-1}(x) + \eta_k D_{X_k} f_{k-1}(x_k)\pi_X\right|_H > (1 + \eta_k)(1 - \varphi(\eta_k))\,S,$$

where $S = \max\{|D_{X_k} f_{k-1}(x_k)|_H, |D_X f_{k-1}(x)|_H\}$. Noting that the left-hand side is $|D_X f_k(x)|_H$, we use that $(x, X) \in D_k$ to estimate that

$$\begin{aligned}
|D_X f_k(x)|_H &> S_k(1 - s_k) \ge |D_{X_k} f_k(x_k)|_H(1 - s_k) \\
&= (1 + \eta_k)(1 - s_k)|D_{X_k} f_{k-1}(x_k)|_H \\
&> (1 + \eta_k)(1 - \varphi(\eta_k))|D_{X_k} f_{k-1}(x_k)|_H.
\end{aligned}$$

Similarly, using the inequalities $S_k \geq (1+\eta_k)(1-s_k)S_{k-1}$ and $(1-s_k)^2 > (1-\varphi(\eta_k))$ we obtain

$$\begin{aligned}
|D_X f_k(x)|_H &\geq S_k(1-s_k) \geq (1+\eta_k)(1-s_k)^2 S_{k-1} \\
&\geq (1+\eta_k)(1-s_k)^2 |D_X f_{k-1}(x)|_H \\
&> (1+\eta_k)(1-\varphi(\eta_k))|D_X f_{k-1}(x)|_H.
\end{aligned}$$

Having verified all assumptions of Lemma 16.2.2 we apply it together with the estimate $S \leq \sqrt{2}$ to get $\|\pi_X - \pi_{X_k}\| \leq 3\eta_k < \frac{1}{5}\alpha_k$, which is the second required inequality, and $\|D_{X_k} f_{k-1}(x_k) - D_X f_{k-1}(x)\| \leq 3\eta_k$. The first required inequality now follows from

$$\begin{aligned}
\|D_X f_k(x) - L_k\| &= \|D_{X_k} f_k(x_k) - D_X f_k(x)\| \\
&\leq \|D_{X_k} f_{k-1}(x_k) - D_X f_{k-1}(x)\| + \eta_k \|(D_{X_k} f_{k-1}(x_k))\pi_{X^\perp}\| \\
&\leq 4\eta_k < \frac{\alpha_k}{5}.
\end{aligned}$$

We still have to show that $\varepsilon(f_k, x, L_k, \xi_k) \leq \varepsilon_k$. Assume for a contradiction that $\varepsilon(f_k, x, L_k, \xi_k) > \varepsilon_k$. Since $\xi_k/\varphi(\varepsilon_k) = \Delta_k \leq \delta_k$, we have

$$\varepsilon_{X_k}(f_k, x, L_k, \xi_k/\varphi(\varepsilon_k)) \leq \varepsilon_{X_k}(f_k, x, L_k, \delta_k) < \alpha_k \leq \varphi(\varepsilon_k).$$

Hence Lemma 16.2.1 with $\varepsilon = \varepsilon_k$ (in which we recall that c_0 and C_0 were replaced by $\varphi(\varepsilon_k)$ and $1/\varphi(\varepsilon_k)$, respectively) provides us with $(y_0, Y) \in \mathcal{D}_2(f_k)$ and $c > \varphi(\varepsilon)$ such that $\|y_0 - x\| < \xi_k/\varphi(\varepsilon_k) = \Delta_k$, $|D_Y f_k(y_0)|_H \geq |L_k|_H + c$ and

$$|(f_k(y_0 + u) - f_k(y_0)) - (f_k(x + u) - f_k(x))| \leq \chi(c)\|u\| \tag{16.12}$$

for every $u \in X_k$.

We show that $(y_0, Y) \in D_i$ for every $i = 0, 1, \ldots, k-1$. (Notice that we do not claim this for the value $i = k$.) Fix an i, $0 \leq i \leq k-1$, and observe that $y_0 \in B(x_k, 2\Delta_k) \subset B(x_i, \Delta_i)$. Recalling by (16.11) that $|L_k|_H \geq S_{k-1} \geq S_i$, we see that

$$\begin{aligned}
|D_Y f_i(y_0)|_H &\geq |D_Y f_k(y_0)|_H - \sqrt{2}\operatorname{Lip}(f_k - f_i) \\
&> |L_k|_H + c - 2\sqrt{2}\eta_{i+1} \geq S_i - 2\sqrt{2}\eta_{i+1} + c.
\end{aligned} \tag{16.13}$$

In particular, $|D_Y f_i(y_0)|_H > S_i - 2\sqrt{2}\eta_{i+1} > S_i(1-s_i)$ by the first condition imposed upon the choice of the numbers η_k in (16.10). It suffices now to show that $(y_0, Y) \in E_i$. Since clearly $(y_0, Y) \in E_0$, we will suppose by induction that $0 < i \leq k-1$ and $(y_0, Y) \in D_{i-1}$. Our aim is to prove that

$$\varepsilon_{X_i}(f_i, y_0, L_i, \delta_i) < \alpha_i.$$

Using that $(y_0, Y) \in D_{i-1}$ and (16.13) with i replaced by $i-1$, we see that

$$S_{i-1} \geq |D_Y f_{i-1}(y_0)|_H \geq S_{i-1} - 2\sqrt{2}\eta_i + c.$$

Consequently, $c \leq 2\sqrt{2}\eta_i$, which implies that $\chi(c) \leq \chi(2\sqrt{2}\eta_i) \leq \frac{1}{5}\alpha_i$ by the third requirement in (16.10). Recalling that both $(x_k, X_k), (x_i, X_i) \in D_i$ we infer from the already proved two inequalities of the present lemma that

$$\|D_{X_k} f_i(x_k) - D_{X_i} f_i(x_i)\| < \frac{\alpha_i}{5} \quad \text{and} \quad \|\pi_{X_k} - \pi_{X_i}\| < \frac{\alpha_i}{5}.$$

Since also $(x, X) \in D_k$ the same argument implies that $\|\pi_X - \pi_{X_k}\| < \alpha_k/5$. We use this to estimate the increment of f at the point x in the direction of X_k. Notice that we do not know that $(x, X_k) \in D_k$, so we have to do it in a slightly roundabout way:

$$
\begin{aligned}
|f_k(x &+ \pi_{X_k} u) - f_k(x) - L_k \pi_{X_k} u| \\
&\leq |f_k(x + \pi_X \pi_{X_k} u) - f_k(x) - L_k \pi_X \pi_{X_k} u| + 2\|\pi_X - \pi_{X_k}\|\|u\| \\
&\leq \alpha_k \|\pi_{X_k} u\| + \frac{2\alpha_k}{5}\|u\| \leq \frac{7\alpha_k}{5}\|u\| \leq \frac{7\alpha_{i+1}}{5}\|u\|.
\end{aligned}
$$

This and (16.12) lead to the conclusion that for $u \in X_i \cap B(0, \delta_k)$,

$$
\begin{aligned}
|f_i(y_0 &+ u) - f_i(y_0) - L_i u| \\
&\leq |f_k(y_0 + u) - f_k(y_0) - L_k u| + \mathrm{Lip}(f_k - f_i)\|u\| + \|L_k - L_i\|\|u\| \\
&\leq |f_k(y_0 + u) - f_k(y_0) - L_k u| + 2\eta_{i+1}\|u\| \\
&\quad + \|L_i - D_{X_k} f_i(x_k)\|\|u\| + \|D_{X_k}(f_i - f_k)(x_k)\|\|u\| \\
&\leq |f_k(y_0 + u) - f_k(y_0) - L_k u| + 2\eta_{i+1}\|u\| + \frac{\alpha_i}{5}\|u\| + \mathrm{Lip}(f_i - f_k)\|u\| \\
&\leq |f_k(y_0 + \pi_{X_k} u) - f_k(y_0) - L_k \pi_{X_k} u| + 2\|\pi_{X_i} - \pi_{X_k}\|\|u\| \\
&\quad + 2\eta_{i+1}\|u\| + \frac{\alpha_i}{5}\|u\| + 2\eta_{i+1}\|u\| \\
&\leq |f_k(x + \pi_{X_k} u) - f_k(x) - L_k \pi_{X_k} u| \\
&\quad + \chi(c)\|u\| + \frac{2\alpha_i}{5}\|u\| + 4\eta_{i+1}\|u\| + \frac{\alpha_i}{5}\|u\| \\
&\leq \frac{7\alpha_{i+1}}{5}\|u\| + \left(\chi(c) + \frac{3\alpha_i}{5} + 4\eta_{i+1}\right)\|u\| \\
&\leq \left(\frac{4\alpha_i}{5} + \frac{8\alpha_{i+1}}{5}\right)\|u\| \leq \frac{9\alpha_i}{10}\|u\|.
\end{aligned}
$$

For $u \in X_i \cap B(0, \delta_i) \setminus B(0, \delta_k)$ we use that

$$\|y_0 - x_k\| \leq 2\Delta_k \leq \frac{1}{4}\beta_k \delta_k \leq \frac{1}{4}\beta_k \|u\|$$

to estimate

$$
\begin{aligned}
|f_i(y_0 + u) - f_i(y_0) - L_i u| &\leq |f_i(x_k + u) - f_i(x_k) - L_i u| + 2\|y_0 - x_k\| \\
&\leq (\alpha_i - \beta_k)\|u\| + \frac{1}{2}\beta_k \|u\| \\
&= \left(\alpha_i - \frac{1}{2}\beta_k\right)\|u\|.
\end{aligned}
$$

These inequalities show that

$$\varepsilon_{X_i}(f_i, y_0, L_i, \delta_i) \leq \max\left\{\frac{9\alpha_i}{10}, \ \alpha_i - \frac{1}{2}\beta_k\right\} < \alpha_i.$$

So $(y_0, Y) \in E_i$, which, as we have already shown, implies that $(y_0, Y) \in D_i$. Having thus proved by induction that $(y_0, Y) \in D_{k-1}$, we infer that

$$S_k(1 - s_k) + \varphi(\varepsilon_k) \leq |D_{X_k} f_k(x_k)|_H + \varphi(\varepsilon_k) \leq |D_Y f_k(y_0)|_H$$
$$\leq |D_Y f_{k-1}(y_0)|_H + \eta_k \sqrt{2} \leq S_{k-1} + \eta_k \sqrt{2} \leq S_k + \eta_k \sqrt{2}.$$

It follows that

$$\eta_k \sqrt{2} \geq \varphi(\varepsilon_k) - S_k s_k \geq \alpha_k - \frac{\varphi(\eta_k)}{\sqrt{2}} \geq \alpha_k - \frac{\eta_k}{\sqrt{2}}.$$

Recalling that $\eta_k < \frac{1}{20}\alpha_k$ we get the desired contradiction. □

16.4   PROOF OF THEOREM 16.1.1

To complete the above arguments, we just need to put the properties of the sequences constructed in the previous section together. Since $(x_j, X_j) \in D_k$ for $j \geq k$, we have $\|x_j - x_k\| < \Delta_k$. The sequence (x_k) converges to some $x_\infty \in H$. Also, for $j \geq k$, the difference $f_j - f_k$ is a linear function of norm at most

$$\text{Lip}(f_j - f_k) \leq \sum_{i=k+1}^{\infty} \eta_i \leq 2\eta_{k+1} < \alpha_{k+1} \leq \varphi(\varepsilon_{k+1}) \leq \varepsilon_k.$$

Hence the sequence (f_j) converges to some function f_∞ and

$$\varepsilon(f_\infty, x_\infty, L_k, \xi_k) \leq \text{Lip}(f_k - f_\infty) + \varepsilon(f_k, x_\infty, L_k, \xi_k)$$
$$\leq \varepsilon_k + \liminf_{j \to \infty} \varepsilon(f_k, x_j, L_k, \xi_k) \leq 2\varepsilon_k.$$

This implies that f_∞ is ε-Fréchet differentiable at the point x_∞ for every $\varepsilon > 0$. Hence f_∞, and also f itself as the sum of a scalar multiple of f_∞ and a linear function, is Fréchet differentiable at x_∞.

16.5   PROOF OF LEMMA 16.2.1

Before the actual proof of Lemma 16.2.1 we state two special cases of integral estimates proved above that will be used here. Their proofs would be slightly simpler than those of the general versions, but not sufficiently so to warrant the repetition. In these lemmas the symbol K denotes a (sufficiently large) constant. First, we need the following special case of the general method of deformation of surfaces from Section 9.3: the function ψ required here is $\psi(u) = \tau\psi_\kappa(u/\tau)$, where ψ_κ comes from Lemma 9.3.2 whose last statement is used with $\theta(u) = |u|^2$.

Lemma 16.5.1. *Given any $\tau > 0$ and $0 < \kappa < 1$, there is a Lipschitz function $\psi \colon \mathbb{R}^2 \longrightarrow \mathbb{R}$ having the following properties.*

(a) $\psi(0) = \tau$,

(b) $\psi(u) = 0$ *whenever* $|u| \geq e^{1/\kappa}\,\tau$,

(c) $|\psi(u)| \leq \tau$ *for all* $u \in \mathbb{R}^2$,

(d) $\|\psi'(u)\| \leq \kappa$ *for almost all* $u \in \mathbb{R}^n$,

(e) $\displaystyle\int_{\mathbb{R}^2} \|\psi'\|^2 \, d\mathscr{L}^2 \leq \frac{K\tau^2}{\log(1/\kappa)}.$

The second result is a special case of two integral estimates from Section 9.5, Corollary 9.5.5 with $n = 2$ and Lemma 9.5.4 with $n = s = 2$. To state it, recall that the Lipschitz constant of g at $u \in \Omega$, where $u \in \Omega$, is the least number $\mathrm{Lip}_u(g) = \mathrm{Lip}_{u,\Omega}(g) \in [0, \infty)$ such that

$$|g(v) - g(u)| \leq \mathrm{Lip}_{u,\Omega}(g)|v - u|$$

for every $v \in \Omega$.

Lemma 16.5.2. *Let $\Omega \subset \mathbb{R}^2$ be either a disk or a square and $g \colon \Omega \longrightarrow \mathbb{R}^2$ a Lipschitz function. Then*

$$\int_\Omega \left(\mathrm{Lip}_{u,\Omega}(g)\right)^4 d\mathscr{L}^2(u) \leq K \mathrm{Lip}(g)^2 \int_\Omega \|g'\|^2 \, d\mathscr{L}^2,$$

and for every $u, v \in \Omega$,

$$|g(v) - g(u)|^3 \leq K \mathrm{Lip}(g)|v - u| \int_\Omega \|g'\|^2 \, d\mathscr{L}^2.$$

Proof of Lemma 16.2.1. To simplify the formulas, we will assume that the constant K from the previous lemmas satisfies $K \geq 1$. We introduce the following parameters. Although we give exact formulas (as this is convenient when recalculating one parameter with the help of another), the main point that should be kept in mind is that, in terms of the parameters from their definition, σ, κ, and c_0 are small and C_1, C_0 are big.

$$\sigma = \frac{\varepsilon^5}{2^{18}},$$

$$\kappa = \exp(-2^9 K^2 \sigma^{-3}),$$

$$c_0 = \min\left\{\frac{\varepsilon}{3}, \frac{\sigma^4}{2^{16} K \pi e^{2/\kappa}}\right\},$$

$$C_1 = \frac{2^7}{\varepsilon^2},$$

$$C_0 = 4C_1 e^{2/\kappa}.$$

We will use a number of inequalities that easily follow from these definitions and the facts that $K \geq 1$ and $\varepsilon \leq 3$ (which is shown below). For example, we will explicitly use in some steps that

$$c_0 < \frac{\sigma}{4}, \quad \kappa < \frac{\sigma}{4}, \quad \frac{4}{\sigma} < e^{1/\kappa}, \quad \text{or} \quad C_1 \geq \max\left\{\frac{2^5}{\varepsilon}, \frac{2^6}{\varepsilon^2}\right\}.$$

Also we may decrease the powers of σ and 2 and leave out K in some estimates of c_0. This agreement indicates that we are not trying to find optimal numerical constants. This is also reflected by making use of powers of 2 only for such constants in almost all occurrences.

All these parameters are determined by the value of $\varepsilon > 0$. In order to define the function $\chi(c)$ it will be convenient to calculate $\varepsilon = \varepsilon(c_0)$ from the third equation above and write c instead of c_0, that is, $\varepsilon = \varepsilon(c)$ as a function of c. Therefore all $\sigma(c)$, $\kappa(c)$, and $c_0(c)$ are now given functions of c. Using this notation, we show that a possible choice for χ is

$$\chi(c) := 8K^{1/4}c^{1/4} + 2e^{-1/\kappa(c)} + 4c_0(c) + 8\kappa(c) + 4\sigma(c) + 4\varepsilon(c).$$

Clearly, $\lim_{t\searrow 0} \chi(c) = 0$ as required. Notice also that the functions $c_0(c)$, $\kappa(c)$, $\sigma(c)$, and $\varepsilon(c)$ are increasing; we will often use this without any further mention because in our (upper) estimates by $\chi(c)$ these functions will be in fact taken at a point $< c$.

Suppose that g, ξ, x, X, and L satisfying the assumptions of the lemma are given. By shifting, we may assume that $x = 0$ and $g(0) = 0$.

We first observe several nonimportant simple estimates. They could have been added to the assumptions of the lemma as their verification in our application is straightforward. However, we prefer to deduce them as this allows us to keep the statement of the lemma (relatively) simple. Since $\|L(y)\| \leq (\text{Lip}(g) + c_0)\|y\|$ for $y \in X$, $\|y\| \leq C_0\xi$, and since $c_0 \leq \frac{1}{3}\varepsilon$, we have $\|L\| = \|L\pi_X\| \leq \text{Lip}(g) + c_0 \leq \text{Lip}(g) + \frac{1}{3}\varepsilon$. The assumption $\varepsilon(g, 0, L, \xi) > \varepsilon$ implies that $\varepsilon < \text{Lip}(g - L)$, and we infer from

$$\varepsilon < \text{Lip}(g - L) \leq \text{Lip}(g) + \|L\| \leq 2\text{Lip}(g) + \tfrac{1}{3}\varepsilon$$

that $\text{Lip}(g) \geq \frac{1}{3}\varepsilon$. As corollaries we immediately get that $\varepsilon \leq 3$ and $\|L\| \leq 2$.

We will consider two cases that correspond to the irregular case and regular case from the previous chapters. Of course, since we do not consider the limit case, we have to distinguish the two cases quantitatively and not only qualitatively as we did before.

It will be convenient to identify X with $\mathbb{R}^2$; in particular, g and L map the Hilbert space H into X.

Case 1 (irregular behavior). *The conclusion of Lemma 16.2.1 holds provided there are $x \in X$ and $y \in H$ with $\|x\|, \|y\| \leq C_1\xi$ such that*

$$\|g(y + x) - g(y) - Lx\| > \sigma(\|y\| + \|x\|).$$

Proof. We first notice that $y \notin X$. Indeed, assuming that $y \in X$, the condition $\|x\| + \|y\| \le 2C_1\xi \le C_0\xi$ would allow us to infer from $\varepsilon_X(g, 0, L, C_0\xi) < c_0$ that

$$
\begin{aligned}
\sigma(\|x\| + \|y\|) &< \|g(x+y) - g(y) - Lx\| \\
&\le \|g(x+y) - L(x+y)\| + \|g(y) - Ly\| \\
&\le c_0\|x+y\| + c_0\|y\| \le 2c_0(\|x\| + \|y\|) \\
&\le \sigma(\|x\| + \|y\|).
\end{aligned}
$$

Denote $w = (y - \pi_X y)/\|y - \pi_X y\|$ and use Lemma 16.5.1 with κ defined above and $\tau = \|y - \pi_X y\|$ to find the function $\psi \colon X \longrightarrow \mathbb{R}$ with the properties (a)–(e).

Let $r = e^{1/\kappa}\|y\|$ and $\Omega = \{u \in X \mid \|u\| \le r\}$. Notice also that the choice of C_0 reveals that $2r \le 2e^{1/\kappa}C_1\xi \le C_0\xi$. Define $\gamma \colon X \longrightarrow H$ by

$$
\gamma(u) = u + \psi(u)w.
$$

and let Y_u be the range of $\gamma'(u)$. Finally, let $L_u = (D_{Y_u}g)(\gamma(u))$ and $h(u) = g(\gamma(u))$. Notice that by (d), $\mathrm{Lip}(h) \le \mathrm{Lip}(\gamma) \le 1 + \kappa \le 2$, and so $\mathrm{Lip}(h - L) \le 4$. Our intention is to find the required pair (y_0, Y) such that $y_0 = \gamma(u_0)$ and $Y = Y_{u_0}$ for a suitable point $u_0 \in \Omega$.

We first relate $|L_u|_{\mathrm{H}}$ and $|h'(u)|_{\mathrm{H}}$: for every $u \in X$,

$$
|L_u|_{\mathrm{H}}^2 \le |h'(u)|_{\mathrm{H}}^2 \le |L_u|_{\mathrm{H}}^2 + \|\psi'(u)\|^2. \tag{16.14}
$$

Indeed, let $e, e^{\perp}$ be an orthonormal basis for X (which may depend on u) such that $\psi'(u)(e^{\perp}) = 0$. Then the vectors $\gamma'(u)(e) = e + \psi'(u; e)w$ and $\gamma'(u)(e^{\perp}) = e^{\perp}$ form an orthogonal basis for Y_u. Hence, denoting by $\widetilde{e}$ the unit vector parallel to $\gamma'(u)(e)$, we have

$$
\begin{aligned}
|h'(u)|_{\mathrm{H}}^2 &= \left\|g'(\gamma(u))(e + \psi'(u; e)w)\right\|^2 + \left\|g'(\gamma(u))(e^{\perp})\right\|^2 \\
&= \|L_u\widetilde{e}\|^2(1 + |\psi'(u; e)|^2) + \|L_u e^{\perp}\|^2 \\
&= |L_u|_{\mathrm{H}}^2 + |\psi'(u; e)|^2\|L_u\widetilde{e}\|^2.
\end{aligned}
$$

The first inequality in (16.14) is now obvious and the second follows by noticing that $|\psi'(u; e)| \le \|\psi'(u)\|$ and $\|L_u\widetilde{e}\| \le \|L_u\| \le \mathrm{Lip}(g) \le 1$.

We will need the following consequence of (16.14). Subtracting $|L|_{\mathrm{H}}^2$ from the second inequality in (16.14), integrating over Ω and using (e) leads to

$$
\begin{aligned}
\int_{\Omega}(|h'|_{\mathrm{H}}^2 - |L|_{\mathrm{H}}^2)\,d\mathscr{L}^2 &\le \int_{\Omega}(|L_u|_{\mathrm{H}}^2 - |L|_{\mathrm{H}}^2)\,d\mathscr{L}^2(u) + \frac{K\tau^2}{\log(1/\kappa)} \\
&= \int_{\Omega}(|L_u|_{\mathrm{H}}^2 - |L|_{\mathrm{H}}^2)\,d\mathscr{L}^2(u) + \frac{\sigma^3\tau^2}{2^9 K}. \tag{16.15}
\end{aligned}
$$

The first statement of Lemma 16.5.2 with $g = h - L$ gives that

$$
\int_{\Omega}\left(\mathrm{Lip}_{u,\Omega}(h - L)\right)^4 d\mathscr{L}^2(u) \le 2^4 K \int_{\Omega}|h' - L|_{\mathrm{H}}^2\,d\mathscr{L}^2. \tag{16.16}
$$

Further,

$$
\int_\Omega |h' - L|_{\mathrm{H}}^2 \, d\mathscr{L}^2 = \int_\Omega \left(|h'|_{\mathrm{H}}^2 + |L|_{\mathrm{H}}^2 - 2\langle h', L \rangle_{\mathrm{H}} \right) d\mathscr{L}^2
$$
$$
= \int_\Omega \left(|h'|_{\mathrm{H}}^2 - |L|_{\mathrm{H}}^2 \right) d\mathscr{L}^2 + 2 \int_\Omega \langle L - h', L \rangle_{\mathrm{H}} \, d\mathscr{L}^2.
$$

Denoting by ν_Ω the outer unit normal vector to Ω we obtain with the help of the divergence theorem

$$
\int_\Omega \langle L - h', L \rangle_{\mathrm{H}} \, d\mathscr{L}^2 = \int_\Omega \operatorname{div} L^*(L - h) \, d\mathscr{L}^2
$$
$$
= \int_{\partial\Omega} L^*(L - h) \cdot \nu_\Omega \, d\mathscr{H}^1
$$
$$
\leq \int_{\partial\Omega} \|L^*\| \, \|h(u) - Lu\| \, d\mathscr{H}^1(u) \leq 4\pi c_0 r^2,
$$

because $\|h(u) - Lu\| \leq c_0 r$ for u with $\|u\| = r$. Combining the last two estimates and recalling that $\pi c_0 r^2 = \pi c_0 e^{2/\kappa} \|y\|^2 \leq 2^{-12} \sigma^3 \|y\|^2 / K$, we have

$$
\int_\Omega |h' - L|_{\mathrm{H}}^2 \, d\mathscr{L}^2 \leq \int_\Omega \left(|h'|_{\mathrm{H}}^2 - |L|_{\mathrm{H}}^2 \right) d\mathscr{L}^2 + \frac{\sigma^3 \|y\|^2}{2^9 K}. \tag{16.17}
$$

To get a lower estimate of the integral on the left, we first notice that

$$
\sigma(\|x\| + \|y\|) < \|g(x + y) - g(y) - Lx\|
$$
$$
\leq \|g(x + \pi_X y) - g(\pi_X y) - Lx\| + 2\|y - \pi_X y\|
$$
$$
\leq \|g(x + \pi_X y) - L(x + \pi_X y)\| + \|g(\pi_X y) - L\pi_X y\| + 2\tau
$$
$$
\leq 2c_0(\|x\| + \|y\|) + 2\tau.
$$

Since $\sigma > 4c_0$ we obtain that $\|x\| + \|y\| < 4\tau/\sigma \leq 4\|y\|/\sigma \leq e^{1/\kappa}\|y\| = r$. Hence both $x + \pi_X y$ and $\pi_X y$ belong to Ω. Now

$$
\|h(\pi_X y + x) - h(\pi_X y) - Lx\|
$$
$$
\geq \|g(y + x) - g(y) - Lx\| - \|\tau w - \psi(\pi_X y + x)w\| - \|\tau w - \psi(\pi_X y)w\|
$$
$$
> \sigma(\|y\| + \|x\|) - |\psi(0) - \psi(\pi_X y + x)| - |\psi(0) - \psi(\pi_X y)|
$$
$$
\geq \sigma(\|y\| + \|x\|) - 2\kappa(\|y\| + \|x\|) \geq \frac{\sigma}{2} (\|y\| + \|x\|).
$$

It follows that the second statement of Lemma 16.5.2 is applicable with $g = h - L$, and we obtain

$$
\int_\Omega \|h' - L\|^2 \, d\mathscr{L}^2 \geq \frac{\|h(\pi_X y + x) - h(\pi_X y) - Lx\|^3}{K \operatorname{Lip}(h - L)\|x\|} \geq \frac{\sigma^3 \|y\|^2}{2^5 K}.
$$

Together with (16.17) this leads to

$$
\int_\Omega \left(|h'|_{\mathrm{H}}^2 - |L|_{\mathrm{H}}^2 \right) d\mathscr{L}^2 \geq \int_\Omega \|h' - L\|^2 \, d\mathscr{L}^2 - \frac{\sigma^3 \|y\|^2}{2^9 K} \geq \frac{\sigma^3 \|y\|^2}{2^6 K}.
$$

Hence in view of (16.15) we conclude that

$$\int_\Omega (|L_u|_H^2 - |L|_H^2) \, d\mathscr{L}^2(u) \geq \int_\Omega (|h'|_H^2 - |L|_H^2) \, d\mathscr{L}^2 - \frac{\sigma^3 \tau^2}{2^9 K} \geq \frac{\sigma^3 \|y\|^2}{2^7 K}. \quad (16.18)$$

In order to place the desired point u_0 far from the boundary of Ω we introduce the following auxiliary function. Let $\zeta \colon \Omega \longrightarrow (0, \infty)$ be defined as

$$\zeta(u) = \begin{cases} 2^8 K c_0 & \text{if } \|u\| \leq r(1 - s_0), \text{ where } s_0 = c_0/\sigma, \\ 2^9 K & \text{otherwise.} \end{cases}$$

Then the choice of c_0 and the equation $r = e^{1/\kappa} \|y\|$ give that

$$\int_\Omega \zeta \, d\mathscr{L}^2 < 2^8 K c_0 \pi r^2 + 2^{10} K s_0 \pi r^2 \leq 2^{-6} \sigma^3 \|y\|^2 + 2^{-6} \sigma^3 \|y\|^2 \leq \frac{\sigma^3 \|y\|^2}{24}.$$

Adding this to (16.16) and observing that $2^4 K$ multiples of the error terms in (16.17) and (16.15) are majorized by $\sigma^3 \|y\|^2 / 24$, we get

$$\int_\Omega \left(\mathrm{Lip}_{u,\Omega}(h - L) \right)^4 + \zeta(u) \, d\mathscr{L}^2(u)$$

$$< 2^4 K \int_\Omega |h' - L|_H^2 \, d\mathscr{L}^2 + \frac{\sigma^3 \|y\|^2}{24}$$

$$\leq 2^4 K \int_\Omega (|h'|_H^2 - |L|_H^2) \, d\mathscr{L}^2 + \frac{2\sigma^3 \|y\|^2}{24} \qquad \text{by (16.17)}$$

$$\leq 2^4 K \int_\Omega (|L_u|_H^2 - |L|_H^2) \, d\mathscr{L}^2(u) + \frac{3\sigma^3 \|y\|^2}{24} \qquad \text{by (16.15)}$$

$$= 2^4 K \left(\int_\Omega (|L_u|_H^2 - |L|_H^2) \, d\mathscr{L}^2(u) + \frac{\sigma^3 \|y\|^2}{2^7 K} \right)$$

$$\leq 2^5 K \int_\Omega (|L_u|_H^2 - |L|_H^2) \, d\mathscr{L}^2(u). \qquad \text{by (16.18)}$$

Hence there is $u_0 \in \Omega$ such that $h'(u_0)$ exists and

$$\mathrm{Lip}_{u_0,\Omega}(h - L)^4 + \zeta(u_0) < 2^5 K (|L_{u_0}|_H - |L|_H)(|L_{u_0}|_H + |L|_H).$$

In particular, the right side is positive, and so, since $|L_{u_0}|_H + |L|_H \leq 3\sqrt{2} \leq 8$,

$$\mathrm{Lip}_{u_0,\Omega}(h - L)^4 + \zeta(u_0) < 2^8 K (|L_{u_0}|_H - |L|_H). \quad (16.19)$$

Since the right side is at most $2^9 K$, we infer that $\zeta(u_0) < 2^9 K$, and so $\|u_0\| \leq r(1 - s_0)$ and $\zeta(u_0) = 2^8 K c_0$. Hence (16.19) gives that

$$c := |L_{u_0}|_H - |L|_H > \frac{\zeta(u_0)}{2^8 K} = c_0.$$

We show that $y_0 = \gamma(u_0)$ and the subspace $Y = Y_{u_0} \in G(H, 2)$ have the required properties. Clearly, $(y_0, Y) \in \mathcal{D}_2(g)$ and

$$\|y_0\| \le \|u_0\| + |\psi(u_0)| \le r(1 - s_0) + \tau \le \frac{1}{2} C_0 \xi.$$

Let $u \in X$. We distinguish the following three cases.

(A) If $u_0 + u \in \Omega$ then (16.19) implies

$$\|h(u_0 + u) - h(u_0) - Lu\| \le \mathrm{Lip}_{u_0, \Omega}(h - L)\|u\| \le 4K^{1/4} c^{1/4}\|u\|.$$

Since $\|u\| \le \|u_0 + u\| + \|u_0\| \le 2r \le C_0 \xi$ and $\mathrm{Lip}(\psi) \le \kappa$,

$$\begin{aligned}
\|g(y_0 &+ u) - g(y_0) - g(u)\| \\
&\le \|h(u_0 + u) - h(u_0) - Lu\| + \|g(u) - Lu\| + \|g(y_0 + u) - g(\gamma(u_0 + u))\| \\
&\le 4K^{1/4} c^{1/4}\|u\| + c_0\|u\| + \|\psi(u_0)w - \psi(u_0 + u)w\| \\
&\le 4K^{1/4} c^{1/4}\|u\| + c_0\|u\| + \kappa\|u\| \le \frac{\chi(c)}{2}\|u\|.
\end{aligned}$$

This is clearly bounded by the required $\chi(c)\|u\|$, but we will need the just established better estimate shortly.

(B) When $u_0 + u \notin \Omega$ and $\|u\| \le \frac{1}{2} C_0 \xi$ we find v on the segment $[u_0, u_0 + u]$ with $\|v\| = r$. It follows that $\|u\| \ge r - \|u_0\| \ge s_0 r$, and so

$$\|v\| \le \frac{\|u\|}{s_0} \quad \text{and} \quad \|u_0 + u\| \le r(1 - s_0) + \|u\| \le \frac{\|u\|}{s_0}.$$

Also

$$\|y_0 - u_0\| = |\psi(u_0)| = |\psi(u_0) - \psi(v)| \le \kappa\|u_0 - v\| \le \kappa\|u\|.$$

The previous case now gives

$$\begin{aligned}
\|g(v) &- g(u_0) - g(v - u_0)\| \\
&\le \|g(y_0 + v - u_0) - g(y_0) - g(v - u_0)\| + 2\|y_0 - u_0\| \\
&\le \frac{\chi(c)}{2}\|v - u_0\| + 2\kappa\|u\| \\
&\le \left(\frac{\chi(c)}{2} + 2\kappa\right)\|u\|.
\end{aligned}$$

Since the vectors $u, v, v - u_0$, and $u_0 + v$ have norm at most $C_0\xi$ (they are bounded by $\|u\|, r, \|u\|$ and $2r \le C_0\xi$, respectively), we can estimate

$$\|g(y_0 + u) - g(y_0) - g(u)\|$$
$$\le 2\|y_0 - u_0\| + \|g(u_0 + u) - g(u_0) - g(u)\|$$
$$\le 2\kappa\|u\| + \|g(u_0 + u) - g(u_0) - g(u)\|$$
$$\le 2\kappa\|u\| + \|g(v) - g(u_0) - g(v - u_0)\| + \|g(u_0 + u) - L(u_0 + u)\|$$
$$\qquad + \|g(v) - Lv\| + \|g(u) - Lu\| + \|g(v - u_0) - L(v - u_0)\|$$
$$\le 2\kappa\|u\| + \left(\frac{\chi(c)}{2} + 2\kappa\right)\|u\| + c_0\big(\|u_0 + u\| + \|v\| + \|u\| + \|v - u_0\|\big)$$
$$\le \left(4\kappa + \frac{\chi(c)}{2} + \frac{2c_0}{s_0} + 2c_0\right)\|u\| = \left(4\kappa + \frac{\chi(c)}{2} + 2\sigma + 2c_0\right)\|u\|$$
$$\le \chi(c)\|u\|.$$

(C) Finally, when $\|u\| \ge \frac{1}{2}C_0\xi$, we have that

$$\|g(y_0 + u) - g(y_0) - g(u)\| \le \|g(y_0 + u) - g(u)\| + \|g(y_0) - g(0)\|$$
$$\le 2\|y_0\| \le 2(\|u_0\| + \tau) \le 4r \le 4e^{1/\kappa}\,C_1\xi$$
$$\le \frac{8C_1 e^{1/\kappa}}{C_0}\|u\| = 2e^{-1/\kappa}\|u\|$$
$$\le \chi(c)\|u\|. \qquad\qquad \square$$

Case 2 (regular behavior). *The conclusion of Lemma 16.2.1 holds provided*

$$\|g(y + x) - g(y) - Lx\| \le \sigma(\|y\| + \|x\|)$$

for every $x \in X$ and $y \in H$ with $\|x\|, \|y\| \le C_1\xi$.

Proof. The assumption $\varepsilon(g, 0, L, \xi) > \varepsilon$ allows us to find $z_0 \in H, \|z_0\| \le \xi$ such that $\|g(z_0) - Lz_0\| > \varepsilon\|z_0\|$. Letting $z = z_0 - \pi_X z_0$, we have

$$\|g(z)\| = \big\|g(z_0) - L(z_0) + \big(g(z_0 - \pi_X z_0) - g(z_0) - L(-\pi_X z_0)\big)\big\|$$
$$> \varepsilon\|z_0\| - \sigma(\|z_0\| + \|\pi_X z_0\|) \ge \frac{\varepsilon}{2}\|z_0\|$$
$$\ge \frac{\varepsilon}{2}\|z\|.$$

We choose an orthonormal basis e_1, e_2 for X such that Le_1 is a non-negative multiple of $g(z)$. Denote

$$\beta = 1 - \frac{\varepsilon^2}{2^6},$$

and let $r > 0$ be such that $\beta^2 r^2 + \|z\|^2 = r^2$. Then $r \ge \|z\| \ge \frac{1}{8}\varepsilon r$. For future use we list a chain of inequalities various pairs of which will be used in various situations:

$$r \le \frac{8\|z\|}{\varepsilon} \le \frac{8\xi}{\varepsilon} \le \frac{1}{4}C_1\xi.$$

The set Ω will be in this case the square

$$\Omega = \{se_1 + te_2 \mid s, t \in [0, r]\} \subset X.$$

Define $\gamma \colon X \longrightarrow H$ by

$$\gamma(s, t) = \beta s e_1 + t e_2 + \frac{s}{r} z$$

and observe that γ is a linear isometry of X onto its range. We denote this range by Y. Hence, letting $L_u = (D_Y g)(\gamma(u))$ and $h(u) = g(\gamma(u))$, we have that

$$h'(u) = L_u \circ \gamma \quad \text{and} \quad |h'(u)|_{\mathrm{H}} = |L_u|_{\mathrm{H}}.$$

Since $r \geq \|z\| \geq \frac{1}{8}\varepsilon r$, the vector $w := g(z)/r$ satisfies

$$1 \geq \|w\| \geq \frac{\varepsilon\|z\|}{2r} \geq \frac{\varepsilon^2}{2^4}.$$

In particular,

$$(1 - \beta)\|Le_1\| \leq 2(1 - \beta) \leq \|w\| \text{ and } (1 - \beta^2)\|Le_1\| \leq 4(1 - \beta) \leq 2\beta\|w\|.$$

We use these inequalities to estimate the Hilbert-Schmidt norm of the linear map $M \in L(X, X)$ given by

$$Me_1 := \beta Le_1 + w \quad \text{and} \quad Me_2 := Le_2.$$

Since Le_1 is a non-negative multiple of w, $|M|_{\mathrm{H}}^2 = (\beta\|Le_1\| + \|w\|)^2 + \|Le_2\|^2$, showing that $|M|_{\mathrm{H}} \leq 4$ and

$$|M|_{\mathrm{H}}^2 = |L|_{\mathrm{H}}^2 + \big((\beta^2 - 1)\|Le_1\| + 2\beta\|w\|\big)\|Le_1\| + \|w\|^2 \geq |L|_{\mathrm{H}}^2 + \|w\|^2. \quad (16.20)$$

Similarly we get

$$\langle M, L \rangle_{\mathrm{H}} = |L|_{\mathrm{H}}^2 + \big((\beta - 1)\|Le_1\| + \|w\|\big)\|Le_1\| \geq |L|_{\mathrm{H}}^2. \quad (16.21)$$

Since $\|\beta s e_1 + t e_2\| \leq 2r \leq C_1\xi$ for every $0 \leq s, t \leq r$, we have the following two estimates.

$$\begin{aligned}
\|h(r, t) - h(0, t) - rMe_1\| &\leq \|h(r, t) - rMe_1 - L(te_2)\| + \|h(0, t) - L(te_2)\| \\
&= \|g(z + \beta r e_1 + te_2) - g(z) - L(\beta r e_1 + te_2)\| \\
&\quad + \|g(te_2) - g(0) - L(te_2)\| \\
&\leq 4\sigma r,
\end{aligned}$$

and, using also $\|\gamma(s, 0)\| \leq 2r \leq C_1\xi$ for $0 \leq s \leq r$,

$$\|h(s, r) - h(s, 0) - rMe_2\| = \|g(\gamma(s, 0) + re_2) - g(\gamma(s, 0)) - L(re_2)\| \leq 4\sigma r.$$

Hence for any operator $T \in L(X, X)$,

$$\int_\Omega \langle h'(u), T \rangle_{\mathrm{H}} \, d\mathscr{L}^2(u) \tag{16.22}$$

$$= \int_\Omega \langle h'(u)(e_1), Te_1 \rangle \, d\mathscr{L}^2(u) + \int_\Omega \langle h'(u)(e_2), Te_2 \rangle \, d\mathscr{L}^2(u)$$

$$= \int_0^r \langle h(r, t) - h(0, t), Te_1 \rangle \, dt + \int_0^r \langle h(s, r) - h(s, 0), Te_2 \rangle \, ds$$

$$\geq r \int_0^r \langle Me_1, Te_1 \rangle \, dt + r \int_0^r \langle Me_2, Te_2 \rangle \, ds - 8\sigma r^2 \|T\|$$

$$= r^2 \langle M, T \rangle_{\mathrm{H}} - 8\sigma r^2 \|T\|. \tag{16.23}$$

With $T = M/|M|_{\mathrm{H}}$ this inequality yields

$$\int_\Omega |L_u|_{\mathrm{H}} \, d\mathscr{L}^2(u) = \int_\Omega |h'(u)|_{\mathrm{H}} \, d\mathscr{L}^2(u) \geq \int_\Omega \langle h', T \rangle_{\mathrm{H}} \, d\mathscr{L}^2 \geq r^2 |M|_{\mathrm{H}} - 8\sigma r^2.$$

By (16.20) and the Cauchy-Schwarz inequality this implies

$$\int_\Omega (|L_u|_{\mathrm{H}}^2 - |L|_{\mathrm{H}}^2) \, d\mathscr{L}^2(u) \geq |M|_{\mathrm{H}}^2 r^2 - |L|_{\mathrm{H}}^2 r^2 - 16\sigma |M|_{\mathrm{H}} r^2$$

$$\geq \|w\|^2 r^2 - 2^6 \sigma r^2 \geq (2^{-8} \varepsilon^4 - 2^{-9} \varepsilon^4) r^2$$

$$= 2^{-9} \varepsilon^4 r^2. \tag{16.24}$$

The inequality (16.23) used with $T = L$ and (16.21) give

$$\int_\Omega |h' - L|_{\mathrm{H}}^2 \, d\mathscr{L}^2 = \int_\Omega (|h'|_{\mathrm{H}}^2 + |L|_{\mathrm{H}}^2 - 2\langle h', L \rangle_{\mathrm{H}}) \, d\mathscr{L}^2$$

$$\leq \int_\Omega |L_u|_{\mathrm{H}}^2 \, d\mathscr{L}^2(u) + r^2 |L|_{\mathrm{H}}^2 - 2r^2 \langle M, L \rangle_{\mathrm{H}} + 2^5 \sigma r^2$$

$$\leq \int_\Omega (|L_u|_{\mathrm{H}}^2 - |L|_{\mathrm{H}}^2) \, d\mathscr{L}^2(u) + \frac{\varepsilon^4 r^2}{2^{10}}. \tag{16.25}$$

Together with the first inequality in Lemma 16.5.2 this implies that

$$\int_\Omega \mathrm{Lip}_{u,\Omega}(h - L)^4 \, d\mathscr{L}^2(u) \leq 2^4 K \int_\Omega \|h' - L\|^2 \, d\mathscr{L}^2$$

$$\leq 2^4 K \int_\Omega (|L_u|_{\mathrm{H}}^2 - |L|_{\mathrm{H}}^2) \, d\mathscr{L}^2(u) + \frac{K\varepsilon^4 r^2}{2^6}. \tag{16.26}$$

We also denote $s_0 = 2^{-18}\varepsilon^4$ and $\widetilde{\Omega} = \{se_1 + te_2 \mid rs_0 < s, t < r(1 - s_0)\}$, and define an auxiliary function $\zeta \colon \Omega \longrightarrow [0, \infty)$ by

$$\zeta(u) = \begin{cases} 2^{11} K s_0 & \text{if } u \in \widetilde{\Omega}, \\ 2^9 K & \text{otherwise.} \end{cases}$$

Then we observe that

$$\int_\Omega \zeta \, d\mathscr{L}^2 < 2^{11} K s_0 r^2 + 2^{11} K r^2 s_0 = \frac{K\varepsilon^4 r^2}{2^6}.$$

Adding this inequality to (16.26), we get with the help of (16.24),

$$\int_\Omega (\mathrm{Lip}_{u,\Omega}(h-L)^4 + \zeta(u)) d\mathscr{L}^2(u) < 2^4 K \int_\Omega (|L_u|_\mathrm{H}^2 - |L|_\mathrm{H}^2) \, d\mathscr{L}^2(u) + \frac{K\varepsilon^4 r^2}{2^5}$$

$$\leq 2^5 K \int_\Omega (|L_u|_\mathrm{H}^2 - |L|_\mathrm{H}^2) \, d\mathscr{L}^2(u).$$

This is the same estimate as in the irregular case, and the rest of the proof is very similar to that case, except of course that Ω and some constants are slightly different. The previous inequality shows that there is $u_0 \in \Omega$ such that $h'(u_0)$ exists and

$$\mathrm{Lip}_{u_0,\Omega}(h-L)^4 + \zeta(u_0) < 2^5 K(|L_{u_0}|_\mathrm{H} - |L|_\mathrm{H})(|L_{u_0}|_\mathrm{H} + |L|_\mathrm{H}).$$

In particular, the right side is positive, and so

$$\mathrm{Lip}_{u_0,\Omega}(h-L)^4 + \zeta(u_0) < 2^8 K(|L_{u_0}|_\mathrm{H} - |L|_\mathrm{H}). \tag{16.27}$$

Since the right side is at most $2^9 K$, we infer that $\zeta(u_0) < 2^9 K$ and so $u_0 \in \tilde{\Omega}$ and $\zeta(u_0) = 2^{11} K s_0 = 2^{-5} K \varepsilon^4$. Hence (16.27) gives that

$$c := |L_{u_0}|_\mathrm{H} - |L|_\mathrm{H} > \frac{\zeta(u_0)}{2^8 K} = \frac{\varepsilon^4}{2^{15}} \geq c_0.$$

We claim that $y_0 = \gamma(u_0)$ and the already defined subspace $Y \in G(H,2)$ have the required properties. To start we notice that, clearly, the pair (y_0, Y) belongs to $\mathcal{D}_2(g)$ and $\|y_0\| = \|u_0\| < 2r \leq C_1 \xi < C_0 \xi$. Let $u \in X$ and distinguish the following three cases.

(A) If $u_0 + u \in \Omega$ then

$$\|h(u_0 + u) - h(u_0) - Lu\| \leq \mathrm{Lip}_{u_0,\Omega}(h-L)\|u\| \leq 4K^{1/4}c^{1/4}\|u\|.$$

Hence, observing that

$$\|u - \gamma(u)\|^2 \leq (1-\beta)^2\|u\|^2 + \frac{\|z\|^2}{r^2}\|u\|^2 = \frac{\varepsilon^2}{2^5}\|u\|^2 \leq \varepsilon^2\|u\|^2,$$

and that $\|u\| \leq \|u_0 + u\| + \|u_0\| \leq 4r \leq C_1 \xi$, we conclude that

$$\|g(y_0 + u) - g(y_0) - g(u)\| \leq \|h(u_0 + u) - h(u_0) - Lu\| + \|g(u) - Lu\|$$
$$+ \|g(y_0 + u) - g(\gamma(u_0 + u))\|$$
$$\leq 4K^{1/4}c^{1/4}\|u\| + \sigma\|u\| + \|u - \gamma(u)\|$$
$$\leq (4K^{1/4}c^{1/4} + \sigma + \varepsilon)\|u\| \leq \chi(c)\|u\|.$$

(B) When $u_0 + u \notin \Omega$ and $\|y_0\| + \|u\| \le C_1\xi$ we use that $u_0 \in \widetilde{\Omega}$ to infer that $\|u\| \ge s_0 r$. Hence we may estimate

$$\|g(y_0 + u) - g(y_0) - g(u)\| \le \|g(y_0 + u) - g(y_0) - Lu\| + \|g(u) - Lu\|$$
$$\le \sigma(\|y_0\| + 2\|u\|) \le \sigma\Big(\frac{2}{s_0} + 2\Big)\|u\|$$
$$= (2\varepsilon + 2\sigma)\|u\| \le \chi(c)\|u\|.$$

(C) Finally, when $\|y_0\| + \|u\| > C_1\xi$ we use that

$$\|u\| \ge C_1\xi - \|y_0\| = C_1\xi - \|u_0\| \ge C_1\xi - 2r \ge \frac{C_1\xi}{2}$$

to infer that

$$\|g(y_0 + u) - g(y_0) - g(u)\| \le \|g(y_0 + u) - g(u)\| + \|g(y_0) - g(0)\| \le 2\|y_0\|$$
$$\le 4r \le \frac{8r}{C_1\xi}\|u\| \le \frac{2^6}{\varepsilon C_1}\|u\| \le \varepsilon\|u\|$$
$$\le \chi(c)\|u\|. \qquad \square$$

Bibliography

[1] E. ACERBI AND N. FUSCO, An approximation lemma for $W^{1,p}$ functions, in *Material instabilities in continuum mechanics* (Edinburgh, 1985–1986), Oxford Sci. Publ., Oxford Univ. Press, New York (1988), 1–5. (Cited on p 88.)

[2] G. ALBERTI, M. CSÖRNYEI, D. PREISS, Structure of null sets, differentiability of Lipschitz functions, and other problems. In preparation. (Cited on pp 2 and 16.)

[3] N. ARONSZAJN, Differentiability of Lipschitzian mappings between Banach spaces, *Studia Math.* 57 (1976), no. 2, 147–190. (Cited on p 15.)

[4] Y. BENYAMINI, J. LINDENSTRAUSS, *Geometric nonlinear functional analysis, Vol. 1*, Colloquium Publications 48, Amer. Math. Soc., 2000. (Cited on pp 2, 3, 5, 12, 13, 14, 15, 16, 23, 24, 31, 33, 34, 75, and 323.)

[5] A. BEURLING, L. AHLFORS, The boundary correspondence under quasiconformal mappings, *Acta Math.* 96 (1956), 125–142. (Cited on p 33.)

[6] J. M. BORWEIN, W. MOORS, Separable determination of integrability and minimality of the Clarke subdifferential mapping, *Proc. Amer. Math. Soc.* 128 (2000), no. 1, 215–221. (Cited on p 37.)

[7] J. M. BORWEIN, D. PREISS, A smooth variational principle with applications to subdifferentiability of convex functions, *Trans. Amer. Math. Soc.* 303 (1987), no. 2, 517–527. (Cited on pp 7, 120, and 121.)

[8] J. M. BORWEIN, J. V. BURKE, A. S. LEWIS, Differentiability of cone-monotone functions on separable Banach space, *Proc. Amer. Math. Soc.* 132 (2004), no. 4, 1067–1076. (Cited on p 223.)

[9] R. D. BOURGIN, *Geometric aspects of convex sets with the Radon-Nikodým property*, Lecture Notes in Math. 993, Springer-Verlag, 1983. (Cited on p 2.)

[10] J. C. BURKILL, U. S. HASLAM-JONES, Notes on the differentiability of functions of two variables. *J. London Math. Soc.* 7 (1932), 297–305. (Cited on p 223.)

[11] J. P. R. CHRISTENSEN, On sets of Haar measure zero in abelian Polish groups, *Israel J. Math.* 13 (1972), 255–260. (Cited on p 13.)

[12] T. DE PAUW, P. HUOVINEN, Points of ε-differentiability of Lipschitz functions from $\mathbb{R}^n$ to $\mathbb{R}^{n-1}$, *Bull. London Math. Soc.* 34 (2002), no. 2, 539–550. (Cited on p 71.)

[13] R. DEVILLE, G. GODEFROY, V. ZIZLER, *Smoothness and renormings in Banach spaces*, Pitman Monographs and Surveys 64, Longman, 1993. (Cited on p 120.)

[14] J. DIESTEL, J. J. UHL, *Vector measures*, Math. Surveys, no. 15, Amer. Math. Soc., 1977. (Cited on p 2.)

[15] M. J. DORÉ, O. MALEVA, A compact null set containing a differentiability point of every Lipschitz function, to appear in Mathematische Annalen. (Cited on pp 2 and 393.)

[16] I. EKELAND, Nonconvex minimization problems, *Bull. Amer. Math. Soc.* 1 (1979), no. 3, 443–474. (Cited on p 6.)

[17] M. FABIÁN, D. PREISS, On intermediate differentiability of Lipschitz functions on certain Banach spaces, *Proc. Amer. Math. Soc.* 113 (1991), no. 3, 733–740. (Cited on p 51.)

[18] S. FITZPATRICK, Differentiation of real-valued functions and continuity of metric projections, *Proc. Amer. Math. Soc.* 91 (1984), no. 4, 544–548. (Cited on p 6.)

[19] T. FOWLER, D. PREISS, A simple proof of Zahorski's description of nondifferentiability sets of Lipschitz functions, *Real. Anal. Exchange* 34 (2008/2009), no. 1, 1–12. (Cited on p 15.)

[20] G. GODEFROY, N. KALTON, G. LANCIEN, Szlenk indices and uniform homeomorphisms, *Trans. Amer. Math. Soc.* 353 (2001), no. 10, 3895–3918. (Cited on p 58.)

[21] P. HOLICKÝ, M. ŠMÍDEK, L. ZAJÍČEK, Convex functions with non-Borel set of Gâteaux differentiability points, *Comment. Math. Univ. Carolin.* 39 (1998), no. 3, 469–482. (Cited on p 34.)

[22] W. B. JOHNSON, J. LINDENSTRAUSS, D. PREISS, G. SCHECHTMAN, Almost Fréchet differentiability of Lipschitz mapping between infinite dimensional Banach spaces, *Proc. London Math. Soc.* (3) 84 (2002), no. 3, 711–146. (Cited on pp 47, 51, 52, 53, 57, 58, 68, 71, 116, 263, and 324.)

[23] W. B. JOHNSON, J. LINDENSTRAUSS, G. SCHECHTMAN, Extensions of Lipschitz maps into Banach spaces, *Israel J. Math.* 54 (1986), no. 2, 129–138. (Cited on pp 88 and 89.)

[24] A. S. KECHRIS, *Classical descriptive set theory*, Graduate Texts in Math. 156, Springer Verlag, 1995. (Cited on pp 109 and 276.)

[25] J. R. LEE, A. NAOR, Extending Lipschitz functions via random metric partitions, *Inventiones Math.* 160 (2005), no. 1, 59–95. (Cited on p 89.)

[26] J. LINDENSTRAUSS, D. PREISS, Almost Fréchet differentiability of finitely many Lipschitz functions, *Mathematika* 43 (1996), no. 2, 393–412. (Cited on pp 47, 50, and 263.)

[27] J. LINDENSTRAUSS, D. PREISS, A new proof of Fréchet differentiability of Lipschitz functions, *J. Eur. Math. Soc.* 2, (2000), no. 3, 199–216. (Cited on pp 43, 44, 47, 51, 204, 219, 222, and 224.)

[28] J. LINDENSTRAUSS, D. PREISS, On Fréchet differentiability of Lipschitz maps between Banach spaces, *Annals of Math.* 157 (2003), no. 1, 257–288. (Cited on pp 4, 6, 72, 96, 151, 170, 198, 199, 222, and 263.)

[29] J. LINDENSTRAUSS, D. PREISS, J. TIŠER, Fréchet differentiability of Lipschitz functions via a variational principle, *J. Eur. Math. Soc.* 12 (2010), no. 2, 385–412. (Cited on pp 47, 127, 222, 224, 225, 248, 277, and 317.)

[30] J. LINDENSTRAUSS, L. TZAFRIRI, *Classical Banach spaces I*, Springer-Verlag, 1977. (Cited on p 320.)

[31] J. LINDENSTRAUSS, L. TZAFRIRI, *Classical Banach spaces II*, Springer-Verlag, 1979. (Cited on pp 48 and 51.)

[32] O. MALEVA, Unavoidable sigma-porous sets, *J. London Math. Soc.* 76 (2007), no. 2, 467–478. (Cited on p 324.)

[33] P. MATTILA, *Geometry of sets and measures in Euclidean spaces*, Cambridge University Press, 1995. (Cited on pp 31, 168, and 207.)

[34] J. MATOUŠEK, E. MATOUŠKOVÁ, A highly non-smooth norm on Hilbert space, *Israel J. Math.* 112 (1999), 1–27. (Cited on pp 33 and 73.)

[35] E. MATOUŠKOVÁ, An almost nowhere Fréchet smooth norm on superreflexive spaces, *Studia Math.* 133 (1999), no. 1, 93–99. (Cited on pp 33 and 73.)

[36] V. D. MILMAN, Geometric theory of Banach spaces II. Geometry of the unit ball, *Uspehi Mat. Nauk.* 26(1971), no. 6(162), 73–149. English translation: *Russian Math. Surveys* 26(1971), no. 6, 79–163. (Cited on p 52.)

[37] R. R. PHELPS, *Convex functions, monotone operators and differentiability*, second ed., Lect. Notes in Math. 1364, Springer-Verlag, 1993. (Cited on pp 69, 186, 188, and 393.)

[38] D. PREISS, Gâteaux differentiable functions are somewhere Fréchet differentiable, *Rend. Circ. Mat. Palermo* (2) 33 (1984), no. 1, 122–133. (Cited on pp 6, 37, 43, and 225.)

[39] D. PREISS, Differentiability of Lipschitz functions on Banach spaces, *J. Functional Anal.* 91 (1990), no. 2, 312–345. (Cited on pp 2, 6, 15, 44, 47, 204, 222, 224, and 262.)

[40] D. PREISS, J. TIŠER, Two unexpected examples concerning differentiability of Lipschitz functions on Banach spaces, GAFA Israel Seminar 92–94, Birkhäuser 1995, 219–238. (Cited on pp 31, 33, 323, and 325.)

[41] D. PREISS, L. ZAJÍČEK, Directional derivatives of Lipschitz functions, *Israel J Math.* 125 (2001), 1–27. (Cited on p 15.)

[42] D. PREISS, L. ZAJÍČEK, Sigma-porous sets in products of metric spaces and sigma directionally porous sets in Banach spaces, *Real Anal. Exchange* 24 (1998/99), no. 1, 295–313. (Cited on p 170.)

[43] E. M. STEIN, J. O. STRÖMBERG, Behavior of maximal functions in $\mathbb{R}^n$ for large n, *Ark. Mat.* 21 (1983), no. 2, 259–269. (Cited on p 168.)

[44] M. TALAGRAND, Deux exemples de fonctions convexes, *C. R. Acad. Sci. Paris Sér. A-B* 288 (1979), no. 8, A461–A464. (Cited on p 34.)

[45] B. S. TSIRELSON, Not every Banach space contains an embedding of ℓ_p or c_0, *Funkcional. Anal. i Priložen.* 8 (1974), no. 2, 57–60. English translation: *Funct. Anal. Appl.* 8 (1974), 138–141. (Cited on p 201.)

[46] Z. ZAHORSKI, Sur l'esemble des points de non-dérivabilité d'une fonction continue, *Bull. Soc. Math. France* 74 (1946), 147–178. (Cited on p 15.)

[47] Z. ZAHORSKI, Sur la première dérivée, *Trans. Amer. Math. Soc.* 69 (1950), 1–54. (Cited on p 6.)

[48] L. ZAJÍČEK, Sets of σ-porosity and sets of σ-porosity (q), *Časopis pro pěst. mat.* 101 (1976), no. 4, 350–359. (Cited on p 170.)

[49] L. ZAJÍČEK, Porosity and σ-porosity, *Real Anal. Exchange* 13 (1987-88), no. 2, 314–350. (Cited on p 33.)

[50] L. ZAJÍČEK, Fréchet differentiability, strict differentiability and subdifferentiability, *Czechoslovak Math. J.* 41(116) (1991), no. 3, 471–489. (Cited on pp 33 and 43.)

[51] L. ZAJÍČEK, On σ-porous sets in abstract spaces, *Abstract and Appl. Analysis* 2005 (2005), no. 5, 509–534. (Cited on p 33.)

[52] W. ZIEMER, *Weakly differentiable functions, Sobolev spaces and functions of bounded variation*, Grad. Texts in Math. 120, Springer Verlag, 1989. (Cited on p 88.)

Index

(d, d_0)-complete, 128
(d, d_0)-lower semicontinuous, 128
Γ-null, 73
Γ_n-null, 73
σ-ideal, 2

almost Fréchet differentiable, 47
arbitrary smoothness, 199
  and Γ-nullness of porous sets, 199
  implied by asymptotic c_0, 199
Asplund space, 3, 29
asymptotic uniform smoothness
  and existence of smooth bumps, 150
asymptotically c_0, 151
  and existence of smooth bumps, 151
  renorming, 155

bump, 144
  convex bump and renorming, 149
  smooth with controlled modulus, 144

cone-monotone function, 223

decreasing, 11
  strictly, 11
defect of regularity, 264
defect of restricted regularity, 264
deformation lemma, 162
derivative
  see also differentiability
  ε-Fréchet, 28
    existence of, 48, 60, 68
  ε-Fréchet, 46
  directional, 9
  Fréchet, 1
  Gâteaux, 1
  upper and lower, 230

upper metric, 332
differentiability
  see also derivative
  almost Fréchet, 47
  Fréchet
    Γ-almost everywhere, 112, 114, 116
    asymptotic, 356
    in direction of, 157
    mean value estimate, 17, 20, 119
    of cone-monotone functions, 223
    of real-valued functions, 222
    of vector-valued functions, 262
    on ℓ^p spaces, 263
    on Hilbert spaces, 262, 392
    regular, in direction of, 158
    set of points is Borel, 36
    upper, 135
  Gâteaux
    Γ-almost everywhere, 75
    almost everywhere, 15
    mean value estimate, 16, 20
    regular, in direction of, 157
  of convex functions
    Γ-almost everywhere, 113
    except σ-porous sets, 29
directed downward, 134
distance function, 31
divergence theorem, 164
downward directed, 134

eccentricity, 165

Fréchet derivative, 1
Fréchet smooth norm, 25
  existence of, 24
function

$\mathcal{C}$-monotone, 223
cone-monotone, 223
Lipschitz, 1
locally affine a.e., 173

Gâteaux derivative, 1
Gauss null, 13
Gaussian measure, 13

Haar null, 13
Hardy-Littlewood operator, 168
Hilbert-Schmidt
    norm, 156
    scalar product, 156

increasing, 11
    strictly, 11
irregular points, *see* regular points

Lipschitz constant
    pointwise, 168
Lipschitz function, 1
locally uniformly convex space, 25
Luxemburg norm $\|(c_i)\|_\omega$, 321

maximal operator, 168
mean value estimate
    for Fréchet derivatives, 17
    for Gâteaux derivatives, 16
    multidimensional, 20
midpoint inequality, 54
modulus
    of smoothness of a function, 133
        asymptotic, 134
        in direction, 134
    of uniform convexity, 51
        asymptotic, 52
    of uniform smoothness, 51
        asymptotic, 52

norm
    Fréchet smooth, 25
    locally uniformly convex, 25
null sets
    Γ-null, 73
    Γ_n-null, 73

$\tilde{\mathcal{A}}$ null, 15
Aronszajn null, 15
Gauss null, 13
Haar null, 13

Orlicz function, 320
Orlicz norm $|(c_i)|_\omega$, 321

perturbation game, 122
porous sets, 10, 169
    Γ-nullness of, 200
    Γ_n nullness of, 185
    σ-, 10, 169
    c-porous, 169
    contained in G_δ, 171, 173
    decomposition property, 170
    directionally, 10, 169
    in direction of
        a family of subspaces, 171
        a subspace, 10, 169
        a vector, 10, 169
        with constant c, 10, 169
    positive, 11
        strictly, 11
pseudometric, 10
pseudonorm, 10

Radon-Nikodým property, 12
regular points, 97
    of convex functions, 97
    of Lipschitz functions, 99
rich family of subspaces, 37
RNP, *see* Radon-Nikodým property

separable determination
    of ε-Fréchet differentiability, 42, 44
    of Γ-nullness, 95
    of σ-porous sets, 41
    of Fréchet differentiability, 43
separable family of subspaces, 172
separable reduction, *see* separable deter-
    mination
shear, 204
slice, 11
    w^*, 11

rank of, 11
smooth bump, 144
   with controlled modulus, 144
smooth norm, *see* Fréchet smooth norm
space
   admitting a smooth bump, 144
   Asplund, 3, 29
   uniformly convex, 48, 51
      asymptotically, 52
      locally, 25
   uniformly smooth, 48, 51
      asymptotically, 52
subdifferential, 23
surface, 73
   see also function
   tangent space to, 74

tangent space, 74

upper ω estimate, 321
upper q estimate, 319
upper Fréchet smooth, 135
upper metric derivative, 332

variational principles
   abstract, 123
   for bimetric spaces, 130
   perturbation game, 122
   standard, 125

Index of Notation

1_A ....indicator function of the set A

$\overline{A}$ ....closure of the set A

conv ....convex hull

diam ....diameter

dist ....distance

div ....divergence (of a vectorfield)

Id ....identity operator or matrix

Lip ....Lipschitz constant

rank ....rank (of a matrix)

sign ....sign function

span ....linear span

Trace ....trace (of a matrix)

$\langle \cdot, \cdot \rangle$ ....scalar product, 156

$\langle \cdot, \cdot \rangle_H$ ....Hilbert-Schmidt scalar product, 156

$\fint_A f(x)\, d\mu(x)$ ....average integral, 11

$\partial f(x)$ ....subdifferential, 23

$|\cdot|$ ....Euclidean norm, 10

$|\cdot|_H$ ....Hilbert-Schmidt norm, 156

$|(c_i)|_\omega$ ....Orlicz norm, 321

$\|\cdot\|_{C^1}$ 73

$\|\cdot\|_k$ 73

$\|\cdot\|_{\leq k}$ 73

$\|\cdot\|_\infty$ ....supremum norm, 10

$\|\gamma\|_{C^1}$ 10

$\|x\|_\mathbb{N}$ 339

$\|(c_i)\|_\omega$ ....Luxemburg norm, 321

α_n ....volume of the unit ball, 11

δ_X ....modulus of convexity, 51

$\overline{\delta}_X$ ....modulus of asymptotic convexity, 52

$\overline{\delta}_X(t, x, Y)$ 54

$\varepsilon(f, x, r, X)$ 34

$\varepsilon(f, x, L, \delta)$ 394

$\varepsilon_Z(f, x, L, \delta)$ 394

$\Gamma(X)$ 73

$\Gamma_n(X)$ 73

$\gamma^{n,s}$ 74

Λ_n .... (2–2) type constant of the maximal operator in $\mathbb{R}^n$, 168

$\bar{\mu}_X(t)$ 54

$\bar{\mu}_X(t,x,Y)$ 54

ω_n 161

π_n .... orthogonal projection on the span of the first n basis vectors, 156

π^n .... orthogonal projection on the span of all but the first n basis vectors, 157

π_V .... orthogonal projection onto V, 156

$\rho(\Omega)$ .... eccentricity, 165

$\rho_\Theta(x;t)$ .... modulus of smoothness of Θ at x, 133

$\rho_{\Theta,Y}(x;t)$ ....ditto in the direction of a subspace Y, 134

$\bar{\rho}_{\Theta,\mathcal{Y}}(x;t)$ ....ditto in the direction of a family $\mathcal{Y}$, 134

$\bar{\rho}_\Theta(x;t)$ .... ditto in the direction of the family of finite codimensional subspaces, 134

$\rho_{\Theta,Y}(T;t) = \rho_{\Theta,\{S\in L(\mathbb{R}^n,X)|S(\mathbb{R}^n)\subset Y\}}(T;t)$ 135

ρ_X .... modulus of smoothness, 51

$\bar{\rho}_X$ ....modulus of asymptotic smoothness, 52

$\bar{\rho}_X(t,x,Y)$ 54

$B_j(x,r)$ .... ball in the first j coordinates, 156

$\text{div}_T\,\varphi(x)$ ....divergence, 339

$\mathcal{D}_2(f)$ 394

$D_j\gamma$ 73

$\underline{D}h(t)$ .... lower derivative, 230

$\underline{D}f(x;v)$ .... directional lower derivative, 329

$\overline{D}h(t)$ .... upper derivative, 230

$\overline{D}f(x;v)$ .... directional upper derivative, 329

$D_V f(x)$ 394

$\overline{D}_u\psi(x)$ .... upper metric derivative, 332

$e_1,\ldots,e_n$ .... standard basis of $\mathbb{R}^n$, 156

$f'(x)$ .... Gâteaux derivative, 9

$f'(x;T)$ .... Fréchet derivative of f in the direction of an operator T, 157

$f'(x;u)$ .... directional derivative, 9

$f'_V(x)$ ....Fréchet derivative of f in the direction of a subspace V, 158

$g^{k,s}$ 106

$g'_n(x)$ ....derivative of with respect to the first n variables, 264

$\text{grad}\,\psi$ ....gradient, 156

$\mathscr{L}^n$ .... outer Lebesgue measure, 11

$\mathscr{L}^{\mathbb{N}}$ .... product Lebesgue measure in $T = [0,1]^{\mathbb{N}}$, 73

$\text{Lip}_n(g)$ .... Lipschitz constant with respect to the first n variables, 264

$\text{Lip}_{u,\Omega}(g)$ .... pointwise Lipschitz constant, 168

$L(X,Y)$ ....space of linear operators, 10

$\mathbf{M}h$ ....Hardy-Littlewood maximal operator, 168

$\mathbf{M}_n h$ .... maximal operator in the direction of the first n variables, 264

$Q_k(t,r)$ 101

$\text{reg}_n\, g(u)$ .... defect of regularity, 264

$\text{reg}_n\, g(u, \Omega)$ .... defect of restricted regularity, 264

$R(f; u, y; \zeta, \rho)$ 98

$T = [0, 1]^{\mathbb{N}}$ 73

$\text{Tan}(\gamma, t)$ .... tangent space, 74

$X \otimes \mathbb{R}^n$ .... tensor product of X and $\mathbb{R}^n$, 157

Ingram Content Group UK Ltd.
Milton Keynes UK
UKHW022323160623
423577UK00008B/746